Flat Panel Display Materials II

MATERIALS RESEARCH SOCIETY
SYMPOSIUM PROCEEDINGS VOLUME 424

Flat Panel Display Materials II

Symposium held April 8-12, 1996, San Francisco, California, U.S.A.

EDITORS:

Miltiadis K. Hatalis
Lehigh University
Bethlehem, Pennsylvania, U.S.A.

Jerzy Kanicki
University of Michigan
Ann Arbor, Michigan, U.S.A.

Christopher J. Summers
Georgia Institute of Technology
Atlanta, Georgia, U.S.A.

Fumiaki Funada
Sharp Corporation
Nara, Japan

MATERIALS
RESEARCH
SOCIETY

PITTSBURGH, PENNSYLVANIA

Single article reprints from this publication are available through
University Microfilms Inc., 300 North Zeeb Road, Ann Arbor, Michigan 48106

CODEN: MRSPDH

Published by:

Materials Research Society
9800 McKnight Road
Pittsburgh, Pennsylvania 15237
Telephone (412) 367-3003
Fax (412) 367-4373
Website: http://www.mrs.org/

Library of Congress Cataloging in Publication Data

Flat panel display materials II : symposium held April 8–12, 1996,
    San Francisco, California, U.S.A. / editors, M.K. Hatalis, J. Kanicki,
    C.J. Summers, F. Funada
        p. cm—(Materials Research Society symposium proceedings ; v. 424)
    Includes bibliographical references and index.
    ISBN 1-55899-327-4
        1. Liquid crystal displays—Materials—Congresses. 2. Thin film
    transistors—Materials—Congresses. 3. Information display devices—
    Materials—Congresses. I. Hatalis, M.K. II. Kanicki, J. III. Summers, C.J.
    IV. Funada, F. V. Series: Materials Research Society symposium
    proceedings ; v. 424.
TK7872.L56F533 1996   96-48483
621.3815'422—dc21   CIP

Manufactured in the United States of America

CONTENTS

*Invited Paper

*Invited Paper

*Invited Paper

PART IV: TRANSPARENT CONDUCTING OXIDES

PART V: FIELD EMISSION DISPLAY MATERIALS AND TECHNOLOGY

*Invited Paper

*Invited Paper

x

PREFACE

This volume contains papers presented at the symposium on flat-panel display materials which was part of the 1996 MRS Spring Meeting held in San Francisco, CA on April 8 to 12. This symposium continued the theme introduced in the first symposium on flat-panel display materials held in 1994. The intent was to cover advances related to materials and processing for the various flat-panel display technologies.

The field of flat-panel displays is experiencing a rapid growth which currently is expanding from initial portable computer applications to include display applications for desktop computers and a wide array of consumer and industrial products. At present, the industrial focus continues to be on active matrix liquid crystal displays (AMLCDs) and that is reflected in this proceeding volume where more than half of the papers are related to AMLCD technology. However, new emerging flat-panel display technologies such as field emission displays (FEDs) and new organic light-emitting display materials are also covered. Part I covers papers related to amorphous silicon thin-film transistor (TFT) technology for AMLCDs. In this part papers presented in a joint session with the "Amorphous Silicon Technology–1996" symposium are also included. Part II includes papers on polycrystalline silicon TFT technology, with topics ranging from low-temperature preparation of polysilicon films to device processing. Part III covers advances in liquid-crystal materials, while part IV includes papers on transparent conducting oxides. Part V covers a range of papers dealing with advances in emerging FED display technology; part VI includes papers on other emissive display materials.

A new feature of this symposium was a tutorial on flat-panel displays presented by J. Kanicki and C. Summers, which provided an excellent introduction to the technical part of the symposium. The organizers of the symposium wish to thank the speakers, the paper authors, and in particular the invited speakers, for their excellent contributions. The assistance provided by the session chairpersons and all of the paper reviewers has contributed to the success of the symposium and is greatly appreciated. We also wish to thank the MRS Staff for their expert editorial assistance in the preparation of this volume.

The financial support provided by DARPA, through the High Definition Systems program, dpiX A Xerox Company, IBM, and SHARP is gratefully acknowledged.

Miltiadis K. Hatalis
Jerzy Kanicki
Christopher J. Summers
Fumiaki Funada

January 1997

MATERIALS RESEARCH SOCIETY SYMPOSIUM PROCEEDINGS

MATERIALS RESEARCH SOCIETY SYMPOSIUM PROCEEDINGS

Prior Materials Research Society Symposium Proceedings available by contacting Materials Research Society

Part I

Amorphous Silicon Thin–Film Transistor Materials and Technology

AMORPHOUS SILICON TFTS AND FLAT PANEL DISPLAYS

T. TSUKADA
Central Research Laboratory, Hitachi, Ltd.

Kokubunji, Tokyo 185, Japan, tsukada@crl.hitachi.co.jp

ABSTRACT

The current status and next-generation technologies for liquid-crystal displays based on amorphous silicon TFTs are described. Panel size, resolution, design rule, and viewing angle are discussed from the standpoint of panel performance, productivity, and product application. The material properties for future TFT/LCDs are specified based on the discussions of design rule.

INTRODUCTION

Design and process technologies based on hydrogenated amorphous silicon (a-Si:H) thin films have led to the development of TFT/LCD panels for use in notebook and nomadic computing machines. For these applications, 10.4-in. VGA (640H x 480V pixels) panels have captured the leading position in the market [1]. However, the market trend is now shifting from VGA to larger (11-12 in.) and higher-resolution (SVGA) panels. This "larger and higher" trend will still keep its pace to XGA and beyond, as shown in Fig.1.

In conventional VGA panels, each square pixel is composed of three RGB dots measuring 110 by 330 μm. The design rule for these panels has been ~10 μm due primarily to the use of premature process technologies. However, the accumulation of technical know-how and production line experience has made it possible to produce larger panels with higher resolution without reducing production efficiency or yield. Therefore, the design rule is now decreasing. This is important because it will keep the aperture ratio of high-resolution (SVGA) panels on the same level as that of conventional VGA panels.

TFT/LCD panels feature flatness, lightweight, low-power consumption, ruggedness, and a high contrast ratio. Application of these panels has broadened recently from conventional computer displays to consumer electronic devices, including TV monitors, video cameras, digital cameras, navigation systems, and projection displays. However, their viewing angle is not as wide as those of CRTs and other emissive displays. For TFT/LCDs to be used in a wider range of products, the viewing angle must be widened. Application to personal digital assistants will also become important in the near future, requiring improvements in display area, resolution, and color.

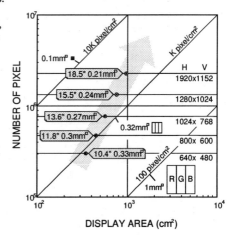

Fig. 1 Number of pixel vs. display area of TFT/LCD panels.

a-Si:H TFT PANELS

The mainstream of TFT/LCD panels (Fig. 2) is now shifting from VGA to SVGA, which has an array format of 800H x 600V.

3

Mat. Res. Soc. Symp. Proc. Vol. 424 © 1997 Materials Research Society

Since SVGA panels range in size from 11 to 13 in., the pixels are reduced in size from 300-330 μm in VGA format to 270-300 μm. This change has been achieved by refining the fabrication technologies for a-Si:H TFT panels. The design rule for these panels has thus been scaled down from the conventional 10 μm to 8 - 5 μm.

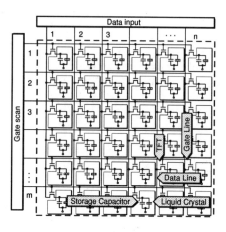

Fig. 2   Configuration of a-Si:H TFT/LCDs.

Design rule

The improvement in performance that is expected from scaled-down MOSFETs in integrated circuits was first discussed by Dennard et al. [2]. They defined dimensionless scaling factor k (>1) and concluded that the delay time and power dissipation of a scaled device are reduced to $1/k$ and $1/k^2$ that of an unscaled device, respectively. They assumed smaller dimensions (t_{ox}, L, and W with a scaling factor of $1/k$, where t_{ox} is the oxide thickness and L and W are the channel length and width of the MOSFET, respectively), a higher doping concentration (k), a lower voltage and current ($1/k$), and a lower capacitance ($1/k$). The discussion of MOSFET scaling in integrated circuits is relatively simple since the device structures are uniformly reduced in size and there are no other exotic devices involved.

TFT/LCD scaling

In TFT/LCDs, however, the situation is not as simple as that of LSIs. TFT scaling has to be considered in conjunction with the liquid- crystal cell specifications. If we assume the liquid-crystal material remains the same as we shift from VGA to SVGA (or even to XGA) format, cell gap d of the twisted-nematic (TN) cells does not change (Δn d is kept constant, where Δn is the anisotropy of the refractive indices). If d is constant and Δε (dielectric anisotropy of liquid-crystal) is kept unchanged (same liquid-crystal material), the capacitance of the liquid-crystal is scaled up by a factor of a, which approximately corresponds to the increase in the aperture ratio in a new pixel design after scaling.

Threshold voltage V_{th} of the TN cell is written as

$$V_{th} = \pi\{[k_{11} + (k_{33} - 2k_{22})/4]/\Delta\varepsilon\}^{1/2}, \tag{1}$$

where k_{11}, k_{22}, and k_{33} are the constants of the splay, twist, and bend stresses of the liquid crystal, respectively. Therefore, V_{th} is independent of d and is determined solely by the material constants. Consequently data voltage V_d applied to the cell is also invariant. To keep the electric field constant in the TFT, it follows that the channel length should also be kept constant. This implies that V_d and L cannot be scaled even if the design rule is reduced by a factor of k. The same applies to the gate voltage V_g and this is equal to saying that the thickness of the gate insulator, t_{SiN}, and the gate capacitance per unit area, C_i, are not scaled since the electric field was assumed constant.

On the other hand, the channel width can be reduced to $1/k$. The drain current thus becomes $1/k$ and the parasitic capacitance C_{gs} is reduced to $1/k^2$ because the overlapping length between the gate and source electrodes can also be scaled to $1/k$.

Two other issues related to scaling are the charging of the pixel capacitance and the gate delay [3]. To charge the pixel capacitance, the TFT drivability has to be taken into account. Time constant τ of the charging behavior is expressed as [4]

$$\tau = C_{px} / [\beta_o (V_g - V_t - V_d)], \tag{2}$$

where C_{px} ($= C_{gs} + C_{lc} + C_{st}$) is the pixel capacitance, where C_{lc} is the capacitance of the liquid-crystal cell and C_{st} is the storage capacitance, $\beta_o = \mu C_i$ (W/L), where μ is the mobility of a-Si:H, and V_t is the threshold voltage of a-Si:H TFT. Since the voltages are not scaled, C_{px} and β_o determine the time constant. Because β_o is reduced to $1/k$, if C_{px} is not reduced, the time constant will be large and the panel operation will suffer from insufficient charge up. However, if C_{px} is reduced appropriately, as will be discussed later, the panel will operate in almost the same way as before scaling.

Gate delay

Gate delay can be estimated from the product of the resistance and capacitance per unit pixel multiplied by the squared number of pixels [5]. The resistance per cell of the gate and drain buslines becomes higher when the width of the busline is narrowed to $1/k$. However, the capacitance per pixel is reduced to $1/k^2$, which is approximated by $C_c + C_{gs} + C_{gd}$, where C_c is the capacitance of the crossed area between two buslines and C_{gd} is the gate-drain capacitance. Therefore, the gate delay per pixel is reduced to $1/k$ by scaling. However, the higher panel resolution and larger panel increase the gate delay by ratio r:

$$r = W' N' / W_i N_i, \tag{3}$$

where W_i and N_i are the width and number of pixels (horizontal) of the starting panel and W' and N' are the width and number of pixels of the target panel. Therefore, the gate delay can be estimated from the ratio, r_g

$$r_g = (1/k) (W' N' / W_i N_i). \tag{4}$$

If the starting panel is VGA ($W_i = 10.4$ in., $N_i = 640$) and the target panel is $W' = 13$ in. and $N_i = 1024$, and if k is assumed to be 2 (design rule reduced from 10 to 5 μm), r_g is calculated to be 1. This means the gate delay is the same for both panels. If the panel is bigger than 13 in. the gate delay increases. When there is no scaling, this gate delay increases with $W'N'/W_iN_i$ ($= r$), which is estimated to be 2 if the starting panel is 10.4-in. VGA and the target panel is 13-in. XGA, as before.

Feedthrough voltage

A dc voltage offset, ΔV, is induced in the node potential (voltage at the pixel electrode) due to stray capacitance C_{gs} of the TFT. This ΔV is given by

$$\Delta V = V_g C_{gs} / (C_{gs} + C_{st} + C_{lc}), \tag{5}$$

where V_g is the amplitude of the gate voltage pulse applied to the gate busline. This offset voltage has an undesirable effect on the performance of the liquid-crystal display, causing flicker, image sticking, and permanent brightness nonuniformities on the panel.

The important implication of smaller TFTs is that the feedthrough voltage due to stray capacitance in the TFTs is reduced to $1/k^2$ (constant thin-film thicknesses). If k is 2, ΔV is reduced to 1/4, a considerable improvement. This will be a step towards the C_{st}-less scheme [1] in which storage capacitance C_{st} is not implemented in the pixel design. Calculation shows that a C_{st}-less scheme can be achieved if parasitic capacitance C_{gs} is reduced to $1/(1+n)$, where n is $C_{st} = nC_{lc}$ where C_{lc} is the capacitance of the liquid-crystal cell. If we set n = 2, which is quite reasonable, the required reduction of C_{gs} is 1/3, which corresponds to a k factor of $\sqrt{3}$. This corresponds to a reduction in the design rule from 10 to 5.8 μm. If the increase in the liquid-crystal capacitance, a, is reasonably assumed to be 1.5, the pixel capacitance after scaling is reduced to around 0.5 for $C_{st} = 0$. Consequently, time constant τ in Eq. (2) remains almost equal after scaling, as discussed above.

The effect of the feedthrough voltage appears not only from its absolute value, but also from the variance in voltage from pixel to pixel (dV). The discussion of dV in Ref. [5] leads us to the conclusion that the complete absence of C_{st} is not a good solution because C_{lc} depends on the voltage. However, the effect of this design rule change should be recognized as quite important.

VOLTAGE SCALING AND LIQUID CRYSTAL

In the preceding section, we discussed the scaling of the panel assuming the liquid-crystal material was the same. Such variables as voltage, channel length, and insulator thickness thus remained constant. However, low-voltage operation is preferred from the standpoint of reducing the power consumption of the driver LSIs. The scaling of the voltage means the scaling of drain voltage V_d, and gate voltage V_g. As a result, channel length L of the TFT and insulator thickness t_{SiN} are also scaled and the intensity of the electric field is kept constant. To scale the voltage to $1/k$, the material constants of the liquid crystal must be redesigned. For threshold voltage V_{th} of the TN cell to be scaled, stress constants k_{ii} and the anisotropy of the dielectric constant, $\Delta\varepsilon$, have to be modified. Even with these modifications, the cell gap d of the liquid crystal is not necessarily scaled because V_{th} is independent of cell gap. The only requirement for d is

$$\Delta n\, d = (\sqrt{3}\,/\,2)\,\lambda, \tag{7}$$

where λ is the wavelength of the light.

Although the cell gap can also be scaled, here we assume the same cell gap to simplify the discussion. The stray capacitance is reduced to $1/k$ in contrast to the $1/k^2$ reduction in the previous section. However, because the voltage is scaled to $1/k$, the offset voltage is scaled to $\sim 1/k^2$ if C_{st} and C_{lc} are not changed. Consequently, the discussion in the previous section applies here in almost the same way.

The discussion of the time constant also follows from before since ß is scaled to k (C_i is scaled to k) and the voltages are scaled to $1/k$. The reduction in C_{px} (if any) simply shortens the time constant. The gate delay discussion does not change.

To make self-consistent the discussions of the previous paragraphs, it is necessary to prepare materials which satisty the scaling conditions. As for liquid cystal, the material constants like anisotropic dielectric constant, constants of stress have to be modified to control the threshold voltage and operating voltage. This is necessary to scale the operating voltages of a-Si:H TFTs.

To scale the gate insulator thickness of TFTs, the gate insulator material has to be chosen from the standpoint of productivity. Since the thinner insulator can cause defects, the film has to be deposited in cleaner environments than before.

CHARACTERISTICS

There have been many efforts to improve the viewing angle of TFT/LCDs. The first approach was the multi-domain scheme [6], in which the TN cells were divided into two domains with their alignment directions perpendicular to each other. The recently reported super-TFT scheme [7] makes use of an in-plane switching mechanism. The viewing angle of super-TFT panels has been improved from ±30° in conventional panels to ±70° with no reversal in the gray-scale colors.

PLATFORMS

Application of TFT/LCDs to such products as WS monitors, TV monitors, navigation systems, and projection TV systems require that they have larger areas to display the larger amount of information. Foldable-display systems have thus been proposed and evaluated [8]. These systems consist of two LCDs whose possible configurations are shown in Fig. 3. One potential application is shown in Fig. 4, where one display panel is used as a tool space and the other as a work area. In this "draw" application, the tool space functions as a palette and the work area functions as a canvas.

Another potential application involves reversing the displays on the two panels so that two people facing each other can easily read the same displayed information.

CONCLUSION

The present status of TFT/LCDs based on hydrogenated amorphous silicon thin films has been reviewed. VGA 10-in. panels are being replaced by SVGA 11-12-in. panels. The design rule is now shifting from conventional 10 μm to 5-7 μm depending on the size and resolution of the panel. Its effects on the panel performances have been discussed.

The viewing angle of TFT/LCDs has been greatly increased by super-TFTs which have recently been introduced. The application of TFT/LCDs will expand greatly due to these advances, and they will be used on a wider range of platforms.

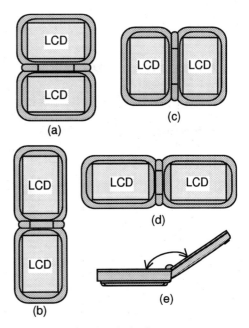

Fig. 3   Foldable display systems.

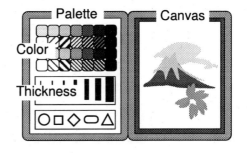

Fig. 4   Draw system.

REFERENCES

1. T. Tsukada, Proc. Intn'l Display Research Conf., 325 (1992).

2. R. H. Dennard, F. H. Gaensslen, H-N Yu, V. L. Rideout, E. Bassous, and A. R. LeBlanc, IEEE J. Solid-State Circuits, SC-9, 256 (1974).

3. T. Tsukada, Mater. Res. Soc. Symp. Proc., 284, 371 (1993).

4. T. Tsukada, <u>TFT/LCD</u>, 1st ed. (Gordon and Breach Science Publishers, New York, 1996), p. 19.

5. ibid, p. 30.

6. K. H. Yang, Proc. Intn'l Display Research Conf., 68 (1991).

7. M. Oh-e, M. Ohta, S. Aratani, and K. Kondo, Proc. Intn'l Display Research Conf. 577 (1995).

8. Y. Kaneko, M. Yamaguchi, H. Matsuya, and T. Tsukada, IEEE Trans. Consumer Electronics, <u>42</u>, 17 (1996).

HIGH DEPOSITION RATE a-Si:H FOR THE FLAT PANEL DISPLAY INDUSTRY

J. HAUTALA, Z. SALEH, J.F.M. WESTENDORP, H. MEILING[1], S. SHERMAN[2] and S. WAGNER[2]
Tokyo Electron America, 123 Brimbal Ave, Beverly MA 01915
[1]Utrecht University, P.O. Box 80000 3508 TA, Utrecht, The Netherlands
[2]Department of Electrical Engineering, Princeton University, Princeton NJ 08544

ABSTRACT

High deposition rates and good quality electrical properties and thickness uniformities over large areas are required for all three films (SiN_x, a-Si:H and n+ a-Si:H) composing the thin film transistors (TFTs) for the AMLCD industry, while maintaining high tool up-time and low particle formation. Generally these conditions have been achieved for most single-panel multichamber PECVD systems; however, it has become increasingly apparent that a compromise is drawn between the TFT mobility and the deposition rate of the a-Si:H layer. Thus it becomes essential to clearly assess the industry requirements for both deposition rates as well as TFT performance for the different device structures used for AMLCDs, and to discover and control these underlying material properties.

The TEL VHF (40/60 MHz) PECVD system produces high quality, low defect density a-Si:H at deposition rates exceeding 1500 Å/min when analyzed by FTIR, CPM, photo and dark conductivity. Even though the low deposition rate a-Si:H exhibits very similar bulk properties, higher mobility TFTs are produced with a-Si:H deposited at lower RF power. Having both a high ion flux and low ion energy in the SiH_4 discharge are likely the most critical conditions for controlling the a-Si:H quality and thus the TFT mobility. Increasing the RF frequency enhances both of these effects, as well as provides a higher deposition rate for a given power density and a higher power threshold for particle/powder formation. For these reasons it is likely a 40/60 MHz plasma will produce better performing TFTs for a given deposition rate when compared with a conventional 13.56 MHz system. Other process conditions such as diluting the SiH_4 in H_2 or Ar also seem to play an important role in the optoelectronic properties of the a-Si:H film and ultimately the TFT performance.

INTRODUCTION

High throughput, high yield and high transistor performance drive the development of Plasma Enhanced Chemical Vapor Deposition (PECVD) systems for Thin Film Transistor Liquid Crystal Display (TFT-LCD) manufacturing. Improving throughput has been approached from several directions including larger panel sizes, batch style or multi-chamber deposition designs, high deposition rates, high etch rate insitu chamber cleaning with long intervals between clean cycles, and efficient robot panel handling/heating/load-lock systems. All these design considerations may influence device performance; however, it is the deposition rate of the a-Si:H that appears to have the largest impact on the ultimate transistor mobilities.

High yield requires excellent film uniformities in thickness, and in electrical and chemical properties extending typically to within 1 cm from the edge of the panel (including a ~ 5 mm exclusion zone of no deposition). Since pinhole densities in the gate nitride film must be at or below 0.01 / cm^2 for high resolution displays, careful attention must be paid to particle generation

9

and control. Many of these requirements are working against each other, such as high deposition rate and low number of cleaning cycles vs low particle generation, and thus both special techniques and compromises must be devised. Increasing the RF frequency from the conventional 13.56 MHz to VHF has demonstrated several clear advantages for the manufacturing of TFTs including higher deposition rates, lower particle generation and superior film properties [1, 2, 3].

In the simplest sense, a good TFT should have a low "off" current, a low threshold voltage and a high "on" current. Most a-Si:H TFTs, especially the standard etch-stopper inverted staggered type with a 500 Å a-Si:H layer, have a very low off current (typically $< 10^{-13}$ A) over a wide range of material deposition and device fabrication conditions. All of the TFTs studied for this paper are of this type. Threshold voltages (V_{th}) should be below 3V to limit power consumption by the TFT array. In our experience, studies involving this parameter demonstrate a large enough scatter in the data (in the 1 to 3V range) to suggest that V_{th} is more sensitive to device preparation conditions than to the film deposition conditions, and no definitive correlations are observed between deposition rates and V_{th}. Achieving a high on current is a more difficult challenge. Aside from device dimensions, the only device parameter that is observed to be correlated with the on current is the field effect mobility, μ, which is proportional to the drain current of the TFT. Although this parameter is also affected by the many processing steps involved in fabricating the TFT array, it seems to be more directly determined by the PECVD film depositions, and primarily the a-Si:H layer. One explanation for the observed dependence of TFT mobility on deposition rate is a correlation between the Urbach energy of the material and the mobility [4]. This study suggests a relationship between the valence and conduction band tail state distribution, and proposes that the decrease in mobility with the a-Si:H deposition rate results from a more broad conduction band tail due to a more disordered silicon network.

According to analysis by Howard [5], mobility is the crucial parameter in determining the limitations in size and resolution of AMLCDs. This study concluded that a-Si:H TFT technology is adequate for high resolution displays up to 40 inches in diagonal, assuming advances in lithography tools, device structures and other processing techniques. With a given tool set and technology base, however, mobility controls "ΔV_p", or change in pixel voltage, which is a major factor in limiting the performance of AMLCDs. When a TFT is shut off the charge in its channel must redistribute itself, and half of it moves through the source electrode to the pixel and any storage or parasitic capacitors connected to this node. This charge changes the voltage on the pixel from the desired data line voltage by an amount ΔV_p, which can be shown to scale as:

$$\Delta V_p \propto L(L = 2\Delta L) r D / \mu \qquad (1)$$

Where L is the length of the gate electrode, ΔL is the overlap between the gate electrode and the source and drain electrodes, μ is the electron mobility, r is the display resolution in units of lines per length and D is the diagonal size of the display. A fixed or known ΔV_p can be accounted for by a voltage on the common electrode or by adjusting the data voltage. The ΔV_p must be kept small, however, to keep driving schemes simple and because non-uniformities in ΔL across the display cannot be easily compensated [5].

As an example of TFT device performance requirements for acceptable display quality using an exposure tool with a 3 µm overlay accuracy (ΔL) to fabricate etch-stopper type TFTs, and with L = 11µm (meaning a 5 µm TFT channel length), in order to make a 10.4 inch VGA (640 x 480) display a linear mobility of 0.45 cm²/Vs is required. Using the same tool set and TFT structure to make a 12.2 inch SXGA (1280 x 1024) display, the mobility must become 1.13 cm²/Vs. As mentioned earlier, other display technology advances such as using a self-aligned

TFT which would reduce L and ΔL to values as low as 5 and 1μm, respectively, would also contribute to allowing for larger displays.

Of all these possible contradictory influences on throughput, yield, and device performance, perhaps the most important compromises must be made between the quality vs deposition rate of the a-Si:H thin film. The deposition rates of both the SiN_x and the a-Si:H layers need to be 1000 Å/min or more, depending on the structure of the transistors, for single panel PECVD systems to be economical. This is not difficult for the SiN_x layer where both high frequency and standard 13.56 MHz deposition systems can more or less just increase the RF power density and adjust gas flows appropriately without any serious costs to film performance. For a-Si:H, however, it has been shown that increasing deposition rate of a-Si:H is directly correlated to lower TFT mobilities for both conventional 13.56 MHz [6, 7, 8, 9], and to a lesser extent for high frequency RF when SiH_4 is diluted in H_2 [10, 11]. It has been proposed that this decrease in device performance for 13.56 MHz deposited material is correlated to an increase in $Si-H_2$ bonds which act as electron traps. These bonds are essentially non-existent in 40 or 60 MHz deposited a-Si:H [10]. In addition, increasing deposition rates by increasing RF power or process pressure results in a correlated increase in the gas phase polymerization of the silane leading to increase in particle incorporated in the film.

Increasing the RF frequency more efficiently couples the electric power into the shi_m disassociation as opposed to vibration excitation. It also has been shown that the peak-to-peak RF voltage decreases as the frequency increases, thus the maximum ion energy at the substrate at 40 MHz would be approximately three times lower than at 13.56 Mhz, even though the ion flux would be actually higher at higher RF frequencies because of the growth of the high energy tail of the electron energy distribution function [2].

The high ion flux plus low energy ion bombardment are considered to be responsible for the maintaining of high quality a-Si:H film optoelectronic and microstructural properties while increasing the deposition rate of VHF reactors. These properties degrade much quicker in conventional 13.56 MHz systems because the ion impact energy becomes too high as the power increases. The ions will induce collision cascades or thermal spikes as the ion energy increases and penetrates further into the growth zone. This is likely to create defects which remain trapped in the bulk of the film [2]. It therefore seems that, although compromises between high deposition rate a-Si:H and TFT performance still have to be made for VHF PECVD systems, much higher deposition rates are possible for a given TFT performance requirement than with conventional 13.56 MHz discharges.

EXPERIMENT

Deposition System

All samples and TFTs were made using a TEL high throughput, multi-chamber, face down, cross-flow VHF production system described in reference [1]. The type of film deposited in each chamber depends on the type of TFT film stack that is required. In order to maximize throughput for an etch-stopper style TFT which requires only ~ 500 Å of a-Si:H and n^+ a-Si:H, but ~ 3000 Å of SiN_x, three chambers would be dedicated to SiN_x and one each for the intrinsic and doped a-Si:H. Two chambers dedicated to both SiNx and intrinsic a-Si:H and one for n^+ a-Si:H would be implemented for the back-channel-etch type TFT where ~ 3000 Å of a-Si:H is required. It is also possible to deposit both SiNx and a-Si:H in a single chamber, however, it is generally preferable to separate the two for purposes of avoiding nitrogen contamination in the intrinsic layer and allowing for different deposition temperatures of the two films.

The process chambers in this system differ from conventional PECVD reactors in three significant ways: (1) VHF RF is used, either 40 or 60 MHz as opposed to the conventional 13.56 MHz, (2) the gas flows across the panels from one long edge to be pumped at the other, as opposed to the conventional showerhead, and (3) the film is deposited face down. The VHF is perhaps the most significant difference for influencing the optoelectronic properties of the films for reasons to be discussed later. The VHF also allows for lower particle generation for a given deposition rate. Both the cross-flow and the face down features are primarily employed to minimize particle contamination in the films. A featureless RF electrode does not have the corners and edges a showerhead provides as sources for micro-cracking and peeling of film, and the face down configuration greatly reduces the possibility of particles to become incorporated in the films. Together they allow for a much longer interval between cleaning cycles and thus a higher throughput.

Film Depositions

All films discussed in this study were deposited in a system similar to the one described above. The samples were deposited at either 60 or 40 Mhz but little difference in the optoelectronic properties of the a-Si:H or performance of the TFTs were detected. All SiN_x films used in the TFT structures were deposited at substrate temperatures between 250 and 325°C and at a rate of 1200 Å/min with a power density of 0.6 W/cm^2. These films have a refractive index of 1.88 +- 0.01 , a breakdown field > 9MV/cm, etch rates < 200 Å/min (for 325°C deposited samples) and < 800 A/min (for 250°C deposited samples) in a 9:1 BOE etch bath.

The a-Si:H films were deposited over a range of power densities from 30 to 600 mW/cm^2 with deposition rates from 100 to 1500 Å/min. Substrate temperatures were held at 250°C, and samples were deposited using pure silane and Ar and H_2 diluted silane in 5 : 1 ratios. The process pressure was 100 mTorr for the pure SiH_4 samples, and 500 mTorr for the diluted depositions.

RESULTS and DISCUSSION

The deposition rate as a function of 40 MHz RF power for films deposited with pure SiH_4 and 5 : 1 dilutions of Ar and H_2 is shown in Figure 1. A different linear dependence on RF power density and deposition rate is demonstrated for each of the gas mixtures. The lower deposition rates observed for the H_2 diluted process are probably due to the competing etch process of the H_2 plasma on the film. The increased deposition rate with the 5 : 1 Ar dilution of SiH_4 is probably due to the increased plasma density the Ar provides, thus allowing for more efficient SiH_4 dissociation. Other process issues such as plasma stability or film thickness uniformity can influence the use of any of these gas recipes for eventual LCD

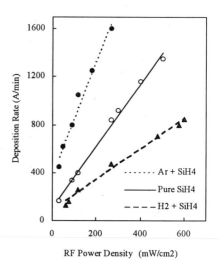

Figure 1. Deposition rate vs 40 MHz RF power. The SiH_4 dilution ratio in Ar and H_2 was 1 : 5.

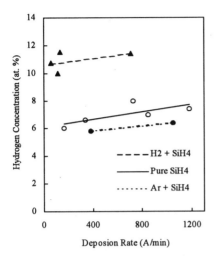

Figure 2. Hydrogen concentration vs deposition rate for samples deposited with pure SiH₄ and diluted 1 : 5 in Ar and H₂. The Hydrogen concentration was determined by FTIR and calibrated to SIMS analysis performed on three of the samples.

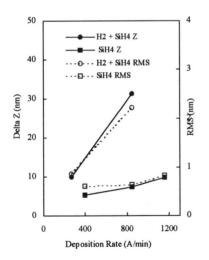

Figure 3. Atomic Force Microscope (AFM) measured surface roughness over a 2 x 2 micron area. The left axis represents the maximum height difference (solid lines) and the right axis is the RMS values over the same area (dotted lines).

manufacturing; however, it is likely that solutions to these and other production type concerns can be found if a clear improvement in the performance of the device is determined.

For all depositions represented in Figure 1, no visible powder was produced in the reactor. Here in lies one of the distinct advantages of VHF discharges over conventional 13.56 MHz : the ability to both produce a higher deposition rate, as well as produce less particles, for a given power density. Dorier et. al. [12] demonstrated that the threshold of gas phase nucleation, or powder formation, in glow discharge SiH₄ plasmas as a function of RF power increases with the plasma frequency. Since for a given RF power density the ion flux, and thus the deposition rate, will be already higher for a VHF plasma it is clear that the particle generation for the 40 MHz reactor should be considerably improved in comparison with the 13.56 MHz systems. Indeed less than 50 particles > 1 μm are measured on 360 x 465 mm panels after more than 10 μm of depositions for both the pure and H₂ diluted SiH₄ films deposited at rates > 1000 Å/min. Particle studies with Ar dilution have not yet been tried.

SIMS analysis show the oxygen, nitrogen and carbon impurity levels at < 5 x 10^{19}cm^{-3} for representative samples. The hydrogen concentration was measured by integrating the rocking mode absorption peak centered at 650 cm^{-1} and calibrated to SIMS measurements performed on three of the samples. All hydrogen appears to be bound as monohydride for all samples represented in Figure 2. The absence of any 2100 cm^{-1} absorption (within the detectability of the FTIR measurement of ~ 1 at. %) indicates that there is very little Si-H₂ in the films and the void fraction is low for all samples studied. Figure 2 shows the hydrogen concentration as a function of deposition rate for the pure and diluted samples. Clearly the H₂ diluted samples have the most hydrogen, however, no clear dependence on RF density or deposition rate is observed. The incorporation of more hydrogen into the films with increasing RF power does not appear to be the source of the decrease in film and device performance for the high deposition rate material.

An increase in surface roughness; how ever, is observed as the RF power density and deposition rate is increased. The AFM measured surface roughness area is shown in Figure 3 for H_2 diluted and pure SiH_4 samples. The delta Z data represent the difference between the maximum and minimum heights measured over a 2 x 2 μm area and the right Y axis represents the RMS of the height differences measured over the same area. The H_2 clearly promotes an uneven surface structure which increases dramatically with higher RF power, perhaps due to the etching behavior of the H_2 plasma. How this might effect the performance of the TFT is unclear, but it is generally accepted that a rough interface between the underlying SiN_x and the a-Si:H does result in an increase of electron traps and a concomitant decrease in the transistor's electron mobilities [13]. The top surface of the SiN_x probably plays a more important role in this regard, but the microstructure of the a-Si:H does also indicate the level of disorder in the film. This effect can also be correlated to the RF frequency. It is known that the higher frequency plasmas result in not only higher ion fluxes for a given power density, but also lower average ion energies which likely result in different (probably smoother) surface microstructure.

The optoelectronic properties were measured for films spanning a wide range of RF power densities for a-Si:H films deposited using pure SiH_4 and 5 : 1 dilutions in Ar and H_2 . The dark and photo conductivities are shown in Figure 4. The photo conductivity values are for AM 1.5 light intensity. The dark conductivities begin to rise for samples deposited at > 800 Å/min, with the corresponding decrease of the photo-to-dark conductivity ratio occurring around the same position as observed in Figure 5. The decrease in the photo-to-dark conductivity is not observed in the one high deposition rate Ar diluted sample, and this may be due to the fact that the Ar diluted films are

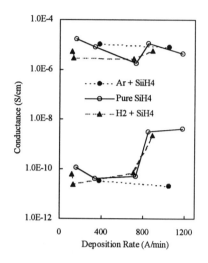

Figure 4. Photo and dark conductivitites vs deposition rate for pure SiH4 and 5 : 1 diluted Ar and H_2 samples. Photoconductivity values were determined using AM 1.5 white light.

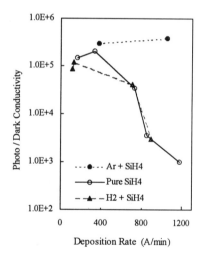

Figure 5. Photo-to-dark conductivity ratio vs deposition rate for samples deposited with pure SiH_4 and diluted with 5 : 1 Ar and H_2. Photoconductivity values were determined using AM 1.5 white light.

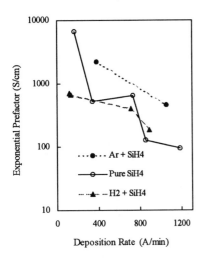

Figure 6. The conductivity exponential prefactor vs deposition rate for pure SiH₄ and 5 : 1 diluted in Ar and H₂ samples.

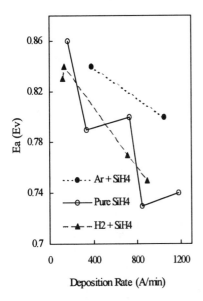

Figure 7. Thermal activation energy vs deposition rate for pure SiH4 and 5 : 1 dilutions in Ar and H2. The activation energy values were determined after the samples were held at 190 C for 30 minutes and then cooled 2 C/min.

deposited at a much lower RF power, as shown in Figure 1.

Similar to Figures 4 and 5, Figures 6 and 7 also indicate a significant decrease in optoelectronic performance at the higher RF powers or deposition rates. Samples were heated at 2°C/min to 190°C, held there for one hour, and then cooled at the same rate to room temperature. The thermal activation energies were obtained from the slope of the Arrhenius plots, and the exponential pre-factor values were taken from the y-axis intercepts of the same plots. The decreasing exponential pre-factor with deposition rate shown in Figure 6 is consistent with the photo-response of these same samples (Figure 5), indicating that a significant fraction of the excited electrons are no longer in extended states but probably in broadened localized band tail states. The continuous decline in the thermal activation energies with increased deposition rate is also be consistent with the broadening of the band tails as well as an increasing and/or shifting of the midgap defect states. It is important to note that even though the optoelectronic performance is decreasing with increasing deposition rate, that even the highest deposition rate materials still represent high quality films that meet the requirements for LCD production.

Defect densities which were determined by the constant photo-current method (CPM) [14] for samples deposited in a TEL 40 MHz CVD system with a 5 : 1 H₂ dilution are shown in Figure 8 [4]. Although the results show a correlation between increasing deposition rate and disorder, the figure indicates all the samples to have suprisingly low defect densities ($< 10^{16}$ cm^{-3}) even for material deposited at > 25 A /sec. The Urbach energies determined by CPM (Figure 8) on the same set of samples exhibit a similar behavior, implying a low but broadening tail state distribution with higher RF power. In fact the Urbach energy correlated surprisingly well with the linear mobility of the TFTs made with identical a-Si:H films. This correlation suggests a relationship between the valence band tail slope, which is measured by the Urbach energy, and the conduction band tail slope which largely controls the TFT field effect mobility. This correlation is

15

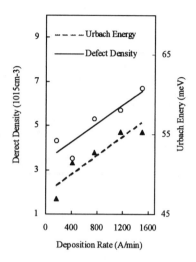

Figure 8. Defect density (solid line) and Urbach energies (dashed line) vs deposition rate. All values were determined by CPM, and all depositions had 5 : 1 dilutions of H2 : SiH4. [4]

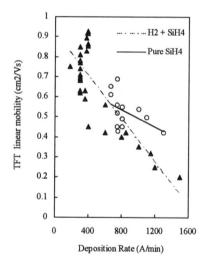

Figure 9. TFT linear mobility vs deposition rate for devices whose a-Si:H layers were deposited with pure SiH4 and diluted with H2. Other deposition parameters and device processing conditions varied slightly for this data set. See text for details.

discussed by Sherman and Wagner in reference [4].

As mentioned in the introduction, the field effect mobility is both the most sensitive and the most significant indicator of the TFT performance. It can be measured in two ways, either as the linear or saturated mobility, referring to the regime of the drain current vs drain characteristic in which the device is operating. To measure the linear mobility, the drain voltage, V_d, is held much smaller than the gate voltage, Vg. For example, V_d = 0.1V, -5V < V_g < 30V; and the drain current is assumed to be proportional to $\mu(V_g - V_{th})$ over most of the sweep. To measure saturated mobility, the gate voltage and drain voltage are held equal and are swept over a similar range as the V_g for the linear mobility. Far above threshold the drain current is assumed to be proportional to $\mu(V_g{}^2)$. The linear mobility is the more relevant measurement for AMLCD application since a typical drive scheme might use 15 volts on the gate while varying the drain voltage between 0 and 5 volts depending on the gray scale required, and thus would never reach the saturated regime. Furthermore, because the pixel charges up exponentially, the majority of the time the TFT is turned on there is very little voltage difference between V_d and V_g (or data line and pixel). The linear mobility is typically half to two thirds the saturated values. Saturated mobility values are still often quoted, and for historical reasons will also be discussed here.

Figures 9 and 10 show the general compromise between deposition rate and TFT saturated and linear mobility. All TFT data presented here represent devices processed at four different locations and using slightly different device structure and dimensions. They were all, however, deposited in TEL 40 or 60 MHz face-down cross-flow CVD tools. Most of the data points represent the average mobility of several devices measured on a single panel. Although most of the a-Si:H deposition parameters such as temperature, gas pressure and gas flows were held constant, much of the scatter in the data of Figures 9 and 10 likely represents the different processing conditions, other than the film depositions, used for fabricating the devices. The general trend, however, does show a compromise between a-Si:H deposition rate and TFT performance as reported elsewhere [6, 7, 8, 9]. In fact Ibaraki et al., [6] show a plot of TFT mobility vs deposition rate for 13.56 Mhz deposited films that exhibits a very similar slope as the pure SiH$_4$ data in Figure 10, only with the mobilities shifted down ~ 30% in comparison to the 40 Mhz data. Figures 9 and 10 s do indicate that high mobility TFTs are possible using pure SiH$_4$ feed gas in the 40 MHz TEL tool at deposition rates > 1000 Å / min. with saturated and linear mobilities > 1.0 and 0.5 cm^2/Vs, respectively.

CONCLUSION

For economical production of AMLCDs, a deposition rate of > 1000 Å/min is required for both the a-Si:H and the SiN$_x$ films, while maintaining high quality optoelectronic properties, good thickness uniformities and low particle generation. This is relatively easier for the SiNx film than for the a-Si:H layer. The TFT performance is closely correlated to the a-Si:H quality which, in turn, has been demonstrated to be inversely correlated to the RF power and/or deposition rate. The data presented in this paper consistently indicate a general increase in material

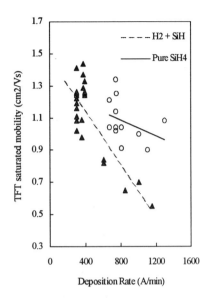

Figure 10. TFT saturated mobility vs deposition rate for devices whose a-Si:H layers were deposited with pure SiH4 and diluted with H2. Other deposition parameters and device processing conditions varied slightly for this data set. See text for details.

disorder with increasing a-Si:H deposition rate; however, a-Si:H films with low defect density (< 10^{16}cm^{-3}), high photo-to-dark conductivity ratios (> 10^4), low hydrogen concentration (< 8 at.%) and high thermal activation energies (>0.78 eV) as well as low particle generation and excellent thickness uniformity (< +/- 6 % with 10 mm exclusion) can still be obtained at deposition rates > 1000 Å/min with the TEL 40 MHz PECVD. Films deposited with pure SiH$_4$ appear to have superior properties for higher deposition rates compared with those implementing a 5 : 1 dilution of H$_2$. This may be due to the significantly lower RF power required for the pure SiH$_4$ process for the same deposition rate. The Ar diluted material requires even less RF power than the pure SiH$_4$ for a given deposition rate and exhibits yet even better optoelectronic properties. The increase in ion flux with a decrease in average ion energy that is produced in VHF plasmas when compared with standard 13.56 MHz reactors enhances the deposition rate while maintaining or possibly improving film quality. With these films the desired values of > 1.0 cm^2/Vs for saturated mobility or > 0.45 cm^2/Vs for linear mobilities for VGA displays can be obtained at deposition rates > 1000 Å/min, which is an approximate 30% improvement over the published data of a similar study implementing a 13.56 Mhz deposition system [6]. Increased demands on the performance of the TFTs will certainly come with SXGA displays, and fully understanding and eventually controlling the mechanism of the a-Si:H film degradation with increasing deposition rate will impact the economic feasibility of large area flat panel active matrix displays.

ACKNOWLEDGMENTS

The authors would like to gratefully acknowledge the TFT processing groups at IBM Research, Yorktown Heights and at AT&T Bell Labs, Murray Hill .

REFERENCES

1.   J.F.M. Westendorp, H. Meiling, J.D. Pollock, D.W. Berrian, A.H. Laflamme Jr., J. Hautala and J. Vanderpot, in *Flat Panel Display Materials*, edited by J. Batey, A. Chiang and P.H. Holloway (Mat. Res. Soc. Symp. Proc. 345, Pittsburgh, PA, 1994) p. 175.

2.   J. Perrin, in <u>*Plasma Deposition of Amorphous Silicon-based Materials*</u>, edited by G. Bruno, P. Capezzuto and A. Madan, Academic Press, 1995, pp.216-18.

3.   H. Curtins, N. Wyrsch, M. Favre and A.V. Shaw, Plasma Chem. Plasma Process. 7 (3). 267, (1987).

4.   S. Sherman and S. Wagner, AMLCDs '95 Workshop, Bethlehem, PA (1995)

5.   W.E. Howard, J. of the SID, 3, #3, 127 (1995)

6.   N. Ibaraki, K. Matsumura, K. Fukuda, N. Hirata, S. Kawamura and T. Kashiro, in 1994 Display Manufacturing Technology Conference (Society for Information Display, Playa del Rey, CA, 1994), pp. 121-2.

7.   P. Roca i Cabarrocas, J. Non-Cryst. Solids 164-166. 37 (1993).

8.   H. Miyashita and Y. Watabe, IEEE Trans. Electron Dev. 41, 499 (1994).

9.   Y. Kuo, in *Amorphous Silicon Technology - 1995*, edited by M. Hack, E.A. Schiff, A. Madan , M. Powell and A. Matsuda, (Mat. Res. Soc. Symp. Proc. 377, Pittsburgh, 1995) p. 701.

10.   H. Meiling, J.F.M. Westendorp, J. Hautala, Z.M. Saleh and C.T. Malone in *Flat Panel Display Materials*, edited by J. Batey, A. Chiang and P.H. Holloway (Mat. Res. Soc. Symp. Proc. 345, Pittsburgh, PA, 1994) p. 65.

11.   S. Sherman, P.Y. Lu, R.A. Gottscho and S. Wagner, in *Amorphous Silicon Technology - 1995*, edited by M. Hack, E.A. Schiff, A. Madan , M. Powell and A. Matsuda, (Mat. Res. Soc. Symp. Proc. 377, Pittsburgh, PA 1995) p.749.

12.   J.L. Dorier, C. Hollenstein, A.A. Howling and U. Kroll, J. Vac. Sci. Technol. A 10 (4) 1048 (1992).

13.   H. Uchida, K. Takechi, S. Nishida and S Kanecko, Jpn. J. Appl. Phys. 30 (12B), 3691 (1991).

14.   M. Vanecek, J. Kocka, J. Stuchlik, Z. Kozisek, O. Stika and A. Triska, Solar Energy Mat., 8, 411 (1983).

SURFACE ROUGHNESS OF SILICON-NITRIDE GATE INSULATORS DEPOSITED IN A 40-MHZ GLOW DISCHARGE

H. MEILING, E. TEN GROTENHUIS, W. F. VAN DER WEG, J. J. HAUTALA[1], AND J. F. M. WESTENDORP[1]
Debye Institute, Utrecht University, P.O. Box 80.000, NL-3508 TA Utrecht, The Netherlands.
[1]Tokyo Electron America, Inc., 123 Brimbal Avenue, Beverly MA 01915, USA.

ABSTRACT

The surface morphology of 40-MHz PECVD SiN_x films is investigated. We report on the correlation between the deposition conditions, bulk properties, and surface roughness of these TFT insulators. The roughness is measured with atomic-force microscopy, AFM. A link will be presented between the AFM properties and the effects of hydrogen dilution during deposition: gas composition and rf-power-density (P) dependences will be discussed. An increase of the surface roughness to 3.7 nm is observed upon H_2 dilution and P increase, ascribed to enhanced ion bombardment of the surface during growth.

INTRODUCTION

Hydrogenated amorphous silicon, a-Si:H, has been the focus point of many research papers in the last twenty years. Apart form solar cells, most studies on devices focus on thin-film transistors, TFTs. Fabrication of these TFTs starts with the coating of the gate metal with an insulator, usually hydrogenated amorphous silicon nitride, a-SiN_x:H or short SiN_x, or silicon dioxide, SiO_2. Then, in the same pumpdown the a-Si:H channel and n^+a-Si:H contact layers are deposited. Charge transport in these devices takes place along the interface of the insulator and semiconductor, through a layer of typically 10 nm thickness. Therefore knowledge of the chemical composition of the interface and the physical and chemical processes that take place at the interface is of crucial importance to the understanding of the device behaviour.

Many papers have been published on the relation between the bulk properties and deposition parameters for plasma-enhanced-chemical-vapour-deposition (PECVD) SiN_x [1–4]. SiN_x can be synthesized from mixtures of SiH_4, NH_3, N_2, and H_2. Sometimes He or Ar dilution is used, e.g. in depositions at low substrate temperatures [5,6]. Recently, deposition of SiN_x was reviewed by Kuo [7] and by Habraken and Kuiper [8]. The key to make good SiN_x insulators, i.e. with low electron trapping rates, is to reduce the amount of Si–H bonds in the material. From mass spectroscopy experiments during deposition from SiH_4 and NH_3 it was deduced that highly insulating nitrides with low defect densities can be made if the creation of $SiH_2(NH_2)_2$, diaminosilane, and the byproduct Si_2H_6 is suppressed [3]. The main precursors are then thought to be $Si(NH_2)_4$, tetra-aminosilane, in which the Si atom is completely saturated with N–H$_2$ aminogroups, and the triaminosilane radical •$Si(NH_2)_3$. Electron trapping rates of films made under these conditions are as low as in PECVD SiO_2.

Several studies have been published about the growth of a-Si:H on various substrates [9,10]. However, a detailed study of the growth process of a-Si:H on SiN_x is virtually non-existent, especially for material made at a higher rf frequency than the conventional 13.56 MHz. Only

19

recently some papers were published on the performance of TFTs with a smooth versus rough interface [11]. It was shown that TFTs with a smooth interface have a higher saturation mobility and show less shift of the threshold voltage upon prolonged gate voltage stress than TFTs with a rough interface. The explanation is that the TFTs with a smooth interface have a lower interface defect density. No details were given as to how the smooth versus rough interfaces were made. In this paper we link the deposition chemistry to the surface roughness for 40-MHz SiN_x films made from $SiH_4/NH_3/H_2$ mixtures. Parameters is the rf power density.

TABLE I:
Process conditions used for SiN_x deposition.

$\nu =$	40 MHz
$T_{sub} =$	320 °C
$\varphi_{SiH4} =$	220 sccm
$\varphi_{NH3} =$	2400 sccm
$\varphi_{H2} =$	2400 sccm
$p =$	100 Pa
$P_{rf} =$	200–1000 W
area =	1674 cm^2

(sccm = standard cm^3/min)

SAMPLE PREPARATION AND EXPERIMENTAL

The SiN_x films were deposited in a PECVD system that was designed for the manufacturing of the active matrix of liquid-crystal displays. High-mobility TFTs can be made on a large surface area in this production tool [12,13]. Previous studies done in this system investigated the effects of deposition-rate variation and H_2 dilution of SiH_4 during channel deposition [14] on TFT properties. In our experiments, variations in the surface morphology of the SiN_x films are induced by changing the rf power and by replacing H_2 with N_2 in the deposition process. The deposition parameters can be seen in Table I. The thickness of all films used in this study amounts to ~300 nm, which is the typical thickness of the insulator in the TFT device.

The measurements of the surface morphology are done using a Nanoscope III (Digital Instruments, Inc.) atomic-force microscope under atmospheric conditions. Microfabricated cantilevers with an integrated Si_3N_4 pyramidal tip are used. The length of the cantilevers is 100 μm, their spring constant 0.6 N/m. All images are acquired without filtering in constant-force mode, with a scan speed of 2 lines/s. Because AFM images represent the interaction of the tip with the real sample topography, they must be interpreted carefully. The acquired image is a convolution of the sample topography and the shape of the pyramidal tip. Only when the tip features are smaller than the features in the film an accurate representation of the roughness is obtained. Inherent to the technique, there is the potential of morphology modification by the tip when the measurement is done in constant-force mode. Especially when SiN_x films are investigated, as compared to e.g. a-Si:H films, the mechanical hardness of the film may be a problem and tip-scraping can be observed, resulting in low-quality images. As a measure for the surface roughness we use the root-mean-square (rms) value, which is the standard deviation of the measured heights within a surface area of 0.5 μm × 0.5 μm. The maximum height difference in this area is obtained by calculating $\Delta Z = Z_{max} - Z_{min}$, with Z_{max} and Z_{min} the maximum and minimum height in the measured area, respectively. The value of ΔZ is sensitive to the presence of particulates, formed during the deposition process, that are incorporated in the film.

Bulk properties of the SiN_x films are deduced from several standard analysis techniques, including measurement of the infra-red absorption spectra, the mechanical stress, and the refractive index n. The hydrogen bonding configuration and the total H concentration in the films is deduced from the FTIR spectra, using the conversion factors of Lanford and Rand [15]. The characteristic absorptions in the FTIR spectra are correlated with the vibrational modes of the chemical bonds as described in Table II.

RESULTS

Depositions under the process conditions of Table I result in a deposition rate between 0.35 and 1.38 nm/s, as can be seen in Fig. 1. In this graph the circles denote the deposition rate and the squares represent the refractive index n. The solid lines are fits through the data. The

TABLE II:
Characteristic FTIR absorption frequencies for PECVD and LPCVD SiN_x films (from [4,16]).

Mode	PECVD		LPCVD	
Si–N stretching	840–875	cm^{-1}	835	cm^{-1}
Si–H stretching	2060–2180	cm^{-1}	2210	cm^{-1}
N–H stretching	3320–3340	cm^{-1}	3330	cm^{-1}
N–H_2 stretching	1550	cm^{-1}		
N–H bending	1180	cm^{-1}		

monotonous increase of the deposition rate with increasing P starts to deviate from a straight line (the dotted line in Fig. 1) when P becomes larger than 350 mW/cm^2. At that P the refractive index has decreased from 1.91 to 1.87, and remains virtually constant with further P increase.

The Si–H and N–H bond densities are plotted in Fig. 2. The open symbols are the results of the calculations for discharges in which H_2 dilution was used, the closed symbols are those for depositions in which the H_2 was replaced by an equal flow rate of N_2. Films with N_2 dilution were made for comparison with conventional SiN_x deposition processes. The circles are the Si–H bond density, the squares represent the N–H bond density. For H_2-diluted discharges the N–H bond density in the film increases with P from 3×10^{22} cm^{-3} to 4.5×10^{22} cm^{-3}. The Si–H bond density remains nearly constant. When H_2 is replaced by N_2 the trends are very similar, only the Si–H content is slightly higher. Replacing H_2 with N_2 also results in a shift to lower energy of the Si–H stretching-mode peak, from 2191 to 2176 cm^{-1}.

In Fig. 3 the results are plotted of measurements of the mechanical stress in the films $vs.$ P, the stress being positive when compressive and negative when tensile. Symbol definition is the same as in Fig. 2. Upon increase of P the stress becomes more compressive for the films deposited with H_2, and becomes less tensile for films from N_2-diluted depositions. Films with a very low tensile stress, i.e. below 100 MPa, can be deposited if N_2 dilution is used at a relatively high P.

Finally, in Fig. 4 we have plotted the results of the AFM measurements. The circles represent the rms value, the squares ΔZ. The surface roughness of the films increases with P. The increase of rms appears to level off at a value of 0.8 nm when $P > 500$ mW/cm^2, but the increase of ΔZ is linear with P and surpasses 3.7 nm. This high value of ΔZ implies that the height differences between peaks and valleys is a considerable fraction (~25 %) of the channel thickness in TFTs.

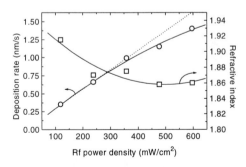

FIG. 1: Deposition rate and refractive index as a function of P, for SiN_x grown from SiH_4, NH_3, and H_2.

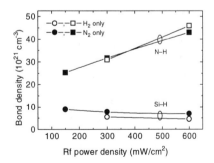

FIG. 2: Hydrogen FTIR bond density versus P for H_2-diluted deposition and N_2-diluted deposition.

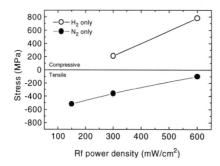

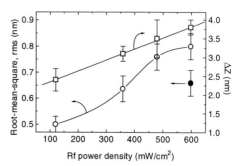

FIG. 3: Stress versus P for SiN$_x$ films made with H$_2$ dilution or N$_2$ dilution.

FIG. 4: AFM surface roughness $vs.$ P. Open symbols are for H$_2$-diluted deposition; the filled symbol is for a typical N$_2$-diluted deposition.

When in the discharge H$_2$ is replaced by N$_2$, the rms value of the roughness of the resulting film is considerably lower, as can be seen from the datapoint represented by the closed circle.

DISCUSSION

Any plasma-CVD process is actually a delicate balance between film growth and etching. If growth prevails a film is deposited, if etching prevails material is removed (or, no deposition takes place!). Various processes occur simultaneously, including dissociation and ionization of source-gas molecules, gas-phase reactions between molecules and radicals, diffusion of radicals to the surface of the growing film, and ion bombardment of the surface due to the potential difference between plasma bulk and substrate electrode. In addition, processes including H elimination at the surface of the film take place. These surface processes are strongly dependent on the substrate temperature and on the energy and flux of the ion bombardment.

The deviation from a straight line of the increase of the deposition rate with increasing P can be explained in terms of this balance between deposition and etching. At low P the dissociation of SiH$_4$ mainly determines the deposition rate, and a Si-rich film with $n > 1.90$ is obtained. Only at higher P the dissociation of H$_2$, i.e. the creation of etchants, starts playing a role because the dissociation energy of H$_2$ (~11eV) is higher than that of SiH$_4$ (~8 eV). In contrast to other reports [3,7] no decrease of the deposition rate at high P is observed. This indicates that the supply of SiH$_4$ is not the rate-limiting step and that deposition still prevails over etching. The decrease of the refractive index n from 1.91 to 1.86 when P increases from 100 to 300 mW/cm^2 implies either a decrease of the Si-N density or an increased H concentration in the film, or both. From Fig. 2 it can be seen that the N-H content increases monotonically with P, also beyond 300 mW/cm^2, while Si-H remains constant. Combining these results we conclude that beyond $P = 300$ mW/cm^2 the Si-N density in the film increases. This is confirmed by the shift of the peak position of the Si-H stretching mode to higher energy when H$_2$ is used instead of N$_2$, due to an enhanced local electronegativity caused by a larger N/Si ratio in the film.

There does not appear to be much difference in bulk properties of SiN$_x$ films deposited from SiH$_4$ and NH$_3$ diluted with H$_2$ versus N$_2$, as far as the H bonding configuration is concerned. However, the stress values of Fig. 3 indicate considerable differences in the Si-N lattice construction. The results of Fig. 3 correspond very well with the trends reported earlier in literature on P-dependence of bulk properties of SiN_x films made with H_2 dilution at 13.56 MHz [17]. Increased ion bombardment of the growing surface leads to a higher N/Si ratio in the film, which induces the stress to change from tensile to compressive [1,17]. This supports the idea that H_2 dilution during deposition (as well as a P increase) results in enhanced creation of positive ions, as compared to N_2 dilution.

We use this enhanced ion flux to explain the trends in the surface roughness of the films as shown in Fig. 4, and to form a model of the roughness creation from $SiH_4/NH_3/H_2$ discharges. Smith et al. [2] suggested for SiN_x deposition at 13.56 MHz from a SiH_4/NH_3 discharge the following: $SiH_4 + 4NH_3 \rightarrow Si(NH_2)_4 + 4H_2$, followed by $3Si(NH_2)_4 \rightarrow Si_3N_4 + 8NH_3$. After condensation of $\bullet Si(NH_2)_3$ radicals and $Si(NH_2)_4$ molecules on the surface of the film, H elimination through NH_3 desorption occurs and a dense Si–N network is formed.

When H_2 dilution is used there are additional highly energetic H^+ ions present. We then suggest the following growth mechanism, see Fig. 5. The white circles represent Si atoms, the grey circles N atoms, and the black dots H atoms. Dangling bonds are represented by a short dash, covalent bonds are represented by long bars. In Fig. 5(a) we present the low-P phase of the growth mechanism from a $SiH_4/NH_3/H_2$ discharge through $Si(NH_2)_4$ and $\bullet Si(NH_2)_3$ precursors, similar to the model of Smith et al., but with the addition of H^+ ions. Upon creation of H^+ ions breaking of weak bonds in the surface-layer of the growing film increases, which will be more pronounced at higher P. This enhances the formation and completion of a dense structure, explaining the observed increase in Si–N density. However, the enhanced ion flux upon H_2 dilution and increased P may also create dangling bonds at the surface of the film. This results in an enhanced sticking coefficient, hence in a reduced surface diffusivity, of growth precursors. Consequently, columnar growth is likely to occur due to a shadowing effect of stuck precursors, as is drawn schematically in Fig. 5(b). The shadowing effect is represented by the dashed lines, the limited surface diffusion of precursors is represented by the crossed-out dashed arrow. The

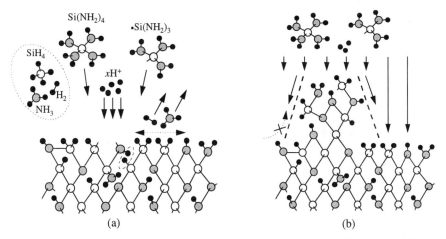

FIG. 5: Model for the formation of surface roughness of SiN_x films deposited from mixtures of SiH_4, NH_3, and H_2. (a) Initial stage, (b) at high rf power density P. White circles are Si-atoms, grey circles N-atoms, and black dots H-atoms.

combination of an increased sticking coefficient and this shadowing effect upon P increase will result in enhanced surface roughness, thus explaining the trends from Fig. 4.

CONCLUSIONS

The increase in surface roughness and changes in bulk properties of SiN_x films upon rf-power increase and H_2 dilution of SiH_4/NH_3 mixtures during deposition can be attributed to enhanced ion bombardment of the film during growth. This is likely to be caused by an increased amount of H^+ ions in the discharge. A reduction in surface roughness is seen upon substitution of H_2 by N_2. A factor of two reduction in the rms value can be obtained by changing the process parameters, stressing the importance of evaluating the surface morphology of these TFT gate insulators.

ACKNOWLEDGEMENT

The research of Dr. Meiling has been made possible by a fellowship of the Royal Netherlands Academy of Arts and Sciences.

REFERENCES

[1] W. A. P. Claassen, W. G. J. N. Valkenburg, M. F. C. Willemsen, and W. M. v. d. Wijgert, J. Electro-chem. Soc. **132** (4), 893 (1985).
[2] D. L. Smith, A. S. Alimonda, C. C. Chen, S. E. Ready, and B. Wacker, J. Electrochem. Soc. **137** (2), 614 (1990).
[3] D. L. Smith, J. Vac. Sci. Technol. A **11** (4), 1843 (1993).
[4] T. J. Cotler and J. Chapple-Sokol, J. Electrochem. Soc. **140** (7), 2071 (1993).
[5] G. N. Parsons, J. H. Souk, and J. Batey, J. Appl. Phys. **70** (3), 1553 (1991).
[6] M. J. Loboda and J. A. Seifferly, J. Mater. Res. **11** (2), 391 (1996).
[7] Y. Kuo, J. Electrochem. Soc. **142** (7), 2486 (1995).
[8] F. H. P. M. Habraken and A. E. T. Kuiper, Mat. Sci. Engineer. **R12** (3), 123 (1994).
[9] H. Shirai, Jpn. J. Appl. Phys. **34** (2A), 450 (1995).
[10] J. R. Abelson, Appl. Phys. A **56**, 493 (1993).
[11] H. Uchida, K. Takechi, S. Nishida, and S. Kaneko, Jpn. J. Appl. Phys. **30** (12B), 3691 (1991).
[12] H. Meiling, J. F. M. Westendorp, J. J. Hautala, Z. M. Saleh, and C. T. Malone, in *Flat Panel Display Materials*, edited by J. Batey, A. Chiang, and P. H. Holloway (Mater. Res. Soc. Proc. **345**, Pittsburgh, PA, 1994), p. 65.
[13] J. F. M. Westendorp, H. Meiling, J. D. Pollock, D. W. Berrian, A. H. Laflamme Jr., J. J. Hautala, and J. Vanderpot, in *Flat Panel Display Materials*, edited by J. Batey, A. Chiang, and P. H. Holloway (Mater. Res. Soc. Proc. **345**, Pittsburgh, PA, 1994), p. 175.
[14] S. Sherman, P.-Y. Lu, R. A. Gottscho, and S. Wagner, in *Amorphous Silicon Technology — 1995*, edited by M. Hack, E. A. Schiff, A. Madan, M. Powell, and A. Matsuda (Mater. Res. Soc. Proc. **377**, Pittsburgh, PA, 1995), p. 749.
[15] W. A. Lanford and M. J. Rand, J. Appl. Phys. **49** (4), 2473 (1978).
[16] J. Campmany, E. Bertran, J. L. Andújar, A. Canillas, J. M. López-Villegas, and J. R. Morante, in *Amorphous Silicon Technology — 1992*, edited by M. J. Thompson, Y. Hamakawa, P. G. LeComber, A. Madan, and E. A. Schiff (Mater. Res. Soc. Proc. **258**, Pittsburgh, PA, 1992), p. 643.
[17] I. Kobayashi, T. Ogawa, and S. Hotta, Jpn. J. Appl. Phys. **31**, 336 (1992).

A NOVEL DEVICE STRUCTURE FOR HIGH VOLTAGE, HIGH PERFORMANCE AMORPHOUS SILICON THIN-FILM TRANSISTORS

A. M. MIRI, P. S. GUDEM, S. G. CHAMBERLAIN and A. NATHAN
Department of Electrical and Computer Engineering, University of Waterloo
Waterloo, Ontario N2L 3G1, Canada

ABSTRACT

Conventional high voltage thin-film transistors (HVTFTs) suffer from performance limitations such as low on-current, V_x shift and large curvature in the linear region of the output characteristics. These limitations are associated with the highly resistive dead region in conventional HVTFT structures. In this paper, we present a novel TFT structure which has a high on-current, improved output characteristics in the linear region, and no V_x shift. The higher on-current and significant improvement in output characteristics allows faster switching. Elimination of the V_x shift leads to more reliable circuit operation. The new structure is based on the conventional low voltage TFT (LVTFT) structure except that it does not suffer from low-voltage breakdown. The low-voltage breakdown of the gate nitride in conventional LVTFTs is perceived to be due to spiking of the drain metallization into the underlying layers which creates regions of very high electric field. In our novel structure, a higher breakdown is achieved by locating the metal contacts away from the gate edge while keeping the necessary drain to gate overlap through a heavily doped microcrystalline layer. Therefore, the new TFT extends the same performance as LVTFTs to high voltage operation. Furthermore, this structure also enhances the yield and reliability by minimizing the common faults in TFTs such as short circuits between gate, source and drain.

INTRODUCTION

Amorphous silicon thin film transistors are widely used as low voltage pixel addressing elements in large area flat panel displays on account of low cost and low processing temperature. However, there are many applications, such as certain ferroelectric liquid crystals and electrographic plotters, that require drive voltages as high as a hundred volts or more [1,2], for which HVTFTs are needed.

Figure 1 shows a schematic cross section of a conventional a-Si:H HVTFT structure. It is also referred to as a gap-type HVTFT structure. The HVTFT can be thought of as an

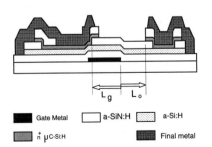

Figure 1: Cross section of a conventional high-voltage amorphous silicon thin-film transistor (HVTFT).

Gate Metal ▢ a-SiN:H ▨ a-Si:H

n^+ μC-Si:H ▨ Final metal

accumulation-mode MOSFET with an offset drain. The a-Si:H region spanned by L_g (Fig. 1) is called the "gated" region which is typical to a conventional LVTFT. It maintains a low electric potential approximately equal to that of the gate electrode. The region of a-Si:H spanned by L_o is referred to as the "offset", "drift", or "dead" region. A high drain voltage is dropped across the offset region thus preventing a large drain potential from appearing across the gate dielectric. This allows high voltage operation without causing breakdown. Figure 2 shows the output characteristics of a HVTFT with $L_g = L_o = 10\mu m$. The operating principles of HVTFTs are elaborated by Martin *et al.* [3,4].

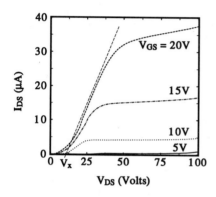

Figure 2: Output characteristic of a HVTFT (Fig. 1) with gate and offset lengths of $10\mu m$. The width of the HVTFT is $200\mu m$.

On examining the output characteristics of conventional HVTFTs, a striking and undesirable feature is the large curvature at small drain voltages. This is similar to the problem of high contact resistance in LVTFTs except that in HVTFTs the curvature is extended over several tens of volts. The other undesirable feature in conventional HVTFTs lies in the shift of output characteristics (I_D vs. V_D) towards higher drain voltages. This is referred to as V_x shift. The value of V_x is obtained by extrapolating the I_D versus V_D curve at the inflection point to zero current (as shown in Fig. 2). The shift in V_x arises from changes in the density of states in the intrinsic a-Si in the offset region near the gate edge [1]. When the transistor is turned off (gate below threshold voltage) and the drain is at a high voltage, a large electric field is created at the edge of the gate. This causes the creation of dangling bonds in this region which would act as electron traps. Thus, when the device is turned on, electrons from the source have to overcome the barrier created by local charges trapped in these defect states, before being injected into the offset region.

Both the above drawbacks (V_x shift and large curvature in the output characteristics) of conventional HVTFTs are due to its offset drain structure. Unfortunately, this is the very feature that is central in turning the LVTFT into a HVTFT structure.

THE HIGH VOLTAGE AND HIGH PERFORMANCE TFT STRUCTURE

Conventional LVTFTs can only withstand drain voltages of up to about 40 volts. Applying a high voltage to the drain causes the breakdown of the gate nitride dielectric at the drain side. However, it is our belief that the low-voltage breakdown of the gate nitride (around 40 volts), observed in LVTFTs, is due to the high electric-field regions which are created by spiking of the top metallization layer into the underlying layers. Experiments indicate that the gate nitride dielectric is affected by the top metallization at the edge of the gate metal line where the dielectric is thinned due to slope coverage limitations. To address the above issues, we propose the following structure as shown in Fig. 3.

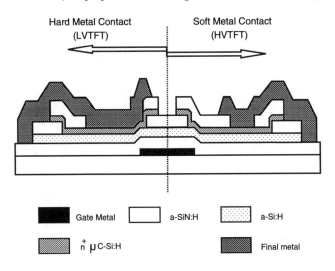

Figure 3: Cross section of our soft contact high-voltage amorphous silicon thin-film transistor.

This structure is exactly the same as our regular LVTFTs [5] except that the physical location of the final metal contact is moved away from the gate region, to prevent overlap between the top and the gate metallizations. However, the position of the actual drain contact, which is defined by the position of the highly conductive n^+ μC-Si:H layer, remains the same. We refer to this as a **"soft contact"** (**SC**) as opposed to the *"hard contact"* where the final metallization layer overlaps the gate line. We refer to our **"soft contact"** TFT structure as **SCTFT**. When the gate is turned on and the accumulation channel is formed, the electrons travel from the source downwards to the channel region and then move along the channel towards the drain. At the drain to gate overlap region, the electrons move upwards towards the n^+ μC-Si:H contact layer and onwards to the drain metallization. The key difference between the SCTFT and LVTFT lies in the separation of the drain metallization from the channel through a series path comprising of a n^+ μC-Si:H layer. Since the resistance of this layer is negligible compared to that of the accumulation channel, there are no adverse effects introduced in the TFT characteristics.

Fabrication of the SCTFT structure is one of the unique features of our recently developed a-Si:H TFT fabrication process [5]. In this process, the position of the metal contacts and n^+ layer are independent of each other. It is precisely this feature that allows fabrication of SCTFT. This is achieved by introducing a third nitride layer after deposition of the heavily doped layer. Contact to the device can be made at any location using the doped layer. The metal contact to the doped layer can then be made at a different location by a via through the third nitride. To the best of our knowledge, this is the only fabrication process that has been reported [5] which can accommodate the soft contact structure. In other TFT fabrication processes, the doped layer is covered by the final metallization and patterned together using a single mask. Therefore, to maintain the essential drain-to-gate overlap will result in the technological drawbacks discussed above. Furthermore, not having to simultaneously pattern the doped and final metallization layers allows a direct metal contact to the a-Si:H layer for realization of Schottky contacts.

EXPERIMENTAL RESULTS AND DISCUSSION

In Fig. 4, we compare two similar TFTs (aspect ratio of $200\mu m/20\mu m$), one with hard contacts and the other with soft contacts for the source and drain. Their characteristics are denoted by the solid and dotted lines, respectively. The inset of Fig. 4 shows the output characteristics of these TFTs in an operating range typical to LVTFTs. As expected, both devices behave identically in this region. As we increase the drain voltage to about 40 V, the characteristics of the LVTFT begin to show a rapid increase in the drain current. This

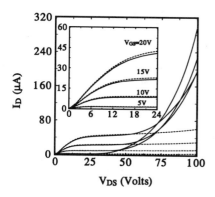

Figure 4: Characteristics of TFTs with hard (solid lines) and soft (dotted lines) contacts. The aspect ratio of the TFT is $200\mu m/20\mu m$.

is due to the breakdown of the gate dielectric. Interestingly, for the same applied voltage, no breakdown is observed for the SCTFT located on the same chip. A variety of measurements performed on devices fabricated with different gate dielectric nitride conditions, showed that the SCTFT consistently gives a very high breakdown voltage. From Fig. 4, we see that the SCTFT can accommodate drain voltages as high as 100 volts without breakdown of the gate dielectric. In fact, no noticeable gate leakage current was observed even up to 120V operation.

In Fig. 5, we plot the gate current versus the drain voltage. We observe that the gate current for the SCTFT is very low (in the range of pico Amps). This is in contrast to the LVTFT (hard contact TFT) which shows a large gate current at drain voltages above 40 V or so. The gate current is negative which indicates current flow to the gate from the drain. In fact, if we compare the behavior of the gate current of the LVTFT (hard contact) with the drain current after breakdown, we see a strong correlation.

We see that in the SCTFT, the gate dielectric can withstand drain voltages as high as 120 volts. In the LVTFT, the very same nitride does not withstand a voltage drop of 50 volts. Since the only difference between the soft and hard contact TFTs is the position of the final metallization relative to the edge of the gate, we attribute the low-voltage breakdown of the gate nitride to the position of the final metallization. At the edge of the gate,

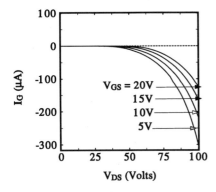

Figure 5: The gate leakage current vs. V_{DS} for two TFTs with hard contact (solid lines) and soft contacts (dotted lines). The aspect ratio of the TFT is $200\mu m/20\mu m$.

the overlying layers have a tensile stress, thus weakening the dielectric material structure. This can result in a magnification of microstructural pinholes in the thin film. Thus, any metal on this edge can penetrate into the underlying layers through these microstructural pinholes. The sharp metal spikes at the edge create regions of very high electric field inside the gate dielectric causing the low-voltage breakdown observed in conventional hard contact LVTFTs. We avoided this technological problem by moving the physical location of the final metallization away from the edge region while keeping the electrical contacts (defined by the n^+ μCSi:H layer) in the same location as with LVTFTs.

Therefore, unlike conventional HVTFTs, the SCTFT structure can operate at high voltages without compromising the device performance. The SCTFT is also more reliable than conventional LVTFTs. Device reliability and yield are two important factors in determining the complexity of the circuit design and the final cost. For example, in active matrix liquid crystal displays (AMLCD), which is one of the biggest application areas of a-Si:H TFTs, it is usual practice to split a pixel into four sub pixels, each with its own driver TFT [6]. This way, the pixel still operates even in the presence of a faulty TFT. The common faults in TFTs are short circuits between gate, source and drain electrodes. All of these faults can be significantly avoided using the SCTFT structure. Here, the risk of drain-to-gate and source-to-gate short circuits is significantly reduced, because there is no overlap between the top metallization and gate metallization layers. In addition, the risk of a short circuit between the source and drain metallizations is decreased due to larger separation.

CONCLUSIONS

A novel TFT structure for high voltage and high performance operation was presented. A "soft contact" structure for the source and drain has been introduced which allows the TFT to withstand high voltages. In low voltage applications, it offers the same performance as conventional LVTFTs with higher reliability and fabrication yield. In high voltage applications, it shows a much superior performance compared to conventional gap-type HVTFTs. The SCTFT does not suffer from the drawbacks of regular HVTFTs, such as a large curvature and V_x shift in its output characteristics. The "soft contact" capability is one of the unique features of our process design which allows independent positioning of the final metallization layer with respect to the doped contact layers.

ACKNOWLEDGMENTS

This work was in part supported by the National Science and Engineering Research Council of Canada (NSERC) and the Information Technology Research Center (ITRC).

REFERENCES

1. R. A. Martin, V. M. D. Costa, M. Hack, and J. Shaw, IEEE Tran. Elect. Dev., **40**, pp. 634-644, March 1993.

2. R. A. Martin, P. K. Yap, J. Shaw, M. Hack, N. Sugiura, and T. Hamano, Technical Digest, IEEE IEDM, pp. 341-344, 1989.

3. J. Shaw, M. Hack, and R. A. Martin, J. Appl. Phys. **69**, pp. 2667-2672, Feb. 1991.

4. R. A. Martin, P. Kein, M. Hack, and H. Tuan, Technical Digest, IEEE IEDM, pp. 440-442, 1987.

5. A. M. Miri, and S. G. Chamberlain, Mat. Res. Soc. Symp. Proc. **377**, pp. 737-742, 1995.

6. J. Kanicki, Amorphous & Microcrystalline Semiconductor Devices: Optoelectronic Devices, Artech House, 1991.

HOT-WIRE DEPOSITED AMORPHOUS SILICON THIN-FILM TRANSISTORS

R.E.I. SCHROPP*, K.F. FEENSTRA*, C.H.M. VAN DER WERF*, J. HOLLEMAN**, and H. MEILING*,
* Debye Institute, Department of Atomic and Interface Physics, Utrecht University, P.O. Box 80 000, 3508 TA Utrecht, The Netherlands
** MESA Research Institute, Department of Electrical Engineering, University of Twente, P.O. Box 217, 7500 AE Enschede, The Netherlands

ABSTRACT

We present the first thin film transistors (TFTs) incorporating a low hydrogen content (5 - 9 at.-%) amorphous silicon (a-Si:H) layer deposited by the Hot-Wire Chemical Vapor Deposition (HWCVD) technique. This demonstrates the possibility of utilizing this material in devices. The deposition rate by Hot-Wire CVD is an order of magnitude higher than by Plasma Enhanced CVD. The switching ratio for TFTs based on HWCVD a-Si:H is better than 5 orders of magnitude. The field-effect mobility as determined from the saturation regime of the transfer characteristics is still quite poor. The interface with the gate dielectric needs further optimization. Current crowding effects, however, could be completely eliminated by a H_2 plasma treatment of the HW-deposited intrinsic layer. In contrast to the PECVD reference device, the HWCVD device appears to be almost unsensitive to bias voltage stressing. This shows that HW-deposited material might be an approach to much more stable devices.

INTRODUCTION

The method of thin film amorphous semiconductor deposition by decomposition of the source gases at a tungsten filament held at a high temperature (Hot Wire Chemical Vapor Deposition — HWCVD) has recently attracted (renewed) attention [1-3]. It is feasible to deposit *device-quality* hydrogenated amorphous silicon (a-Si:H) with a very low hydrogen content (down to < 1 at.-%). Thus far, reasonably low density of states (DOS) figures could not be obtained in amorphous silicon materials made by plasma deposition (Plasma Enhanced Chemical Vapor Deposition — PECVD) at hydrogen concentrations lower than 6 - 8 at.-%. Moreover, the saturated DOS resulting from a prolonged stressing period with non-equilibrium free carrier concentrations (such as light soaking conditions) has been shown to be lower than that of conventional PECVD layers [2]. Therefore, the new types of amorphous silicon obtainable by HWCVD are interesting for application in (photo-)electronic devices. The incorporation of these new materials in solar cells is currently being investigated [4-7], however, this requires an extensive effort in improving the compatibility between doped and undoped layers in a sandwich type structure.

This paper deals with the application of Hot-Wire deposited a-Si:H in thin film transistors (TFTs). The initial purpose of this work is to demonstrate the possibility of utilizing this material in devices. Furthermore, it is investigated whether the improved stability of the materials correlates with an improved reliability for the TFTs. Vice versa, the TFT is an excellent diagnostic tool for the study of the metastable behaviour of amorphous silicon because Fermi-level shifts can be induced in the dark by applying a gate bias. This can now be performed in device quality layers with a low hydrogen content, which should lead to

31

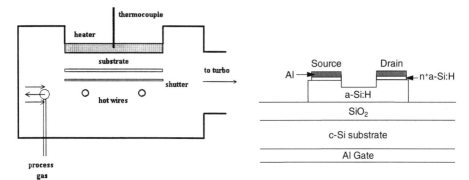

Figure 1: Schematic drawing of the HW deposition chamber.

Figure 2: Schematic cross-section of the inverted-staggered TFTs studied.

further insight in the relation between the creation of metastable states and the hydrogen content.

HW deposition offers a number of important technological advantages. First, a high deposition rate can be obtained (up to 50 Å/s) [8] for device-quality amorphous silicon deposition. Second, all photolithography and etching steps for these HW deposited layers can be carried out as for any conventional thin film device. Third, merely by tuning the deposition parameters in the gas flow and pressure regime, HW deposition is also suitable for the deposition of poly-Si thin films at a relatively low temperature.

Possible concerns about HW deposition are the difficulty of achieving uniform films, frequent breakage of the wire, and the risk of tungsten incorporation in the deposited film. Whereas the first problem can be resolved by proper design of the deposition geometry, we have found that the latter two do not represent real problems. In our experience, using proper procedures, the lifetime of the wire until breakage allows for the deposition of films with an accumulated thickness of 35 - 40 μm. Conventional PECVD systems normally require maintenance after this amount of deposited material, or earlier. No tungsten could be detected in films that were deposited at a sufficiently high filament temperature.

EXPERIMENTAL

Hot-Wire deposition chamber

A schematic drawing of the deposition chamber is shown in Fig. 1. This chamber is part of a multichamber ultra-high vacuum (UHV) system (PASTA) originally delivered as a PECVD system by Elettrorava S.p.A. under license from MVSystems, Inc. We have adapted one of the chambers to accommodate for a Hot-Wire deposition assembly. The substrates can be up to 10 cm × 10 cm or 4 inch diameter, which are fairly large dimensions compared to HW laboratory systems currently in use. The substrates can be heated to a maximum temperature of 550°C by heater elements outside the vacuum in a heater well. We selected substrate temperatures of 300°C and 450°C, to achieve a "standard" hydrogen content (9 at.-%) and a low hydrogen content (5 at.-%), respectively. All films in this study are purely amorphous. Some typical properties of the HW-deposited materials are listed in Table 1. The source gas is pure silane (SiH₄) and is decomposed at the surface of one

Table 1: Typical properties of the HW-deposited layers as used in the TFT structures. The hydrogen content and the microstructure parameter R* have been determined from Fourier Transform InfraRed spectroscopy; the ambipolar diffusion length has been determined by the steady state photocarrier grating technique (SSPG).

Deposition temperature (°C)	300	450
Deposition rate (Å/s)	16	20
Hydrogen content (at.-%)	9	5
Microstructure parameter R*	0.18	< 0.1
Ambipolar diffusion length (nm)	76	115

or more tungsten wires mounted parallel to each other. The wire is resistively heated by a DC current to a temperature in excess of 1800°C, to ensure catalytic gas dissociation at a high rate. The gas flow is chosen as to prevent depletion of the silane at a deposition rate of 15-20 Å/s (a factor 10 faster than conventional PECVD deposition rates). The pressure has been optimized in order to induce the correct rate of gas phase reactions between the wires and the substrate. Note that the ambipolar diffusion length could be further optimized.

TFT preparation technology

The sample structure is shown in Fig. 2. For this initial research project we used thermally oxidized (100) single-crystal highly doped p-type wafers as a substrate. The gate oxide thickness was 1700 Å. At the rear side the oxide was stripped and the bare surface was aluminum coated to serve as a gate contact. The active semiconductor layers were made using HWCVD from 100 % silane at a substrate temperature of either 300°C or 450°C. Reference devices were made by conventional rf PECVD at a substrate temperature of 194°C from 100 % silane at a discharge frequency of 13.56 MHz. The thickness of the intrinsic amorphous silicon layers was 200 - 250 nm. This was followed by the deposition of 50 nm of n-type amorphous silicon, for all devices by conventional rf PECVD at 194°C from a silane/phosphine/hydrogen mixture. Without prior patterning the structures were first metallized by thermal vacuum evaporation of aluminum.

Next, the common back-etch procedure was used to define source and drain contacts as well as the channel dimensions. This involves wet etching of the aluminum followed by accurate dry etching of the n^+ region between the contacts.

Surface treatment and adhesion

In the course of this project films have been deposited on Corning 7059 glass, on bare c-Si, and on thermally oxidized Si wafers. It has been observed that the a-Si:H films flake off locally on the latter two types of substrates. Visual inspection using an optical microscope revealed the possibility of gas inclusion and disruptive escape of the gas during deposition. From depth profilometry, it became clear that the depth of the craters can vary and that it can also be just a fraction of the film thickness. Interface treatments used to improve the adhesion, such as sputtering with an Ar plasma and wet cleaning with a hydrofluoric solution, have had only a limited effect. It was concluded that the film rupture is due to H_2 gas formation due to heating up of the film by the radiation from the wires during deposition. Consequently, films with a low H content appeared to stick better and to form only very small craters. The method that did completely avoid rupture of

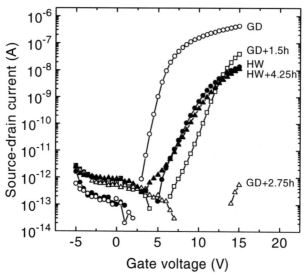

Figure 3: The transfer characteristics for a TFT made with a standard PECVD a-Si:H layer (open symbols) and for a TFT made with a 5 at.-% hydrogen content HWCVD a-Si:H layer (closed symbols). For each TFT, the set of curves represents the effect of various durations of voltage bias stressing at +25 V.

the films with a low as well as a high hydrogen content was to let the temperature of the substrate equilibrate in the chamber prior to deposition by exposing it to the heat of both the substrate heater and the heated wire in the presence of a gas flow.

RESULTS & DISCUSSION
Transfer characteristics

Figure 3 shows the drain current I_d versus the gate voltage V_g, at a small source-drain voltage ($V_{sd} = 0.5$ V) for a TFT made with a standard PECVD a-Si:H layer (open symbols) and for a TFT made with a 5 at.-% hydrogen content HWCVD a-Si:H layer (closed symbols). For each TFT, the set of curves represents various stages of voltage bias stressing. It can be seen that the PECVD reference device is a state-of-the-art TFT with a 2×10^6 ON/OFF switching ratio and a threshold voltage of 6.4 V, and that it has good quality hole-blocking source-drain contacts. The field-effect mobility μ_{FE} is 1.0 cm^{2}V^{-1}s^{-1}, as determined from the saturation regime. This shows good control of the basic technology.

The HWCVD device has a somewhat smaller, but sufficient, switching ratio in excess of 1×10^5 and a higher threshold voltage. The mobility, however, is orders of magnitude lower ($\mu_{FE} = 0.001$ cm^{2}V^{-1}s^{-1}). Presumably, the interface treatment is not yet optimized, although this does not seem to have a large effect on the switching ratio.

Output characteristics

Figure 4 shows the drain current I_d versus the source-drain voltage V_{sd}, for various

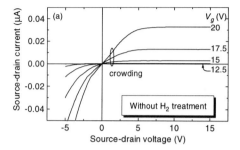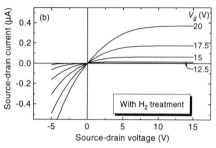

Figure 4: The output characteristics of two TFTs with a different i/n$^+$ interface treatment; (a) no special procedure leading to current crowding effects, and (b) using a H$_2$ plasma treatment on the HW-deposited i-layer prior to deposition of the PECVD n$^+$-layer.

gate voltages (output characteristics), of two types of HWCVD devices. At first, as shown in Fig. 4a, some current crowding was observed, which did not occur in the reference devices. Usually in staggered electrode configurations, this effect is observed [9] in "thick" TFTs or in TFTs with poorly injecting contacts. Therefore, the origin of the current crowding was thought to be due to the i/n$^+$ interface at which hydrogen depletion could lead to a defective layer with a high resistance. The procedure for the device in Fig. 4a was to leave the Hot-Wire deposited i-layer in the deposition chamber under a SiH$_4$ flow until it had cooled down far enough to transfer it to the PECVD n$^+$-chamber for deposition of a plasma-deposited n$^+$-layer. The procedure that eliminated the current crowding effect entirely in Fig. 4b, included an extra H$_2$-plasma treatment at the temperature used for the n-layer. Apparently, the HW-deposited layer has been effectively rehydrogenated in order to reduce the resistive barrier near the i/n$^+$ interface.

Stability

Both the PECVD and the HWCVD devices have been subjected to a continuous bias-voltage stress of +25 V for several hours (see Fig. 3). The HWCVD device is a TFT with a rehydrogenated i/n$^+$ interface. The PECVD device showed to be very unstable and suffered from a threshold voltage shift ΔV_g of 4 V and 11 V, after 1.5 h and 2.75 h of stress, respectively. The HWCVD device showed hardly any threshold voltage shift although it was stressed for a longer time (for 4.25 h). This demonstrates that the TFTs incorporating a HW-deposited, low H content, active layer are much less prone to threshold voltage shifts than TFTs with conventional device-quality PECVD layers. Although the initial transfer characteristics of the HWCVD devices need improvement, the characteristics after bias-stressing are superior to those of PECVD devices. Since only a small number of TFTs has been made using HWCVD, we expect to be able to also improve the as-deposited characteristics.

CONCLUSIONS

We presented the first TFTs with an intrinsic a-Si:H layer deposited by HWCVD. The exposure of the gate dielectric to heat in the deposition chamber is different than for conventional PECVD, which was found to be a source for adhesion problems. This issue can however be resolved by a proper predeposition treatment. Resistive effects leading to current crowding in the output characteristics could be resolved by a proper post-deposition treatment (H_2 plasma) of the HW deposited layer. The transfer characteristics of thus fabricated HWCVD TFT are almost completely insensitive to voltage bias stressing, whereas those of the reference PECVD TFT show a threshold voltage shift of 11 V after 2.75 h at a gate voltage of 25 V. This work shows that TFTs with a HW-deposited a-Si:H layer are feasible and that they are potentially much more stable.

ACKNOWLEDGEMENTS

We gratefully acknowledge the technical assistance of M.H.H. Weusthof of the MESA Research Institute, University of Twente. The research of H. Meiling has been made possible by the Royal Netherlands Academy of Arts and Sciences (KNAW). We thank J. Wallinga and D. Knoesen for confirming that the W contamination is below the detection limit of EDX.

REFERENCES

1. A.H. Mahan, J. Carapella, B.P. Nelson, R.S. Crandall, and I. Balberg, J. Appl. Phys. **69**, 6782 (1991).
2. R.S. Crandall, A.H. Mahan, B.P. Nelson, M. Vanecek, and I. Balberg, AIP Conf. Proc. **268**, 81 (1992).
3. A.H. Mahan, B.P. Nelson, S. Salamon, and R.S. Crandall, J. Non-Cryst. Solids **137 & 138**, 657 (1991).
4. B.P. Nelson, E. Iwaniczko, R.E.I. Schropp, A.H. Mahan, E.C. Molenbroek, S. Salamon, and R.S. Crandall, Proc. 12th International E.C. Photovoltaic Solar Energy Conference 1994, Eds. R. Hill, W. Palz, and P. Helm (H.S. Stephens and Associates, 1994) p679-682.
5. E. Iwaniczko, B.P. Nelson, E.C. Molenbroek, R.E.I. Schropp, R.S. Crandall, and A.H. Mahan, AIP Conf. Proc. **306**, 458 (1993).
6. A.H. Mahan, B.P. Nelson, E. Iwaniczko, Q. Wang, E.C. Molenbroek, S.E. Asher, R.C. Reedy, Jr., and R.S. Crandall, 13th NREL PV Program Review Meeting, 1995.
7. P. Papadopulos, A. Scholz, S. Bauer, B. Schroeder, and H. Oechsner, J. Non-Cryst. Solids **164 - 166**, 87 (1993).
8. E. Molenbroek, E.J. Johnson, and A.C. Gallagher, 13th European Photovoltaic Solar Energy Conference, 23-27 October 1995, Nice, France, Eds. W. Freiesleben, W. Palz, H.A. Ossenbrink and P. Helm (H.S. Stephens and Assoc., 1994) p319-322.
9. M.J. Powell and J.W. Orton, Appl. Phys. Lett. **45**, 171 (1984).

NANOSTRUCTURE AND ELECTRICAL PROPERTIES OF ANODIZED AL GATE INSULATORS FOR THIN-FILM TRANSISTORS

Toshiaki Arai, Yasunobu Hiromasu, and Satoshi Tsuji
IBM Yamato Laboratory, 1623-14 Shimo-tsuruma, Yamato-shi, Kanagawa 242, Japan

ABSTRACT

A novel anodic oxidation method for aluminum thin-film on the glass substrate is proposed for the fabrication of a gate insulator in thin-film transistor (TFT). A proposed rear-positioned cathode makes for much more smooth surface with an average surface roughness of under 4 nm in comparison with conventional cathode positions. The anodic oxidation conditions were comprehensively investigated. The current density, ethylene glycol concentration, and cathode position were chosen as the elements of anodic oxidation condition matrix, and all these conditions affected the film formation. The surface morphology and nanostructure of the anodized films were characterized with an atomic force microscopy (AFM) and cross-sectional transmission electron microscopy (TEM). Films formed at higher ethylene glycol concentrations of over 50% and higher current densities of over 0.9 mA/cm^2 exhibited higher breakdown electric fields of over 7 MV/cm, and lower leakage currents. These films had relatively smooth surfaces and took on the shapes of the underlying aluminum. In contrast, films formed at lower ethylene glycol concentrations of under 50% and lower current densities of under 0.9 mA/cm^2 had rough surfaces and microvoids. A high concentration of ethylene glycol increases the cost of mass production, however, the proposed rear-positioned cathode method can be achieved even when the concentration of ethylene glycol is under 50%.

INTRODUCTION

Bottom-gate thin-film transistors (TFTs) have been widely investigated for use in active-matrix liquid crystal displays (AMLCDs) [1, 2]. In recent years, much effort has been devoted to developing low-resistivity gate bus lines, to meet the need for large, high-resolution LCDs [3-5]. Aluminum (Al) and its alloys are remarkably good materials for gate bus lines, because of their low resistivity, low cost, high adhesion, and superior patternability. But aluminum easily forms hillocks and whiskers during heating processes such as chemical vapor deposition (CVD) film formation, and causes defects as a result of short-circuits and leakage between the gate and the upper electrodes. Al_2O_3 ,which is formed by anodic oxidation at room temperature, is therefore used as a protective layer against hillock formation. It also functions as a gate insulator, and the electrical properties of Al_2O_3 are therefore very important. The surface flatness of the film is also important, because the next layer takes on the shape of the underlying surface [6]. The properties of Al_2O_3 have been investigated, and the effects of electrolyte and annealing after the formation of the film have been reported [7, 8].

To obtain an anodically oxidized film with a high breakdown voltage and low leakage current, we focused on the anodic oxidation condition. The current density, the ethylene glycol concentration of the electrolyte, and the position of the cathode were chosen as the elements of the anodic oxidation condition matrix. The breakdown voltage and leakage current of the anodized film were studied as electrical properties, while the surface morphology and nanostructure of the anodized film were studied by tapping-mode atomic force microscopy (AFM) and cross-sectional transmission electron microscopy (TEM).

TFT AND CAPACITOR STRUCTURE

Figure 1 shows a schematic cross-sectional view of a pixel. First, the aluminum used in the gate bus line and the bottom electrode of the storage capacitor (Cs) are patterned. Al_2O_3 is

37

then formed on the surface of the aluminum by anodic oxidation. Next, silicon nitride (SiN_x) is formed on the Al_2O_3 by the plasma-enhanced CVD method to secure the properties of the interface between the gate insulator and the amorphous silicon (a-Si) layer.

Electrically, Al_2O_3 is required to have a high breakdown voltage and a low current leakage, since it is used as the gate insulator. Breakdown leads to the destruction of the TFT and the Cs, and current leakage reduces the value of the on-off current ratio of the TFT and the retention ability of the Cs. Physically, Al_2O_3 is required to have surface flatness, because the next layer takes on the shape of the underlying surface [6]. If the Al_2O_3 film has a rough surface, the surfaces of the SiN_x and a-Si layers will be rough, and the field effective mobility of the a-Si layer will be decreased.

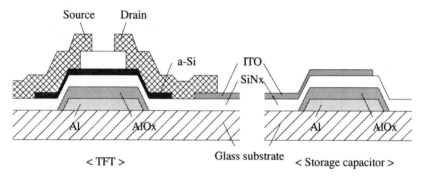

Figure 1.   Schematic cross-sectional view of a TFT and storage capacitor.

EXPERIMENT

A schematic diagram of the anodic oxidation system is shown in Figure 2.

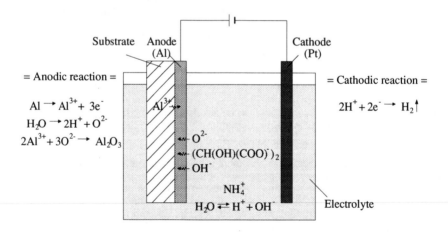

Figure 2.   Schematic diagram of the anodic oxidation system.

The electrolyte is a mixture of 3 wt% tartaric acid solution and ethylene glycol. Its pH is kept at 6.5-7.5 by addition of an ammonium solution. The temperature is kept at 20°C. The anode is a 350 nm layer of aluminum sputtered on a 125×125 mm sheet of OA-2 glass. The cathode is a platinum-coated titanium mesh with the same area as the anode.

Overall, the reaction can be explained as follows:

$$2Al + 3H_2O \rightarrow Al_2O_3 + 6H^+ + 6e^- \qquad (1)$$

When a positive bias is applied to the anode, H_2O is electrolyzed into hydrogen ions and oxygen ions. Hydrogen ions each receive an electron from the cathode and are released as H_2 gas. Oxygen ions react with the aluminum ions created by the release of electrons from aluminum. At the same time, anions such as OH^- and $(CH(OH)(COO)^-)_2$ are attracted to the Al_2O_3 film by the electric field [7]. The attraction is much weaker than in the case of a strong acid electrolyte such as phosphoric acid. The bias increases at a constant current density (CC-mode) and stops at a fixed voltage. The current then decreases at a constant voltage (CV-mode) until this reaction ends.

The concentration of ethylene glycol was examined at 33%, 50%, 67%, and 75%, and the current density in CC-mode was examined at 0.3, 0.9, 1.5, and 2.1 mA/cm^2. The fixed voltage in CV-mode was 150 V and the total reaction time was 30 min. Cathode position both in front of and behind the anode (Figure 3) were examined. This report gives no data for a position beside the anode, but they would be rather similar to the data for a position behind the anode. The side position is suitable for mass production, because the bath capacity can be greatly reduced in this position.

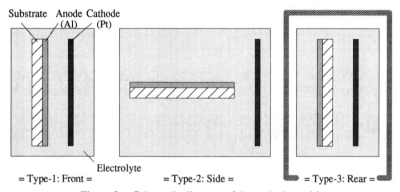

Figure 3. Schematic diagram of the cathode position.

The surface morphology was observed by means of a tapping-mode AFM (Digital Instrument-Nanoscope-3). A cross-sectional view was observed by TEM after a sample had been prepared by focused ion beam (FIB) etching. The electrical properties of the sample were measured after the fabrication of a 1 mmϕ molybdenum (Mo) / Al_2O_3 / Al capacitor. The breakdown voltage and leakage current were measured by means of a curve tracer (Tektron-K213). The rate of the voltage increase was 100 V/sec. The capacitance was measured by means of a computer-controlled HP-4140B, 1737 system. The thickness of Al_2O_3 was estimated from the capacitance and the dielectric constant of 9.1, and was confirmed by the cross-sectional TEM view.

RESULTS AND DISCUSSION

Figures 4 and 5 show the breakdown electric field and the leakage current for a changing concentration of ethylene glycol and current density. They show that a current density of at least 0.9 mA/cm^2 is required and that increasing the current density above 0.9 mA/cm^2 has little effect. They also show that the concentration of ethylene glycol should be over 50%, and that a higher concentration of ethylene glycol gives a higher breakdown voltage.

The surface morphology observed by AFM varied markedly with the concentration of ethylene glycol and the current density. The surfaces of films with low breakdown voltages were very rough, and deep crater-like holes were observed in some places (Figure 6). In contrast, the surfaces of films with high breakdown voltages were very smooth (Figure 7).

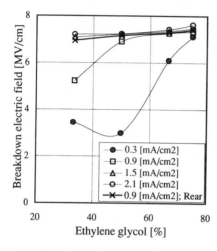

Figure 4. Breakdown electric field vs. concentration of ethylene glycol and current density.

Figure 5. Leakage current vs. concentration of ethylene glycol and current density.

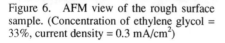
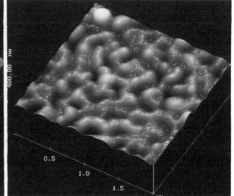

Figure 6. AFM view of the rough surface sample. (Concentration of ethylene glycol = 33%, current density = 0.3 mA/cm^2)

Figure 7. AFM view of the smooth surface sample. (Concentration of ethylene glycol = 75%, current density = 1.5 mA/cm^2)

Figure 8 shows the average roughness (Ra) of these films. The concentration of ethylene glycol strongly affects the Ra. The current density should not be too low. A 75% concentration of ethylene glycol and a 0.9-1.5 mA/cm^2 current density were the best conditions for Ra.

Figures 9 and 10 show cross-sectional TEM views of these films. Figure 9 shows films with rough surfaces containing many voids. However, the surface of the aluminum is very smooth. This means that the surface roughness of the anodized film is related to the formation of the voids. Figure 10 shows films with smooth surfaces containing no voids and no grain boundaries. This indicates that an extremely regular film can be obtained by selecting good anodic oxidation condition. The surface morphology of the anodized film takes on the shape of the aluminum with slight emphases. The undulations of the film originate in the grain boundaries of the aluminum.

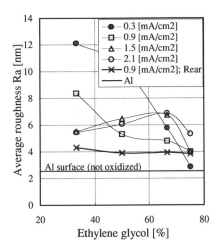

Figure 8. Average roughness (Ra) vs. concentration of ethylene glycol and current density.

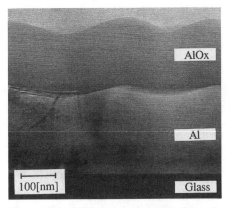

Figure 10. TEM cross-sectional view of the smooth surface sample. (Concentration of ethylene glycol = 75%, current density = 1.5 mA/cm^2)

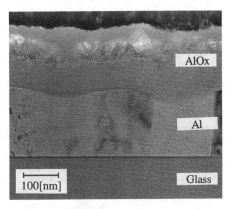

Figure 9. TEM cross-sectional view of the rough surface sample. (Concentration of ethylene glycol = 33%, current density = 0.3 mA/cm^2)

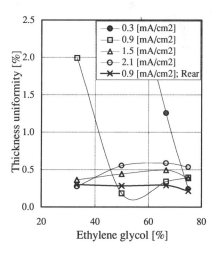

Figure 11. Thickness uniformity vs. concentration of ethylene glycol and current density.

The concentration of ethylene glycol was found to have a strong effect on the quality of the anodized film. However, ethylene glycol is very expensive and is not suited for mass production. We therefore examined the cathode position, with the aim of changing the anion's behavior in the electrolyte. The results are shown in Figures 4, 5, and 8 by solid lines. The data for the rear position are superior to those for the frontal position. We were concerned that the uniformity of the film might be worse than for the frontal position, but it was actually rather better. Figure 11 shows the thickness average of the Al_2O_3. The anodized film has quite good uniformity in comparison with a CVD film of over 5% uniformity. When the current density was 0.9-1.5 mA/cm^2, the uniformity was under 0.5% and the average roughness was under 7 nm, in spite of the frontal position of the cathode. In the case of a rear-positioned cathode, the uniformity was under 0.3 % and the roughness average was under 4 nm. With a rear-positioned cathode, the concentration of ethylene glycol can be reduced to under 50%, and the cost of mass production can be very greatly reduced.

CONCLUSION

We investigated the morphology and nanostructure of anodized film by AFM and cross-sectional TEM. We also demonstrated a novel technique to fabricate high quality anodized film. Our results show that a film with an fairly smooth surface less than 4 nm in average roughness can be obtained by positioning the cathode to the rear of the anode. The cathode position had an extremely strong effect on the surface roughness and uniformity of the anodized film. A high concentration of ethylene glycol and a high current density affected the improvement of breakdown voltage and the leakage current of the anodized film. A high concentration of ethylene glycol increases the manufacturing cost; however, the proposed rear-positioned cathode method can be achieved even when the concentration of ethylene glycol is under 50%. An understanding of the rear-positioned cathode effect may thus prove beneficial in future mass production of thin-film transistors.

ACKNOWLEDGMENTS

The authors are grateful to S. Hirano, H. Matino, and K. Furuta of IBM Display Technology for constructive and valuable advice, and to M. McDonald of IBM Yamato Laboratory for his many contributions to this paper.

REFERENCES

1. M. Dohjo, T. Aoki, K. Suzuki, M. Ikeda, T. Higuchi, and Y. Oana, in SID International Symposium Tech. Dig., p.330 (1988).
2. Y. Kanemori, M. Katayama, K. Wakazawa, H. Kato, K. Yano, Y. Furuoka, Y. Kanatani, Y. Ito, and M. Hijikigawa, in SID International Symposium Tech. Dig., p.408 (1990).
3. T. Tsukada, MRS Symposium Proceedings v.284 p.371-382 (1993).
4. M. Yamamoto, I. Kobayashi, T. Hirose, S. M.Bruck, N. Tsuboi, Y. Mino, M. Okafuji, and T. Tamura, '94 SID International Display Research Conference Digest,142-145 (1994).
5. P. M. Fryer, L. Jenkins, R. John, Y. Kuo, A. Lien, R. Nywening, B. Owens, L. Palmateer, M. E. Rothwell, J. Souk, J. Wilson, S. Wright, J. Batey, and R. Wisnieff, '94 SID International Display Research Conference Digest, 146-149 (1994).
6. C. R. M. Grovenor, <u>Microelectronic Materials</u> (Institute of Physics, Oxford, England, 1992), p.67.
7. C. F. Yeh, J. Y. Cheng, and J. H. Lu, Jpn. J. Appl. Phys., Vol. 32, 2803-2808 (1993).
8. K. Ozawa, K. Miyazaki, and T. Majima, J. Electrochem. Soc., Vol. 141, No. 5, 1325-1333 (1994).

HIGH-RATE DEPOSITED AMORPHOUS SILICON NITRIDE FOR THE HYDROGENATED AMORPHOUS SILICON THIN-FILM TRANSISTOR STRUCTURES

Tong Li, Chun-ying Chen, Charles T. Malone*, and Jerzy Kanicki
Dept. of Electrical Engineering and Computer Science
Center for Display Technology and Manufacturing
The University of Michigan, Ann Arbor, MI 48109
*Optical Imaging Systems, Inc., Northville, MI 48167

ABSTRACT

Hydrogenated amorphous silicon nitride thin films, prepared in a large area plasma-enhanced chemical vapor (PECVD) deposition system utilizing high-rate deposition technique, have been characterized by various techniques. Experimental data obtained from this study are presented and compared to low-rate deposited PECVD films. Special attention has been devoted during this study to the difference between high- and low-rate deposited samples. The amorphous silicon (a-Si:H) thin-film transistors (TFTs) based on high-rate PECVD materials have been fabricated and characterized. The evaluation of a-Si:H TFTs indicates a good electrical performance which is comparable to its low-rate PECVD materials counterparts.

INTRODUCTION

High-rate deposition of different materials is the key factor in determining the throughput of hydrogenerated amorphous silicon (a-Si:H) thin film transistors (TFTs) in active-matrix liquid crystal display (AMLCD) manufacturing. Many attempts to achieve the high deposition rate without degrading film's electrical quality have, thus, been accomplished up to now [1-3]. Furthermore, the feasibility of high-rate a-Si:H TFTs' fabrication has also been investigated recently [4,5]. In order to establish the correlation between material quality and device performance for high-rate deposition technology, a comprehensive study of the materials and their impacts on TFTs' performance is needed. Besides, the comparison between different properties of the high- and low-rate deposited materials and a-Si:H TFT performance is needed to understand the effects of the deposition rates on materials and consequent TFTs' quality.

In this work, we have characterized the chemical, physical, optical, electrical properties of high-rate PECVD hydrogenated amorphous silicon nitride (a-SiN$_x$:H) thin films, and compared with that of the low-rate PECVD a-SiN$_x$:H. Some correlations between various experimental parameters, and between high- and low-rate a-SiN$_x$:H's properties have been established during this study.The electrical performance of a-Si:H TFTs fabricated from high-rate PECVD materials have also been evaluated to establish a relationship between materials' quality and device performance.

EXPERIMENTAL

Hydrogenated amorphous silicon nitride thin films were deposited in a large area 13.56 MHz plasma enhanced chemical vapor deposition system (AKT CVD 1600) at the temperature ranging from 250 to 350 °C. The details about the deposition conditions for all films evaluated during this study cannot be disclosed. However, for example the film high#1 given in Table I was deposited under the following conditions: silane flow = 50 to 150 sccm, ammonia flow = 300 to 700 sccm, nitrogen flow = 2500 to 3500 sccm, pressure = 1 to 1.5 Torr and r.f. power = 1.5 kW. For film thickness and etch rate measurements, a step was etched in the film with NH$_4$HF/HF buffered oxide etch (BOE), at room temperature. The step heights were measured using Dektak profiler.

Optical data were taken by Cary 45 spectrophotometer from ultraviolet-visible (UV-Vis) spectra obtained for the thin films deposited on quartz substrate. KaleidaGraph was used to perform interference fringe correction and corresponding calculation of the absorption coefficient for different films. The obtained results are tabulated in Table I.

The quantitative analysis of total hydrogen content and film stoichiometry were performed by Rutherford Backscattering Spectroscopy (RBS) and Secondary Ion Mass Spectroscopy (SIMS), and the results are also given in Table I. Quantification of the hydrogen content in the a-SiN$_x$:H by SIMS technique was accomplished by using a relative sensitivity factor derived from a-SiN standard of known composition. The most important bond densities, at room temperature, were estimated from infrared (IR) spectra, measured by Bio-Rad FTS-40 FTIR; bond densities were calculated using Win IR software.The total hydrogen densities for different samples deduced from IR spectra were compared with the RBS and SIMS results, and are given in Table II.

In order to observe hydrogen evolution behavior, films on both double-polished wafer and quartz were thermally annealed in a special high-temperature cell. The temperature stability was ±1 °C below 300 °C and ±3 °C above 300 °C. The annealing was carried out at different temperatures for 30 min. The variations of bond densities as well as optical bandgap were recorded during the thermal-annealing.

The electrical properties are evaluated by metal-insulator-semiconductor (MIS) and metal-insulator-metal (MIM) structures, respectively. In MIS structures, aluminium was deposited on a-SiN$_x$:H films which were deposited on both n- and p-type {100} silicon wafers. In MIM structures, on the other hand, aluminium and chromium were used as top- and bottom-contacts, respectively. A Keithley 590 CV analyzer was used to record the capacitance(C)-voltage(V) characteristics.

Finally, the electrical performance of a-Si:H TFTs fabricated from high deposition rate PECVD materials were characterized using Hewlett Packard 4156A semiconductor analyzer. The fabrication of a-Si:H TFT is described in [12].

Table I. Properties of high- and low-rate PECVD a-SiN$_x$:Hs

Sample#	R_{dep} (nm/min)	BOE (nm/min)	Eg (eV)	N/Si	C_H (at.%)	B (cm*eV)$^{-1/2}$
high#1	165	66	5.1	1.5	20.6	699.3
high#2	60	116	5.4	1.6	22.4	831.7
high#3	420	116	3.8	1.1	24.0	518.1
high#4	208	90	4.9	1.5	22.6	680.3
high#5	166	118	4.5	1.4	23.8	645.2
low#1	-	-	5.2	1.6	17.0	831.7

RESULTS AND DISCUSSION

A. Optical Properties

Fig. 1 illustrates the semilog plot of optical absorption coefficient (α) as a function of photon energy (E) for high- and low-rate PECVD a-SiN$_x$:H thin films. In this figure the high-rate deposited film with a N/Si (=x) atomic ratio of 1.6 is compared with the low-rate deposited film having x of 1.6. At $\alpha < 10^4$ cm^{-1} there is an exponential dependence of α on photon energy (valence-band) [6]:

$$\alpha \approx \exp\left(\frac{E}{E_0}\right) \qquad (1)$$

where E_0 is the Urbach edge, which is merely the slope of the curve around $10^4 cm^{-1}$. The Urbach edges, being the measure of the steepness of the valence-band tail, are similar for both films, of which the high-rate film has a value of 341 meV and the low-rate film has a value of 321 meV.

At $\alpha > 10^4$ cm^{-1}, absorption is believed to take place between extended valence- and conduction-band states, and is often described by Tauc's expression [6]

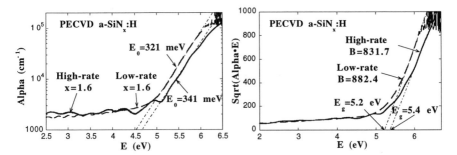

Fig. 1 Optical absorption coefficient as function of photon energy for high- and low-rate PECVD a-SiN$_x$:H having x of 1.6. E_0 is the Urbach edge of the valence bandtail.

Fig. 2 $(\alpha\hbar\omega)^{1/2}$ as a function of photon energy for high- and low-rate PECVD a-SiN$_x$:H thin film having x of 1.6. The unit of B is $(cm \cdot eV)^{-1/2}$.

$$(\alpha\hbar\omega)^{1/2} = B(\hbar\omega - E_g) \qquad (2)$$

where h is Plank's constant, ω is the wavenumber, and B is a proportionality constant. Equation (2) represents, in fact, the tangential slope at $\alpha > 10^4$ cm^{-1} in Fig. 2, in which $(\alpha\hbar\omega)^{1/2}$ is plotted as a function of $\hbar\omega$ for low- and high-rate PECVD films. The Tauc optical bandgap (E_g) can be determined from this figure at $(\alpha\hbar\omega)^{1/2} = 0$. Fig. 2 shows that the high- ($x = 1.6$) and low-rate ($x = 1.55$) PECVD films have E_g of 5.39 and 5.22 eV, and B-values of 831.7 and 882.4 $(cm \cdot eV)^{-1/2}$, respectively. The Tauc optical bandgap are plotted in Fig. 3 as a function of film's

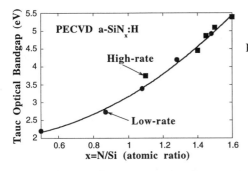

Fig. 3 Tauc optical bandgap as a function of film stoichiometry for high- and low-rate PECVD a-SiN$_x$:H thin films.

stoichiometry for both high- and low-rate PECVD films. They clearly follow the same trend indicating that the a-SiN$_x$:H film optical bandgap measurements can be used for both low- and high-rate PECVD films to estimate film's stoichiometry.

Note that both Fig. 1 and 2 were interference fringes corrected using fringe-matching method [7]. That is an artificial interference fringe pattern was created to match the pattern of the measured data. Those two patterns were then subtracted from each other to yield a reliable data for further analysis.

B. Buffered Oxide Etching Rate

The room temperature etching rates in buffered oxide etch (BOE) for various samples, having N/Si ratio between 1.1 and 1.6 are listed in Table I. The etching rate increases with the hydrogen content, in general, for the same hydrogen content BOE of high-rate PECVD films is very comparable to that of the low-rate PECVD films indicating that the chemical nature of both films is very similar.

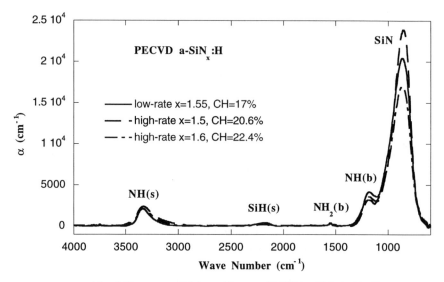

Fig. 4 IR spectra of high- and low-rate PECVD a-SiN$_x$:H thin films.

<u>C. Film Microstructure</u>

Typical baseline corrected IR spectra for high- and low-rate PECVD films are shown in Fig. 4. From these spectra the following vibrational peaks can be identified [8]: N-H$_2$ (3450 cm^{-1}), N-H (3340 cm^{-1}) and SiH (2160 cm^{-1}) stretching modes; NH$_2$ (1550 cm^{-1}) and N-H (1180 cm^{-1}) bending modes; Si-N (890 cm^{-1}) asymmetric stretching mode. The curves of high- and low-rate deposited films show an overall similarity for similar films stoichiometry and hydrogen contents. However, the scale-up curves shown in Fig. 5 illustrate the subtle peak shape discrepancies for both films. For N-H(s) mode (Fig. 5a), high-rate films have a larger α than low-rate films, but peaks are located at about the same location (3340 cm^{-1}); N-H$_2$(s) mode which is hidden in the main peaks at around 3450 cm^{-1}, when differentiated out, show a closed location and shape for both high- and low-rate films; for Si-H(s) mode (Fig. 5b), on the other hand, low-rate film has a larger α with the peak location at around 2160 cm^{-1}, while the high-rate films have smaller αs with the peak location at around 2210 cm^{-1}; for N-H$_2$(b) mode (Fig. 5c), all the films have a similiar peak shape and location; all curves of N-H(b) mode (Fig. 5d), again, have similar peak shape and location, but low-rate peak has larger α than high-rate peak, which is the opposite observation to that made for N-H(s) mode (the discrepancy is consistent to all samples characterized during this study, although only three curves of those are illustrated here). This observation is very important in determination of the hydrogen content from IR spectra which is usually determined by the summation of the integrated peak areas multiplied by a correction factor associated with different strectching modes present in IR spectra [9]. From our study it is clear that high-rate PECVD film have different correction factors from that of conventional low-rate PECVD film. The hydrogen content determination from only stretching mode will, thus, create errors unless the system is calibrated for the films deposited at various rates. Therefore, a peak intensity summation over all vibratonal modes would yield a more accurate result. This result will be shown in next subsection.

The microstruction analysis of the film based on IR spectra indicates that high-rate PECVD materials have a better microstructure than low-rate PECVD samples, i.e. the Si-H concentration in high-rate is lower than in low-rate films.This implies a more stable film, and this point will be discussed below.

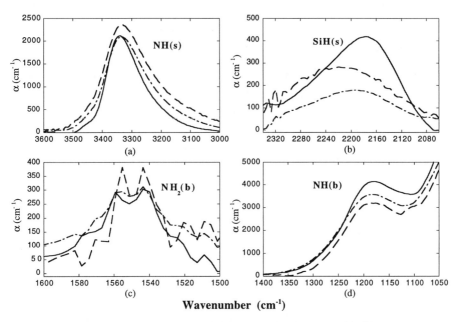

Fig. 5 Scale-up spectra for high- and low-rate PECVD a-SiN$_x$:H thin films. The lengend is the same as that in Fig 4.

D. Bond Density

The bond densities shown in IR spectra are calculated using the following equations[11]

$$\text{bonds / cm}^3 = K(\text{bond})_{\text{mode}} \int\limits_{\text{bond}} \left(\frac{\alpha(\omega)}{\omega} \right) d\omega \tag{3}$$

where ω is the wavenumber of the peak, and K is the proportionality constant. In this calculation following constants have been used [11]

$$K(NH)_{bonding} = 2.1 \pm 0.2 \times 10^{19} \, cm^{-2}$$
$$K(NH)_{stretching} = 1.2 \pm 0.3 \times 10^{20} \, cm^{-2}$$
$$K(NH_2)_{bonding} = 3.8 \pm 0.4 \times 10^{20} \, cm^{-2}$$
$$K(NH_2)_{stretching} = 5 \pm 0.5 \times 10^{20} \, cm^{-2}$$
$$K(SiH)_{stretching} = 1.4 \times 10^{20} \, cm^{-2}$$

The calculated mean values for bond densities are listed in Table II. Comparison between RBS and SIMS data shows a good agreement with each other. However, most of IR results are slightly smaller than RBS and SIMS data. This is reasonable because only N-H, N-H$_2$, and Si-H bending

47

and stretching modes are involved in this calculation. Thus, the difference (about 15%) between IR and RBS/SIMS data is attributed to: (a) other existing vibration modes (though not significant) which are not included in our IR calculation, and (b) possible error in the K-values given above.

Table II. Bond densities for high- and low-rate PECVD a-SN$_x$:Hs deduced from IR spectra, and comparisons among the results of IR, RBS, and SIMS.

Bond ID	Bond density ($\times 10^{22}$/cm^3)					
	high#1	high#2	high#3	high#4	high#5	low#1
NH(b)	0.182	0.208	0.197	0.188	0.176	0.220
NH(s)	1.151	0.944	0.340	0.903	0.643	0.901
SiH	0.110	0.056	1.058	0.360	0.702	0.213
NH$_2$(b)	0.181	0.180	0.176	0.203	0.242	0.160
NH$_2$(s)	0.135	0.090	0.061	0.126	0.101	0.085
H$_{IR}$	1.778	1.478	1.834	1.780	1.863	1.579
H*$_{RBS}$	1.90	1.78	1.77	2.23	2.13	1.71
H*$_{SIMS}$	2.0	1.8	-	2.2	2.3	-

* Density of hydrogen atoms ($\times 10^{22}$/cm^3).

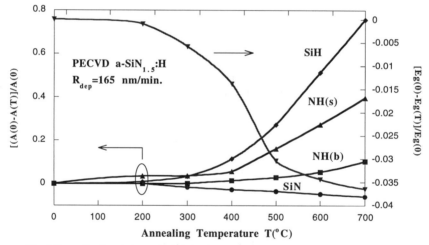

Fig. 6 Normalized variations of bond densities (left axis), and normalized variation of Tauc optical bandgap as functions of thermal annealing temperature for high-rate deposited PECVD a-SiN$_{1.5}$:H thin film.

E. Hydrogen Evolution

Fig. 6 illustrates hydrogen evolution from the high-rate PECVD a-SiN$_x$:H thin film. The bond density variations normalized to their room temperature values, and Tauc optical bandgap variation normalized to its room temperature value, are plotted as functions of thermal annealing temperature on the left- and right-hand sides, respectively. Note that for the simplicity, the variations of N-H$_2$ bonds are treated as part of that of N-H bonds. Based on the experimental data we have found that the Si-H bond dissociates faster than N-H bonds indicating that Si-H bond is thermally less stable than N-H bonds. The density of Si-N bond increases as a result of hydrogen evolution from Si-H and N-H bonds. This bond formation from passivation between silicon and nitrogen dangling bonds is confirmed by an increase optical bandgap with thermal-annealing.

The slopes of bond density variation, in this case, are much smaller than their low-rate PECVD counterparts [10]. At 700 $^{\circ}C$, there is still a large N-H and Si-H content in the film whereas in the low-rate case, almost all the Si-H bonds are dissociated. In addition, the N-H stretching mode here dissociates at a faster rate than bending mode, whereas in low-rate case, both N-H stretching and bending modes dissociate at almost the same rate. Consequently, it can be concluded that this high-rate PECVD a-SiN$_x$:H is thermally more stable than its low-rate PECVD counterpart, although they both follow the same hydrogen evolution trend.

F. Electrical Properties

High frequency (1 MHz) capacitance (C) - applied voltage (V) characteristics, at room temperature, of the Al/ a- SiN$_{1.5}$:H/c-Si (MIS) structure is shown in Fig. 7. The orientation of crystalline silicon substrate is p-type {100}. The C-V characteristics were swept from inversion to accumulation and then back to initial (inversion) regions, as indicated by the up and down arrows in the figure. This characteristics shows a fairly small C-V hysteresis. From the hysteresis ΔV, and mean voltage value V* defined as $\dfrac{V_{max} - V_{min}}{2}$ the effective trap, and fixed charges can be deduced. The estimated number of effective trap and fixed charges are tabulated in Table III. Furthermore the dielectric constant ε is also derived from this C-V characteristics which is also listed in Table III. The comparatively small stretch-out of the C-V characteristics indicates a small number of interfacial traps located at a-SiN$_x$:H/c-Si interface.

In addition, high-frequency C-V characteristics for some samples using Al/ a- SiN$_x$:H/Cr (MIM) structure have also been used to derive dielectric constants which are compared with the values obtained for MIS structures. The results are given in Table III. The MIM structure yields a larger dielectric constants than MIS structures. This discrepancy might be caused by measuring error, and/or error in the film thickness measurements.

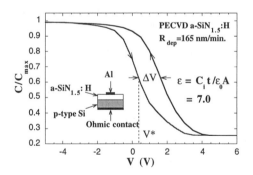

Fig. 7 C-V characteristics for Al/a-SiN$_{1.5}$:H/ c-Si (MIS) structure.

Table III. Electrical properties of high-rate PECVD a-SiN$_x$:H thin films.

Sample #	ΔV (V)	V* (V)	N_{eff} (cm^{-2})	N_{fixed} (cm^{-2})	ε_{MIS}	ε_{MIM}
high#1	1.25	0.33	5.65×10^{11}	1.49×10^{11}	7.0	-
high#2	1.75	-3.95	5.09×10^{11}	9.14×10^{11}	7.2	7.3
high#3	5.75	-5.76	1.66×10^{12}	1.66×10^{12}	6.5	-
high#4	-	-	-	-	6.9	7.2
high#5	-	-	-	-	6.6	-

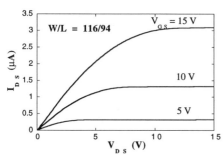

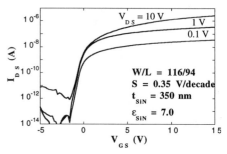

Fig. 8 Characteristics of drain-source current (I_{DS}) versus drain voltage (V_{DS}) for gate voltage (V_{GS}) of 5, 10, 15 V, respectively.

Fig. 9 Transfer characteristics for V_{DS} of 0.1, 1, 10 V, respectively.

G. TFT Characteristics

The back-channel etched a-Si:H TFTs were fabricated using high-rate PECVD materials. The deposition rates for a-SiN$_{1.5}$:H, a-Si:H, and n$^+$ a-Si:H are 190, 50, and 60 nm/min, respectively.

The channel width and length of typical a-Si:H TFTs are 116 and 94 μm, respectively. The gate to source/drain overlap in our TFT was about 3 μm. More details about TFT's parameters can be found elsewhere [12].

Fig. 8 shows the drain-source current (I_{DS}) versus drain voltage (V_{DS}) characteristics for various gate voltages (V_{DS}). At small drain voltages (below 2 V), the current crowding effect is absent, which indicates a good electrical quality of source/drain contacts in our TFT. Fig. 9 gives the TFT transfer characteristics for various V_{DS}. The ON-OFF current ratio exceeds 10^6 and the OFF-current is less than 10^{-12} A; the gate leakage current was below the detection limit of HP4145. This a-Si:H TFT with low OFF-current satisfies the requirement of low-leakage current needed for AMLCDs. In linear region ($V_{DS} = 0.1$ V) the subthreshold slope (S) is about 0.35 V/decade. This slope is fairly sharp compared to that of low-rate a-Si:H TFTs, and sharpness implies a fairly low density of deep-gap states localized at or near the a-SiN:H/a-Si:H interface.

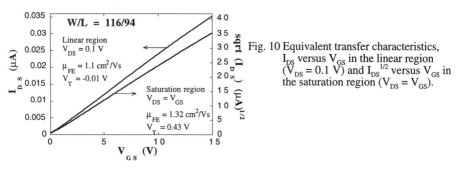

Fig. 10 Equivalent transfer characteristics, I_{DS} versus V_{GS} in the linear region ($V_{DS} = 0.1$ V) and $I_{DS}^{1/2}$ versus V_{GS} in the saturation region ($V_{DS} = V_{GS}$).

The equivalent transfer characteristics, I_{DS} versus V_{GS} in the linear region ($V_{DS}=0.1$V) and $I_{DS}^{1/2}$ versus V_{GS} in the saturation region ($V_{DS}=V_{GS}$), are shown in Fig. 10. The mobility and threshold voltage (V_T) were calculated using the gradual channel approximation [13]. The field-effect mobilities extracted from Fig. 10 are 1.32 and 1.1 cm^2/Vs in the saturation and linear regions, respectively. The difference in mobility values is most likely associated with some nonlinearity present in these equivalent transfer characteristics. The fairly high mobility value is in good agreement with a low value of the subthreshold slope deduced from Fig. 9. The threshold volages obtained from Fig. 10 are about 0.43 and -0.01 V in the saturation and linear regions, respectively. The low V_T indicates that the interface states and fixed charges localized at or near a-Si:H/a-SiN:H interface are small.

CONCLUSION

In conclusion we have found that the high-rate PECVD a-SiN$_x$:H thin films have, overall, comparable chemical, physical, optical, and electrical properties to their low-rate PECVD counterparts. The a-Si:H TFTs fabricated using a high-rate deposition PECVD technique show also a good TFT electrical performance compared to that of conventional TFTs fabricated from low-rate PECVD materials. Therefore, it is concluded that this high-rate PECVD deposition technique can be used to replace conventional low-rate PECVD technique in the production of a-Si:H TFT-LCDs.

ACKNOWLEDGEMENT

This work within Center for Display Technology and Manufacturing at the University of Michigan was supported by AFOSR/ARPA through Multidisciplinary University Research Initiative (MURI) under the contract number F49020-95-1-0524. PECVD of different materials was carried out at OIS Inc. by D. Hagerty and J. Gordon. SIMS was done by S. Novak at Evans East.

REFERENCES

1. A. Shah, J. Dutta, N. Wyrsch, K. Prasad, H. Curtins, F. Finger, A. Howling, and Ch. Hollenstein, Mat. Res. Soc. Symp. Proc. <u>258</u> 15 (1992).
2. F. Finger, U. Kroll, V. Viret, and A. Shah, W. Beyer, X. M. Tang, J. Weber, A. Howling, and Ch. Hollenstein, J. Appl. Phys. <u>71</u> 5665 (1992).
3. M. Heintze, R. Zedlitz, and G. H. Bauer, J. Phys. D: Appl. Phys. <u>26</u> 1781 (1993).
4. H. Meiling, J. F. M. Westendorp, J. Hautala, Z. M. Saleh, and C. T. Malone, Mat. Res. Soc. Symp. Proc. <u>345</u> 65 (1994).
5. S. Sherman, P. Y. Lu, R. A. Gottscho, and S. Wagner, Mat. Res. Soc. Symp. Proc. <u>377</u> 749 (1995).
6. W. Fuhs in <u>Amorphous & Microcrystalline Semiconductor Devices Volume II</u>, edited by J. Kanicki, Artech House (1992), p. 1-8.
7. O. S. Heavens, <u>Optical Properties of Thin Solid Films, Butterworths</u>, London, 1955.
8. J. Kanicki, and P. Wagner, Electrochem. Soc. Proc. <u>87-10</u> 261 (1987).
9. W. A. Lanford, M. J. Rand, J. Appl. Phys. <u>49</u> 2473 (1978).
10. T. Li, J. Kanicki, M. Fitzner, and W. L. Warren, AMLCDs '95, 123, Lehigh (1995).
11. E. Bustarret, M. Bensouda, M. C. Habrard, J. C. Bruyere, S. Poulin, S. C. Gujrathi, Phys. Rev B <u>38</u> 38 (1988).
12. C. Y. Chen, and J. Kanicki, IEEE Elec. Dev. Lett. (submitted).
13. S. Kishida, Y. Naruke, Y. Uchida, and M. Matsumura, Jpn. J. Appl. Phys. <u>22</u> 511(1983).

An Amorphous Silicon Thin Film Transistor Fabricated at 125°C by dc Reactive Magnetron Sputtering

C. S. McCormick, C. E. Weber, and J. R. Abelson, Coordinated Science Laboratory and the Department of Materials Science and Engineering, University of Illinois, Urbana IL 61801.

Abstract

We deposit hydrogenated amorphous silicon-based thin film transistors using dc reactive magnetron sputtering at a substrate temperature of 125°C, which is low enough to allow the use of plastic substrates. We characterize the structural properties of the a-Si:H channel and a-SiN$_x$:H dielectric layers using infra-red absorption, thermal hydrogen evolution, and refractive index measurements, and evaluate the electrical quality using capacitance-voltage and leakage current measurements. Inverted staggered thin film transistors made with these layers exhibit a field effect mobility of 0.3 cm^2/V-s, a I_{on}/I_{off} ratio of 5 x 10^5, a sub-threshold slope of 0.8 V/decade, and a threshold voltage of 3 V.

Introduction

The hydrogenated amorphous silicon (a-Si:H) based thin film transistor (TFT) is the dominant switching element in active matrix liquid crystal flat panel displays. Thin film transistors are usually deposited by plasma enhanced chemical vapor deposition (PECVD) using silane to produce the a-Si:H channel, silane-ammonia to produce the silicon nitride (a-SiN$_x$:H) dielectric, and silane-phosphine to produce the n$^+$ a-Si:H contacts. There is a strong industrial interest in fabricating TFTs on plastic substrates, which requires lowering the deposition temperature to below 150°C. However, PECVD is currently unable to produce high quality TFTs at temperatures below ~250°C [1,2]. The main problem is that the films become very hydrogen-rich at low temperatures, ie., the stoichiometry is largely determined by thermal mechanisms at the film growth surface [3].

DC reactive magnetron sputtering (RMS) has been shown to deposit excellent quality hydrogenated amorphous silicon [4] and silicon nitride layers [5]. A crystal silicon target is sputtered in a working gas of (Ar+H$_2$) or (Ar+H$_2$+N$_2$), respectively. For doped contact layers, a doped silicon target is used [6]. RMS offers the potential for low temperature deposition since the film composition can be controlled via the reactive gas partial pressures, and the translational energy of the sputtered species produces a dense and smooth film microstructure. In addition, RMS is an environmentally "green" and low cost method since no hydride gases are employed [7]. RMS has been previously shown to deposit high quality TFTs at 300 °C [8].

Experimental

In this work, thin film transistors are deposited by RMS in a load-locked, UHV deposition system with water and oxygen background pressures less than 10^{-9} Torr [4,9]. The system has three interconnected chambers for depositing the channel, dielectric, and doped layers. Anode biasing and fringing magnetic fields are arranged to minimize ion bombardment of the substrates [10]. We deposit films under low total pressure conditions (Ar+N$_2$ $\leq$ 3 mTorr) to ensure a dense homogeneous microstructure. The substrate temperature is fixed at 125°C for all layer depositions.

Results and Discussion

We performed an initial optimization of the growth conditions using single layer characterizations. The best a-Si:H layers are produced at $P(H_2)$ = 0.4 mTorr and $P(Ar)$ = 1.5 mTorr. They have monohydride dominant bonding as measured using infra-red absorption: the ratio of the 2000 cm^{-1} (monohydride) mode to the 2100 cm^{-1} (dihydride) mode peak heights is ~ 5. The dark conductivity is 4 x 10^{-12} (Ωcm)$^{-1}$ and photoconductivity at 100 mW/cm^2 white light illumination is 5 x 10^{-7} (Ωcm)$^{-1}$. The Tauc optical bandgap is 1.81 eV. These results indicate that the sputtered a-Si:H films deposited at 125°C have quality comparable to PECVD films deposited at higher temperatures.

We deposit the nitride layer under two different conditions depending on the application. For the encapsulation (back channel) layer, which does not have to sustain high electric fields, we grow under high total pressure conditions (Ar+N$_2$ $\geq$ 4 mT). This produces a-SiN$_x$:H with a columnar microstructure that has a fast wet etch rate, which is convenient for device processing. For the dielectric layer, we grow under low total pressures (Ar+N$_2$ < 3 mTorr) to produce a dense film with minimum defect density. Optimized nitride films are grown at $P(N_2)$ = 1.3 mTorr, $P(Ar)$ = 1.0 mTorr, and $P(H_2)$ = 0.7 mTorr. These films have an refractive index of ~ 1.88 which is within the 1.85-1.90 range that has been shown to yield high quality TFTs [11]. Infra-red absorption shows dominant nitrogen-hydrogen bonding. Using literature values for oscillator strengths [12], the hydrogen content is found to be 3-4x10^{22} /cm^3. Thermal evolution spectra show three distinct peaks centered at 250, 500, and 900 °C. The peaks at 500°C and 900°C are similar to those found in nitrides sputtered at higher temperature (230°C) [5], and to PECVD nitrides [13,14]. The small low temperature peak is unique to the 125 °C nitrides. Current leakage through the nitride is < 5x10^{-8} A/cm^2 under a 2 MV/cm electric field. High frequency (100 kHz) capacitance-voltage measurements have been made on Al/a-SiN$_x$:H/c-Si structures. We bias sweep from +1 to -1 MV/cm at 0.05 V/sec. Although the a-SiN$_x$:/c-Si interface is not found in a TFT, these measurements yield information about the quality of the nitride. Optimized nitride films show a flat band voltage and hysteresis of approximately -1 V and 2 V, respectively. This indicates that the sputtered nitrides have low bulk trapped charge and are of good electrical quality.

Figure 1 shows a schematic cross section of the inverted staggered TFT fabricated in this study using a three mask process as follows [15]. The cromium gate is evaporated onto the glass substrate and patterned; a trilayer of a-SiN_x:H/a-Si:H/a-SiN_x:H is deposited; the TFT active areas are patterned; source/drain contacts are patterned; aluminum is evaporated and lifted off the back nitride. A 2000 Å gate dielectric layer, 1000 Å channel layer, and 2500 Å encapsulation layer are used.

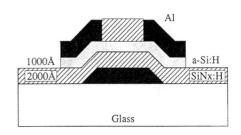

Fig. 1 Cross sectional view of a-Si:H TFT used in this experiment

Figure 2 shows the transfer characteristic for a TFT deposited at 125°C. The threshold voltage is 3 V, subthreshold slope is 0.8 V/decade, and the I_{on}/I_{off} current ratio is 5 x 10^5. The field effect mobility as calculated from the slope of the $(I_{ds})^{1/2}$ vs V_g plot in the saturation region is 0.3 cm^2/V-s. Figure 3 shows the family of I_{ds} vs V_{ds} curves at different V_g for the same TFT. For this device, no n^+ contact layer was deposited in order to reduce mask steps in processing; the additional source/drain resistance is evident by the offset in the turn on near $V_{ds} = 0$.

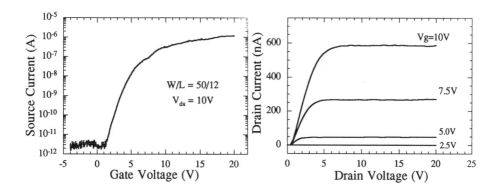

Fig. 2 Transfer characteristic of TFT deposited at 125° C

Fig. 3 The family of I_{ds} vs V_{ds} Curves at different V_g for the TFT used in this experiment

Conclusions

In summary, we have deposited high-quality TFTs at a substrate temperature of 125 °C using dc reactive magnetron sputtering. To our knowledge, this sets a low-temperature record for a quality device and opens the possibility of device fabrication on plastic substrates. We anticipate that the device performance can be further optimized by careful attention to the a-SiN$_x$:H / a-Si:H interface, eg., tailoring the hydrogen content by varying the hydrogen partial pressure in the sputtering discharge. The combination of low substrate temperature (compatible with photoresist) and sputter deposition (which is highly directional [16]) also opens new possibilities for simplified device processing involving lift-off of the a-SiN$_x$:H or a-Si:H layers. And the sputtering process is environmentally friendly and low-cost due to the absence of hydride gases.

Acknowledgments

This work was supported through the ARPA AMLCD flat panel display initiative through a subcontract from Intevac and a University Partnership Award from IBM. The authors would like to thank Dr. Verle Aebi and Dr. Gary Davis of Intevac for supplying the ARPA TFT mask set.

REFERENCES

1. M.S. Feng, C.W. Liang, and D. Tseng, J. Electrochemical Soc. **141**, 1040 (1994).
2. W. Liao, C. Lin, and S. Lee, Appl. Phys. Lett. **65**, 2229 (1994).
3. J. Perrin, in Plasma Deposition of Amorphous Silicon-Based Materials, ed. by G. Bruno, P. Capezzuto, and A. Madan (Academic Press), **177** (1995).
4. M. Pinarbasi, N. Maley, A.M. Myers, and J.R. Abelson, Thin Solid Films **171**, 217 (1989).
5. C.S. McCormick, C.E. Weber, and J.R. Abelson, in preparation.
6. Y.H. Liang, S.Y. Yang, A. Nuruddin, and J.R. Abelson, Mat. Res. Soc. Symp. Proc. **336**, 589 (1994)
7. J.L. Crowley "PECVD vs. DC Magnetron Sputtering for TFT Fabrication: A cost of ownership Analysis," (Intevac Report, 1993).
8. A. Kolodziej and S. Nowak, Thin Solid Films **175**, 37 (1989).
9. J.R. Abelson, "[Sputtered] Hydrogenated Amorphous Silicon, Silicon Carbide, and Micro-crystalline Silicon," in Handbook of Thin Film Deposition, (Institute of Physics, 1995) p. X2.2:1.
10. J.R. Doyle, A. Nuruddin, and J.R. Abelson, J. Vac. Sci. Tech. A **12**, 886 (1994).
11. Y. Kuo, J. Electrochemical Soc. **142**, 186 (1995).
12. A. Morimoto, Phys. Stat. Sol. B **63**, 715 (1983).
13. M. Fitzner, J. R. Abelson, and J. Kanicki, Mat. Res. Soc. Sym. Proc. **258**, 649 (1992).

14. T. Li, J. Kanicki, M. Fitzner, W.L. Warren, AMLCD 1995 Workshop Proceedings, Lehigh University, (1995).
15. Mask set designed for the ARPA Active Matrix Flat Panel Display Program and Supplied by the Intevac Corp.
16. A. Nuruddin, J.R. Doyle, and J.R. Abelson, J. Appl. Phys. **76**, 3123 (1994).

IMPROVED a-Si:H TFT PERFORMANCE USING a-Si$_X$N$_{1-X}$ / a-Si$_X$C$_{1-X}$ STACK DIELECTRICS

G. LAVAREDA, E. FORTUNATO, C. NUNES CARVALHO AND R. MARTINS
Faculdade de Ciências e Tecnologia - Departamento de Ciência dos Materiais, FCT/UNL
Centro de Excelência de Microelectrónica e Optoelectrónica de Processos, CEMOP/UNINOVA
Centro de Física Molecular, CFM/UTL.
Quinta da Torre, 2825 Monte da Caparica - PORTUGAL

ABSTRACT

In this paper we present a study on the electrical characteristics (conductivity, σ and relative dielectric constant, ε_r) of amorphous silicon nitride (a-Si$_X$N$_{1-X}$) and carbide (a-Si$_X$C$_{1-X}$) films deposited by PECVD, used as dielectric materials in TFT devices, aiming to select the most adequate alloy that lead to improve device performances. Besides that, double stack a-Si$_X$N$_{1-X}$/a-Si$_X$C$_{1-X}$ structures were developed and applied as dielectric layers on TFTs, whose performances show to be superior to those ones using single silicon nitride or silicon carbide as dielectric.

INTRODUCTION

Amorphous silicon TFTs are under development since the pioneer work of the Dundee group [1], for several switching/addressing applications such as LCDs [2] and image sensors [3]. One of the main device components is the gate dielectric that besides being highly isolated has to present the required surface morphology (low surface density of states).

Amorphous silicon nitride has been widely used as a dielectric material [4], because of its typical very low conductivity. However, due to its coordination number, nitrogen makes an intrinsically stressed alloy with silicon, leading to a defective interface with the amorphous silicon active layer [5]. In order to search other dielectric materials, with a better interface behavior, amorphous silicon TFTs with a-Si$_X$C$_{1-X}$ gate dielectric have been studied [6]. The data reported show a significant decrease on threshold voltages (< 1V) inspite of the moderate transconductances presented by such devices ($\sim 3 \times 10^{-8}$ A.V^{-2}). The only disadvantage of the a-Si$_X$C$_{1-X}$ is related to its low critical electric field, leading to low breakdown voltages. This limitation is responsible for the undesirable low I$_{ON}$/I$_{OFF}$ current ratios presented by those TFTs. In the present work, a similar study on a-Si$_X$N$_{1-X}$ alloys for TFT applications was made, aiming to compare the main electric characteristics of those films and their correlation with the performances exhibited by the corresponding TFTs. Finally an a-Si$_X$N$_{1-X}$ /a-Si$_X$C$_{1-X}$ stack dielectric TFT was produced and tested, whose performances are superior than the ones presented by the single-layer insulator TFTs.

EXPERIMENTAL DETAILS

1. Preparation and Characterization of a-Si$_X$C$_{1-X}$, a-Si$_X$N$_{1-X}$ and a-Si:H

The insulating alloys as well as the semiconductor films were prepared on glass substrates coated with ITO, using the PECVD technique. The production parameters used are listed in Table I, where the silane gas concentration [SiH$_4$] on the mixtures used varied from 10% to 60%. The insulator thickness was determined by a Sloan Dektak IIA apparatus.

Table I - Production Parameters of the dielectric and semiconductor layers.

	a-Si$_X$C$_{1-X}$	a-Si$_X$N$_{1-X}$	a-Si:H	a-Si:H(N+)
Total Gas Flow (SCCM)	25	40	15	15
Deposition Pressure (mTorr)	500	500	400	300
Deposition Temperature (°C)	200	230	200	200
r.f. Power (W)	20	24	12	12

After the deposition of the insulator, a top metal conductor (Al) were evaporated (2000 Å thick), to make a capacitor structure, to determine the values of the relative dielectric constant, ε_r, using a standard capacitor area of 0,0707 cm^2. The film conductivity measurements was performed in the same structure, at room temperature, by means of a digital multimeter (Keithley 617).

2. Preparation and Characterization of the TFTs

The TFTs were produced with a bottom-gate structure (figure 1) with a W/L channel ratio of 68. Both insulators (a-Si$_X$N$_{1-X}$ and a-Si$_X$C$_{1-X}$) were prepared using a 20% ratio of NH$_3$/SiH$_4$ or CH$_4$/SiH$_4$, respectively. The device characteristics was measured by a HP4145B I(V) curve tracer. The I(V) curves plotted are I$_{DS}$=f(V$_{DS}$,V$_{GS}$) (conductance characteristics) and $\sqrt{I_{DS}}$ =f(V$_{GS}$), for a constant V$_{DS}$ (transconductance characteristics), where V$_{DS}$ and V$_{GS}$ means the drain-to-source and the gate-to-source voltages, respectivelly, and I$_{DS}$ is the drain current. The values of the field effect mobilities, μ_{FE} and the threshold voltage, V$_T$, are inferred from the tranconductance characteristics, by means of linear regression, taking y=$\sqrt{I_{DS}}$ and x=V$_{GS}$, and have been used as a figure of merit of the device performances.

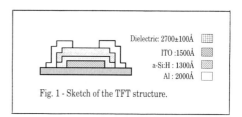

Fig. 1 - Sketch of the TFT structure.

Dielectric: 2700±100Å
ITO :1500Å
a-Si:H : 1300Å
Al : 2000Å

RESULTS AND DISCUSSION

1. Dielectric and amourphous silicon electric properties

Silicon carbide and silicon nitride alloys

The conductivity (σ) of a-Si$_X$C$_{1-X}$ alloys and its dependence on [SiH$_4$] is shown in figure 2. The data reveal that σ is independent from [SiH$_4$], in the range of 10%<[SiH$_4$]<40%, having a significant increase for [SiH$_4$]>40%. Nevertheless, when using this dielectric in TFTs, it has been shown that the best device performances are achieved when [SiH$_4$]=20% [4], since to this concentrations, higher breakdown voltages are achieved than the ones obtained using [SiH$_4$]=40%. The data shown in figure 3 (plot of ε_r versus [SiH$_4$]) reveal a linear dependence of ε_r on [SiH$_4$], where a gradual change of the predominant behaviour, from amorphous carbon ($\varepsilon_r \sim$ 5.5, in diamond-like structures [7]) to amorphous silicon ($\varepsilon_r \sim$ 11.3), is observed.

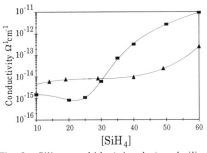

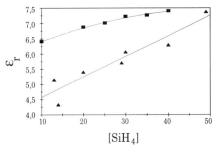

Fig. 2 - Silicon carbide (triangles) and silicon nitride (squares) conductivities as a function of [SiH₄]

Fig. 3 - Silicon carbide(triangles) and silicon nitride (squares) relative dielectric constant as a function of [SiH₄]

The σ versus [SiH$_4$] dependence for silicon nitride is also shown in figure 2. The data reveal a minimum on σ, at about [SiH$_4$]=20%. This minimum is normally reported in the near-stoichiometric (Si$_3$N$_4$) nitrides [8]. From figure 2, we conclude that silicon nitride is better insulator than silicon carbide, in the range 10%<[SiH$_4$]<30%. In figure 3 we plot the dependence of ε_r on [SiH$_4$]. The data show a monotonic increase of ε_r with [SiH$_4$]. Here, once more, we can infer the near-stoichiometric composition of the nitride alloy, comparing the ε_r value for the minimum conductivity ($\varepsilon_r \sim 6.9$) with the values published by other authors [9] ($\varepsilon_r \sim 7$). From the results above achieved, the [SiH$_4$] conditions selected to produce the TFT using both type of dielectrics studied, are the ones for which [SiH$_4$]=20%.

Amorphous Silicon Properties

The semiconductor used in the TFT channel was intrinsic (non-doped) hydrogenated amorphous silicon (a-Si:H), while the Source and Drain zones were phosphorous doped hydrogenated amorphous silicon, a-Si:H(N+). The electric and optical characteristics of both semiconductors are presented in Table II.

Table II - Properties of the semiconductor layers.

	Conductivity, σ ($\Omega^{-1}cm^{-1}$)	Optica Gap, E_{OP} (eV)	Thermal Activation Energy, ΔE (eV)
a-Si:H	2.7×10^{-9}	1.74	0.88
a-Si:H(N+)	5.2×10^{-5}	1.60	0.35

2. Thin-Film Transistor characteristics

Silicon Carbide dielectric TFTs

The transconductance and the conductance characteristics of the TFTs produced using Si$_x$C$_{1-x}$ dielectric are shown on figures 4 and 5. The breakdown voltage of the dielectric is low (between 10V and 15V), limiting so the range of values analysed (between 0V and 5V).

Nevertheless, the TFTs transconductance is 2.5×10^{-7} A/V^2, leading to field-effect mobilities (μ_{FE}) of 0.352 cm^2/(V.s). As we can see in figure 4, the threshold voltage, $V_T \sim 0.86$V, while the I_{ON}/I_{OFF} ratio is about 500 (I_{ON} measured for $V_G = V_D = 5$V).

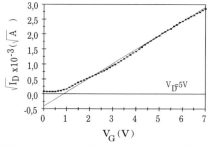

Fig. 4 - Transconductance characteristic of a silicon carbide dielectric TFT.

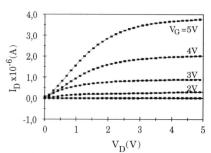

Fig. 5 - Conductance characteristic of a silicon carbide dielectric TFT.

Silicon Nitride dielectric TFTs

The TFTs made using Si_XN_{1-X} as dielectric presents a high breakdown voltage, allowing application of higher voltages, up to 40V. As we can see in figure 6, the $V_T \sim 13$V and the transconductance is 5.3×10^{-8} A/V^2, with $\mu_{FE} = 0.129$ cm^2/(V.s). The corresponding I_{ON}/I_{OFF} ratio is about 3500 (I_{ON} measured for $V_G = V_D = 40$V).

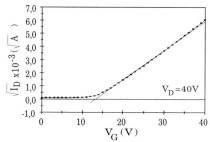

Fig. 6 - Transconductance characteristic of a silicon nitride dielectric TFT.

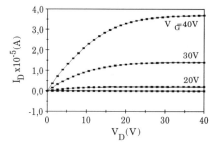

Fig. 7 - Conductance characteristic of a silicon nitride dielectric TFT.

From the above results we conclude that silicon nitride based TFTs presents better field effect properties due to the higher ε_r values resistivity and breakdown voltages than the ones exhibited by silicon carbide based TFTs. On the other hand, silicon carbide TFTs have improved transport properties, essentially due to the better interface with amorphous silicon.

In order to take advantage of the complementary characteristics presented by these layers, we propose a new dielectric structure, composed by a Si_XN_{1-X} /Si_XC_{1-X} stacked layers, to be used as dielectrics, in TFT devices (where the interface with amorphous silicon is made by silicon carbide).

Silicon Nitride/Silicon Carbide stacked dielectric TFTs

The TFTs produced using Si_xN_{1-x}/Si_xC_{1-x} stacked layers as dielectrics (72% of total dielectric thickness is Si_xN_{1-x}), it presents the same insulator brakdown voltage than Si_xN_{1-x} dielectric TFTs, allowing so the use of high test voltages. The tansconductance measured from the characteristics shown in figure 8 is 6.4×10^{-8} A/V^2 [$\mu_{FE} = 0.168$ cm^2/(V.s)] and $V_T \sim 7.2$V. The I_{ON}/I_{OFF} ratio is about 7600 (I_{ON} measured for $V_G = V_D = 40$V), about twice larger than the same value recorded in silicon nitride based TFTs.

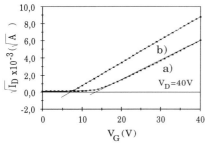

Fig. 8 - Transconductance characteristics of an Si_xN_{1-x} dielectric TFT (a), and Si_xN_{1-x}/Si_xC_{1-x} stacked dielectric TFT (b).

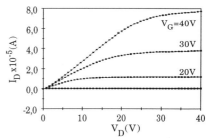

Fig. 9 - Conductance characteristic of a Si_xN_{1-x}/Si_xC_{1-x} stacked dielectric TFT.

CONCLUSIONS

In the present study we presented the main advantages in using a Si_xN_{1-x}/Si_xC_{1-x} dual-layer dielectric instead of the usual single nitride layer, lowering the threshold voltages in 45%, and increasing the I_{ON}/I_{OFF} ratio in about 120%, while the field effect mobilities are improved in 30%. This improvement is due to the combination of the better volume properties of the nitride layer with the better interface properties exhibited by silicon carbide, with amorphous silicon. These data reveal that more stable and high performance TFTs can be produced, by proper matching of the dielectrics, layers structure and composition with the characteristics of the a-Si:H layer used to produce the channel (of the device). It is important to determine the influence of the thickness of the individual stacked layers and their number on the final device performances, encouraging so the prosecution of our work, concerning this proposed new stacked structure to produce highly stable and reliable TFTs, for active matrice applications.

REFERENCES

1. P. G. LeComber, W. E. Spear and A. Ghaith, Electron. Lett. **15**, 179 (1979).
2. K. Suzuki, <u>Flat Panel Displays Using Amorphous and Microcrystalline Semiconductor Devices</u> in: "Amorphous and Microcrystalline Semiconductor Devices", **Vol I**, Artech House, Boston, ed. Kaniki (1991), p. 77.
3. A. Madan, M. P. Shaw, "The Physics and Applications of Amorphous Semiconductors", Academic Press (1988), p. 310.

4. A.R. Grant, P. D. Persons, R. F. Kwasnick and G. E. Possin, Mat. Res. Soc. Symp. Proc, **297**, 883 (1993).
5. C. Van Berkel, <u>Amorphous Silicon Thin Film Transistors: Physiscs and Properties</u> in: "Amorphous and Microcrystalline Semiconductor Devices", **Vol II**, Artech House, Boston, ed. Kaniki (1992), p. 421.
6. G. Lavareda, E. Fortunato, C. N. Carvalho, R. Carrapa and R. Martins, Int. Semicond. Device Res. Symp. Proc. **1**, 161 (1993).
7. Walter Lambrecht, Mat. Res. Soc. Symp. Proc., **339**, 565 (1994).
8. X. Xu and S. Wagner, <u>Phisics and Electronic Properties of Amorphous and Microcrystalline Silicon Alloys</u> in : "Amorphous and Microcrystalline Semiconductor Devices", **Vol II**, Artech House, Boston, ed. Kaniki (1992), p. 118.
9. J. Kanicki, Mat. Res. Soc. Symp. Proc., **219**, 363 (1991).

AMORPHOUS SILICON TFTs ON STEEL-FOIL SUBSTRATES

S. D. THEISS and S. WAGNER
Princeton University, Department of Electrical Engineering, Princeton, New Jersey 08544

ABSTRACT

We describe the successful fabrication of device-quality a-Si:H thin-film transistors (TFTs) on stainless-steel foil substrates. These TFTs demonstrate that transistor circuits can be made on a flexible, non-breakable substrate. Such circuits could be used in reflective or emissive displays, and in other applications that require rugged macroelectronic circuits.

Two inverted TFT structures have been made, using 200 μm thick stainless steel foils with polished surfaces. In the first structure we used the substrate as the gate and utilized a homemade mask set with very large feature sizes: L = 45 μm; W = 2.5 mm. The second, inverted staggered, structure used a 9500 Å a-SiN$_x$:H passivating/insulating layer deposited on the steel to enable the use of isolated gates. For this structure we used a mask set which is composed of TFTs with much smaller feature sizes. Both TFT structures exhibit transistor action. Current-voltage characterization of the TFTs with the inverted staggered structure shows typical on/off current ratios of 10^7, leakage currents on the order of 10^{-12} A, good linear and saturation current behavior, and channel mobilities of 0.5 cm^2/V·sec. These characteristics clearly identify the TFTs grown on stainless steel foil as being of device quality.

INTRODUCTION

The principal substrate for thin-film electronics to date has been glass. The transparency of glass allows light penetration through the TFT back-planes to the liquid crystal of active-matrix liquid crystal displays. The flatness and comparative chemical inertia of glass makes it a useful substrate for thin-film circuits that have little tolerance for roughness and for contamination. However, glass has several disadvantages in regards to its use as a display back-plane in portable device electronics, among them its fragility and weight. Breakage of the display glass is said to be the most important failure mechanism of portable display products. Furthermore, for many portable applications one also would like to reduce the weight that is associated with present glass substrates [1].

In this paper we report the first successful demonstration of steel-foil substrates for amorphous silicon thin-film transistors (TFTs). The toughness of steel opens the possibility of making non-breakable TFT back-planes, and of greatly reducing the substrate thickness and weight. While the opacity of steel limits the TFT circuit applications to reflective or emissive displays, the steel surface can be modified by coatings to achieve a wide range of optical properties. Since we have made no attempts to optimize the substrate or film parameters for these TFTs, the relatively high quality of the devices described in this paper is very encouraging. This suggests that growth of

TFTs on metal-foil substrates has the potential to be a very robust process, easily transferable to industrial applications.

EXPERIMENTAL

The stainless steel substrates used in these experiments are of the type used in industry for flexible amorphous silicon solar cells [2]. They are bright-annealed, grade 430 stainless steel, 200 μm thick, polished on one side, and possess an rms surface roughness of 0.1 μm. We use 3x3 sq. in. plates. We first investigated the feasibility of making TFTs in a configuration in which the entire substrate serves as a gate electrode, as shown in Fig. 1. We call this the plate-gate configuration. For the second configuration we coated the substrate with an a-SiN$_x$:H insulating layer before fabricating bottom gate TFTs in the inverted staggered configuration, shown schematically in Fig. 2. This we call the isolated-gate configuration. The bottom gate TFTs were fabricated using a 4-level mask process, utilizing an integrated TFT mask set designed at Penn State, under the auspices of the ARPA-EPRI collaboration, and produced in the Nanofabrication Laboratory at Cornell University. This mask set was designed to be used for TFT characterization, and contains TFTs with a wide range of W/L ratios and other parameters. All a-Si:H and a-SiN$_x$:H layers were deposited in a three chamber plasma enhanced chemical vapor deposition (PECVD) system, in which undoped a-Si:H, n$^+$ a-Si:H, and a-SiN$_x$:H are deposited in separate chambers provided with diode electrode configuration. The deposition system also includes a separate, heatable load-lock chamber which eliminates the need for long pump-down times. All depositions were done using various levels of RF power at a frequency of 13.56 MHz. Metallization was done by thermal evaporation of chromium in a separate system.

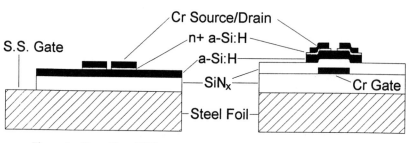

Figure 1. *Plate-Gate TFT.*   Figure 2. *Isolated-Gate TFT.*

The plate-gate configuration was made with the smallest number of steps possible to seek a feasibility demonstration. The gate dielectric, active channel, and ohmic contact layers for the plate-gate configuration were all deposited using the same conditions as for the isolated-gate configuration and are not described separately. Since the substrates were shipped in the state in which they are used for solar cell deposition, no further cleaning was deemed necessary. Before placing the substrates into the deposition system, they were first rinsed in de-ionized water and blown dry with dry N$_2$ to remove any dust particles. The samples were heated in the load-lock chamber to the desired deposition temperature and then transferred to the dielectric deposition chamber. The process parameters for the different TFT layers are summarized in Table 1 below. We first deposited a 9500 Å thick insulating a-SiN$_x$:H layer and returned the sample to the load-lock to cool. We then removed the sample from the deposition system and evaporated and patterned 1500 Å thick chromium gate lines on top of the insulating nitride. We returned the sample to the deposition system and, without breaking vacuum, deposited the 3200 Å a-SiN$_x$:H gate-dielectric layer, the 1600 Å thick active layer of undoped a-Si:H, and the 500 Å thick n$^+$ a-Si:H ohmic contact layer. All depositions described in this paper were done at a pressure of 500 mTorr. Next we removed the sample from the deposition system and used reactive-ion etching (RIE) with a CF$_4$/O$_2$ gas mixture to define the TFT structure and expose the gate pads. Finally, we evaporated and patterned the chromium sources and drains and used RIE to remove the n$^+$ material between the source-drain contacts. An optical micrograph of one of the finished isolated-gate TFTs is shown in Figure 3.

Layer	Gas Flows (sccm)	Temperature (°C)	RF Power (mW/cm^2)
a-SiN$_x$:H	4 : SiH$_4$ 100 : NH$_3$	310	22
a-Si:H	50 : SiH$_4$	250	20
n$^+$ a-Si:H	44 : SiH$_4$ 6 : PH$_3$	260	16

Table 1. *PECVD Process Parameters.*

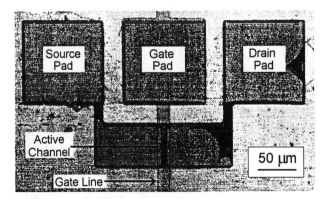

Figure 3. *Optical micrograph of an isolated-gate TFT on a stainless-steel substrate.*

RESULTS

The transfer and output characteristics of an isolated-gate TFT with W=50 μm and L=10 μm are shown in Figures 4 and 5. They clearly demonstrate the high quality of the TFTs described in this paper. In Fig. 4, at a drain voltage of 10 V, the off current in this device is around 0.5 pA and the on current at a gate voltage of 20 V is around 5 μA, resulting in an on/off current ratio of 10^7. The output characteristics shown in Fig. 5 are typical of device quality TFTs grown on glass substrates, with high output drain currents. All TFT measurements were done with the source electrode grounded.

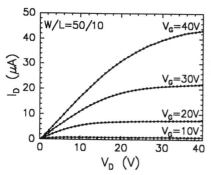

Figure 4. *Transfer characteristics of an isolated-gate TFT.*

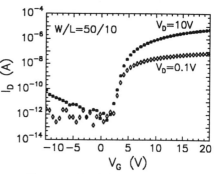

Figure 5. *On/Off characteristics of the isolated-gate TFT of Fig. 4.*

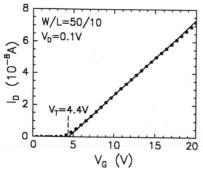

Figure 6a. *Subthreshold characteristics of the isolated-gate TFT of Figs. 4 and 5*

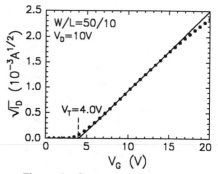

Figure 6b. *Saturation characteristics of the isolated-gate TFT of Figs. 4 and 5.*

68

Figures 6a and 6b show the drain current plotted against the gate voltage for the two drain voltages shown in Fig. 5. These plots are used to determine the threshold voltage (V_T) and field-effect mobility (μ_{FE}) from the following equations [3]:

$$I_{D\,(linear)} = \mu_{FE} C'_{SiN} \frac{W}{L}\left[(V_G - V_T)V_D - \frac{V_D^2}{2}\right] \qquad (\text{small } V_D) \qquad (1)$$

$$I_{D\,(saturation)} = \mu_{FE} C'_{SiN} \frac{W}{2L}(V_G - V_T)^2 \qquad (\text{large } V_D) \qquad (2)$$

where C' is the capacitance per unit area of the a-SiN$_x$:H, and W and L are the channel width and length, respectively, of the TFT. The capacitance for our silicon nitride layers has not been measured yet, but from its optical properties and its thickness (3200 Å), we would anticipate it to lie somewhere in the range of 1.8-1.9×10^{-8} F/cm^2. This range of values comes from using a relative dielectric constant of 6.5-7, which is typical of gate nitride films grown under our deposition conditions [4,5]. From the plots, we see that both Eqs. (1) and (2) are well obeyed in their respective regions. In the linear current regime (Fig. 6a), we obtain a threshold voltage and mobility of 4.4 V and 0.5 cm^2/V·sec, respectively. In the saturation current regime (Fig. 6b), we obtain a threshold voltage and mobility of 4.0 V and 0.5 cm^2/V·sec, respectively, in excellent agreement with the values obtained in the linear current regime.

DISCUSSION

We have demonstrated that it is possible to grow device quality a-Si:H TFTs on stainless steel foils. Preliminary tests indicate that devices fabricated in this fashion are highly rugged and continue to function after undergoing significant mechanical stresses, including macroscopic bending and falling to the ground from heights of several feet. This suggests that these devices are highly suitable for applications in which durability are important. With appropriate surface treatment of the steel substrate to enhance its reflectivity, this technology could easily be incorporated into displays for both civilian and military applications in which reflective liquid-crystal or emissive displays are already in use.

SUMMARY

We have fabricated the first demonstration of device quality a-Si:H TFTs on stainless steel foil substrates. Current-voltage characterization of the inverted staggered TFTs in the isolated-gate configuration show typical a-Si:H TFT behavior, with low reverse-bias leakage currents on the order of 10^{-12} A, and on/off current ratios of 10^7. Plots of the drain current versus gate voltage in both the linear and saturated current regimes show the expected behavior and give threshold voltages in the range of 4 V and field-effect mobilities of 0.5 cm^2/V·sec.

ACKNOWLEDGMENTS

We would like to acknowledge the assistance of Helena Gleskova for helpful discussions and for aid with TFT characterization. We would also like to thank Xixiang Xu and the United States Solar Corporation (USSC) for providing us with the substrate material used in this experiment. The multichamber deposition system used in this experiment was funded by the Electric Power Research Institute (EPRI). This research is supported by Motorola and by ARPA under Air Force grant number F33615-94-1-1464, managed by WPAFB.

REFERENCES

1. D. M. Moffatt, MRS Bulletin, **21**(3), 31 (1996).

2. P. Nath and M. Izu, Conf. Rec. 18th IEEE Photovoltaic Specialist Conf., Las Vegas, Oct. 21-25, 1985. IEEE, New York, 1985, p. 939.

3. S. M. Sze, Semiconductor Devices, (Wiley, New York, 1985), p. 206.

4. M. Gupta, V. K. Rathi, R. Thangaraj, O. P. Agnihotri and K. S. Chari, Thin Solid Films, **204**, 77 (1991).

5. S. Sherman (private communication).

a-Si:H TFTs PATTERNED USING LASER-PRINTED TONER

HELENA GLESKOVÁ*[1], S. WAGNER[1] AND D.S. SHEN[2]
[1] Princeton University, Department of Electrical Engineering, Princeton, NJ 08544
[2] University of Alabama in Huntsville, Department of Electrical and Computer Engineering, Huntsville, AL 35899

ABSTRACT

We fabricated top-gate amorphous silicon thin-film transistors (a-Si:H TFTs) on alkali-free glass foil, for the first time using laser-printed toner for the patterning of each layer. The toner for the first mask level was applied by feeding the glass foil through a laser printer, and for the following mask levels from patterns laser-printed on transfer paper. The transistors have off currents from $\sim 10^{-12}$ A to $\sim 10^{-11}$ A and on-off current ratios of $\sim 10^6$. Thus we have demonstrated a technology for the patterning of TFT circuits by printing.

INTRODUCTION

The flat-panel display industry introduced a new trend in semiconductor technology -- scaling-up replaced miniaturization. The semiconductor processing tools developed for integrated circuit fabrication were scaled up to accommodate the size of the active matrix. Newly developed mass-production sputtering and dry-etching systems [1] and also steppers can handle substrates up to 40 x 50 cm^2, but they are correspondingly expensive. We need a revolution of the manufacturing technology of integrated circuits to reduce the cost of future wall-size displays. In our effort to develop a large-area integrated thin-film circuit technology for such consumer products, we have replaced photolithography with electrophotographic printing. Electrophotography, a mature technique used in laser printers and photocopiers, already allows to print with a resolution of 1800-2400 dpi, leading to a design rule of 11-15 μm, with further reduction in sight [2]. Commercial large-area photocopiers handle three-feet wide rolls of paper. Combining a laser printer with a computer offers a simple means for the generation of toner masks and their dynamic correction. The toner masks can be directly applied to semiconductor, insulator, and metal surfaces [2] without any further patterning.

We reported earlier that hydrogenated amorphous silicon (a-Si:H) can be deposited on flexible glass foil and can be patterned using an etch mask of xerographic toner [2]. Recently, we have shown that a-Si:H TFTs can be made by using toner masks for the patterning of all layers [3]. In this paper we report improved electrophotographic processing that provides TFTs with improved electrical performance.

EXPERIMENTAL PROCEDURES

TFTs were fabricated on 50-μm thick alkali-free glass foil (Schott # AF45). The TFTs have a top-gate staggered structure that requires three different patterning steps. These include direct laser printing [2] for the first mask level, and transfer of toner masks [4] for the higher levels. The three toner patterns are shown in Fig. 1.

The channel length and width of the TFTs are 100 μm and 1mm, respectively. The width of the gate electrode is 250 μm. The entire process sequence is shown in Fig. 2. First, a ~ 100 nm thick Cr layer is thermally evaporated over the clean glass substrate. Then a ~ 100 nm thick (n+) a-Si:H layer is deposited on the Cr using rf plasma enhanced chemical vapor deposition (PECVD). The positive source-drain toner pattern (see Fig. 1a) is printed on the (n+) a-Si:H layer by feeding the glass foil through a commercial 600 dpi laser printer. After printing, the toner is further baked at 120°C for 1 hour in air. During this additional baking the sintered toner clusters fuse together and create a continuous layer of polymer toner. Using this mask, the (n+) a-Si:H layer is etched in KOH solution [2]. Then the toner is stripped in an ultrasonic bath of PRX 100 stripper [5] heated to 80°C. The Cr layer is etched in CAN etch [6] at room temperature. In this case the patterned (n+) a-Si:H layer serves as a mask. The 170-nm-thick channel layer of undoped a-Si:H and the 300-nm-thick gate insulator layer of silicon nitride (SiNx) are deposited in a three-chamber PECVD system using rf excitation.

Because with our off-the-shelf laser printers we cannot register two successive layers, we use the transfer paper technique [4] for the second and higher mask levels. The negative gate pattern (see Fig. 1b) is printed on a special transfer paper using the above mentioned laser printer. The printed side of the transfer paper and the glass substrate are brought in contact and are aligned under an optical microscope. Applying the proper amount of heat and normal force causes the toner to stick to the substrate. Following a water soak the transfer paper is peeled off the toner. Unfortunately, when the toner is transferred in this way, it sticks well to semiconductor and insulator layers only if they are coated with photoresist. Using the toner as a UV-mask the photoresist is exposed to UV-light and is developed. A ~ 100-nm-thick layer of Al is thermally

(a) Positive source-drain mask.

(b) Negative gate mask.

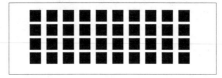

(c) Positive mask for contact hole opening and transistor separation.

Fig. 1. Toner masks for TFT fabrication.

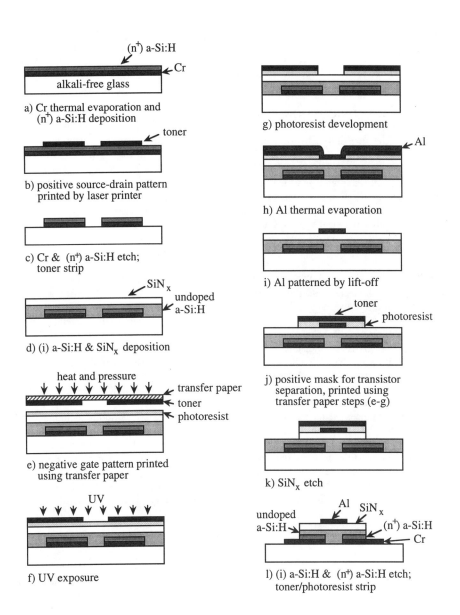

Fig. 2. TFT process sequence.

evaporated and the gate electrode is patterned by lift-off with PRX 100 stripper. In the final step, which includes source/drain contact hole opening and transistor separation, we again use the transfer paper technique. We print a positive etch mask (see Fig. 1c) and then etch the SiN_x in 10% HF. The undoped and (n^+) a-Si:H are etched in AZ 300 T stripper heated to ~ 130°C which simultaneously strips the toner/photoresist layer.

RESULTS AND DISCUSSION

The characteristics of a transistor fabricated as described above are shown in Fig. 3. Fig. 3(a) shows the dependence of the source-drain current I_{ds} on the gate voltage V_{gs} for V = 10 V. The off-current is 1 x 10^{-12} A and the ratio of the on-off current is ~ 10^6. The transistors have a relatively high threshold voltage which is also visible in Fig. 3(b). This figure shows the source-drain current as a function of source-drain voltage V_{ds} for eight different gate voltages ranging from 0 to 45 V. A magnified photograph of that particular transistor is shown in Fig. 4.

In the saturation regime the source-drain current is given by:

$$I_{ds} = (W/2L)\, C_{SiN}\, \mu_n\, (V_{gs} - V_t)^2$$

where L is the channel length, W the channel width, C_{SiN} the capacitance of the gate insulator, μ_n the electron mobility, and V_t the threshold voltage. Plotting $(I_{ds})^{1/2}$ versus V_{gs} gives a value of ~ 12 V for the threshold voltage and a value of 8.32 x 10^{-9} F V^{-1} s^{-1} for the ($C_{SiN}\, \mu_n$) product. Because $C_{SiN} = \varepsilon_o\, \varepsilon_r\, /\, d_{SiN}$, a value of 2.07 x 10^{-8} F cm^{-2} is calculated for the capacitance of the nitride, assuming ε_r = 7. The corresponding calculated mobility is 0.40 cm^2 V^{-1} s^{-1}. This a reasonable value given our immature technology. On the same glass substrate we also fabricated transistors from which the (n^+) a-Si:H layer was removed on purpose. Their characteristics look very similar to those of TFTs with the (n^+) a-Si:H layer. This suggests that the n^+ layer does not affect the contact between the Cr source-drain metallization and the undoped a-Si:H, probably because of the relatively low saturation current densities.

Our current process sequence is more complicated than the laser printing technique itself requires. This is because we have no tool available at present that allows us to align second and higher toner levels in the laser printer. Therefore, we need to rely on the transfer paper technique, which introduces to the process sequence the photoresist and associated processing steps. Alignment of each toner level with the patterned substrate directly in the electrophotographic printer will eliminate the photoresist and the transfer paper steps (steps e, f, and g in Fig. 2). Such alignment tools are close to commercialization, but are not yet available in off-the-shelf printers.

SUMMARY

We fabricated amorphous silicon thin-film transistors for the first time in a process where all pattern definition steps use electrophotographic toner masks. While the process needs improvements, it already combines the potential for a drastic reduction of process steps with the advantage of electronic pattern generation. We view this combination as a powerful tool for developing large-area electronics.

This work is supported by ARPA through WPAFB under Contract F33615-94-1-4448 and by EPRI. We thank S. D. Theiss for the help with the SiN_x depositions, and Schott Corporation for providing the glass foil substrates.

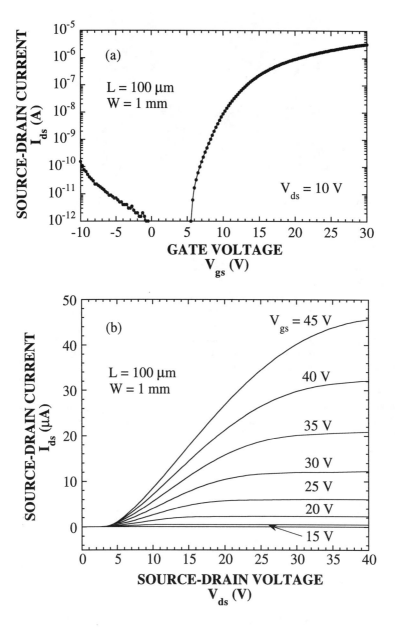

Fig. 3. Source-drain current as a function of a) gate voltage for V_{ds} = 10 V, b) source-drain voltage for eight different gate voltages. The source electrode is grounded.

100 μm

Fig. 4. Magnified photograph of the transistor whose characteristics are shown in Fig. 3.

REFERENCES

* On leave from the Department of Solid State Physics, Comenius University, 84215 Bratislava, Slovakia.

1. R. Gardner, J. Kiyota, I. Suguira, T. Hori, H. Nakamura, S. Ishibashi, T. Kuroda, H. Takei, H. Kawamura, Y. Ohta and K. Nakamura, in 1996 Display Manufacturing Technology Conference, Digest of Technical Papers, San Jose, CA, Feb. 6-8, 1996, (Society for Information Display, Santa Ana, CA, 1996) pp. 60-70.
2. H. Gleskova, S. Wagner and D.S. Shen, IEEE Electron Device Lett. **16**, 418 (1995).
3. H. Gleskova, R. Könenkamp, S. Wagner and D.S. Shen, IEEE Electron Device Letters - to be published.
4. H. Gleskova, S. Wagner and D.S. Shen, in Amorphous Silicon Technology - 1995, edited by M. Hack, E.A. Schiff, A. Madan, M. Powell and A. Matsuda (Mat. Res. Soc. Proc. **377**, Pittsburgh, PA, 1995) pp.719-724.
5. Silicon Valley Chemlabs, Inc., 245 Santa Ana Court, Sunnyvale, CA 94086.
6. Foto Chemical Systems, Inc., Wayne, NJ 07474-3188.

INFLUENCE OF THE DENSITY OF STATES AND SERIES RESISTANCE ON THE FIELD-EFFECT ACTIVATION ENERGY IN a-Si:H TFT

Chun-ying Chen and Jerzy Kanicki
Department of Electrical Engineering and Computer Science, Center for Display Technology and Manufacturing, The University of Michigan, Ann Arbor, MI 48109

ABSTRACT

We have proposed a new two-dimensional simulation model, which takes into account the density of states of hydrogenated amorphous silicon (a-Si:H) and temperature-dependence of the source/drain series resistances (R_S), to explain the dependence of the activation energy (E_{act}) of drain-source current (I_{DS}) on gate-source bias (V_{GS}) in a-Si:H thin-film transistors (TFTs). We found that the influence of series resistance cannot be ignored, else an overestimated E_{act} will result. The results of our simulation are in agreement with experimentally observed saturation of the E_{act} at higher V_{GS}.

INTRODCTION

The hydrogenated amorphous silicon (a-Si:H) thin-film transistor (TFT) has been widely used as a switching device in active-matrix liquid-crystal display because of its low-temperature and large-area deposition capability. The drain-source current (I_{DS}) of a-Si:H TFT for a given gate-source bias (V_{GS}) is measurement temperature dependent with an activation energy, E_{act}. Since E_{act} depends not only on V_{GS}, but also on the density of states of both a-Si:H layer and a-Si:H/gate insulator interface [1,2], its value at higher V_{GS} has been widely used as an indicator of a-Si:H TFT electrical performance. Indeed, some researchers have reported a correlation between the density of states and E_{act} [3,4]. However, the analytical models used to explain the variations of E_{act} with V_{GS} have assumed zero series resistance (R_S). In a real a-Si:H TFT, the source/drain series resistance is temperature-dependent and non-negligible, and will result in discrepancy between the analytical model and experimental data.

In this paper, we present a two-dimensional model which can explain the variation of the activation energy with V_{GS}. The temperature-dependence of the source/drain series resistance was considered in our simulation model. From our simulation results, we found that the E_{act} will be overestimated if the source/drain series resistance is not negligible.

INFLUENCE OF THE DENSITY OF STATES

The inverted-staggered back-channel etched a-Si:H TFT structure used in our simulation and experiment has 0.25 μm-thick amorphous silicon nitride (a-SiNx:H) as a gate insulator, 0.2 μm-thick a-Si:H as a semiconductor and 0.07 μm-thick n$^+$ a-Si:H as a source/drain contact layer. In this simulation, all the materials were assumed to be homogeneous. The distribution of density of states was also assumed to be measurement temperature independent ranging from 300 to 420 K. A two-dimensional Semicad Device simulation program based on finite-element method and solving Poisson's and electron / hole continuity equations was used to analyzed the physics and influence of the density of states and source/drain series resistance on the evolution of E_{act} as a function of V_{GS}. The model of the density of states of a-Si:H in our case is described by two exponentially decreasing band-tail states near the mobility edge to mid-gap and by gaussian distributed deep-gap states in the center of gap. The conduction-band-tail states are acceptor-like traps which are neutral when empty, and become negatively charged when occupied by electrons. The valence band-tail states are donor-like states which are neutral when occupied by electrons and positively charged when empty. The nature of deep-gap states depends on the charges localized at different energy levels [5]. For simplicity, the deep-gap states were modeled by two gaussian-like distributions representing the donor-like and acceptor-like traps. Analytical expressions for the density of states used in our simulation can be expressed by

$$g_a(E) = g_{ta}\exp[(E-E_C)/E_a] + d_a\exp[(E-\lambda_a)^2/\sigma_a^2] \qquad (1)$$
$$g_d(E) = g_{td}\exp[(E_V-E)/E_d] + d_d \exp[(E-\lambda_d)^2/\sigma_d^2] \qquad (2)$$

where g_a, g_d are the densities of acceptor-like and donor-like states in the gap respectively; g_{ta}, g_{td} are the densities of the acceptor-like and donor-like-tail states at $E=E_C$ and $E=E_V$ respectively; d_a, d_d are the peak values of gaussian distribution of acceptor-like and donor-like deep-gap states respectively; λ_a, λ_d are the mean energies of Gaussian distributions of acceptor-like and donor-like states respectively; σ_a, σ_d are the standard deviations of Gaussian distributions of acceptor-like and donor-like states respectively. The more detailed information about the parameters used in our modeling were reported elsewhere [6].

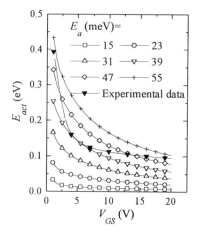

Fig. 1. Simulated evolution of the activation energy of I_{DS} versus V_{GS} for various conduction-band-tail slopes. The deep-gap states and interface states are assumed to be negligible in this calculation.

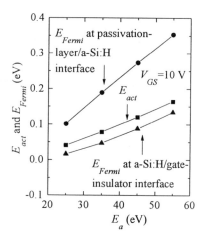

Fig. 2. Simulated evolution of the activation energy and Fermi level as a function of conduction band-tail slopes for V_{GS}=10V.

To establish the dependence between the a-Si:H conduction-band-tail slopes (E_a) and E_{act} in a-Si:H TFT, first the a-Si:H TFT I_{DS}-V_{GS} characteristics at V_{DS}=0.1V were simulated at different temperatures ranging from 300 to 420 K. Then the E_{act} was calculated from the Arrhenius plot, i.e., the slope of log(I_{DS}) versus 1/T plot at a given V_{GS}. The simulated evolution of E_{act} as a function of V_{GS} for various conduction-band-tail slopes is shown in Fig. 1. An experimental result is also given in this figure. At a given V_{GS}, a larger E_{act} has been achieved for a larger E_a.

Fig. 2 shows the variations of the E_{act} and the relative Fermi level (E_{Fermi}) as a function of conduction-band-tail slope at V_{GS}=10 V. The E_{Fermi} is calculated from the difference between the conduction-band edge and the Fermi-level position at T=0 K. The increase of E_a has the effect of increasing total gap densities, and gives rise to a larger value of E_{Fermi} on both a-Si:H/gate insulator and passivation layer/a-Si:H interfaces, and also causes the increase of E_{act}.

As V_{GS} increases, the Fermi-level position at the a-Si:H/a-SiNx:H interface approaches closer to the conduction band edge, and E_{act} becomes smaller, Fig. 3. In Fig. 2 and 3, the values of E_{act} are between the values of E_{Fermi} related to both a-Si:H/a-SiNx:H interfaces. This is because the activation energy represents the average effective barrier height for current conduction in the a-Si:H TFT which is controlled by bulk a-Si:H and both interfaces mentioned above. However, E_{Fermi} values calculated for both a-Si:H/a-SiNx:H interfaces are associated with the Fermi-level position on these interfaces.

The influence of the deep-gap states on E_{act} is different from that of the conduction-band-tail slope described above. Fig. 4 shows the influence of the peak value of the gaussian distributed acceptor-like deep-gap states (d_a) on the variation of E_{act} with V_{GS}. The increase in d_a causes E_{act}-V_{GS} curve to shift to a higher V_{GS} values. From our simulation result, onset V_{GS} voltage at which the curve shifts is corresponding to the TFT threshold voltage. Above threshold voltage the TFT performance is not affected by d_a [6] and, consequently, the E_{act}- V_{GS} curve does not change, Fig. 4.

Even the best optimization of our simulation solely based on the variation of the band-tail and deep-gap states cannot explain the experimental values of E_{act} shown in Fig. 1. To explain the discrepancy between experimental and theoretical data, the influence of source/drain series resistance on a-Si:H TFT characteristics has to be considered.

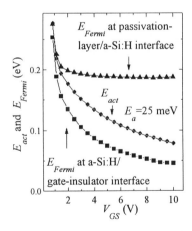

Fig. 3. Simulated variations of E_{act} and E_{Fermi} as a function of V_{GS} for E_a=25 meV.

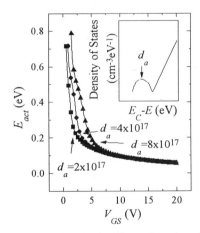

Fig. 4. Simulated effect of peak value of deep-gap states on the evolution of E_{act} with V_{GS}. The conduction-band-tail slope was kept constant at 25 meV.

INFLUENCE OF THE SOURCE/DRAIN SERIES RESISTANCE

In a-Si:H TFT, the source/drain series resistance (R_S) consists of source/drain metal/n$^+$ a-Si:H contact resistance (R_C), P-doped (n$^+$) a-Si:H film resistance (R_f), intrinsic a-Si:H bulk resistance (R_b) and parasitic resistances associated with a-Si:H TFT design [7]. To investigate the role of the R_S in a-Si:H TFTs, an 8-μm-channel-length a-Si:H TFT was used for analysis. Fig. 5 shows I_{DS} versus V_{GS} in linear region (V_{DS}=0.1V) for different measurement temperatures. To extract R_S, the experimental data were best fitted by using Levenberg-Marquardt algorithm [8] to the analytical equations based on gradual channel approximation [9]:

$$I_{DS} = \mu_{FE}(W/L)C_i[(V_{GS}-I_{DS}R_S) - V_T](V_{DS}-2I_{DS}R_S) \tag{3}$$

where C_i (2.5×10^8 F/cm^2) is the insulator capacitance, L (8 μm) and W (100 μm) are the channel length and width, respectively. The I_{DS} versus V_{GS} curve fittings are also shown in Fig. 5. The extracted R_S is temperature-dependent with an activation energy 0.23 eV, Fig. 6. The curve fitting yields $R_S \approx 134$ kΩ and $\mu_{FE} \approx 0.5$ cm^2/Vs at T=27 °C. The channel resistance (R_{ch}) defined at ($V_{GS}-V_T$)=20V is given by: L/($20\times W\mu_{FE}C_i$); $R_{ch}\approx320$ kΩ at 27 °C. Hence, the total source/drain series resistances, which is defined as the sum of source and drain series resistances, R_S^T =2R_S=268 kΩ, accounts for 46% of total TFT resistance, R_T=2R_S +R_{ch}=588 kΩ, at 27 °C. The extracted R_S is 23 kΩ, μ_{FE} is 1.7 cm^2/Vs and R_{ch} at ($V_{GS}-V_T$)=20V is 94 kΩ at 100 °C. Therefore, the contribution of R_S to total TFT resistance drops from 46 to 32 % when the measurement temperature increases from 27 to 100 °C. For a smaller ($V_{GS}-V_T$)=5V, the contribution of R_S to total TFT resistance is 17 % at 27°C. In summary, R_S contribution is higher to R_S/R_T ratios for higher ($V_{GS}-V_T$) or lower temperature values. This non-negligible and temperature-dependent R_S will affect the extraction of a-Si:H TFT activation energy.

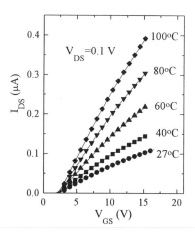

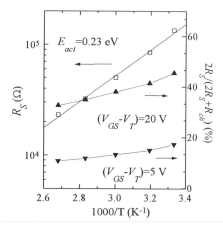

Fig. 5. Experimental data for I_{DS} versus V_{GS} in linear region (V_{DS}=0.1V) for different measurement temperatures. The solid symbols are the best curve-fitted data to equation 3.

Fig. 6. Variation of series resistance as a function of 1/T for V_{DS}=0.1 V. The up and down triangle symbols represent the ratio of R_S to the total resistance of a-Si:H TFT for ($V_{GS}-V_T$)=20 and 5 V, respectively.

From the previous discussion, it is clear that the temperature dependence of source/drain series resistance can not be ignored. Hence, to create an accurate model for E_{act} of a-Si:H TFT, the temperature-dependent source/drain series resistance has to be considered in the simulation model. In our two-dimensional simulation program, a-Si:H bulk resistance and resistances associated with a-Si:H TFT design were already previously considered and described in [10]. However, the exact models for R_C and R_f need to be established for the complete model of source/drain series resistance.

In general, R_C depends on the n$^+$ a-Si:H film resistivity and metal work function [11]. From the transmission line model measurement [12], the values of R_C and R_f can be measured experimentally. The experimental data are shown in Fig. 7 for Mo/n$^+$ a-Si:H structure having the thickness of n$^+$ a-Si:H film on the order of 70 nm. The R_f is extracted from the slope of this curve and the intersection of x-axis gives the transfer length, L_T. R_C is given by $R_C = R_f \times L_T^2$. At room temperature R_C and R_f are about 0.25 Ωcm^2 and 313 Ωcm, respectively. Fig. 8 shows the variation of R_C and R_f with temperatures. The measured activation energies of R_C and R_f are 0.20 and 0.18 eV, respectively. These experimental values were used in our simulation described below.

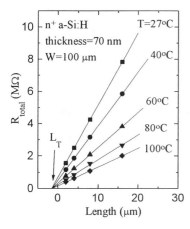

Fig. 7. Variation of total resistance as a function of the distance between two contact pads at different temperature in the transmission line structure. L_T is the transfer length; $L_T \approx 1.5$ μm at room temperature.

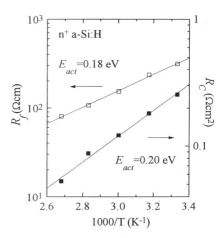

Fig. 8. Variation of n$^+$ a-Si:H film resistivity and n$^+$ a-Si:H/Mo contact resistance with measurement temperature. The straight lines are the best fits to the experimental data shown in this figure.

Fig. 9 shows the evolution of E_{act} as a function of V_{GS} for a-Si:H TFTs with 8 and 64 μm-length channels. The TFT activation energy at a given V_{GS} for 8-μm-channel-length is larger than that for 64 μm-channel-length. This difference in E_{act} for both devices is associated with influence of R_S, which is more important for a short- than for a long-channel-length a-Si:H TFT. This difference in E_{act} will be enhanced at higher V_{GS} values because the R_S contribution will be higher to R_S/R_T ratios at higher V_{GS} values. Simulated values of E_{act} based on two-dimensional simulation taking into consideration the temperature-dependent source/drain series resistance are also shown in this figure. The conduction-band-tail slope used in this simulation is 41 meV and the peak value of deep-gap states is 10^{17} cm^{-3}eV^{-1}. The R_C and R_f are 0.25 Ωcm^2 and 313 Ωcm, respectively; their activation energies are 0.2 and 0.18 eV respectively. The simulation results show a good agreement with experimental data. Hence, the combination of models of density-of-states in a-Si:H and temperature-dependent source/drain series resistance can correctly explain the behavior of E_{act} with V_{GS} in a-Si:H TFTs. A simulation result assumed zero contact resistance is also shown in Fig. 9. In this case the E_{act} values are lower than those obtained for non-zero contact resistance. Hence the measured E_{act} will be overestimated for a-Si:H TFT with non-negligible contact resistance.

CONCLUSIONS

The influence of temperature-dependent source/drain series resistance on the I_{DS} activation energy in a-Si:H TFT has been analyzed. The source/drain series resistance consists of source/drain metal/n$^+$ a-Si:H contact resistance, n$^+$ a-Si:H film resistance, intrinsic a-Si:H bulk resistance and resistances associated with a-Si:H TFT design. This source/drain series resistance is temperature-dependent and can not be neglected. In our simulation of the variation of E_{act} with V_{GS}, not only the density of states of a-Si:H but also the series resistances were taken into consideration. Our simulation results show a good agreement with the experimental data. We also found that E_{act} will be overestimated if the series resistance is ignored. This phenomenon will be more obvious for either a short-channel-length TFT or a long-channel-length TFT having high source/drain contact resistances.

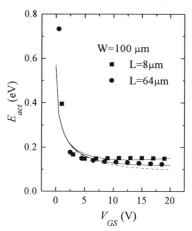

Fig. 9. The experimental and simulated data for 8 and 64 μm channel-length a-Si:H TFT. The symbols are experimental data. Solid lines represent model taking into consideration the source/drain contact resistances. Dash line represents the model with zero source/drain contact resistance.

ACKNOWLEDGEMENTS

This research was supported by the Center for Display Technology and Manufacturing at the University of Michigan.

REFERENCES
1. S. K. Kim, K. S. Lee and J. Jang, J. Korean Phys. Soc., **28**, S116 (1995).
2. N. Lustig, J. Kanicki, R. Wisnieff and J. Griffith, Mat. Res. Soc. Proc., **118**, 267 (1988).

3. B. A. Khan and R. Pandya, IEEE Trans. Elec. Dev., **37**, 1727 (1990).

4. B. S. Bae and C. Lee, J. Appl. Phys., **68**, 3439 (1990).

5. C.-T Sah, Proc. IEEE, 55 (1968).

6. C.-Y. Chen and J. Kanicki, Proc. AM-LCD '95, 46 (1995).

7. J. Kanicki, F. R. Libsch, J. Griffith and R. Polastre, J. Appl. Phys., **69**, 2339 (1991).

8. W. H. Press, S. A. Teukolsky, W. T. Vetterling and B. P. Flannery, Numerical Recipes in C, (Cambridge University Press 1992).

9. M. Shur, Physics of Semiconductor Devices, (Prentice-Hall, New Jersey, 1990), p. 361.

10. C.-Y. Chen and J. Kanicki, Digest of AM-LCD '95, 145 (1995).

11. J. Kanicki, E. Hasan, J. Griffith, T. Takamori and J. C. Tsang, Mat. Res. Soc. Proc., **149**, 239 (1989).

12. G. K. Reeves and H. B. Harrison, IEEE Electron Dev. Lett., **3**, 111 (1982).

Investigation of the off-current in amorphous silicon thin film transistors for SiO₂ and SiNₓ gate insulators

Jeong Hyun Kim, Woong Sik Choi, Chan Hee Hong and Hoe Sup Soh
LCD R&D center, LG Electronics Inc., Anyang, Kyungkido, 430-080, Korea

ABSTRACT

The off current behavior of hydrogenated amorphous silicon (a–Si:H) thin film transistors (TFTs) with an atmospheric pressure chemical vapor deposition (APCVD) silicon dioxide (SiO₂) gate insulator were investigated at negative gate voltages. The a–Si:H TFT with SiO₂ gate insulator has small off currents and large activation energy (E_a) of the off current compared to the a–Si:H TFT with SiNₓ gate insulator. The holes induced in the channel by negative gate voltage seem to be trapped in the defect states near the a–Si:H/SiO₂ interface. The interface state density in the lower half of the band gap of a–Si:H/SiO₂ appears to be much higher than that for a–Si:H/SiNₓ.

INTRODUCTION

Hydrogenated amorphous silicon (a–Si:H) thin film transistors (TFTs) have been used   as switching devices of commercial active matrix liquid crystal displays (AMLCDs). The negative gate voltage is applied to the a–Si:H TFTs almost all the time while the LCD is working. The off currents of the TFTs should be sufficiently low so that the data information remains on a pixel until next data signal is applied. Therefore, the off current is one of the most important factors to get a high performance display with no flicker. The flicker can be arisen by the degration of the liquid crystal holding ratio (pixel capacitance). Several mechanisms responsible for the off current of the polysilicon TFTs have been suggested using junction theories [1-4], i.e., thermal emission and field emission (tunneling) in the drain depletion region. The behavior of the off current in the a–Si:H TFT has been characterized similar to that in the polysilicon TFTs [5][6]. However, there are few reports on the behavior of off current of a–Si:H TFTs which have atmospheric pressure chemical vapor deposition (APCVD) silicon dioxide (SiO₂) as the gate insulator.

In this work, the off currents in the negative gate voltage region were compared between the TFTs with a–Si:H/SiO₂ and with a–Si:H/SiNₓ interfaces. The off currents of the TFT with SiO₂ gate insulator were found to be much lower than those for the TFT with SiNₓ gate insulator.

EXPERIMENT

Figure 1 shows a cross sectional view of the a–Si:H TFT with a-Si:H/SiO₂ interface. The gate electrode of Cr (150nm thick) was deposited and patterned on a glass substrate. The 300nm thick silicon dioxide layer as a gate insulator was deposited by APCVD. The O_2/SiH_4 gas flow ratio and the substrate temperature to deposit SiO_2 film were 20 and 430℃, respectively. The N_2 plasma interface treatment was carried out to decrease the interface state density between a-Si:H and SiO₂ [7]. The 140nm thick a-Si:H as an active layer and the 40nm thick P-doped (n⁺) a-Si:H to make ohmic contact between active layer and drain/source electrodes were successively deposited on SiO₂ layer by plasma enhanced chemical vapor deposition (PECVD). The n⁺ a-Si:H layer between drain and source electrodes was removed by the plasma etching after forming of drain/source electrodes of Cr. Finally, a passivation layer of SiN_x was prepared to protect the back channel from the contamination (not shown). For the comparison of the off current, the a-Si:H TFTs with SiN_x(300nm) gate insulator were fabricated. The SiN_x gate insulator was deposited by PECVD at 320℃. And the formation of junction between a-Si:H layer and source/drain electrode was made the same condition for both TFTs with SiO₂ and with SiN_x gate insulators. The width/length ratio of the TFTs is 32um/8um. The transfer characteristics (I_d-V_g) of TFTs were measured using a HP4145B Semiconductor Parameter Analyzer.

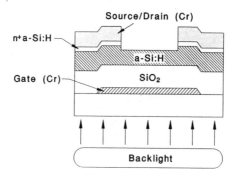

Fig. 1 A cross sectional view of the a-Si:H TFT

RESULTS AND DISCUSSION

Figure 2 shows the depth profile of Si, O, H and N for the interface region between a-Si:H and APCVD SiO₂ by the secondary ion mass spectrometry (SIMS). A large amount of N atoms was observed at a-Si:H/SiO₂ interface after the N_2 plasma interface treatment. In the previous work, the presence of Si-N bond at the interface was confirmed from an electron spectroscopy for chemical analysis (ESCA). The N_2 plasma interface treatment on SiO₂ gate insulator improves the performance of a-Si:H TFT significantly [7][8].

Figure 3 shows the comparison of the transfer characteristics of a-Si:H TFT for both SiN_x and SiO₂ gate insulators. For a-Si:H TFT with SiN_x gate insulator, it was explained that the off current for the negative gate voltage

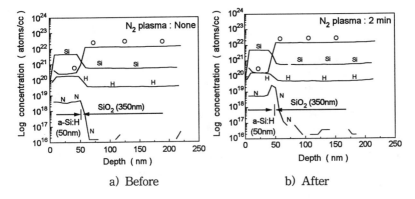

a) Before   b) After

Fig.2   SIMS depth of profile for the interface plasma treatment

was limited by the hole injection from the drain junction in the dark [5]. In addition, the off current under backlight illumination was the hole current limited by two components i.e. the channel conductance by the hole accumulation at low negative gate voltage region and the conductance of the drain junction region by the photogeneration at high negative gate voltage region [9]. However a-Si:H TFT with SiO_2 gate insulator had lower off current about one order of magnitude under the back light illumination. It could be explained that the low backlight-induced off current of TFT in the negative gate voltage region was attributed to large density of states at the interface between a-Si:H and SiO_2 gate insulator. Therefore, the hole trapping occurs in the defect states near the a-Si:H/SiO_2 interface.

Figure 4 shows the comparison of the I_d-V_g characteristics of a-Si:H

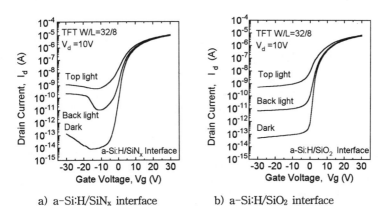

a) a-Si:H/SiN_x interface   b) a-Si:H/SiO_2 interface

Fig.3   Transfer (I_d-V_g) characteristics of a-Si:H TFTs under the dark and the light illumination

TFTs for both a-Si:H/SiO₂ and a-Si:H/SiNₓ interfaces. The other TFT device with SiO₂ 250nm thick gate insulator was prepared to meet the same gate capacitance so that the induced charge in the channel had the same value. Therefore the on current of TFTs were almost the same for both SiNₓ and SiO₂ gate insulators. The off currents in a-Si:H TFTs increase with the negative gate voltage and with the drain voltages (V_d), where the dominant carriers for the off currents are holes. The off currents of the TFTs with a-Si:H/SiO₂ interface were lower than those of the TFTs with a-Si:H/SiNₓ interface for the same negative gate voltage and the same drain voltages. In the vicinity of zero gate voltage, the behavior of off currents between the TFTs with SiNₓ and with SiO₂ gate insulators were found to be quite different. To understand the behavior of off currents, the temperature dependence was measured.

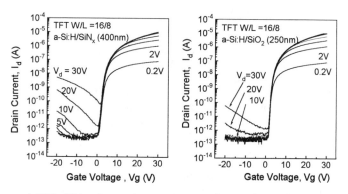

a) SiNₓ (400nm) gate insulator   b) SiO₂ (250nm) gate insulator

Fig.4 Drain-to-Source dependence of the off current characteristics of a-Si:H TFTs for the same gate capacitance

Figure 5 shows the temperature dependence of the I_d-V_g characteristics of the a-Si:H TFTs with SiNₓ and with SiO₂ gate insulators at V_d=20V. The temperature dependence is different between the a-Si:H TFTs with SiNₓ and with SiO₂ gate insulators even though the transfer curves at higher temperature above 125℃ have a similar shape by the thermal emission. In the TFTs with a-Si:H/SiNₓ interface, the transfer curves (I_d-V_g) have a V shape and the off current increases slightly with temperature. It seems to be that the off current for SiNₓ gate insulator shows the field dependence, i.e. more field-induced tunneling. The transfer curves in the TFTs with a-Si:H/SiO₂ interface have more or less round shape and the off current increases rapidly with temperature. The temperature dependence of the current can be explained by the trap state density at the a-Si:H/SiO₂ interface.

Figure 6 shows the activation energy, Ea, of off current for negative gate voltages. Experimentally, the activation energy can be obtained from the slope of log(I_d) versus $1/k_BT$, where k_B is the Boltzman constant, at constant drain voltage (V_d). The E_a for the TFT with a-Si:H/SiNx interface is 0.06eV at V_d = 20V. This value of E_a is quite small compared with the midgap energy. Thus it could be considered that the off currents are determined by the field emission, i.e. the holes are injected into the channel from the drain junction.

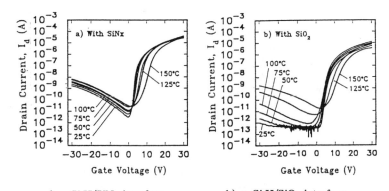

a) a-Si:H/SiN$_x$ interface   b) a-Si:H/SiO$_2$ interface

Fig.5   Temperature dependence of I_d–V_g characteristics of a-Si:H TFTs

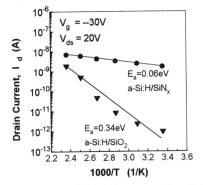

Fig.6   Activation energy of the off current of a-Si:H TFTs
for both SiN$_x$ and SiO$_2$ gate insulators

However, for the TFT with a-Si:H/SiO$_2$ interface, the E_a has a relatively high value (0.34eV) at the high electrical field even though the drain junction between drain electrode and a-Si:H layer was formed as a same condition for both TFTs with SiN$_x$ and with SiO$_2$ gate insulators. Thus, the off currents seem to be dominated by the hole accumulation in the channel and these holes are trapped in the defect states near the a-Si:H/SiO$_2$ interface. As a result, there is different mechanism of off current between the TFTs with a-Si:H/SiN$_x$ and with a-Si:H/SiO$_2$ interfaces. The off current for the negative gate voltage is dominated by considering field emission at the drain junction and trap state density near the interface. The detailed study on the off current characteristics of a-Si:H TFT with APCVD SiO_2 gate insulator will be done in the future.

CONCLUSION

In summary, the behavior of off current of a-Si:H TFTs with SiO_2 gate insulator were investigated in the negative gate voltage region. The TFTs with a-Si:H/SiO_2 interface have lower off currents and higher activation energy compared to the TFTs with a-Si:H/SiN_x interface. The dominant mechanism for the lower off currents seems to be caused by the hole trapping in the defect state near the interface rather than by the hole injection from the drain junction for the negative gate voltage.

ACKNOWLEDGMENTS

The authors would like to thank the director Mr. C. R. Lee for encouragement during this work and Dr. J. K. Yoon of LG Electronics Inc., and Prof. J. Jang of Kyung Hee Univ. for many helpful comments and suggestions.

REFERENCES

[1] J. G. Fossum, A. Ortiz-Conde, H. Schijo, and S. K. Banerjee, IEEE Trans. Electron Devices, vol. 32, No. 9, pp.1878 (1985)

[2] K. Ono, T. Aoyama, N. Konishi and K. Miyata,IEEE Trans. Electron Devices, vol. 39, No. 4, pp.792 (1992)

[3] A. Rodriguez, E. G. Moreno, J. F. Nijs and R. P. Mertens, IEEE Trans. Electron Devices, vol. 40, No. 5, pp. 938 (1993)

[4] A. W. De Groot, G. C. McGonigal, D. J. Thomson, and H. C. Card, J. Appl. Phys., vol. 55, pp. 312 (1982)

[5] G. E. Possin, Mat. Res. Soc. Proc. Vol. 219, 327 (1991)

[6] T. Globus, M. Shur and M. Hack, Mat. Res. Soc. Symp. Proc., Vol. 258, pp. 1013 (1992)

[7] J. H. Kim, E. Y. Oh and C. H. Hong, J. Appl. Phys. Vol. 76(11), pp.7601, (1994)

[8] S. K. Kim, K. S. Lee, J. H. Kim, C. H. Hong and J. Jang, Int'l workshop on AMLCD, Tokyo, Japan, pp. 12 (1994)

[9] J. K. Yoon, Y. H. Jang, B. K. Kim, H. S. Choi, B. C. Ahn and C. Lee, J. Non-Crystalline Solids, 164-166, 747 (1993)

PHYSICS OF BELOW THRESHOLD CURRENT DISTRIBUTION IN a-Si:H TFTs

H. C. SLADE, M. S. SHUR*, S.C. DEANE**, AND M. HACK‡
* University of Virginia, Thornton Hall, Charlottesville, VA 22903-2442, U.S.A.
** Philips Research Laboratory, Redhill, Surrey, RH1 5HA, U.K.
‡ dpiX, 3406 Hillview Ave., Palo Alto, CA 94304-1345, U.S.A.

ABSTRACT

We have examined the material properties and operation of bottom-gate amorphous silicon thin film transistors (TFTs) using temperature measurements of the subthreshold current. From the derivative of current activation energy with respect to gate bias, we have deduced information about the density of states for several different transistor types. We have demonstrated that, in TFTs with thin active layers and top nitride passivation, the current conduction channel moves from the gate insulator interface to the passivation insulator interface as the transistor switches off. Our 2D simulations clarify these experimental results. We have examined the effect of bias stress on the transistors and analyzed the resulting reduction in the subthreshold slope. Based on these results, we have extended our analytic amorphous silicon TFT SPICE model to include the effect of bias stress.

I. INTRODUCTION

In the last decade, the market for hydrogenated amorphous silicon (a-Si:H) devices has grown to become second only to that of crystalline silicon. Much of the credit is due to the success of a-Si:H thin film transistors (TFTs) in large area electronics, particularly in active matrix liquid crystal displays (AMLCDs), where the TFT acts as a pixel switching element. Although the mobility in a-Si:H is low, the TFTs have a very low OFF current and a high ON-TO-OFF current ratio, necessary to maintain a large number of grey levels, reduce leakage, and to eliminate crosstalk in the display [1].

One of the concerns for device designers is the lack of stability of the amorphous silicon, since a-Si:H is strongly affected by metastable localized electronic states in the energy gap. The changes in these localized states under gate bias stress and the effect of these changes on device performance are of particular interest for device design, since they can have a severe effect on the TFT characteristics.

Our goal is to use a-Si:H TFTs as tools to examine the physics of the amorphous network and incorporate our understanding into an analytic, physics-based device model for circuit simulators, such as SPICE. In this work, we link material characterization and device physics and examine the operation and material properties of different transistor types using temperature measurements of the TFT gate transfer characteristics. We first use activation energy measurements of the subthreshold current to analyze the density of states. In order to demonstrate the validity of this approach, this analysis is applied to TFTs with nitride gate insulators and to those with thermal oxide insulators, and to n- and p-channel nitride TFTs. We then use two different transistor structures to explore the current conduction paths above and below threshold. 2D simulations are employed to aid in the interpretation of our experimental results. From bias stress experiments on TFTs fabricated at two different fabrication facilities, we analyze the effect of gate bias stress tests on the density of states and the subsequent degradation of the transistor subthreshold current. In our analytic device model, we account for these changes by increasing the density of states and including an additional current component. We use this approach for modeling bias stress on the TFTs.

Mat. Res. Soc. Symp. Proc. Vol. 424 © 1997 Materials Research Society

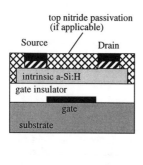

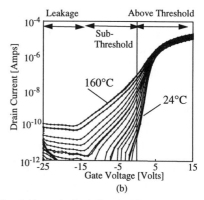

Figure 1: (a) Inverted staggered a-Si:H TFT structure. Top nitride passivation is deposited for display-quality TFTs. (b) Gate transfer characteristics at varying temperatures (data points are connected for clarity). Subthreshold current is temperature activated.

II. EXPERIMENTAL PROCEDURE

Traditionally, field effect experiments have provided information about the density of states (DOS) in the energy gap of a-Si:H (for example [2,3]). In these experiments, charge induced into the device channel by the gate bias (V_g) fills the localized energy states leading to a shift in the Fermi level (E_F). The channel current increases accordingly, providing information which can then be related back to the DOS. Activation energy (E_A) measurements are a useful method of following current variation on a logarithmic scale [4]. Since the subthreshold TFT current is temperature activated, activation energy measurements of the subthreshold current can provide information about the a-Si:H bandgap. The position of the Fermi level (E_F) at the silicon-nitride interface is linked to the activation energy $E_A = E_C - E_F$ (n-channel) or $E_A = E_C - E_F$ (p-channel), where E_C is the bottom of the conduction band in the bulk a-Si:H.

In [4] we proposed a differential technique, similar to that described in [5], which uses the derivative of the below-threshold current activation energy with respect to gate bias to analyze the DOS. The differential technique of DOS analysis is useful in determining features of the DOS deep in the band gap that cannot be easily resolved by more conventional field effect DOS techniques. However, the differential DOS technique cannot resolve the magnitude of such features and, as we show in this work, is affected by the charge distribution in the device. We have used the differential method to examine the DOS variation for several different transistor types. We have also found that activation energy measurements can be used to establish current conduction paths in TFTs.

The majority of the devices used in this study were inverted staggered devices and were fabricated at Philips Research Laboratory (PRL) in Redhill, U.K. Additional low-leakage TFTs used in the bias stress tests were fabricated for this work at Xerox Palo Alto Research Center (PARC), U.S.A. The fabrication details have been described elsewhere [6,4]. The activation measurements on the PRL devices were performed using a Hewlett Packard HP4140B picoammeter/dual voltage source and a temperature controlled chuck. The sample was measured in a light tight metal box and contacted through tungsten-tipped microprobes. Before each test, the TFT was annealed at 240°C for 10 minutes in order to bring the sample back to an initial, unstressed condition. In order to minimize temperature stress on the device, the activation energy measurements were taken in the range 35-75°C. Measurements of the Xerox PARC devices are similar and are described in [7]. Typical device structure and gate transfer characteristics are shown in Fig. 1.

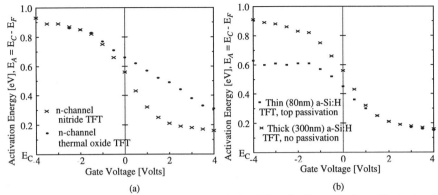

Figure 2: (a) Comparison of n-channel a-Si:H TFTs with nitride and thermal oxide gate insulators. The density of states can be related to the variation of activation energy with gate bias. The reduced slope of the thermal oxide TFT indicates a higher DOS in the upper part of the bandgap. (b) Comparison of nitride n-channel TFTs with thin and thick intrinsic layers. At negative gate biases (below threshold), the thick i-layer device conducts in the bulk a-Si:H. In comparison, current conduction takes place at the top nitride interface for the thin i-layer device.

We performed a comparative study of n-channel TFTs with nitride gate insulators, those with thermal oxide gate insulators, and p-channel nitride TFTs. We compared current conduction paths in the device using n-channel TFTs with a thick (300nm) intrinsic a-Si:H layer (i-layer) and no top passivating insulator to TFTs with a thin (80nm) intrinsic layer and top nitride passivation. These latter TFTs underwent bias stress tests in order to see the effect on the activation energies.

III. RESULTS AND ANALYSES

Differential Density of States Analysis

The comparison of the activation energies as a function of gate bias (V_g) for TFTs with a nitride gate insulator (threshold voltage, $V_T \sim 3V$) and with a thermal oxide insulator ($V_T \sim 6V$) is given in Fig. 2(a). It is important to note that these devices are both fabricated with a thick (300nm) intrinsic a-Si:H active layer and have no top passivating insulator. The activation energies for the nitride TFT indicate that, at negative gate biases, the Fermi level is located deep in the band gap, at about 0.85eV below the conduction band. As the gate bias is increased positively, the Fermi level moves closer to E_C, through a region of lower DOS until it reaches the tail states (approximately 0.15 eV below E_C) and moves into the above threshold regime. The high DOS in this region pins the Fermi level in the tail states, as indicated by the reduced slope of the E_A vs. V_g curve. The activation energies for the thermal oxide TFT show that the density of states is higher in the top of the bandgap than for the nitride TFT. The shallower slope of the E_A vs. V_g curve indicates that the Fermi level is moving more slowly through the larger DOS in the thermal oxide TFTs and therefore will enter the tail states at a higher gate voltage (larger V_T) than for the nitride TFTs. This is confirmed by field effect DOS analyses and the prediction of the defect pool model [8]. This is also in agreement in the activation energy measurements of a p-channel nitride, [9]. For p-channel devices, we found that the slope of the E_A vs. V_g plot indicates a high DOS in the lower part of the gap for the nitride transistors, in agreement with previous analyses.

Current Conduction Paths

Referring again to Fig. 2(a), we notice that for both the nitride and thermal oxide TFTs, the Fermi level is pinned at negative gate biases at 0.85eV below the conduction band. This corre-

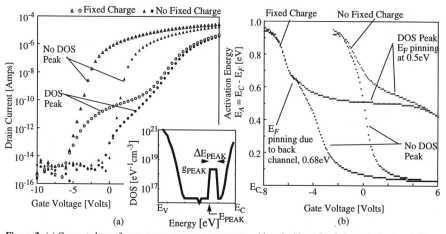

Figure 3: (a) Computed transfer characteristics for a-Si:H TFTs with and without fixed charge in the top nitride and with and without a peak in the DOS. (b) Activation energies calculated as a function of gate bias obtained from a 2D simulation of an a-Si:H TFT with and without fixed charge in the top nitride passivation and with and without a peak in the DOS. A fixed charge of $2 \times 10^{11} cm^{-2}$ causes the Fermi level to show a pinning at approximately 0.68 eV. The effect of a peak in the DOS is shown by a strong Fermi level pinning near 0.5 eV.

sponds to the bulk Fermi level position in amorphous silicon. This shows that, at negative gate biases, the conduction has moved away from the gate insulator interface and takes place in the bulk semiconductor. Fig. 2(b) shows the $E_A(V_g)$ curves for two n-channel nitride TFTs both fabricated at PRL: one with a 300 nm intrinsic a-Si:H layer and no top passivating nitride and one with a thin, 80 nm i-layer and a top passivating nitride. As we noted before, at negative gate biases, current conduction in the 300 nm i-layer device occurs in the bulk of the a-Si:H. However, for the 80 nm i-layer TFT, the active layer is so thin that the band-bending due to the fixed insulator charge extends throughout the entire layer. In this case, the conduction moves away from the gate nitride interface to the passivating nitride interface (a "back" channel), where the bands are pinned at approximately 0.6 eV at fabrication [10]. This has been confirmed by 2D simulations in which leakage currents are caused by conduction in a channel away from the gate electrode, caused by a combination of surface states at the top nitride interface and the fixed charge in the top passivation dielectric [11].

In order to further examine the effect of the back channel, the thin active layer transistor was light soaked using a Schott 1500T tungsten halogen light supply which focused white light onto the top of the transistor at an intensity of approximately 1.6 Wcm^{-2}. The light soaking creates a large number of additional localized states in the band gap *and* increases the interface states at the top nitride interface, thus changing the band bending at the surface. A 6-hour light soak to the top of the TFT shifted the device threshold voltage from 3V to approximately 10V, due to the large number of states created under the light soaking conditions. This was also reflected by a severely reduced slope in the $E_A(V_g)$ curve (See [9] for more details). More importantly, however, the Fermi level pinning at negative gate biases increased to 0.7 eV, which suggests that the band-bending at the top nitride interface has decreased due to the creation of interface states and agrees with the existence of the back channel.

Bias stressing of TFTs increases the density of dangling bonds in a-Si:H TFTs [12]. We applied positive and negative bias stress (75°C, $V_g = \pm 20$V, 3 hours) to the thin active layer TFT. As discussed in [9]. The threshold voltage shifted as expected due to state creation in the a-Si:H under bias stress: V_T increased after positive stress and decreased after negative stress. For both the unstressed and stressed cases, the voltage at which the current reached a minimum corresponded to the voltage at which the Fermi level became pinned at 0.6 eV. However, the bias

94

stress did not alter the energy of the Fermi level pinning, since no additional states were created in the top nitride. This agrees with the concept of the back channel, since bias stress applied to the gate electrode should not significantly affect the back channel.

2D simulations were performed to verify the experimental results. In Fig. 3(a) transfer characteristics for a-Si:H TFTs with and without fixed charge in the top nitride and with and without a peak in the DOS are shown. The fixed charge shifts the transfer characteristics laterally; a DOS peak creates a shoulder in the characteristics. The combination of a peak and fixed charge shifts the characteristics negatively *and* increases the shoulder. In Fig. 3(b) the activation energies calculated as a function of gate bias obtained from a 2D simulation of an a-Si:H TFT with and without fixed charge in the top nitride passivation and with and without the peak in the DOS are shown. A fixed charge of $2 \times 10^{11} \text{cm}^{-2}$ causes the Fermi level to exhibit a pinning be pinned at approximately 0.68 eV. This corresponds to conduction at a back channel. The effect of a peak in the DOS is shown by an additional, more severe, Fermi level pinning, near 0.5 eV. These simulations confirm our conclusions based on the experimental data.

<u>Analytic TFT Modeling: Stress Effects</u>

In [7], we have presented the results of bias stressing of TFTs fabricated at Xerox PARC. We found that a region of reduced slope exists in the transfer characteristics for these TFTs (shown by the experimental points in Fig. 4(a)). This region increases considerably with applied bias stress. Examination of the activation energies reveals that the Fermi level pinning occurs well before the device reaches a minimum current; in fact, in all cases, the Fermi level pinning occurs at the start of this reduced slope region. The 2D simulations (Fig. 3) have demonstrated that a large DOS near the top nitride interface may play a role at higher gate voltages before the back channel pinning begins. This "peak" in the DOS causes the Fermi level to become pinned before the current reaches the back channel. In the previous section, we describe the bias stressing of the PRL devices to support the existence of the back channel. However, for the Xerox PARC TFTs, the results suggest that *both* the increase in the DOS and the back channel conduction play a role in the E_F pinning. (This is described in detail in [7].) Therefore, we can reproduce the structure in the subthreshold slope of the TFTs by incorporating a large, flat peak in the DOS.

We have developed a temperature-dependent analytical model for amorphous silicon TFTs, presented in [13, 14]. We have expressed the current-voltage characteristics in the above- and below-threshold regimes in a unified manner and have included a description of the leakage regime. Our analysis of the density of states from the activation energy experiments and from 2D simulations allow us to describe the effect of bias stress in terms of an additional current term, I_{peak}, which corresponds to the peak in the DOS. The relevant parameters are shown in the insert in Fig. 3: the width of the peak in energy (ΔE_{peak})and the magnitude of the peak (g_{peak}). These can be related to the region in the transfer characteristics of reduced subthreshold slope. Shown in Fig. 4(a) are the measured gate transfer characteristics for the Xerox TFTs for both the unstressed device and after positive bias stress. The modeled data without inclusion of the peak current is also shown. After the bias stressing, the discrepancy between modeled and actual data becomes more severe. In Fig. 4(b), we demonstrate the effect of the DOS peak included into the model. A larger peak in the DOS for the positively stressed TFT than for the unstressed device reproduces the effect of the bias stress. The parameters necessary for modeling the stress effects are extracted from the transfer characteristics in a straightforward manner.

V. CONCLUSIONS

Activation energy measurements of TFTs are a useful method of estimating the density of localized states in amorphous silicon. These measurements can also clearly pinpoint the current conduction paths in the device. Activation energy measurements on TFTs with different gate insulators and of n- and p-type confirm this analysis. In addition, measurements of TFTs with thin and thick intrinsic a-Si:H layers show that the current conduction moves away from the gate insulator interface as the device is switched off. For thick i-layer devices with no top passivating

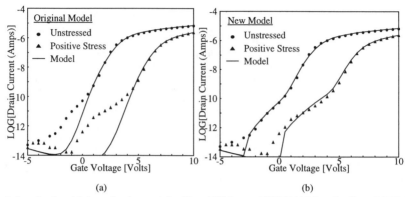

Figure 4: Actual and modeled transfer characteristics (V_{ds} = 10V) for an a-Si:H TFT fabricated at Xerox PARC afte no applied bias stress (circles) and after positive bias stress (triangles). (a)Stress effects not into the model. (b) Stres effects modeled by a flat peak in the density of states. Relevant parameters include the width of the peak in energ: and the magnitude of the peak.

insulator, the conduction moves into the bulk a-Si:H. For thin i-layer devices with a top nitride passivation layer, the conduction takes place at the top nitride interface, in a "back" channel. Further testing using bias stressing and light soaking supports this conclusion. Using activation energy measurements and 2D simulations, we have attributed the reduction in the subthreshold slope for the Xerox PARC TFTs after applied stress to an increase in the density of states. This causes the Fermi level to be pinned before the conduction reaches the back channel. We have included these results in our analytical SPICE model for a-Si:H TFTs.

ACKNOWLEDGEMENTS

This work has been supported by Advanced Projects Research Agency under contract #MDA 972-91-K-0002 (Project Monitor, Dr. G. Henderson) and by the National Science Foundation. Ms. Slade would like to thank the ARCS Foundation for fellowship support and The Electrochemical Society for the F. M. Becket Memorial Award. We are also grateful to Dr. W.I. Milne and the Thin Film Laboratory at Cambridge University for their support and advice on this project. Much appreciation is extended to Rene Lujan at Xerox PARC for sample preparation.

REFERENCES

[1] M.J. Thompson, *J. Non-Crystal. Sol.,* **137&138**, 1209, 1991.
[2] M.J. Powell, *Phil. Mag. B.*, **43**(1), pp. 93-103, 1981.
[3] M. Grunewald, P. Thomas, D. Wurtz, *Phys. Stat. Sold. B*, **100**, k139, 1980.
[4] T. Globus, H.C. Slade, M.S. Shur, M. Hack, *Mat. Res. Soc. Proc.*, **336**, 823, 1994.
[5] G. Fortunato, D.B. Meakin, P. Migliorato, P.G. LeComber, *Phil. Mag. B*, **57**(5), 573, 1988.
[6] P.N. Morgan, Ph.D. Dissertation, Cambridge University, 1994.
[7] H.C. Slade, M.S. Shur, M. Hack, *Elec. Chem. Soc. Conf. Proc.*, **94-35**, 207, 1994.
[8] S.C. Deane and M.J. Powell, *J. Appl. Phys.*, **74**(11), 6655, 1993.
[9] H.C. Slade, *Elec. Chem. Soc. Interface*, **4**(4), Winter, 1995.
[10] G.E. Possin, *Mat. Res. Soc. Proc.*, **219**, 327, 1991.
[11] M. Hack, H. Steemers, and R. Weisfield, *Mat. Res. Soc. Proc.*, **258**, 949, 1992.
[12] M.J. Powell, C. van Berkel, A.R. Franklin, S.C. Deane, W.I. Milne, *Phys. Rev. B*, **45**, 4160, 1992.
[13] M.S. Shur, M.D. Jacunski, H.C. Slade, M. Hack, *J. Soc. Info. Displays*,**3**(4), 223,1995.
[14] H.C. Slade, M.S. Shur, M. Hack, *Mat. Res. Soc. Conf. Proc.*, **377**, 725, 1995.

ENERGY LEVELS OF DEFECTS IN a-Si:H FROM OPTICAL AND ELECTRICAL CHARACTERISTICS

T. GLOBUS
EE Department, University of Virginia, Charlottesville, VA 22903, tg9a@Virginia.edu

ABSTRACT

Two novel characterization techniques for hydrogenated silicon thin films have been recently proposed which show promise in providing critical feedback for evaluating materials and monitoring the device fabrication process. The first technique is the optical interference spectroscopy for a quick non-destructive measurement of absorption coefficient and refractive index spectra of amorphous- and poly-Si thin films in a wide range of the incident photon energies (0.5-3.5 eV) [1]. By using this technique, the absorption related to defects in the subgap energy region has been determined for device quality thin films. The second technique is the novel version of the field effect conductivity (FEC) method for the direct density-of-states (DOS) determination from analysis of thin film transistor (TFT) quasi-static transfer characteristics [2]. This sensitive, fast, and easy to use, method makes it possible to resolve fine-scale features in the midgap DOS of a-Si:H. In this work, data from two methods of spectroscopy are analyzed together. Very close correlation of results is demonstrated which provides a unique opportunity to identify midgap defect states and to understand the fundamental physics of hydrogenated silicon films. The energy map of defect states in the upper half of a-Si:H bandgap is presented. These results permits to use TFT transfer characteristics and optical interference technique measurements as effective tools to control the quality of TFT manufacturing process.

INTRODUCTION

Over the past two decades, considerable efforts were made in order to find efficient, reliable, and nondestructive methods for hydrogenated silicon thin film characterization. The objective was to provide control of material quality for device fabrication and at the same time yield valuable insight into the fundamental physics of these materials. This work was motivated by the variety of applications of a-Si:H based materials for large area microelectronic devices, such as thin film transistor arrays for active-matrix liquid-crystal flat panel display technology and terrestrial-based large area photovoltaics including solar cells and linear line-sensors. The materials properties which have to be probed include the surface morphology and interfacial roughness, thin film thickness, grain size and crystallinity degree, alloy composition and hydrogen content. Not less important, although much more difficult for real-time monitoring, are measurements of many key electronic properties of bulk material, such as the band gap, the concentration and energy of localized defect states, and the doping efficiency. It is well known that the localized defects and their energy states distribution control most of a-Si:H properties as well as device performance including the form of static current-voltage characteristic and the transient characteristics of TFTs.

Currently, the electronic states within the mobility gap of hydrogenated silicon are investigated using photoconductivity experiments, charge transport via space-charge-limited-current, and device characteristics of thin-film field-effect transistors. The main problem with these experiments is that an unambiguous interpretation of experimental data is often made difficult by a large number of simplifying assumptions about both the material properties and the measurement procedures.

Several defect energies were determined by different methods. Chen et al. obtained the energy of a neutral intrinsic defect $E (D_i^0)=1.1$ eV from the conduction band E_C, and the energy of defect with an electron $E (D_i^-)=0.6$ eV using photomodulation spectroscopy [3]. The threshold energy about 0.25 eV was obtained from photoinduced midgap absorption [4] and was interpreted in terms of hole trap level near the valence band edge. Broad defect related transitions peaking at

0.9 eV and 1.35 eV are well known from the luminescence [5]. The results obtained from an analysis of photoluminescence spectra [6] are consistent with a distribution of D$^-$ levels with peaks at 0.95 eV below E_c and a peak of D$^+$ levels at 0.55 eV below E_c.

Optical measurements are among the most powerful methods of materials characterization. Knowledge of subband absorption is important to characterize the defect density of states. However, the conventional transmission technique, as well as many other methods for measurement subband absorption, have serious limitations due to interference effects.

In [1], the alternative "interference technique" has been developed. The method is based on new self-consistent data-analysis algorithms for simultaneously measured optical transmission, T, and specular reflection, R. Recently, the variation of this technique was applied to films of a-Si:H and alloys [8, 9] and was found effective in the photon energy range above the optical gap. The absorption coefficient has been determined from the ratio T/(1-R), which almost completely eliminates disturbance from the interference effect. This, however, limits the sensitivity of the interference method which is potentially close to the sensitivity of modulation spectroscopy. The "interference technique" presented in [1] uses exact interference equations for the system of a film on a substrate [10, 11]. These account not only for the interference in the film but also for the absorption inside the substrate and for the reflection from both substrate surfaces. In this work, we demonstrate that the "interference technique" can be used as a tool to measure subgap absorption of hydrogenated silicon intrinsic layers. We present experimental results and detailed analysis of optical characteristics of non-doped a-Si:H films near and below the fundamental edge. New structures and singularities near the mobility gap are demonstrated and discussed. In the subgap energy range, our technique resolves structures related to absorption due to defects. We demonstrate that the interference technique has a high potential for detailed and fast control of band gap and defect density of states.

Field effect conductance methods are among the most commonly used for the density of states determination in the bandgap of a-Si:H. In spite of all efforts, important questions about the presence of structures in the midgap DOS and the ability of the FEC method to resolve fine-scale features are still controversial issues. The lack of reliable methods to determine the flat band voltage is one of the important sources of uncertainties. A novel version of the field effect conductivity method for the direct DOS determination has been recently developed [2]. This sensitive, and easy to use method, provides the detailed information on the defect density in the gap of hydrogenated Si. The ability of this technique to extract parameters of the TFT, including the flat band voltage, from the experimental dependence of the surface potential on the gate voltage, and to resolve fine-scale features in the midgap DOS of a-Si:H has been demonstrated. There are two wide bands of localized states in the upper part of the gap which are separated by a region with lower density of states. One of these bands is the band of tail states and the second is the band of the deep, almost midgap localized states in the range of 0.3- 0.6 eV from the conduction band E_c . The dependence of the DOS on the energy is close to exponential in the range 0- 0.15 eV from E_c , while deeper in the gap, the DOS can be described as a series of localized defect bands with values in the range 10^{17} cm^{-3} eV^{-1} - 2×10^{18} cm^{-3} eV^{-1}.

The development of new, highly sensitive methods [1,2] which are totally independent of the existing techniques provides a new approach in order to establish a more detailed description of the density of localized states in amorphous and polycrystalline silicon. In this work, we present new results from these two methods of optical and electrical spectroscopy which we analyze together.

EXPERIMENT

The transmission and reflection spectra are measured with Cary 5E Spectrophotometer using a two-beam scheme with specular reflection attachments which ensure a near normal (~8 °) angle of incidence on the sample. We used undoped samples of device quality a-Si:H thin films with hydrogen content ~15 % (Sample #1) and with lower hydrogen content ~ 10% (Sample #2), and the TFT fabricated from material with 10% of hydrogen. The experimental techniques have been described in [1,2]. The detailed structure of a-Si:H TFTs used for FEC measurements has been described earlier [12].

RESULTS

Optical results

Figure 1 demonstrates absorption of two samples with different hydrogen content near the fundamental edge. The absorption of one of these samples in the subgap energy range is shown in Fig.2. Our measurements indicate new structures in the absorption spectrum of a-Si:H. These are peaks and steps near the fundamental edge (see Fig. 1) and in the subgap region (Fig.2). The form of absorption near the edge can be used for material quality control. The near edge structure can only be observed in high quality film with low defect concentration. As subgap absorption related to defects is rising, the near edge structure becomes less pronounced and disappears. The absorption in the subgap can vary significantly and can give detailed information regarding defects and impurities including their concentration and energy position in the gap. Note the increased accuracy of the new method. The accuracy of traditional methods is not sufficient to detect these fine structures.

A number of mechanisms have been proposed for behavior which is visible near the edge and subgap regions of Figs.1, 2. These regions can be explained in terms of localized state-localized state transitions, transitions involving localized states and the bands, and a broadened exciton response [13, 14]. We can expect that transitions of electrons from localized states into one of the band (conduction or valence) give the major contribution in the subgap absorption. Fig. 3 shows two major paths which include transitions 1 from the states below the Fermi level, E_F, filled by electrons, into conduction band states, and transitions 2 from the valence band states into empty defect states above E_F. The probability of such kind of transition in a-Si can be much higher than in crystalline material due to the lack of momentum conservation. The bulk thermal activation energy for device quality a-Si:H deposited by PECVD method, E_a, is known from literature [15]. It gives us the estimates of the Fermi level position in the gap relative to the conduction band, E_F, as 0.68 eV and 0.76 eV. Thus, we can expect transitions of type 1 in the energy range between E_F and E_g, and transitions of type 2 from the valence band at energies above $E_g - E_F \approx 0.95$ eV.

The threshold energy E_1 in the well known Tauc dependence ($K \hbar\omega$)$^{1/2}$=B ($\hbar\omega$- E_1) is usually interpreted as the optical gap E_g [16]. Here, K is the absorption coefficient and $\hbar\omega$ is the photon energy. The value of E_g depends on the hydrogen content. However, large additional absorption near the edge makes it difficult to obtain accurate values of the optical gap. Fig. 4 demonstrates the Tauc dependencies for two a- Si films after subtraction of background absorption, which was taken in a first approximation as a constant near the edge. This procedure gives value of E_g equal to 1.64 eV and to 1.70 eV for a-Si films with hydrogen content 10% and 15 %, respectively.

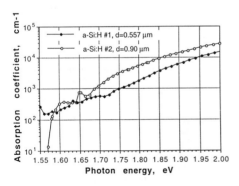

Fig. 1. The absorption coefficient of a-Si:H near the edge.

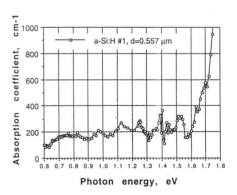

Fig. 2. The details of the a-Si:H absorption spectrum in the subgap region.

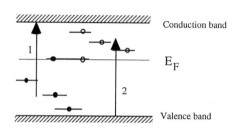

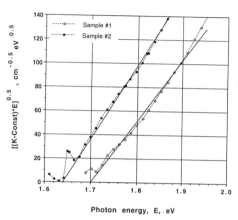

Fig. 3. Schematic illustration of subgap optical transitions.

Fig. 4. The Tauc plot for absorption coefficient after subtraction of near edge background absorption.

In the subgap absorption spectrum of samples #1 demonstrated in Fig.2, there are several bands related to different defect states. We do not presently understand exactly the type of electron transitions which can provide the observed sharpness of peaks. This might reflect the complicated nature of deep defects.

Results from FEC method

Fig 5. demonstrates the density of states for undoped a-Si:H [2]. The energy scale is related to the conduction band edge. The dependence of the DOS on the energy is close to exponential in the range 0- 0.15 eV from the conduction band. This energy range corresponds to the Urbach tail in the absorption spectrum near the edge ($\hbar\omega$= 1.55-1.7 eV in Figures 1 and 2). Deeper in the gap, the DOS can be described as a series of localized defect bands with values in the range 10^{17} cm^{-3} eV^{-1} - 2×10^{18} cm^{-3} eV^{-1}. We suppose that peaks in the DOS serve as indicators of the energy position of the involved defects relatively to the conduction band.

Analysis of results

Characteristic energies for the peaks in the DOS, E_t, are listed in column 1 of the table I. The estimated energy for transitions to the second band E_g-E_t are presented in column 2 for the sample #1 (with E_g= 1.70 eV), and in the column 4 for the sample #2 (with E_g= 1.64 eV). Characteristic energies for the most pronounced peaks E_p in subgap absorption for these two samples are shown in columns 3 & 5, respectively. Peaks in the DOS at energies 0.65 eV and 0. 75 eV have been determined from measurements at elevated temperatures (+80˚C). Comparison of peak positions E_g-E_t and E_p shows very good correlation between these two data sets.

Thus we can conclude that both techniques are able to detect transitions related to defects, and we can use the positions of the peak to generate the map of the most pronounced localized defect states in the gap. These one electron levels of defects are demonstrated in the Fig. 6. As we can see, all observed peaks in absorption listed in the Table I belong to the transitions of type 2 and can not be normally detected by the constant photocurrent technique because of the very low mobility of holes. The other important result of this analysis is that the energy of optical transition from the valence band into localized states above the Fermi level changes with changing Eg. Thus these

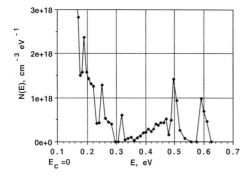
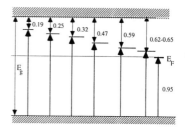

Fig.5. The density of states in a-Si:H TFT.   Fig. 6. One electron levels of defects.

Table 1

| FEC peaks, E_t, eV | Sample #1 | | Sample #2 | |
	(Eg=1.71 eV) E_g-E_t, eV	Optical peaks, Ep, eV	(Eg=1.65 eV) E_g-E_t, eV	Optical peaks, Ep, eV
1	2	3	4	5
–	–	–	–	1.65
–	–	1.64	–	1.61
0.19	1.52	1.52	1.46	1.48
0.25	1.46	1.45	1.40	1.41
0.32	1.39	1.40	1.33	1.32
0.47	1.24	1.24	1.18	1.17
0.5	1.21	–	1.15	–
0.59	1.12	1.13	1.06	1.07
0.65*	1.06	1.06	1.0	0.98
0.75*	0.96	0.96	0.9	0.9

defects in the upper half of the gap originate from the conduction band.

We suppose that the observed structure in the absorption spectrum near the fundamental edge (1.6-1.7 eV) does not correspond to one electron defect states and is due to exciton states.

CONCLUSIONS

Two characterization techniques have been developed for evaluation of the energy distribution of deep defects in high quality a-Si:H thin films and TFTs from the analysis of optical and electrical spectroscopy measurements. Results from two techniques perfectly correlate. The energy map of defect states in the upper half of a-Si:H bandgap is presented. Further work with p-type TFTs will hopefully give information regarding defects in the lower half of the gap.

ACKNOWLEDGMENTS

The author would like to thank Dr. C. Seager, Dr. G. Gildenblat and Dr. S. Fonash for many helpful and stimulating discussions and Dr. M. Hack of Xerox Parc for supplying the samples of a-Si:H thin films and TFTs used in this study.

REFERENCES

1. T. Globus, G. Gildenblat, S. Fonash, MRS Proc., **406**, p.313, 1996.

2. T. Globus, B. Gelmont, Q.L.Sun, R.J. Mattauch, to be published in 1996
   MRS Spring Meeting Proc., Symp. A, San Francisco, CA, 1996.

3. L. Chen, J. Tauc, J. Kocka, and J. Stuchlik, Phys. Rev. **B46**, p. 2050 (1992).

4. P. O'Connor and J. Tauc, Phys. Rev. **B25**, 2748 (1982).

5. R. A. Street, Phys Rev.**B21**, p.5775 (1980).

6. R. A. Street, and D. K. Biegelsen, Solid St. Commun. **33,** 1159 (1980).

7. K. Pierz, W. Fuhs and H. Mell, Philosophical Magazine B, 63, 123 (1991).

8. Y. Hishikawa, N.Nakamura, S. Tsuda, S. Nakano, Y. Kishi, and Y. Kuwano, Jpn. J. Appl. Phys. **30**, 1008 (1991).

9. N. Maley, Jpn. J. Appl. Phys. **31**, 768 (1992).

10. J. F. Hall and W. F. C. Ferguson, J. Opt. Soc. Am. **45**, 714 (1955).

11. T. R. Globus, B. L. Gelmont, K. I. Geiman, V. A. Kondrashov, and A. V. Matveenko, Zh. Exsp. Teor. Fiz. **80**, 1926 (1981) [Sov. Phys. JETF **53 (5)**, 1000 (1981)].

12. T. Globus, H. C. Slade, M. Shur, and M. Hack, MRS Proc. **336**, 823 (1994).

13. N.G. Tarr, D.L. Pulfrey, and P.A. Iles, J. Appl. Phys. **51**, 3926 (1980).

14. J. Tauc, in Optical Properties of Solids, Plenum Press, New-York, p.128 (1969).

15. J.Kanicki in Amorphous & Microcrystalline Semiconductor Devices, Vol. 2, Ed. J.Kanicki, Artech House, Boston-London, p.195 (1992).

16. W. Fuhs, in Landolt-Bornstein, Numerical Data and Functional Relationships in Science and Technology, New Series, edited by O.Madelung, M. Schultz, H. Weiss (Springer-Verlag Berlin-Heidelberg, Ney-York, Tokyo 1985), Band 17, p.43.

LOW TEMPERATURE GROWTH OF NANOCRYSTALLINE SILICON FROM SiF$_4$ + SiH$_4$

YU CHEN, M. TAGUCHI* AND S. WAGNER
Princeton University, Department of Electrical Engineering, Princeton, New Jersey 08544

ABSTRACT

The electrical conductivity of nc-Si films grown from SiF$_4$ and H$_2$ with constant arsenic doping rises from 10^{-5} to 10 Scm^{-1} as the thickness rises from ~ 0.1 to 1 μm. This variation demonstrates the strong influence of film structure on conductivity. We show that the conductivity of undoped nc-Si films of constant thickness can be varied by adding SiH$_4$ to the SiF$_4$ and H$_2$ source gas.

INTRODUCTION

Nanocrystalline silicon may be defined as material that ranges from barely measurable grain size to a size that is comparable to the electron scattering length (at 300K) and the smallest space charge widths typically found in crystalline silicon devices. These measures place the grain size between ~ 2 and ~ 50 nm. Because nanocrystalline silicon (nc-Si) can be made by the deposition techniques used for hydrogenated amorphous silicon (a-Si:H), it is an attractive candidate for large area electronics. Its deposition temperatures typically lie between 200 and 350°C. Therefore, it can be deposited on glass substrates. When its grain size is comparable to the electron scattering length, nc-Si might exhibit electron and hole mobilities closer to those found in crystalline silicon than those in a-Si:H. Therefore, nc-Si devices may perform better than a-Si:H devices. Such considerations have been the motive behind experiments to make solar cells [1,2] and thin-film transistors [3,4] from nc-Si.

Nanocrystalline silicon can be made from silane or from silicon fluoride, SiF$_4$. In both cases, the necessary deposition conditions are explained best by a near-balance between deposition and etching. When grown on glass substrates or on a-Si:H, the structure of nc-Si films typically changes in the growth direction from amorphous to increasingly granular [5]. This variation can affect the electrical properties. Fig. 1 provides a drastic example by showing how the conductivity of a number of nc-Si:As samples varies with film thickness. The films had been grown from a single, highly arsenic-doped SiF$_4$ source, with small (factor of 2) variations of the SiF$_4$:H$_2$ gas flow ratio and at substrate temperatures from 250 to 290°C. The electrical conductivity, which is averaged over the film thickness, is seen to vary by six orders of magnitude, reflecting the effect of film structure on dopant activation and carrier transport. While Fig. 1 stands for highly doped material, it is plausible that the electrical properties of undoped nc-Si also vary with film thickness. Understanding this variation, making it reproducible, controlling or suppressing it, and making use of it will be a major task in the introduction of nc-Si

---

Premanent address: Sanyo Electric New Materials Res. Ctr., 1-18-13 Hashiridani, Hirakata, Osaka 573.

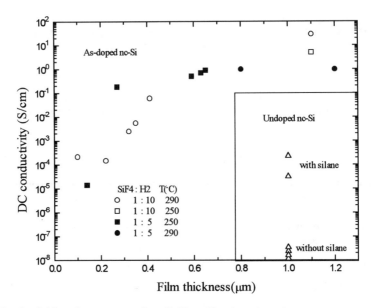

Fig.1. Conductivities of two groups of nc-Si films. The data along the curve to the left and top stand for samples grown under nearly identical conditions (see inset) and with constant As-doping. They demonstrate the effect of structure as the films become increasingly crystalline with rising thickness. The data point at the lower right show the range of conductivities of undoped nc-Si films of constant, 1 μm thickness obtained by adding SiH_4 to the standard $SiF_4 + H_2$ source gas.

as a device material.

Our goal for the growth of *undoped* nc-Si from glow discharges in mixtures of SiF_4, SiH_4 and H_2 is the control of the conductivity at any given thickness.

EXPERIMENTS

The nc-Si films were deposited by plasma-enhanced chemical vapor deposition in a three-chamber diode system with d.c. excitation. Before deposition we pumped down to a pressure of ~ 10^{-7} torr (~ 10^{-3} Pa). We cleaned the heated Corning 7059 substrates for 5 minutes in a 100 mWcm^{-2} hydrogen plasma. The film deposition conditions are listed in Table 1.

We evaluated film thickness, crystalline structure, H and F bonding, impurity content, optical gap, subgap optical absorption spectra, and dark conductivity and its thermal activation energy.

The film thickness was determined at steps with a Dektak surface profiler and from interference fringes in optical transmission spectra measured with a Hitachi U-3410 visible and near-infrared spectrophotometer. The crystallite sizes of the films were estimated from X-ray diffraction spectra measured using the Cu K$_\alpha$ 1.54 Å line. The crystal volume fractions were

Table 1. Ranges of deposition conditions used for all samples, and for the undoped samples of Fig.1.

		All samples			Fig.1
Substrate temperature	(°C)	250	--	290	290
Pressure	(mTorr)	750	--	900	750
SiF_4 flow rate	(sccm)	6	--	10	10
SiH_4 flow rate	(sccm)	0	--	1	0 or 1
H_2 flow rate	(sccm)	30	--	50	50
Power density	(mWcm^{-2})	90	--	300	90

estimated from Raman scattering spectra measured using a 488 nm Ar$^+$ laser source at ~ 65° off normal incident angle. At this wavelength, which corresponds to a photon energy of 2.5 eV, we measured optical absorption coefficients of ~ 1×10^5 cm^{-1}. Therefore the penetration-and-return probe depth for vertical incidence is approximately 0.05 μm. The infrared absorption spectra were measured with a Nicolet Fourier-transform spectrophotometer. The impurity content of the films was determined by Auger electron spectroscopy. The optical (Tauc) gap was evaluated from the transmission spectrum. The subgap optical absorption was measured by photothermal deflection spectroscopy (PDS). The dark conductivity measurements were carried out with two co-planar gap electrodes of evaporated chromium (gap length 0.5 mm, width 7.5 mm) in vacuum, at temperatures between R.T. and 107°C.

RESULTS AND DISCUSSION

> *Thickness.* The deposition rates were determined by dividing film thickness by growth time. The rates were found to vary from 0.3 Å/s to 1 Å/s. The low growth rate reflects the near-balance between etching and deposition.

> *Crystal size.* X-ray diffraction spectra show the highest intensity in the (220) orientation (Fig. 2). Crystallite sizes from 10 to 30 nm were estimated from the full width at half maximum of the X-ray peaks, using Scherrer's equation [6]. Raman scattering spectra of the TO phonon band (Fig. 3) were taken from the front and through the substrate sides of the films. Fig. 3 illustrates the change in film structure over a thickness of 0.3 μm. The bottom of the film, next to the substrate, is amorphous, because the spectrum is broad and peaks at 480 cm^{-1}. The top produces the sharp crystalline peak at 515 cm^{-1}.

> *H and F bonding.* The IR spectra show that most of the H is bonded as SiH_2 and most of the fluorine as SiF_2.

> *Impurities.* The IR spectra also show a pronounced Si-O peak at 1090 cm^{-1}, much of which in our experience reflects a surface oxide. Auger electron spectroscopy of the bulk of one sample, reached by sputtering, showed an oxygen concentration of 0.5 at.% and a carbon concentration of 8 at.%. We plan to follow up with more impurity analysis.

> *Optical gap.* The Tauc gaps lie between 1.9 and 2.0 eV.

> *Subgap absorption spectrum.* The subgap optical absorption spectra typically exhibit two regimes, in each of which the optical absorption coefficient α rises exponentially with photon energy. These regimes meet at hv $1.6 \approx$ eV, where $\alpha \approx 100$ cm^{-1}. In the steeper regime above

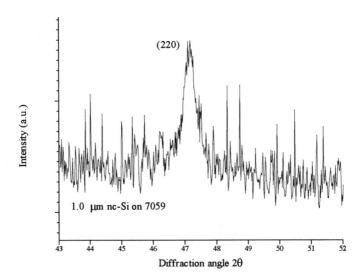

Fig. 2. X-ray diffraction spectrum showing the dominant (220) reflection of a film grown from SiF_4, H_2, and SiH_4.

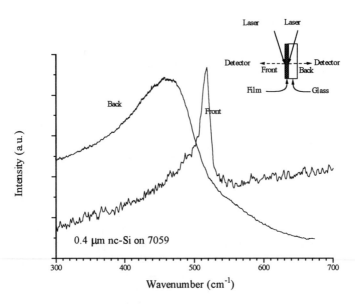

Fig. 3. Raman spectra of a film grown from SiF_4, H_2, and SiH_4. The two spectra were taken from the top and through the glass substrate. The probing depth is ~ 0.05 μm.

1.6 eV α rises to 10^5 cm^{-1} at 2.4 eV. In the less steep regime below 1.6 eV α drops to ~ 10 cm^{-1} at 0.8 eV.

> *Conductivity.* For films of a given thickness, the conductivity can be varied widely by changing the SiH$_4$:SiF$_4$ ratio. A standard set of growth parameters, using only SiF$_4$ and H$_2$ is listed as the last column of Table 1. 1 μm thick films grown with these parameters have dark conductivities of 10^{-8} Scm^{-1} (see lower right of Fig. 1). When 1 sccm of SiF$_4$ is added without any other change, the conductivity becomes 10^{-4} Scm^{-1} (Fig.1). The corresponding thermal activation energies are 0.6 and 0.3 eV. Such results show that the conductivity of undoped nc-Si at a given film thickness can be varied over a wide range by simple variations of the growth parameters.

CONCLUSION

The conductivity of nc-Si films is strongly influenced by the structure of the film. This is demonstrated by growing As-doped nc-Si with constant As doping and essentially constant deposition conditions, so that the principal variable is the film thickness. We show that it is possible to vary the conductivity of undoped nc-Si by adding SiH$_4$ to the standard SiF$_4$ and H$_2$ source gas composition.

ACKNOWLEDGEMENTS

We thank Henning Sirringhaus and Prof. A. Kahn for the Auger spectra, Daojing Wang for help with the Raman spectra, and Chiajen Lai for help with the X-ray spectra. This work is supported by ARPA under Grant #F33615-94-1-1464 managed by WPAFB, and by EPRI.

REFERENCES

1. J. Meier, R. Flückiger, H. Keppner and A. Shah, Appl. Phys. Lett. **65**, 860 (1994).
2. J. Meier, S. Dubail, R. Flückiger, D. Fischer, H. Keppner and A. Shah, in Conf. Record 1994 IEEE 1st World Conf. Photovoltaic Energy Conv., Waikoloa, Hawaii, December 5-9, 1994. (IEEE, Piscataway, NJ 1994), p.409.
3. T. Nagahara, K. Fujimoto, N. Kohno, Y. Kashiwagi and H. Kakinoki, Jpn. J. Appl. Phys. **31**, 4555 (1992).
4. J. Jang, H.J. Lim and B.Y. Ryu, in Soc. Info. Display Internat Symp. Digest vol. XXVI, (SID, Santa Ana, CA 1995), p.325.
5. M. Nakata, PhD thesis, Tokyo Institute of Technology, Department of Electronic Chemistry, March 1992.
6. B.D. Cullity, Elements of X-ray diffraction, Addison-Wesley, Reading, MA, 1967, p.99.

HYDROGEN BEHAVIOR IN PECVD NITRIDE
BY SiH$_4$ & ND$_3$ DURING RTA

S. S. He, V. L. Shannon, & T. Nguyen,
Sharp Microelectronics Technology, Inc.,
Camas, Washington USA

Abstract

PECVD silicon nitride was deposited by silane and deuterium ammonia. Silicon rich and nitrogen rich silicon nitride were deposited by varying the ratio of the SiH$_4$/DH$_3$. From FTIR, we found that the wave numbers of SiD and ND shifted lower when compared to SiH and NH bond in the NH$_3$ nitride. In Si-rich nitride, both Si-H and Si-D bonds were found, which is different from N-rich nitride, where only an ND bond was found. Most of the hydrogen in NH(D) comes from the ammonia during PECVD deposition. We found that some of the deuterium exchanges its bonding to silicon from the initial bonding to nitrogen during a rapid thermal annealing process.

Introduction

a-Si:H TFT is a thin film device. There is a trend to reduce the thickness of the a-Si:H channel in order to reduce the leakage current and light sensitivity. When the thickness of the a-Si:H channel is reduced to below 500Å, the TFT becomes a surface device. Due to its superior interface, PECVD silicon nitride is used in a-Si:H TFT LCDs as its gate dielectrics and/or etching stopper [1, 2]. A very thin layer of a-Si:H channel is surrounded with thick silicon nitrides. The high quality of the nitride should have good interface to the a-Si:H channel, good insulation properties, and good stability. There is a lot of hydrogen in the PECVD silicon nitride due to its low temperature deposition process [3]. The hydrogen is bonded to silicon and nitrogen. The hydrogen can be as high as 40 at.% [3-5], so that the stoichiometric of the film is achieved from three elements, silicon, nitrogen and hydrogen. We can no longer consider the hydrogen as an impurity, such as oxygen, in most nitrides. Even the Si-H and N-H bonds in the nitride are stronger than that of O-H bonds in silicon oxide. The hydride bond can be broken and exchange its bonding during the prolonged TFT operation or with a changed thermal environment [4-6]. The changes can introduce some stability problems for the TFTs during operation, which, in turn, changes the gray level and contrast ratio of the LCDs.

The total hydrogen concentration of the PECVD nitride depends on the ratio of silane and ammonia flow, deposition temperature, and deposition power during deposition. The hydrogen comes from two sources, silane and ammonia. To use deuterium ammonia as its nitrogen source

Mat. Res. Soc. Symp. Proc. Vol. 424 ©1997 Materials Research Society to deposit PECVD nitride will separate the hydrogen from the silane and ammonia. In this research, we study where the hydrogen in the PECVD comes from and how the hydrogen changes during a rapid thermal process.

Experiment

We deposited silicon nitride by using SiH_4 as the silicon source and NH_3 and ND_3 as the nitrogen source on a Si wafer. In the process, we varied the ratio of silane and ammonia flow. In this way, Si-rich and N-rich film were deposited. The deposition temperature was set to 400°C. Table 1 shows the deposition conditions. A low frequency plasma power was used to reduce surface charges during deposition.

<table>
<tr><td colspan="2" align="center">**Table 1**
PECVD Deposition Conditions</td></tr>
<tr><td>SiH_4</td><td>0.3-0.5 slm</td></tr>
<tr><td>N_2</td><td>1.6 slm</td></tr>
<tr><td>ND_3</td><td>4.0-4.2 slm</td></tr>
<tr><td>HF Power</td><td>450W</td></tr>
<tr><td>LF Power</td><td>550W</td></tr>
<tr><td>Pressure</td><td>2.6 Torr</td></tr>
</table>

<table>
<tr><td colspan="2" align="center">**Table 2**
RTA Conditions</td></tr>
<tr><td>Temperature</td><td>500-900°C</td></tr>
<tr><td>Increments</td><td>50°C</td></tr>
<tr><td>N_2</td><td>On</td></tr>
<tr><td>Time</td><td>30 seconds</td></tr>
</table>

Some of the films were annealed in a rapid thermal process. During RTA, nitrogen flowed to protect the surface from oxidation. Table 2 shows the rapid thermal annealing conditions. FTIR was tested before and after a rapid thermal annealing process. SiD, SiH, SiN, NH and ND peaks of the nitride and their integrated area were compared before and after RTA. Hydrogen Forward Spectrum was tested for the nitrides before and after RTA to get the total concentration of the hydrogen and deuterium. From the HFS, we know that only small amount of hydrogen/deuterium is diffused out from nitride film during RTA below 650°C. PECVD silicon nitrides with NH_3 ammonia nitride were deposited as references. Those samples were annealed in RTA also.

Results

Figure 1 is the FTIR spectrum of silicon nitride by deuterium and hydrogen ammonia. From the bottom to the top of the spectrum, (a) is a nitrogen rich nitride by NH_3, (b) standard nitride by NH_3, (c) Si-rich nitride by NH_3, (d) N-rich nitride by ND_3, (e) standard nitride by ND_3, and (f) Si-rich nitride by ND_3. Standard nitride, from NH_3 and ND_3, is called because its reflective index is close to 2, which is the reflective index for thermal nitride. Table 3 lists the film's properties and the FTIR results.

The deposition conditions were modified to deposit 8000Å of nitride film. From FTIR we found that when ND_3 replaced the NH_3 as its nitrogen source, hydride peaks, SiD and ND, were shifted to a lower wave number, since the deuterium is heavier than hydrogen. For Si-rich nitride, both SiH and SiD bonds were found from FTIR, so that not only the hydrogen in silane contributes to silicon hydride bonding, but also the deuterium from ammonia. For N-rich nitride, only ND was significant. The NH bonds were very small. Most of hydrogen in NH bonds were carried from ammonia, not from silane. This is an indication that even in the plasma there are reactive precursors from elements (N, H and Si) and hydrides (NH, NH_2, SiH, SiH_2 or SiH_3), only the hydrides can give a significant contribution to nitride film's growth. Otherwise, we should find both NH and ND in the N-rich nitride.

The position of the SiD was increased when more ammonia flowed. The material becomes less dense in this case. When the surrounding environment becomes light, replacing silicon by H and D or N, the Si-H(D) bonds shifted to a higher wave number. When the NH(D) bond increased, the shoulder of SiN was increased also and the SiN peak shifted to a higher wave number since there are two modes for the nitride [7,8]. We can see the changes in Figure 1.

Table 3
Properties of Silicon Nitride by Varying Ratios of Silane & Ammonia

	R.I.	SiH Area	SiD Area	ND Area	NH Area
Si-rich by ND_3	2.50	2.09	1.36	0.38	0.04
Standard Nitride by ND_3	2.06	0.88	3.39	0.95	0.22
Ni-rich by ND_3	1.94	0.33	2.83	2.31	0.27
Si-rich by NH_3	2.12	6.13	None	None	0.99
N-rich by NH_3	1.94	2.34	None	None	5.29

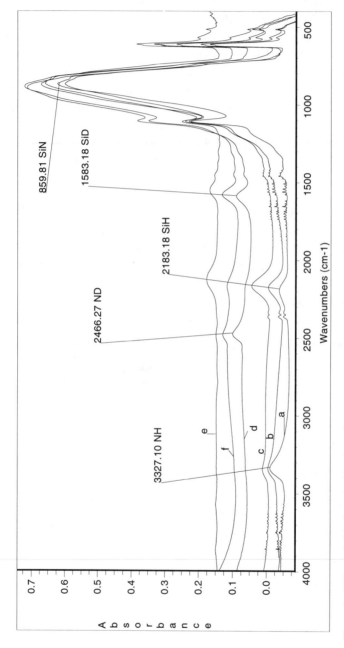

Figure 1. Silicon nitride deposited by deuterium and hydrogen ammonia. Deposition conditions of the spectrums from top to bottom are N-rich (H), standard (H), Si-rich (H), N-rich (D), standard (D), and Si-rich (D).

RTA Effect

Figures 2 and 3 show the integrated SiD and ND bonds in FTIR. After RTA, we found that SiD increased from deposition to its highest value around 650-700°C, then the SiD peak deceased. This behavior is the same as we have found from NH_3 silicon nitride during RTA [4, 5]. During low temperature RTA to the thick, >5000Å, nitrides for a very short time, the hydrogen (deuterium) in the films should be a constant, since there is no hydrogen (deuterium) source during RTA. The hydrogen bond and deuterium bond increase should come from inside the nitride. We believe that the hydrogen (deuterium) is from released hydrogen (deuterium) from N-H(D) bonds by the following reaction:

$$Si\text{-}Si + ND == SiN + SiD. \tag{1}$$

Since the thermal energy of RTA is small and the energy could not break the N-H(D) bond directly, the RTA is an energy source to exchange hydrogen (deuterium) from NH(D) to SiH(D). And, the final configuration is more stable than before [4-6]. When the annealing temperature goes higher, we found that out diffusion of the hydrogen (deuterium) will dominate. As the films lost total hydrogen, both SiH(D) and NH(D) bonds are decreased.

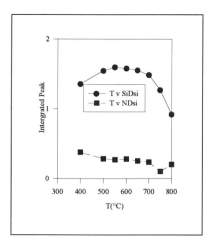

Figure 2. SiD and ND bond integrated area changes with RTA temperature for Si-rich silicon nitride by ND_3 ammonia.

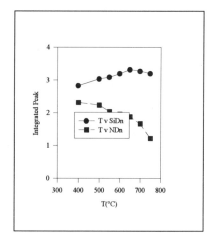

Figure 3. SiD and ND bond integrated area changes with RTA temperature for N-rich silicon nitride by ND_3 ammonia.

Conclusions

1. The silicon nitride deposited by ND_3 ammonia showed that its ND and SiD bond shifted to a lower wave number when compared to SiH and NH bond in the hydrogen nitride.

2. For Si-rich nitride, both SiD and SiH bonds were found. However, only the ND bond is significant to N-rich nitride.

3. This is an indication that hydride bonds dominate the surface growth of the PECVD nitride, since most of the nitrogen bonds to deuterium in N-rich nitride, not to hydrogen.

4. During RTA, some of the hydrogen and deuterium exchanges its bonding from nitrogen to silicon.

References

1. Y. Kuo, J. Electrochem. Soc., Vol. 142, No. 7, July 1995, p. 2486.

2. N. Ibaraki, T. Shimano, F. Fukuda, K. Matsumura, K Suzuki, H. Toeda and O. Takikawa, Conf. Record 91 Int. Display Research Conf., San Diego, CA, p. 97-100.

3. J. Robertson, Phil. Mag., B63, p.47 (1991).

4. S.S. He and V.L. Shannon, MRS Vol. 377, p. 337.

5. S.S. He and V. L. Shannon, Proceedings of the Fourth ICSICT '95, Beijing, PRC.

6. H.J. Stein, V. A. Wells, R. E. Hamphy, J. Electrochem. Soc. 126, 1750, (1979).

7. D. V. Tsu, G. Lucovsky and M.J. Mantini, Phy. Rev., V. 33, No 10, p. 7069.

8. G. Lucovsky, J. Yang, S.S. Chao, J.E. Tyler, and W. Crubatyj, Phy. Rev. B, V. 28, No. 6, p. 3234.

CHARACTERISTICS OF XeCl EXCIMER-LASER ANNEALED INSULATOR

KEUN-HO JANG, HONG-SEOK CHOI, JAE-HONG JUN, JHUN-SUK YOO AND MIN-KOO HAN
Department of Electrical Engineering, Seoul National University, San 56-1, Shinrim-dong, Kwanak-ku, Seoul 151-742, Korea

ABSTRACT

The laser annealing effects on the TEOS (Tetra-Ethyl-Ortho-Silicate) oxide of MOS (Al/TEOS/n+ Silicon) structures was investigated with different initial oxide conditions, such as breakdown field. The breakdown field increased upto the 170 mJ/cm^2 with increasing laser energy density and decreased at 220 mJ/cm^2. It is considered that the increase of breakdown field is originated from the restore of strains which exist mainly at the metal/oxide interface.

INTRODUCTION

Laser-induced crystallization method has been successfully utilized to fabricate the low-temperature polysilicon (poly-Si) films for the application to the thin film transistors (TFTs) [1]. It is essential that the electrical characteristics of poly-Si TFTs depend not only on the active poly-Si layer but also on the quality of insulator. However, most researches have been focused on the laser-induced crystallization of amorphous silicon (a-Si) films , while the laser annealing effects on the characteristics of insulators, such as Si_3N_4 and SiO_2, has been reported little.

The excimer laser-induced dopant activation is widely used for the formation of source/drain contacts in the fabrication of low-temperature poly-Si TFTs [2]. In this case, the gate insulator layer is exposed to the thermal energy produced by the laser absorption at gate electrodes, which may influence the characteristics of gate insulator. In this paper, we have prepared MOS (Metal/Oxide/Silicon) structure samples with TEOS (Tetra-Ethyl-Ortho-Silicate) oxides, and investigated the laser annealing effects on the characteristics of oxides with respect to the initial film quality of breakdown fields.

EXPERIMENTAL

We prepared MOS (Al/TEOS/n+Si wafer) structure samples for the laser annealing experiments. 1000 Å of TEOS oxide was deposited by rf PECVD (Plasma Enhanced Chemical Vapor Deposition) at the 390 ^{o}C. The aluminum was deposited by sputtering and patterned into a circle electrode with 300 μm in diameter for the measurements of breakdown fields. Before the aluminum deposition, the oxide layer was exposed to SF_6* radical damage with PECVD in order to change the initial film quality of breakdown fields. The exposure time of SF_6* radical damage was 0, 30 or 60 seconds.

In laser annealing, XeCl excimer laser was used and the annealing energy densities were 0, 105, 150, 170 and 220 mJ/cm^2.

RESULTS AND DISCUSSION

The schematic diagrams of conventional process in laser-induced dopant activation and the

115

structure of sample used in our works are shown in Fig. 1. The annealing energy of dopant activation is around 200 mJ/cm^2. In dopant activation step (Fig. 1b), the gate oxide is exposed to the laser-induced thermal energy which may change the characteristics of oxide, such as breakdown field and oxide/silicon interface states.

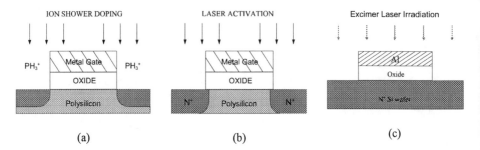

ION SHOWER DOPING   LASER ACTIVATION   Excimer Laser Irradiation

(a)   (b)   (c)

Fig. 1. The schematic diagrams of conventional process in laser-induced dopant activation (a), (b) and the sample structure used in our work (c).

We have measured the breakdown fields with different laser annealing energy density of 0, 105, 150, 170 and 220 mJ/cm^2. Figure 2 shows the breakdown characteristics of oxide film which were damaged by the SF$_6$ * during 60 seconds. We evaluated soft breakdown and abrupt breakdown fields which was shown in Fig. 3 as a function of laser energy density.

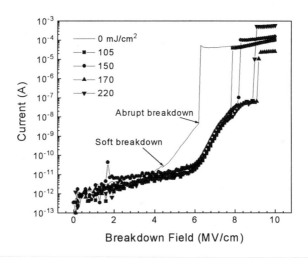

Fig. 2. Current vs. Breakdown fields with different laser annealing energy density. TEOS oxide layer was exposed to the damage of SF$_6$* during 60 seconds.

In Fig. 3, the breakdown fields of initial film (0 mJ/cm^2) shows large deviation due to the possible spatial nonuniformity of SF$_6$* flow. The abrupt breakdown, which take places mainly in the mechanically weak points such as micro-void and cracks, increase with laser energy density and decrease in excess of 170 mJ/cm^2. The soft breakdown , mainly related to the trap sites [3], also exhibits a similar trends compared to the abrupt breakdown. The increase of breakdown fields may be resulted from the damage recovery mainly at the metal/oxide interface.

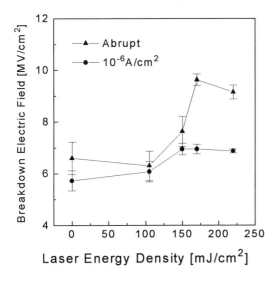

Fig. 3. Characteristics of breakdown field in SF$_6$* damaged TEOS oxide film as a function of laser energy density. The damage exposure time is 60 seconds.

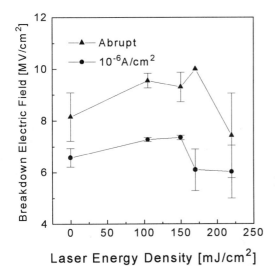

Fig. 4. Characteristics of breakdown field in SF$_6$* damaged TEOS oxide film as a function of laser annealing density. The damage exposure time is 30 seconds.

Figure 4 shows the breakdown field with 30 seconds of SF$_6^*$ exposure. Breakdown field also increase around 170 mJ/cm^2. Figure 5 shows the breakdown field with no SF$_6^*$ exposure. The breakdown field almost unchanged with laser energy density, and the breakdown field of initial film (0 mJ/cm^2) shows higher value than that of SF$_6^*$ treated films. It is showed that the annealing effects on the metal/oxide interface is not noticeable with no damage of SF$_6^*$.

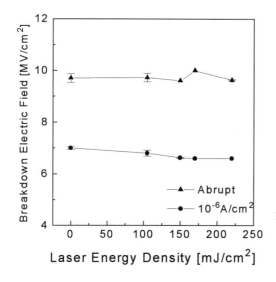

Fig. 5 Characteristics of breakdown voltage in TEOS with no SF$_6$* damage.

CONCLUSION

We have investigated the laser-annealing effects on the insulator in MOS structure samples. The increase of breakdown field with laser energy density is mainly originated from the restore the strains which exist at the metal/oxide interface.

ACKNOWLEDGMENT

The part of this work has been supported by HAN (G-7) project.

REFERENCES

1. K. Shimizu, O. Sugiura, and M. Matsumura, Jpn. J. Appl. Phys. **29**, L1775 (1990).

2. A. Slaoui, F. Foulon, and P. Siffert, J. Appl. Phys. **67**, 6197 (1990).

3. I. Brodie, and P. Richard, Proceedings of the IEEE **82**, 1006 (1994).

MATERIAL PROPERTIES OF SIPOS FILMS PREPARED BY XeCl EXCIMER LASER ANNEALING OF a-Si:N$_x$ FILMS

Hong-Seok Choi, Jae-Hong Jun, Keun-Ho Jang and Min-Koo Han
Department of Electrical Engineering, Seoul National University, San 56-1, Shinrim-dong, Kwanak-ku, Seoul 151-742, Korea

ABSTRACT

The material properties of laser-annealed a-Si:N$_x$ films were investigated. The a-Si:N$_x$ films for laser-annealing were deposited by rf plasma enhanced chemical vapor deposition (PECVD) with NH$_3$ and SiH$_4$ gas mixtures. At the 0.35 of NH$_3$/SiH$_4$ ratio, the optical band-gap was abruptly increased to 2.82 eV from 2.05 eV by laser-annealing which indicates that Si-N bonding comes to be notable at that ratio. The electrical conductivity showed the maximum value of 4×10^{-6} S/cm at the 0.11 of NH$_3$/SiH$_4$ ratio where the grain growth and the increase of Si-N bonding are optimized for the enhancement of electrical conductivity. The σ_P/σ_D ratio which is related to the defects states for photo generation centers was decreased with increasing NH$_3$/SiH$_4$ ratio. Our experimental data showed that the optical band gap and electrical conductivity of laser-annealed a-Si:N$_x$ films were dominantly affected by the NH$_3$/SiH$_4$ ratio at the 250 mJ/cm^2 of laser-annealing energy density.

INTRODUCTION

The high electrical mobility is required for the operation of polycrystalline silicon (poly-Si) thin film transistors (TFTs) as driving elements in liquid crystal displays (LCDs). Laser-annealing method is successfully used to fabricate the low-temperature poly-Si TFTs for its realization of high electrical mobility [1,2]. However, the leakage current of poly-Si TFTs can be increased with increasing mobility, which may cause to the degradation of image through the change of pixel voltage. It is well known that the leakage current of poly-Si TFTs is caused by the resistive current through the active-bulk layer or field-emission current near the drain junction where the electric field is high [3,4]. The former can be reduced by the use of thin active layer which increases the resistance of active layer and the latter can be reduced by the modifications of device structure such as active gate, LDD, offset-gate structures which reduce the drain junction field [5-6]. However, the abrasion or fluctuation of the laser-annealed films may be happened as the film thickness decreased due to the decrease of heat capacitance, which leads to the limitation for the use of thin active layer presumably below 500 Å. In LDD or offset-gate structures, the additional masks are required for the device fabrication which cause to the decrease of production yield. We propose the double active layers of amorphous silicon (a-Si)/ a-Si:N(:O)$_x$ films for laser-annealing to reduce the both resistive and field-emission currents. The thin a-Si layer can be used by retaining the heat capacitance of double active layers and the field-emission carriers can be reduced by the increase of optical band-gap in the semi-insulating polycrystalline silicon (SIPOS) film.

In this paper, we have prepared SIPOS films by XeCl excimer laser (λ=308 nm) annealing of a-Si:N$_x$ films and investigated the material properties of SIPOS films such as optical band-gap

Mat. Res. Soc. Symp. Proc. Vol. 424 © 1997 Materials Research Society

and electrical conductivity for the application to the poly-Si TFTs with double active structure of poly-Si/SIPOS.

EXPERIMENT

a-Si:N$_x$ film was deposited on the glass (Corning 7059) and silicon wafer by rf (13.56 MHz) PECVD system at the power of 50 W, deposition pressure of 1 Torr and substrate temperature of 370 °C with gas mixtures of SiH$_4$ and NH$_3$. We fixed the total flow rate of gas mixture at 20 sccm and varied NH$_3$/SiH$_4$ ratio from 0 to 1. The aluminum electrode was deposited by sputtering and patterned into coplanar structure to measure the electrical conductivity. The channel length and width were 80 μm and 300 μm, respectively. The laser-annealing was performed by XeCl (λ= 308 nm) excimer laser at the 250 mJ/cm^2 of laser energy density and 20 °C of substrate temperature. We evaluated the absorption coefficients and optical band-gap from ultra-violet (UV) spectroscope data, and investigated the change of optical band-gap before and after the laser-annealing.

RESULTS AND DISCUSSION

The melting depth of laser absorption is related to the absorption coefficient, α, of the laser-annealing films and characteristic wave length of laser beam. Figure 1 shows the absorption coefficients of a-SiN$_x$ films as a function of photon energy with different ratio of NH$_3$ to SiH$_4$ flow rates. The absorption coefficient was evaluated from UV reflectance and transmittance data. The absorption coefficient at 4 eV (λ= 308 nm) is decreased with increasing the NH$_3$/SiH$_4$ ratio due to the evolution of Si-N bonds. We extracted the optical band-gap and skin depth from absorption data which was shown in Fig. 2.

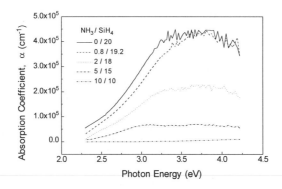

Fig. 1. The absorption coefficients of a-Si:N$_x$ films as a function of photon energy with different ratio of NH$_3$ to SiH$_4$ flow rates.

In Fig. 2, the optical band-gap is almost unchanged upto the 0.35 of NH₃/SiH₄ ratio and increase in excess of 0.35. The skin depth shows 1 μm at 1 of NH₃/SiH₄ ratio which is thicker than the thickness of initial film. At this case, the abrasion was happened in laser-annealing by the explosive absorption at a-SiN$_x$ /glass interface where the absorption is high due to the possible existence of defects.

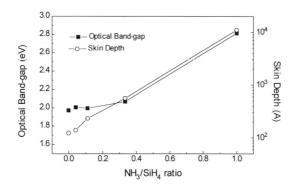

Fig. 2. The optical band-gap and skin depth of initial a-SiN$_x$ films as a function of NH₃/SiH₄ ratio. Optical band-gap was evaluated by the extrapolation of $\alpha^{-1/2} \cdot eV^{-1/2}$ vs eV plots.

Figure 3 shows the optical band-gap and conductivity of SIPOS films annealed by 250 mJ/cm² of laser energy density. Upto the 0.11 of NH₃/SiH₄ ratio, optical band-gap is almost unchanged. However, the optical band-gap shifts up about 0.15 eV from that of the initial films, which may be resulted from the O₂ introduction from atmosphere in laser-annealing. At 0.35 of NH₃/SiH₄ ratio, large increase of optical band-gap happens due to the activation of Si-N bonding by laser-absorption thermal energy while the change of optical band-gap in initial film comes to be notable at 1 of NH₃/SiH₄ ratio. The electrical conductivity shows a peak value at 0.11 of NH₃/SiH₄ ratio. In laser-annealing, the grain growth is enhanced by the increase of the melting and solidification times [7] which generally results in the improvement of electrical conductivity. The melting and solidification times can be increased by employing low-thermal conductivity material, such as a-Si:N$_x$. Therefore the electrical conductivity is increased upto the 0.11 of NH₃/SiH₄ ratio through the possible increase of the melting and solidification times. The decrease of electrical conductivity in excess of 0.11 of NH₃/SiH₄ ratio may be resulted from the increase of Si-N bonds which shows high resistivity. However, it is considered that the laser energy density of 250 mJ/cm² is not enough to promote grain growth by fully melting because of the low value below 10⁻⁷ S/cm in the film without NH₃ (NH₃/SiH₄ = 0).

Figure 4 shows the photo to dark conductivity, σ_P/σ_D as a function of NH₃/SiH₄ ratio and the value of σ_P/σ_D decreases with increasing NH₃/SiH₄ ratio. The value of σ_P/σ_D is related to the defects states which is responsible for the photo generation centers. It may be explained that the defect states are reduced by the promotion of crystal network through the increase of melting and solidification time.

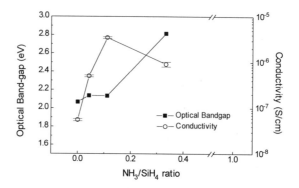

Fig. 3. The optical band-gap and conductivity of laser-annealed SIPOS films as a function of NH₃/SiH₄ ratio.

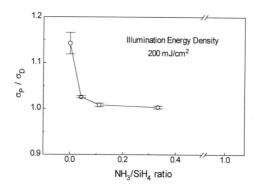

Fig. 4. Photo to dark conductivity, σ_P/σ_D ,as a function of NH₃/SiH₄ ratio.

CONCLUSION

We have investigated the optical band-gap and electrical conductivity of laser-annealed a-Si:N$_x$ films. The optical band-gap was more increased in SIPOS films than that in a-Si:N$_x$ films by the laser-induced thermal energy. The electrical conductivity showed the maximum value of 4×10^{-6} S/cm at the 0.11 of NH₃/SiH₄ ratio where the grain growth and the increase of Si-N bonding are optimized for the enhancement of electrical conductivity. However, the conductivity still remains low value compared with the typical value ($\sim 10^{-3}$ S/cm) of laser-annealed poly-Si films due to the evolution of Si-N bonding. The SIPOS films may be effective to reduce the leakage current of the poly-Si TFTs with poly-Si/SIPOS double-active-structure, and the enhancements of electrical conductivity of poly-Si layer may be also predictable through grain growth by employing the low-thermal conductivity material of SIPOS layer.

ACKNOWLEDGMENT

The part of this work has been supported by HAN (G-7) project.

REFERENCES

1. Y. Miyata, M. Furuta, T. Yoshioka and T. Kawamura, Jpn. J. Appl. Phys. **31**, 4559 (1992).

2. C. D. Kim and M. Mastumura, IEEE Trans. Electron Devices **43**, 576 (1996).

3. B. A. Khan and R. Pandya, Mater. Res. Soc. Symp.Proc. **106**, 353 (1988).

4. J. G. Fossum, A. O. Conde, H. Shichijo and S. K. Banerjee, IEEE Trans. Electron Devices **32**, 1878 (1985).

5. C. M. Park, B. H. Min, K. H. Jang, J. H. Jun and M. K. Han, Extented Abstracts of Solid-State Electron Devices and Materials, 656 (1995).

6. M. Yazaki, S. Takenaka and H. Ohshima, Jpn. J. Appl. Phys. **31**, 206 (1992).

7. M. O. Thompson, G. J. Galvin and J. W. Mayer, Phys. Rev. Lett. **52**, 2360 (1984).

Part II

Polycrystalline Silicon Thin–Film Transistor Materials and Technology

LOW TEMPERATURE POLYSILICON MATERIALS AND DEVICES

D. PRIBAT, P. LEGAGNEUX, F. PLAIS, C. REITA, F. PETINOT and O. HUET
Thomson CSF, LCR, Domaine de Corbeville, 91404, Orsay, France.

ABSTRACT

In this paper, we essentially discuss the material aspects of low temperature ($\leq 600°C$) polysilicon technologies. Emphasis is put on the properties of polysilicon films, depending on the way they are obtained. Solid phase crystallisation as well as pulsed laser crystallisation processes are presented in some detail, together with thin film transistor characteristics. Although not yet stabilised and despite uniformity and reproducibility problems, laser crystallisation will probably end up being the technology of choice for the manufacture of large area electronics products, because it allows the fabrication of devices exhibiting superior properties, with a reduced thermal budget.

I- INTRODUCTION

Thin films of semiconductors have long raised interest for display, sensors or photovoltaic applications [1]. Amongst these materials, polycrystalline silicon (poly-Si) is probably the most attractive for several reasons. First, it is a single element material, compared for instance to cadmium selenide (CdSe) or even to hydrogenated amorphous silicon (a-Si:H). Second, it exhibits superior transport and stability properties, particularly in comparison to a-Si:H and third, it can produce both n and p-type thin film transistors (TFTs), which opens up the possibility of realising complementary metal-oxide-semiconductor (CMOS) devices and circuits. The original driving force behind poly-Si development has been the integration of the addressing circuits that drive liquid crystal displays. However, recent work has shown that the use of poly-Si in replacement of a-Si:H for pixel switching could lead to a reduction of the mask count in the manufacture of active matrix liquid crystal displays (AMLCDs) even when the drivers are external instead of being integrated [2].

In this paper, we shall discuss the particular properties of poly-Si (and associated devices), depending on the way it is synthesised; we shall restrict ourselves to low temperature material and processes, corresponding to glass-compatible operations. Also, we shall only consider chemical vapour deposition processes (CVD) for the preparation of the silicon films, such processes being by far the most commonly used. Although poly-Si can be obtained by direct deposition from the gas phase, the preferred method used so far for its synthesis has been the crystallisation of an amorphous precursor. Because it is a relatively simple and attractive process (easily amenable to mass production), solid phase or furnace crystallisation (around 600°C) has been extensively studied. However, over the past few years, increased attention has been paid to laser crystallisation, because it has allowed the fabrication of very high quality devices, with mobility values reaching those of their silicon on insulator (SOI) counterparts. Particular emphasis will be put on the latter process since we believe we are at crossroads in the technological evolution concerning the crystallisation of amorphous silicon (a-Si).

II- DEPOSITION OF THE AMORPHOUS PRECURSOR

As quoted in the introduction and with reference to figs. 1 and 2, poly-Si can be directly deposited from the gas phase (direct-poly), provided the temperature and pressure are adequately chosen [1]. However, since crystalline growth lends itself to faceting, direct-poly films tend to exhibit rough surface morphologies. This is detrimental to carrier transport, particularly in field effect devices, where inversion or accumulation layers are confined within a few nanometres beneath the interface [3]. To our knowledge, the best result (in term of field effect mobility) obtained so far using direct-poly films is 28 cm^2/Vs, obtained on a p-type TFT [4]. Although this result is worth of consideration, higher mobility values can be obtained when poly-Si is synthesised by solid phase crystallisation (SPC) of an amorphous precursor, because the surface of the films preserves the smooth and even morphology which is characteristic of amorphous deposits.

Fig. 1 summarises the crystalline state (and texture when appropriate) of silicon deposits obtained by pyrolysis of SiH4 over a wide range of temperatures and pressures. Schematically, decreasing the pressure or increasing the temperature tends to favour crystalline growth, because in either case the surface diffusion coefficient of Si adatoms is increased.

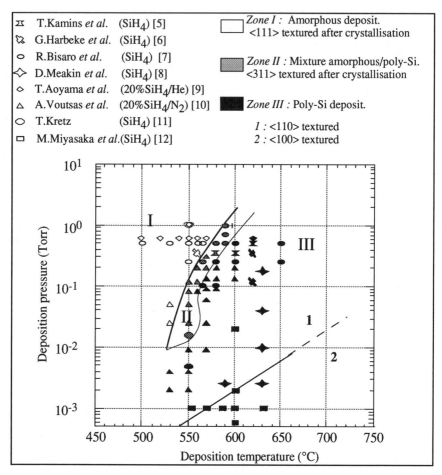

Fig. 1 : Crystalline state and structure of as-deposited silicon films as a function of deposition temperature and deposition pressure, when the gas precursor is SiH4. For each author, we have indicated whether SiH4 was diluted or not.

It appears that for pressures ranging between a few 10^{-2} and 1 Torr, which are typical conditions for low pressure chemical vapour deposition (LPCVD) processes, a-Si can be obtained in the temperature range between ~ 530 and 570°C. However, if the temperature and pressure are kept low, the deposition rate drops to unacceptably small values. Therefore, for practical reasons, SiH4 pyrolysis is usually performed in the Torr pressure range and at a temperature around 550°C. In such conditions, the deposition rate is around 5 nm.min[-1]. Note that there exists a region (zone II on fig. 1) where the deposits consist of mixed phase Si, in the form of crystallites embedded in the amorphous matrix. Interestingly, the use of this kind of deposit can reduce the solid phase crystallisation time, by enhancing the nucleation rate [13].

Over the last few years, the use of Si_2H_6 as gas precursor in replacement of SiH_4 has been investigated. In the same way as for SiH_4, fig. 2 shows the state diagram for silicon films deposited by pyrolysis of Si_2H_6. Although fig. 2 is not as documented as fig. 1, because the use of Si_2H_6 is quite recent, here again three distinct regions are observed, corresponding to amorphous, mixed phase or polycrystalline material. We have shown recently that as for the case of SiH_4, the use of mixed phase films resulted in an appreciable decrease of the crystallisation time [19].

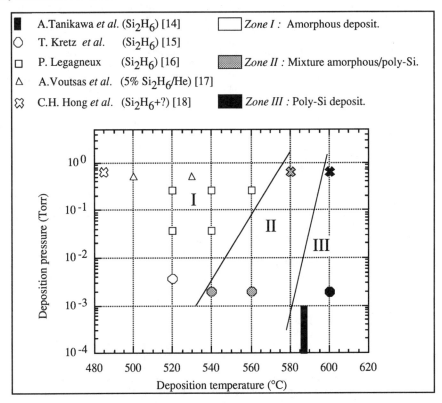

Fig. 2 : Same as fig. 1 but with Si2H6 as gas precursor.

III- SOLID PHASE CRYSTALLISATION

The solid phase crystallisation of an a-Si film takes place by nucleation of crystalline clusters which grow spontaneously when a critical size is reached; this corresponds to the situation where the free energy gained by converting a-Si to the crystal becomes larger than the surface energy corresponding to the creation of the interface between the crystalline cluster and the amorphous matrix. It must be emphasised that in order for nucleation and growth to occur, bond breaking must take place in the amorphous material, so as to allow for atom rearrangement.

Nucleation has been shown to occur heterogeneously at the substrate/a-Si interface [20, 21] and the critical size of the crystal clusters has been estimated around 2-4 nm at a (corrected) temperature of ~ 650°C [21]. Also, it appears that twinning occurs at a very early stage of crystallisation, probably during nucleation [21]. This could be explained by the fact that in a-Si, the 6-membered rings are not organised in chair or boat configurations; hence, upon nucleation, depending on the local bonding environment and the local bond angle distortions of the continuous random network, some rings will go on a chair configuration, whereas some others will adopt the boat configuration, thus initiating twinning [22].

Once the cluster has reached the critical size, growth proceeds by displacement of atoms from the amorphous phase to the crystalline phase, through the interface. The growth mechanism is very similar to that of solid phase epitaxy (SPE) which has been extensively studied. In the atomistic model of SPE [23], the a-Si/crystal interface consists of (111) terraces separated by [1$\bar{1}$0] ledges, which minimises the number of interfacial dangling bonds. To be considered part of the crystal, an atom in the amorphous phase must establish two undistorted bonds with atoms at crystalline sites; this favours growth on ledges and their quick motion, particularly in the [$\bar{1}$12] direction, where the atoms of the amorphous phase can attach to the crystal one by one [23, 24]. Growth proceeds by dangling bond generation on the ledge (bond breaking, corresponding to most of the activation step which is experimentally observed) and propagation along it [25]. The growth mechanism can also be viewed as a defect-assisted kink motion along the ledges [26]. Lu *et al.* have shown that a single bond breaking event can recrystallise up to 200 Si atoms [27].

The common feature of SPC poly-Si films is their very high twin density. As suggested above, these can be generated upon nucleation of the crystalline clusters [22]. However, as shown by Drosd and Washburn [24], twinning can also occur during growth, upon ledge formation on atomically flat (111) crystallographic planes, because on such planes, a cluster of three atoms is needed to fulfil the criteria of two undistorted bonds per atom of the crystal phase (or alternatively, to fulfil the criteria of completion of 6-membered rings). As there are two ways of placing such clusters on the (111) surface, corresponding to boat-type or chair-type rings (with indeed a fault energy for the former situation), twins are easily nucleated, again depending on the local bonding environment and bond angle distortions (i.e., when the fault energy is smaller than the energy needed to bend back or break a bond in order to locally achieve the right chair-type configuration). Also, Drosd and Washburn have shown that the presence of twins facilitates growth by providing favourable attachment sites for the atoms of the amorphous phase, at the boundary of the twin with the matrix. This results in a local growth enhancement which tends to favour the formation of extended twin bands. Since in poly-Si, the critical crystalline clusters which initiate growth are certainly bounded by (111) planes for energetic reasons, twinning by the above mechanism is very likely, resulting in the well known highly twinned structure of the SPC films [22].

Fig. 3 shows plan view TEM micrographs of ~ 100 nm thick poly-Si films which have been deposited and crystallised under various conditions. In fig. 3a, the gas precursor for a-Si deposition was SiH$_4$, whereas in figs. 3b-d, Si$_2$H$_6$ was used instead. Concerning the latter, the deposition pressure has been varied over three orders of magnitude, between 2 mTorr and 2 Torr. Crystallisation was carried out at 580°C during 65h for the high pressure deposits (SiH$_4$ as well as Si$_2$H$_6$) and for only 5h for the low pressure ones. The larger grains are obtained when a-Si is deposited at high pressure by pyrolysis of Si$_2$H$_6$. This corresponds to a situation where nucleation in the a-Si films is difficult, as we have measured an activation energy for the nucleation rate (E_n) of 4.78 eV in such films [28]. Hence, as few grains are nucleated, they can grow farther before impinging on their neighbours. Table I compares the various nucleation and growth activation energies (E_n and E_g) measured (see refs. 19 and 28 for details) on the films shown in fig. 3. Note that the grain sizes (fig. 3) scale well with the values of ΔH^* which are closely related to the nucleation barriers. The activation energy for the growth rate is similar for the various deposits, except for the high pressure Si$_2$H$_6$ case. We believe that this is due to a weaker strength of the Si-Si distorted bonds in a-Si, due to a higher strain, thus rendering easier the breaking of interfacial bonds upon growth. Actually, using Raman spectroscopy (measurements of the half-width of the TO-like band [29, 30, 31]) we have deduced larger

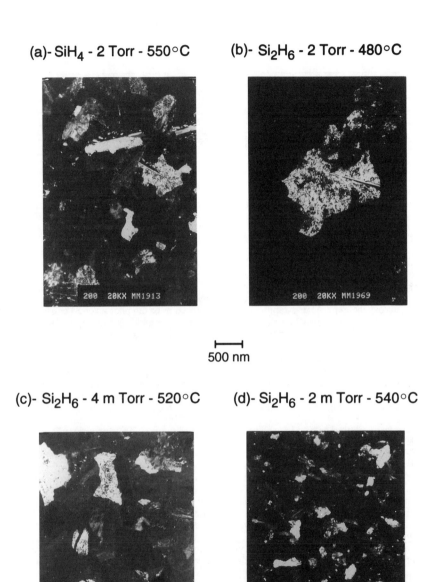

(a)- SiH$_4$ - 2 Torr - 550°C (b)- Si$_2$H$_6$ - 2 Torr - 480°C

⊢——⊣
500 nm

(c)- Si$_2$H$_6$ - 4 m Torr - 520°C (d)- Si$_2$H$_6$ - 2 m Torr - 540°C

Fig.3: Plan view TEM micrographs of 100 nm-thick poly-Si films corresponding to various deposition conditions for the a-Si precursor. Crystallisation was performed at 580°C.

average bond-angle distortions for the high pressure Si_2H_6 films than for their SiH_4 counterparts [32].

Deposition conditions			E_n	E_g	$\Delta H^* = E_n - E_g$
Gaz	T (°C)	p (Torr)	(eV)	(eV)	(eV)
SiH_4	550	2	3.27	2.67	0.60
Si_2H_6	480	2	4.78	2.29	2.49
Si_2H_6	520	4 10^{-3}	3.74	2.69	1.05
Si_2H_6 *	540 *	2 10^{-3} *	2.89	2.65	0.24

Table I : Comparison of the crystallisation parameters of different amorphous films as a function of the deposition conditions. (*) : mixed phase material

The results concerning TFT fabrication and characteristics from such SPC films are presented in section V.

IV- PULSED-LASER CRYSTALLISATION

Initiated by soviet scientists [33], pulsed-laser annealing was first used to activate ion-implanted dopants in crystalline Si (c-Si) and remove the corresponding lattice damage. During irradiation of semiconductor materials with photons of energy $h\nu > E_{gap}$ (where E_{gap} is the enrgy gap of the semiconductor), absorption of the laser energy takes place by excitation of electron-hole pairs across the gap. After a rapid thermal equilibration of the photoexcited carrier system ($\sim 10^{-14}$ s), the energy is transferred to the lattice by phonon emission, on a picosecond time scale [34]. For the particular case considered here of a-Si irradiation by UV wavelength lasers (e.g. excimer lasers with pulse duration $\tau \sim 30$ ns), the absorption depth of the laser light is of the order of 10^{-6} cm and the heat diffusion length in a-Si over the pulse duration, $(D\tau)^{1/2}$ (D is the thermal diffusivity in a-Si), amounts to $\sim 10^{-5}$ cm at elevated temperature. In these conditions, it is clear that most of the energy deposited during the laser pulse will be localised within the first 100 nm of the irradiated film, whereas the underlying substrate will remain practically thermally unaffected (see e.g. ref. 35 for transient temperature measurements during pulsed-laser irradiation).

Pulsed-laser melting of a-Si is very peculiar, because the melting temperature T_a of a-Si is about 225°C below that of c-Si [36]. In normal heating conditions, a-Si crystallises before it melts. However, when pulsed-laser heating is used, the heat supply is so fast that crystallisation is by-passed and a-Si is directly melted by a first order phase transition. As a consequence, liquid Si (l-Si) obtained in this way can be severely undercooled relatively to the equilibrium melting temperature of c-Si. If this undercooling is maintained for long enough (see below), nucleation events take place in the liquid, which trigger explosive crystallisation of the underlying a-Si material [36]. Explosive crystallisation of a-Si has been thoroughly studied by Lowndes and coworkers [37, 38] who have shown that nucleation in the undercooled liquid occurs within 10 ns after the onset of surface melting, i.e. before the end of the laser pulse in the conditions they used. The latent heat of crystallisation released by the nucleation and growth events in the highly undercooled surface melt (primary melt) raises the temperature of the liquid above T_a, which induces melting of the contiguous a-Si; a thin liquid layer is formed underneath the nucleated crystallites (buried liquid layer), whose downwards motion is driven by the difference in the latent heat of crystallisation between a-Si and c-Si ($L_a = 0.7 \ L_c$). Initial velocities around 12-14 ms^{-1} have been measured for buried liquid layers [37], which means that for the laser parameters considered here, the primary melt will propagate into the fine grain poly produced earlier by explosive crystallisation [39].

With the above remarks in mind, fig. 4 schematically shows the various transformations that can occur upon laser irradiation of a-Si films with pulse duration of 30-50 ns. In fig. 4b, the laser is on, explosive crystallisation has been triggered and the downwards propagation of the buried liquid layer leaves behind fine grain poly material. Later in the laser pulse (fig. 4c) the buried layer has crossed the film throughout, and the primary melt front penetrates the fine grain

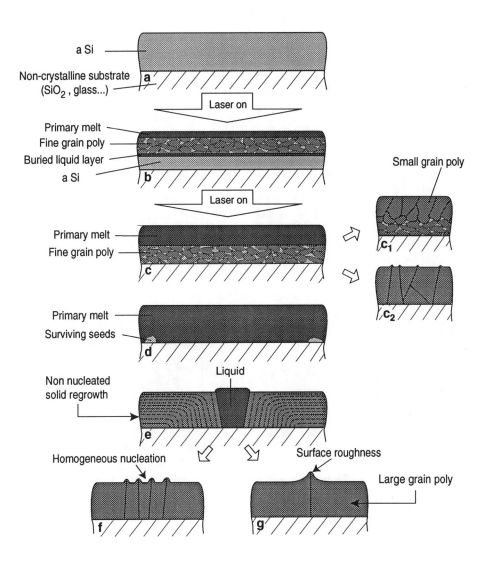

Fig.4: Schematic representation of the various phase transformations
occuring upon pulsed-laser irradiation of an a-Si film.

material, remelting it. At this point, if the laser energy density is not high enough to remelt the entire fine grain poly throughout, solidification will occur, and small grain poly will grow non nucleated, seeded on the fine grain material synthesised earlier by explosive crystallisation; the resulting structure is shown in fig. 4c$_1$. If, on the other hand, the fine grain material is completely melted and the liquid allowed to cool down after the end of the laser pulse, here again, large undercoolings will be reached (depending on the thermal leak through the substrate), before homogeneous nucleation takes place in the melt, leading to the formation of small grain poly [40]; the resulting structure is shown in fig. 4c$_2$. Figs. 4d-g represent a special situation, where the laser energy density is such that the fine grain poly resulting from explosive crystallisation is not totally melted and some surviving seeds sparsely distributed at the interface can act as (non nucleated) regrowth centres during cooling of the liquid. The grains grow laterally around the seeds, until they impinge on each other and grain sizes of several microns can be reached; homogeneous nucleation is prevented for a while between the growing grains, because the latent heat released by the crystallisation process warms up the liquid around the growing grains (fig. 4g). However, if the surviving seeds are sufficiently apart, homogeneous nucleation will indeed be triggered between the grains (the thermal diffusion length in l-Si is < 1 µm for a melt duration of 50 ns), leading to the situation schematically shown in fig. 4f, where small grain poly is found between the large grains. This particular regime of grain growth is often referred to as super lateral growth (SLG) [41, 42].

Fig. 5 shows a plan view SEM micrograph of poly-Si material obtained in the SLG regime after irradiation (10 shots at 580 mJ.cm^{-2}) of a 80 nm-thick a-Si film with a XeCl excimer laser. The crystallised film has been "secco-etched" before observation and the micrograph illustrates the situation depicted in fig. 4f, where homogeneous nucleation has occurred between two large grains (only a small part of one of the large grains is visible on the left hand side of the picture). Note that the grain in the centre of the picture is more than 4 µm in diameter and that there are few intra grain defects, in comparison to SPC material.

$\vdash$ **1 µm** $\dashv$

Fig. 5 : SEM view of the "secco-etched" surface of a 80 nm-thick poly-Si film laser-crystallised in the SLG regime (picture by courtesy of G. Fortunato)

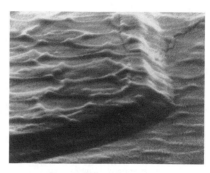

$\vdash$ **0.2 µm** $\dashv$

Fig. 6 : Surface roughness of a 80 nm-thick poly-Si film laser-crystallised in the SLG regime (see text)

Due to its peculiar mechanism, the SLG regime corresponds to a narrow processing window. Hence, a very good laser beam uniformity as well as pulse to pulse reproducibility is needed, if the beam is to be scanned over large areas. Sophisticated beam homogenisers are necessary, which end up costing more than the lasers. Furthermore, as far as scanning is concerned, the beam overlap region is another source of non-uniformity, since at the edges (even for a top-hat profile) there is always an energy gradient, which induces a variety of situations of partial melting and explosive crystallisation. The original material is thus modified and the laser irradiation conditions are no longer optimised in this particular region. The use of very high power excimer lasers capable of delivering up to 45 J per shot and working on a step and repeat mode might be a solution for the future [43].

Another problem of pulsed-laser crystallisation is illustrated on fig. 6, which shows a SEM micrograph (tilted plan view) of the unetched surface of a 80 nm-thick film, again crystallised in the SLG regime. Clearly the surface is very rough, with a high density of protuberances whose height can reach 30-40% of the film thickness. These protuberances decorate the grain boundaries, particularly at places where three grains meet. As shown by Lowndes et al. [37], who have used Cu tracing and Z-contrast STEM analysis, the protuberances are due to the laterally colliding growth fronts, which tend to expel the liquid out of the solidifying film. This roughness is of course detrimental to TFT quality, particularly for top gate structures (see below) because of the high electric field values that can develop at the tip of the protuberances, leading to injection into the oxide.

However, despite the above problems and drawbacks, laser crystallisation is just too attractive to be left aside, as it is a very low thermal budget process, and superior device properties are obtained, due to a much better crystalline quality of the poly-Si films (see below). All the major industrial players in the field of AMLCDs are heavily involved in the optimisation of this process and there is little doubt that its industrialisation will come soon.

As can be seen on fig. 5, the characteristic lateral growth length of the SLG regime is of the order of 2 μm (~ half the grain diameter), whereas as stated above, the thermal diffusion length in l-Si is below 1 μm . It is therefore tempting to induce lateral thermal gradients after melting, in order to control lateral solidification and hence the location of grain boundaries. Since grain boundaries are well known to affect the transport properties in TFTs, the control of their location should allow to fabricate devices in (pre-registered) monocrystalline regions and improve their characteristics [44]. Lateral thermal gradients can be established through the use of patterned antireflective stripes (of e.g. SiO_2 or Si_3N_4) which modulate the absorbed laser energy in the film. Using this technique, Kim and Im have obtained large elongated grains with almost parallel grain boundaries [45]. Although the above authors claim the novelty of the process, it must be emphasised that Colinge et al. have already used it in 1982 to grow silicon films on insulator with a cw argon laser [46].

Still building on the characteristic lateral growth length of the SLG regime, Im and Sposili have recently obtained high quality Si films by sequential lateral solidification [47]. They project a mask pattern on the substrate, consisting of an array of narrow stripes (~ 5 μm width), which allows parallel processing. Between two consecutive laser shots, the substrate is translated over a distance slightly smaller than the characteristic lateral growth length, so that each new molten stripe is laterally seeded on large grain material previously obtained. The crystallised films only exhibit grain boundaries parallel to the direction of translation. A second passage in a perpendicular direction would probably yield practically monocrystalline material.

V- THIN FILM TRANSISTORS

The TEM pictures of fig. 3 clearly show that the extended defects in SPC poly-Si films are not just located at grain boundaries. The situation is more favourable for laser-crystallised material, but still, there exits intra grain defects, in the form of twins, stacking faults and dislocations. Moreover, in either case, there are also point defects, corresponding to dangling bonds and distorted or reconstructed bonds, particularly in the regions of grain boundaries. All these defects, whether extended or punctual, induce a continuum of deep states in the energy gap of the semiconductor, which can be occupied by carriers, depending on the position of the Fermi level. Although the distribution of these states is not spatially uniform, because there are regions of good material adjacent to regions which are much more disorganised, it has been shown that the properties of poly-Si devices can be adequately described by assuming an effective density of states (DOS), uniform over the whole material, as in a-Si:H [48]. Measurements of the DOS can be performed by direct analysis of the field-effect characteristics of TFTs in the linear regime [48, 49, 50], or by analysis of the temperature dependence of the field effect [51]. The latter method is more accurate, but also more time consuming. Fig. 7 shows typical DOS distributions for SPC as well as for laser-crystallised poly-Si materials, measured by the Hirose method [49], using plasma-hydrogenated TFTs. Clearly, and as anticipated from the above remarks, the laser-crystallised material is of much better quality, since it exhibits more than one order of magnitude less defects around mid-gap than its SPC counterpart. Of course, this will induce profound effects, not only on carrier mobility, but also on the threshold voltages and sub-threshold slopes of TFTs, the latter properties being of prime importance for logic circuits. Finally we would like to point out that despite the differences in grain size that can be observed on fig. 3 for the different SPC materials (depending on the deposition and crystallisation conditions), the measured DOS after hydrogenation remain essentially the same for all the films, with a mid-gap density in the high 10^{17} cm^{-3}.eV^{-1}. This is consistent with the above observation that the defects are not only located at grain boundaries, but also within the grains. Note that before hydrogenation, the DOS value around mid-gap can be one order of magnitude higher (7 10^{18} to 10^{19} cm^{-3}.eV^{-1}).

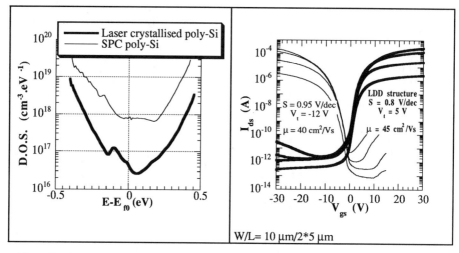

Fig. 7 : Density of states of laser crystallised and SPC poly-Si materials

Fig. 8 : Typical transfer characteristics of SPC transistors. Vds = ± 0.1; 1; 5 and 10 V

The most widely used TFT structure is the top gate, self-aligned type (see e.g. ref. 1), where the source and drain contacts are obtained by ion implantation, using the gate as a mask. This structure is also referred to as the coplanar type and is similar to that of the SOI MOSFET. The staggered type structure still has the gate electrode on top, but it uses highly doped deposited films for the source and drain (S/D) regions. Finally, the inverted staggered structure (or bottom gate structure) is identical to that of a-Si:H TFTs, with the gate electrode being deposited and patterned before the gate dielectric and active poly are laid down. Again here, the S/D regions can be doped by ion implantation, using for instance photoresist to mask the channel. Alternatively, they can be deposited and subsequently etched back over the channel. For this latter operation, an etch stopper previously deposited (SiO_2 or Si_3N_4) can be used, in order to avoid etching the channel, particularly if very thin poly-Si films are of concern. Note that the active poly-Si material (i.e. the channel material) is usually undoped.

When ion implantation is used for S/D doping, some kind of thermal treatment is needed to activate the dopants and cure the implantation damage. Laser activation is of course one possibility [52]. However, when SPC films are concerned, dopant activation is usually carried out thermally, around 600°C, in a furnace. This processing step can severely increase the thermal budget, particularly if classical implantation is used. Actually, for phosphorous implantation to the usual doses of a few 10^{15} cm^{-2}, the poly-Si films can be amorphised (depending on their thickness, on the implantation energy, on the TFT structure and on whether the implantation is carried out through the gate dielectric), and subsequent crystallisation must proceed through a nucleation step which is usually long, as already discussed in section III. Typically, 15 to 25 hours annealing may be needed, which, added to the crystallisation time, may induce an unacceptable deformation of the glass substrate. In order to overcome this problem, the implantation parameters must be carefully adjusted, so as to avoid the complete amorphisation of the film. In such conditions, nucleation is by-passed and crystallisation proceeds by solid phase "epitaxial" regrowth, which is quite fast, as regrowth rates between ~ 0.5 and 11 nm.min^{-1} have been measured at 550°C for crystal orientations between (111) and (100) [53]. Ion shower implantation machines have been developed (see e.g. ref. 52 and 54), which can work at very low energy (below 10 keV), thus avoiding amorphisation and even promoting some kind of beam-activated regrowth, because of a high flux of hydrogen ions due to the fact that PH$_3$ is used as source gas, without mass separation [55].

Fig. 8 shows the transfer characteristics of coplanar TFTs fabricated in SPC poly-Si material. For these particular devices, we have used a mixed phase precursor, deposited to a thickness of 120 nm by pyrolysis of disilane at 540°C and at a pressure of 2 mTorr. SPC was carried out at 580°C during 10 hours. Although 5 hours are sufficient to obtain full crystallisation (see section III and fig. 3d), we have shown in the past that a prolonged annealing resulted in improved structural as well as electrical properties for SPC poly-Si films [56]. The gate dielectric (SiO$_2$) was deposited (also to a thickness of 120 nm) by distributed electron cyclotron resonance (DECR) plasma enhanced CVD, at floating temperature, using a mixture of pure SiH$_4$ and O$_2$ [57]. The S/D regions were implanted through the oxide to doses of 2 10^{15} cm^{-2}, with energies of 90 keV for P (in such conditions, the bottom of the Si film is not amorphised) and 35 keV for B. For the n-type device, a lightly doped drain (LDD) structure was used, in order to reduce the well known exponential dependence of the leakage current on gate and drain bias (this dependence clearly appears on the p-type device). The implantation anneal was performed at 580°C during 8 hours. This is more than necessary, but here again, a prolonged anneal results in lower sheet resistivities for the S/D regions, together with slightly improved device performances. After passivation and metallisation, the devices were hydrogenated at 300°C during 2 hours in a PECVD-type plasma. Still for the n-type device, the apparent carrier mobility is lower than that we have obtained in the past (on a SPC poly-Si material similar to that of fig. 3b) [58], because of the series resistance effect due to the LDD regions. Actually, we emphasise that the various SPC materials shown in fig. 3 give very similar TFT properties (mobility, threshold voltage and subthreshold slope), which is consistent with the similarity of the DOS measurements quoted above.

Fig. 9 displays the transfer characteristics of coplanar TFTs fabricated in 80 nm-thick poly-Si films crystallised with a pulsed excimer laser (τ ~ 30 ns) in the SLG regime [59]. Crystallisation was performed in the scanning mode, with a beam overlap of 90% and a substrate temperature around 350°C. Again, the gate dielectric is SiO$_2$, deposited by DECR PECVD. Activation of the implanted dopants was performed at 580°C during 8 hours. The devices were hydrogenated in the same conditions as above. Clearly, the TFT characteristics are very good, with all parameters within less than a factor of two from their SOI counterparts [57, 60, 61]. Similar device properties have recently been reported by Kuriyama et al. [62], and Sameshima et al. [63].

For the sake of completeness, we show on fig. 10 the transfer characteristics of TFTs fabricated in poly-Si films irradiated with a single shot from a 45 J XeCl laser delivering pulses with 150 ns duration [64]. Compared to our former experiments with a similar laser [43], we have improved the mobility and subthreshold slope values by a factor of two. We emphasise that so far, the maximum grain size that we have obtained with such long pulses is relatively small (~ 500 nm) and that the SLG mechanism probably does not apply. However, the grain size is much more uniform.

We have not discussed here the reliability aspects of poly-Si TFTs and in particular hot carrier effects, which induce device instabilities. However, detail on this matter can be found in a recent review paper [65].

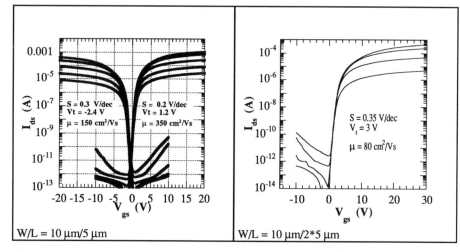

Fig. 9 : Transfer characteristics of TFTs crystallised with multishot laser irradiation (10 shots) Vds = ± 0.1, 0.3, 1,3 and 6 V

Fig. 10 : Transfer characteristics of a n-type TFT crystallised with single shot laser irradiation (45 J per pulse over ~ 100 cm^2). Vds = 0.1, 1, 5 and 10 V

VI- CONCLUSION

We have presented in this paper the two main processes used for the synthesis of polycrystalline silicon thin films, namely solid phase crystallisation and pulsed laser crystallisation of amorphous precursor deposits. SPC with a moderate thermal budget (18 hours at 580°C, if a mixed phase precursor is used) yields TFTs with transport characteristics far superior to their a-Si:H counterparts. However, the subthreshold slopes and threshold voltages remain high, which is a drawback for the fabrication of efficient logic circuits. On the other hand, pulsed laser crystallisation allows the fabrication of high quality TFTs, which have to be compared to their SOI counterparts rather than to a-Si:H devices. In spite of the various problems that are not yet solved, concerning uniformity, pulse to pulse reproducibility, surface roughness and also maintenance, there is little doubt that pulsed laser crystallisation will end up in polysilicon facilities, producing the required amount of plates per hour.

ACKNOWLEDGEMENTS

We would like to thank G. Fortunato and R. Carluccio of CNR in Roma, for the multishot laser irradiations as well as for numerous discussions. Thanks are also due to P. Boher of SOPRA for the single shot laser irradiations.

REFERENCES

1. See e. g. D. Pribat, F. Plais, P. Legagneux and C. Reita, in <u>Science and Technology of Thin Films,</u> edited by F.C. Matacotta and G. Ottaviani (World Scientific, 1995), p. 293.
2. S. Inoue, M. Matsuo, K. Kitawada, S. Takenada, S. Higashi, T. Ozawa, Y. Matsueda, T. Nakazawa and H. Ohshima, Proceedings of the 15th international Display Research Conference (Asia Display' 95), p. 339.
3. See e.g. J.R. Brews in <u>Silicon Integrated Circuits</u>, Part A, Applied Solid State Science, Supplement 2A, edited by Dawon Kahng (Academic Press, 1981), p. 1.

4. H.C. Lin, H.Y. Lin, C.Y. Chang, T.F. Lei, P.J. Wang and C.Y. Chao, Appl. Phys. Lett. **63**, p. 1351 (1993).
5. T.I. Kamins, M.M. Mandurah and K.C. Saraswat, J. Electrochem. Soc. **125**, p. 927 (1978).
6. G. Harbeke, L. Krausbauer, E.F. Steigmeier, A.E. Widmer, H.F. Kappert and G. Neugebauer, J. Electrochem. Soc. **131**, p. 675 (1984).
7. R. Bisaro, J. Magariño, N. Proust and K. Zellama, J. Appl. Phys. **59**, p. 1167 (1986).
8. D.B. Meakin, N.A. Economou, P.A. Coxon, J. Stoemenos, A. Lowe and P. Migliorato, Appl. Surf. Science **30**, p. 372 (1987).
9. T. Aoyama, G. Kawachi, N. Konishi, T. Susuki, S. Okajima and K. Miyata, J. Electrochem. Soc. **136**, p. 1169 (1989).
10. A.T. Voutsas and M.K. Hatalis, J. Electrochem. Soc. **139**, p. 2659 (1992).
11. T. Kretz, Thesis, University of Strasbourg, 1993.
12. M. Miyasaka, T. Nakazawa, W. Itoh, I.Yudasaka and H. Ohshima, J. Appl. Phys. **74**, p. 2870 (1993).
13. A.T. Voutsas and M.K. Hatalis, Appl. Phys. Lett. **63**, p. 1546 (1993).
14. A. Tanikawa and T. Tatsumi, J. Electrochem. Soc., p. 2848 (1994).
15. T. Kretz, D. Pribat, P. Legagneux, F. Plais, O. Huet and M. Magis, Mat. Res. Soc. Symp. Proc. **345**, p. 123 (1994).
16. P. Legagneux, unpublished.
17. A.T. Voutsas and M.K. Hatalis, J. Electrochem. Soc. **140**, p. 871 (1993).
18. C. H. Hong, C. Y. Park and H. J. Kim, J. Appl. Phys. **71**, p. 5427 (1992).
19. T. Kretz, D. Pribat, P. Legagneux, F. Plais, O. Huet and R. Bisaro, Jpn. J. Appl. Phys. **34**, L660 (1995).
20. R. Bisaro, J. Margarino, Y. Pastol, P. German and K. Zellama, Phys. Rev. B **40**, p. 7655 (1989).
21. R. Sinclair, J. Morgiel, A. S. Kirtikar, I. W.Wu and A. Chiang, Ultramicroscopy **51**, p. 41 (1993).
22. J. L. Maurice, D. Pribat, T. Kretz, P. Legagneux and F. Plais, to be published.
23. F. Spaepen and D. Turnbull in <u>Laser Annealing of semiconductors,</u> edited by J. M. Poate and J. W. Mayer (Academic Press, 1982), p. 15.
24. R. Drosd and J. Washburn, J. Appl. Phys. **53**, p. 397 (1982).
25. F. Spaepen and D. Turnbull in <u>Laser-Solid Interactions and Laser Processing</u>, edited by S. D. Ferris, H. J. Leamy and J. M. Poate (American Institute of physics, 1979) p. 73.
26. J.S. Williams and R.G. Elliman, Phys. Rev. Lett. **51**, p. 1069 (1983).
27. G. Q. Lu, E. Nygren and M. Aziz, J. Appl. Phys. **70**, p. 5323 (1991).
28. T. Kretz, R. Stroh, P. Legagneux, O. Huet, M. Magis and D. Pribat, in <u>Solid State Phenomena</u>, Vol. 37-38, edited by H. P. Strunk, J. H. Werner, B. Fortin and O. Bonnaud (Scitec Publications, 1994), p. 311.
29. J.S. Lannin, Physics Today, July 1988, p. 28.
30. D. Beeman, R. Tsu and M. F. Thorpe, Phys. Rev. B **32**, p. 874 (1985).
31. W. C. Sinke, S. Roorda and F. W. Saris, J. Mater. Res. **3**, p. 1201 (1988).
32. J. F. Morhange and T. Kretz, unpublished.
33. E. I. Shtyrkov, I. B. Khaibullin, M. M. Zaripov, M. F. Galyatudinov and R. M. Bayazitov, Sov. Phys. Semicond. **9**, p. 1309 (1976).
34. see e.g. R. F. Wood and G. E. Jellison, in <u>Pulsed Laser Processing of Semiconductors,</u> Semiconductors and Semimetals, Vol. 23, edited by. R. F. Wood, C. W. White and R. T. Young (Academic Press, 1984), p. 165.
35. T. Sameshima, M. Hara and S. Usui, Jpn. J. Appl. Phys. **28**, L2131 (1989).
36. M. O. Thompson, G. J. Galvin, J. W. Mayer, P. S. Peercy, J. M. Poate, D.C. Jacobson, A. G. Cullis and N. G. Chew, Phys. Rev. Lett. **52**, p. 2360 (1984).
37. D. H. Lowndes, S. J. Pennycook, G. E. Jellison, S. P. Withrow and D. N. Mashburn, J. Mater. Res. **2**, p. 648 (1987).
38. D. H. Lowndes, S. J. Pennycook, R. F. Wood, G. E. Jellison, and S. P. Withrow, Mat. Res. Soc. Symp. Proc. **100**, p. 489 (1988); R. F. Wood and G. A. Geist, Phys. Rev. Lett. **57**, p. 873 (1986).

39. K. Murakami, O. Eryu, K. Takita and K. Masuda, Phys. Rev. Lett. **59**, p. 2203 (1987).
40. S. R. Stiffler, M. Thompson and P. S. Peercy, Phys. Rev. Lett. **60**, p. 2519 (1988).
41. H. J. Kim, J. S. Im and M. O. Thompson, Mat. Res. Soc. Symp. Proc. **283**, p. 703 (1993).
42. J. S. Im and H. J. Kim, Appl. Phys. Lett. **64**, p. 2303 (1994).
43. F. Plais, P. Legagneux, C. Reita, O. Huet, F. Petinot, D. Pribat, B. Godart, M. Stehle and E. Fogarassy, Microelect. Eng. **28**, p. 443 (1995).
44. see e.g. N. Yamauchi, J. J. J. Hajjar and R. Reif, IEEE Trans. Electron. Dev. ED **38**, p. 55 (1991).
45. H. J. Kim and J. S. Im, Mat. Res. Soc. Symp. Proc. **358**, p. 903 (1995).
46. J. P. Colinge, E. Demoulin, B. Bensahel and G. Auvert, Appl. Phys. Lett. **41**, p. 346 (1982).
47. J. S. Im and R. S. Sposili, (MRS Bulletin, March 1996), p. 39.
48. G. Fortunato and P. Migliorato, Appl. Phys. Lett. **49**, p. 1025 (1986).
49. M. Hirose, M. Tanaguchi and Y. Osaka, J. Appl. Phys. **50**, p. 377 (1979).
50. S. Hirae, M. Hirose and Y. Osaka, J. Appl. Phys. **51**, p. 1043 (1980).
51. G. Fortunato, D. B. Meakin, P. Migliorato and P. Lecomber, Phil. Mag. B **5**, p. 573 (1988).
52. A. Mimura, G. Kawachi, T. Aoyama, T. Suzuki, Y. Nagae, N. Konishi and Y. Mochizuki, IEEE Trans. Elect. Dev. **40**, p 513 (1993); also see A. Mimura, this conference.
53. L. Csepregi, E. F. Kennedy, J. W. Mayer and T.W. Sigmon, J. Appl. Phys. **49**, p. 3906 (1978).
54. K. Mamuso, M. Kunigita, S. Takafuji, N. Nakamura, A. Iwasaki and M. Yuki, Jpn. J. Appl. Phys. **29**, L2377 (1990).
55. A. Yoshinouchi, A. Oda, Y. Murata, T. Morita and S. Tsuchimoto, Jpn. J. Appl. Phys. **33**, p. 4833 (1994).
56. J. L. Maurice, J. Dixmier, T. Kretz, P. Legagneux, F. Plais and D. Pribat in Solid State Phenomena, Vol. 37-38, edited by H. P. Strunk, J. H. Werner, B. Fortin and O. Bonnaud (Scitec Publications, 1994), p. 335.
57. N. Jiang, M.C. Hugon, B. Agius, T. Kretz, F. Plais, D. Pribat, T. Carriere and M. Puech, Jpn. J. Appl. Phys. **31**, L1404 (1992).
58. F. Plais, P. Legagneux, T. Kretz, R. Stroh, O. Huet and D. Pribat, Mat. Res. Soc. Symp. Proc. **345**, p. 87 (1994).
59. P. Legagneux, F. Petinot, O. Huet, F. Plais, C. Reita, D. Pribat, R. Carluccio, A. Pecora, L. Mariucci and G. Fortunato, submitted for publication.
60. F. Plais, O. Huet, P. Legagneux, D. Pribat, A.J. Auberton-Hervé, and T. Barge, paper presented at the IEEE International SOI Conference (Tucson, October 3-5, 1995), to be published.
61. K. R. Sarma, C. Rogers and C. Chanley, SID '94 Digest , p. 419 (1994).
62. H. Kuriyama, T. Nohda, Y. Aya, T. Kuwahara, K. Wakisaka, S. Kiyama and S. Tsuda, Jpn. J. Appl. Phys. **33**, p. 5657 (1994).
63. T. Sameshima, A. Kohno, M. Sekiya, M. Hara and N. Sano, Appl. Phys. Lett. **64**, p. 1018 (1994).
64. We have used an experimental system manufactured by Sopra and consisting of three coupled 15 J lasers (Sopra : 28, rue Pierre Joigneaux, 92270 Bois-Colombes, France).
65. G. Fortunato, paper presented at POLYSE'95 : Polycrystalline Semiconductors - Physics, Chemistry and Technology, September '95, Gargano, Italy, to be published.

LOW ENERGY ION DOPING TECHNOLOGY
FOR POLY-Si TFTS

A. Mimura, M. Nagai, Y. Shinagawa

Hitachi Research Laboratory, Hitachi Ltd.
7-1-1 Omika-cho, Hitachi-shi, Ibaraki-ken, 319-12 Japan
E-mail : amimura@hrl.hitachi.co.jp

ABSTRACT

Ion doping technology using a bucket-type ion source to fabricate low temperature poly-Si TFTs is presented. Low energy and high ion density conditions are applied to realize the short doping time and resist mask doping for large area glass substrates. A novel CMOS doping technology with one resist mask process is proposed using compensation and inversion of the shollow doped layer. In combination with excimer laser crystallization and activation below 450 ℃, high performance CMOS TFTs are produced.

INTRODUCTION

Low-temperature poly-Si TFTs offer a promising route to realizing large area microelectronics devices. In particular, with their high performance, poly-Si TFTs may be used for novel system integration on display panels. Producing low-temperature poly-Si TFTs on glass substrates, however, takes additional technologies compare to the a-Si TFT process, including laser crystallization, gate insulator SiO_2 deposition, and ion doping and activation[1].

Low-temperature crystallization of Si films by excimer laser has made great progress and high performance poly-Si TFTs have been developed. During the first stage of research, small sized square beams were used. By optimizing the conditions of Si film deposition, Si film thickness, laser irradiation conditions and substrate temperature, electron field effect mobilities of 440-640 cm^2/Vs were achieved[2][3]. Recently, linear beams with lengths of 150-360 mm were developed for high through put annealing[4][5]. High quality gate insulator SiO_2 films with good step coverage was deposited by PECVD in the same way as the a-Si TFT process using a TEOS-O_2 system[1].

Non-mass-separation ion doping was also developed for large glass substrates. Mass production machines with an RF ion source are applied for large glass substrates. Activation is carried out by self-activation or furnace anneling.

An application of a bucket-type ion source for ion doping and laser activation at room temperature for large glass substrates has also been put forth[6][7]. This doping technology has been utilized to realize large area poy-Si TFT substrates below 450 ℃[1].

In this paper, low energy and high speed ion doping technology with a high ion density ion source and excimer laser activation are reviewed and discussed. Fundamental characteristics of low energy doping below 5 kV and laser activation are described. To realize resist mask doping for CMOS coplanar TFTs, doping conditons are optimized. Novel technologies for a one mask CMOS doping is tried using compensation and inversion of a shallow doped layer.

EXPERIMENTS AND RESULTS

Low Energy Ion Doping and Laser Activation

Fig.1 shows the TFT structures during ion doping. Today, to get fine poly-Si TFT-LCDs, ion implantation is utilized to penetrate the gate insulator by applying a high acceleration voltage such as 80 kV (Fig.1(a)). In this process, the glass substrate temperature is raised by ion beam energy. Substrate cooling or moderate long time doping with a low ion beam density is necessary. A high heat-resisting polyimide mask is also applied. In this paper, direct doping with a low acceleration voltage and high ion beam density is carried out for a bare source, drain and gate Si layers to realize photoresist mask doping (Fig.1(b)).

Fig.2 plots sheet resistance dependence on doping time, doping gas, and activation methods for Si films doped by using a small size doping apparatus with a 160 mm diameter ion source[6]. In ion implantation at 30 kV, activation was possible by thermal annealing at 600 °C. But in ion doping with low energy, the dopant was not sufficiently activated and laser activation at room temperature was introduced. In the first stage, hydrogen diluted PH_3 gas was used for dopant gas. But too long doping time caused an increase in sheet resistance due to Si film etching by the hydrogen gas. Then, helium diluted dopant gas was selected. Sheet resistance below 1000 $\Omega/\square$ was obtained at a doping time of 10 s and the TFTs operated successfully[6]. Thus, the possibility of a very short doping time was shown using bucket-type ion source.

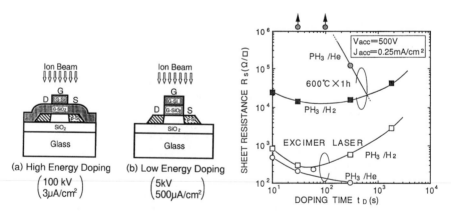

Fig.1 TFT Structures in Two Doping Processes

Fig.2 Effect of Doping Time on Sheet Resistance

Fig.3 shows a schematic side view of the doping apparatus developed for the 350mm by 450mm glass substrates. The arc discharge chamber consists of two tungsten filaments for thermal electron source, permanent magnets, and three-layer ion extraction electrodes. A high-density and uniform discharge plasma is formed by thermally emitted electrons and the line cusp field of the permanet magnets. The ion beam is extracted from the arc discharge chamber and accelerated and irradiate onto a glass substrate without mass-separation. The bucket ion source with a size of 100mm by 400mm was attached on the bottom of the doping chamber. The glass substrate can be transported horizontally in its longitudinal direction through the beam. It is supported by an aluminum holder and irradiated from below.

Table 1 lists the doping conditions for phosphorus and boron. The bucket-type ion source realizes a very high ion density 0.5 mA/cm^2 which is about two orders of magnitude higher than that of a conventional ion doping apparatus[8]. Doping time is defined as the time required to pass a 100mm length through the ion source. It is controlled by the substrate transportation speed. Table 2 lists laser activation conditions. To realize low temperature activation and low sheet resistance, XeCl excimer laser annealing is carried out in the vacuum chamber at room temperature.

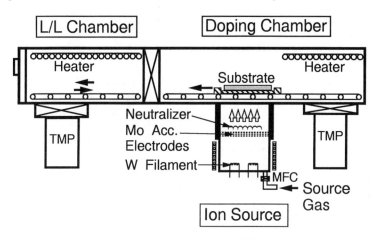

Fig.3 Side View of Ion Doping Apparatus for
Large Glass Substrates.
Ion Beam Size is 100mm × 400mm

Table 1 Ion Doping Conditions	
Ion beam size	100mm × 400mm
Ion beam density	0.5 mA/cm^2
Acceleration voltage	1.0 - 5.0 kV
Source gas	PH3, B2H6(1%)/He
Irrdiation time	6 - 45 s

Table 2 Laser Activation Conditions	
Laser energy	150 - 250 mJ/cm^2
Beam Size	8mm × 8mm
Shot number	10 - 40 shots
Sub. temperature	25℃
Pressure	2.7Pa

Fig.4 plots ion beam intensity and sheet resistance profiles in the substrate lateral direction. The area of high ion beam intensity indicated a rather low sheet resistance because of the high phosphorus content. A low enough sheet resistance below 1000 $\Omega/\square$ was obtained for 6, 10, 45 s doping times. The sheet resistance profile in the substrate longitudinal direction (substrate transportation direction) was very uniform indicating that very uniform doping was possible when ion beam and substrate transportation were stable[9].

Doping Mechanism

Fig.5 plots the concentrations of phosphorus before and after laser activation for doping of 1 kV for 45 s and 5 kV for 6 s. In the former doping, the as-doped profile showed a very high surface concentration was achieved. When the activation laser power was increased, the profile became lower and flatter, and total phosphorus content decreased slightly. The results implied there was evaporation of phosphorus by high energy laser irradiation. A high concentration profile was still obtained in the 5 kV, 6 s doping. Total content was almost constant after laser activation because the projection range was deep and evaporation was supressed.

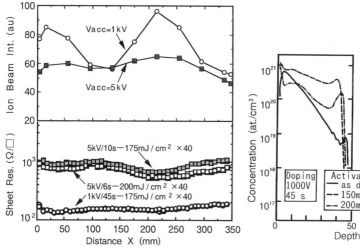

Fig.4 Ion Beam Intensity Profiles and
Sheet Resistance of Phosphorus
Doped Si Films

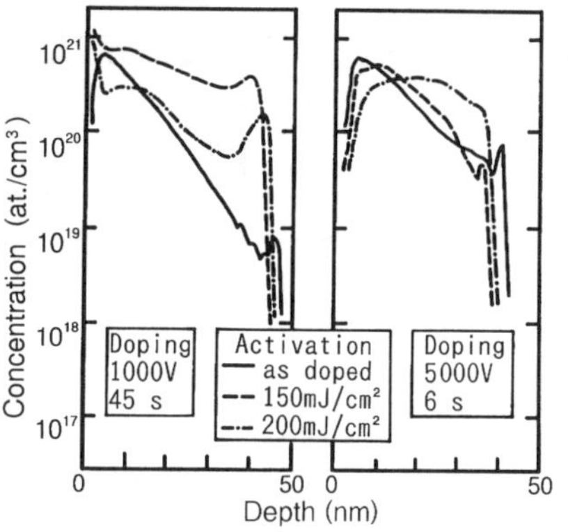

Fig.5 Phosphorus Concentration
Profiles measured by SIMS

For a very low energy, long time doping (1 kV, 45 s), partial peeling of the thin deposited layer was observed by SEM. The peeled layer may be the deposited phosphorus layer or the heavily doped silicon layer. For slightly higher energy and shorter time doping (5 kV, 6 s), peeling was not observed. Only riges from melted Si appeared[10].

Fig.6(a) shows XPS profiles of phosphorus and Si in as-doped Si film. Binding energy profiles were measured by step etching with low energy Ar ions. Etching depth after 1 min was about 1 nm. Phosphorus and its oxide were clearly observed. This indicated the presence of a deposited thin phosphorus layer and surface oxidation. The results implied that the Si was excessively doped and there was a possibility that a shorter doping time would reduce deposition. SiO_2 was also observed, indicating the undamaged natural oxide. As in Fig.6(b) after laser activation, only a trace of phosphorus oxide was observed and the natural silicon oxide was almost all destroyed. This indicated that the deposited phosphorus was diffused into the Si and partially evaporated. In the 5 kV, 6 s doping, XPS analysis showed that the amout of deposited phosphorus decreased and natural oxide was almost completely destroyed.

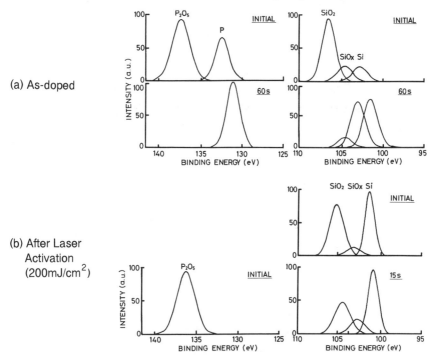

(a) As-doped

(b) After Laser
Activation
($200mJ/cm^2$)

Fig.6 XPS Profiles of Phosphorus and Si
(Doping : 1000V, 45s)

It is important to understand the doping mechanism to optimize doping conditions. Fig.7 illustrates a doping model for low energy (1 kV) ion doping and laser activation. After ion doping, the thin phosphorus layer is deposited and natural oxide partialy remains. The deposited phosphorus layer acts as a diffusion source during laser activation. Part of the phosphorus is evaporated. It appears that removing the natural oxide before doping promotes effective diffusion of phosphorus into Si. A slightly higher energy (5 kV) doping eliminates the natural oxide so that doping is not affected by the natural oxide.

Fig.8 shows the effect of surface cleaning before ion doping (1 kV) on sheet resistance. Chemical etching using HF solution slightly reduced sheet resistance. After wet cleaning, natural oxide should be produced again before sample loading into the doping chamber. Ion beam surface treatment followed by ion doping in the same doping chamber greatly reduced the sheet resistance. Hydrogen ion beam irradiation clearly reduced the sheet resistance implying the effect of both reduction and ion bombardment. Helium ion irradiation was more effective to increase doping efficiency and to reduce sheet resistance. Low energy helium ion beam should have a moderate ability to remove natural oxide.

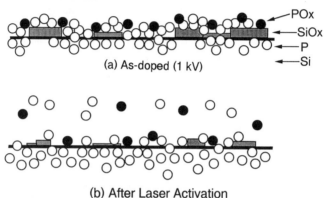

(a) As-doped (1 kV)

(b) After Laser Activation

Fig.7 Schematic Model of Low Energy Doping and Laser Activation

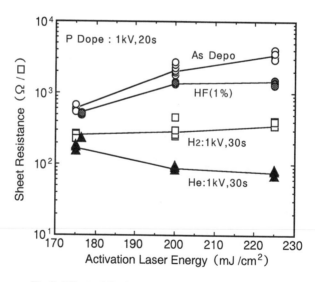

Fig.8 Effect of Surface Cleaning before Ion Doping
on Sheet Resistance

Photoresist Mask Doping

In this paper, resist mask doping for CMOS TFTs was studied. A slightly higher energy (5 kv) doping is necessary to avoid reduction of surface concentration of dopant in the resist removing process. Fig.9 shows the surfaces of a resist mask after 5 kV, 10s doping and resist removing process. There was no damge in the resist for 5 kV doping of phosphorus or boron, and the resist was successfully removed using the conventional O2 ashing and wet process.

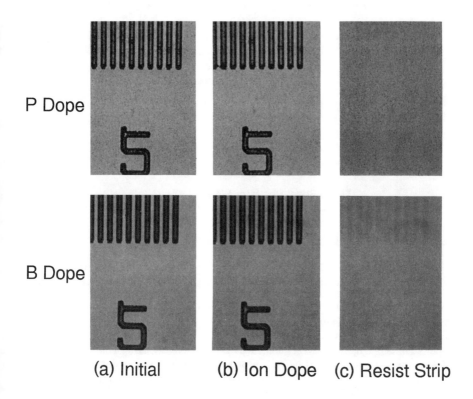

(a) Initial   (b) Ion Dope   (c) Resist Strip

Fig.9 Resist Mask Doping at 5 kV for 10s

Fig.10 plots sheet resistance of Si films doped by using the resist mask process with the 5 kV short time doping. A low enough sheet resistance below 1000 $\Omega/\square$ was obtained by optimizing the activation laser energy. It was confirmed that 6 s doping was sufficient to operate poly-Si TFTs[1]. Thus, resist mask short time doping for the CMOS process is possible.

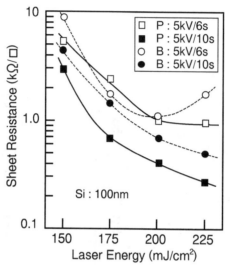

Fig.10 Sheet Resistance in Resist Mask
Doping Process at 5 kV.

Application to CMOS Poly-Si TFTs

Fig.11 illustrates the coplanar CMOS TFT process on 100mmx100mm glass substrates using the conventional two mask doping process and the novel one mask doping process. In the two mask doping process, first a 50 nm thick Si film was deposited by LPCVD at 450°C using Si_2H_6 on the underlayer PECVD SiO_2. After laser crystallization using XeCl excimer laser with a energy of 400 mJ/cm^2, poly-Si film was patterned into islands. A gate insulator of SiO_2 and gate poly-Si with thickness of 100 nm were deposited by PECVD and LPCVD, respectively. After gate patterning, the p-channel region was protected by photoresist and phosphorus was doped at 5 kV for 10 s in the n-channel region. After removing the resist, the n-channel region was covered by resist and boron was doped into the p-channel region. Laser activation was carried out at 200 mJ/nm^2 after removing the resist. Hydrogenation was done at 300 °C for 60 min after passivation by PECVD SiO_2 and metallization.

In the one mask CMOS doping process, first helium ions were irradiated at 1 kV, for 30 s to remove natural oxide and phosphorus was continuously doped at 1 kV for 20 s. Then the n-channel region was protected by resist. Oxgen ashing was applied for the p-channel region for 6 min in the RIE machine to remove the high surface concentration layer. Then boron was doped at 5 kV for 20 s to invert the doped area from n-type to p-type. Reduction of the phosphorus concentration by RIE ashing completed the inversion.

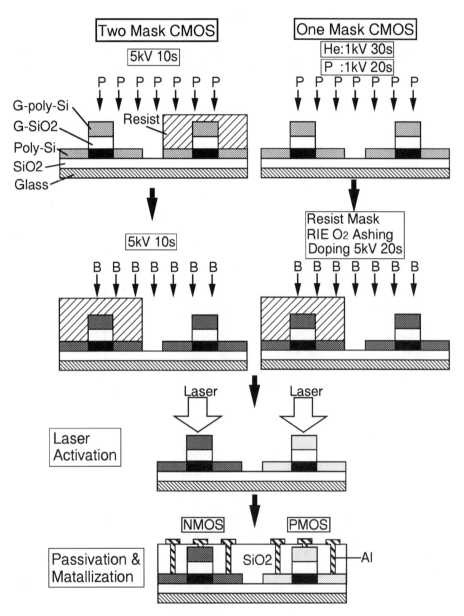

Fig 11. Two Mask and One Mask Doping Processes
for CMOS TFTs

Fig.12 shows the Vg-Id characterics of TFTs fabricated by one and two mask CMOS doping process. Both one mask and two mask processed TFTs had high peformance characteristics, though mobility was slightly lower for the p-channel TFT and the off current was slightly higher for n-channel TFT produced by the one mask process. Thus, the novel one mask doping is a promising way to simplify the CMOS doping process using a large area high speed ion doping machine.

In devices fabricated using this resist mask doping process, it was confirmed that off-set gate structure reduced off current effectively and the 21-step CMOS ring oscillator operated at 5.2 MHz at Vdd of 18 V. These results also showed a possibility of getting driver integartion and low off curent pixel TFTs in low-temperature large area poly-Si TFT-LCDs.

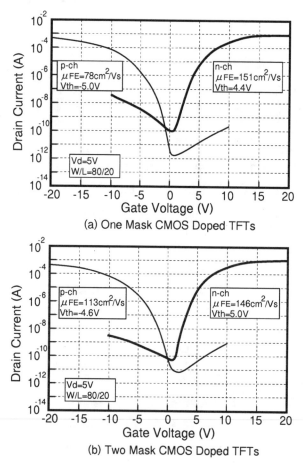

(a) One Mask CMOS Doped TFTs

(b) Two Mask CMOS Doped TFTs

Fig. 12 Vg-Id Characteristics of CMOS TFTs
by One and Two Mask Doping Processes

CONCLUSIONS

Technology for low-energy ion doping with non-mass-separation and laser activation for low temperature poly-Si TFTs was developed. The bucket-type ion source realized very high ion density. Short, 6-10 s doping times at 1-5 kV were applicable to producing source and drain of the TFTs. These moderate doping conditions made resist mask doping possible for CMOS TFTs. Surface cleaning with low energy helium ions before ion doping was employed to improve doping efficiency. A novel CMOS doping technology with a one resist mask process was also developed using conpensation and inversion of the shallow doped layer. High performance poly-Si TFTs were obtained by using ion doping in conbination with laser crystallization and laser activation below 450°C. These technologies appear promising for realizing low-temperature poly-Si TFT-LCDs on large glass substrates.

ACKNOWLEDEMENTS

The authors would like to thank Mr. I. Ikuta and Mr. H. Suga for thier help in TFT device fabrication, and Mr. N. Konishi and Mr. K. Kinugawa for their support of this work.

REFERNCES

[1] A. Mimura, Y. Mikami, K. Kuwabara, Y. Mori, M. Nagai, Y. Nagae, and E, Kaneko, Proc. of 1994 Int. Display Res. Conf., p.126(1994)
[2] H. Kuriyama, T. Nohda, Y. Aya, T. Kuwahara, K. Wakisaka, S. Kiyama, and S. Tsuda, Jpn. J. Appl. Phys., 33, p.5657(1994)
[3] A. Kohono, T, Sameshima, N. Sano, M. Sekiya, and M. Hara, IEEE Trans. Electron Devices, 42, p.251(1995)
[4] S. Chen, Y. H. Sun, P. Mei, and J. B. Byce, Proc. The 2nd Int. Display Workshop, p.15(1995)
[5] H. Hayashi, M. Kunii, N. Suzuki, Y. Kanaya, M. Kuki, M. Minegishi, T. Urazone, M. Fijino, T. Noguchi, and Kanagawa, IEDM 95 Digest, p.829(1995)
[6] A. Mimura, G. Kawachi, T. Aoyama., T. Suzuki, Y. Nagae, N. Konishi, and Y. Mochizuki, IEEE Trans. Electron Devices, 40, p.513(1993)
[7] G. Kawachi, T. Aoyama, T. Suzuki, A. Mimura, Y. Ohno, N. Konishi, Y. Mochizuki, and K. Miyata, Jpn. J. Appl. Phys., p.L2370(1990)

[8] H. Ohshima, T. Hashizume, M. Matsuo, S. Inoue, and T, Nakazawa, Proc. of 1993 SID Int. Symp. p.387(1993)
[9] A. Mimura, M. Nagai, Y. Mimkami, K. Kuwabara, Y. Mori, Y. Nagae, and E. Kaneka, Proc. Int. Symp. Sputtering and Plasma Processes, p.293(1994)
[10] A. Mimura, M. Nagai, Y. Mikami, K. Kuwabara, Y. Mori, Y. Nagae, and E. Kaneko, Proc. 1995 Int. Display Res. Conf., p.329(1995)

ELECTRICAL PERFORMANCE OF AN N-CHANNEL TFT
FABRICATED ON A QUARTZ SUBSTRATE BY ZMR

Si-Woo Lee[#], Seung-Ki Joo, Byung-Il Lee and In-Gon Lim[*]
Electronic Material Laboratory, Dept. of Metall. Eng., Seoul Nat'l Univ., Seoul, 151-742,
*Device and Material Laboratory, LG Electronics Research Center, Seoul, 137-140
#Current Address : Basic Research Team, Samsung Electronic Co. Ltd., 449-900

ABSTRACT

Single crystalline silicon thin films were formed on a transparent quartz substrate by zone melting recrystallization for the application of flat panel display. The recrystallized film shows a perfect (100) texture and the main defects are subgrainboundaries. High residual tensile stress, which can cause the film cracking, can be successfully eliminated by post-annealing. The n-channel thin film transistors fabricated on a recrystallized film showed excellent device characteristics : carrier mobility ~ 420cm^2/Vsec, subthreshold slope ~ 100mV/dec and off-state leakage current < 0.2pA.

INTRODUCTION

LCD(Liquid Crystal Display) is widely used for the large area FPD(Flat Panel Display) panel such as a notebook computer or a slim type portable TV. It consists of a-Si (amorphous silicon) TFT's (Thin Film Transistor) on a glass substrate as pixel transistors and external driving circuits. A-Si, however, has so low carrier mobility about 1cm^2/Vsec and driving current that it has a limitation for the application of large area, high resolution and high contrast panels such as engineering workstations. P-Si (poly-crystalline silicon) TFT can be a candidate for a large area display material because it has a relatively higher mobility (~100cm^2/Vsec) than a-Si TFT so that the driving circuit TFT's can be integrated on peripheries of a panel simultaneously, as well as enhanced contrast and resolution. However, it shows higher off-state leakage current than a-Si TFT, which is known to be due to the electron-hole pair generation at the grain boundaries[1]. It needs higher process temperature than a-Si TFT.

On the other hand, x-Si (single crystalline silicon) shows very high carrier mobility as well as very low leakage current compared with a-Si or p-Si since it has no defects for leakage source. So, it can be applicable for both pixel switching TFT and peripheral driving circuit TFT. Since device size can be significantly reduced to few μm high contrast, high resolution and high aperture ratio can be achievable, which is applicable for HMD (Head Mounted Display), projection TV and view finder of the portable video recorder. However, it is difficult to fabricated a single crystalline silicon film on a transparent amorphous substrate.

Recently, some approaches, such as silicon-on-glass[2] or wafer bonding and etch-back[3], have been demonstrated. In silicon-on-glass process, panel devices are formed on a conventional SOI(Silicon-on-Insulator) wafer and transferred to a glass substrate by separation of

TFT film from the wafer. 1.7-inch diagonal projection panel has been demonstrate. In wafer bonding and etch-back process, single crystalline wafer is bonded with the quartz substrate at high temperature and etched back to desired thickness. The fabricated TFT showed high carrier mobility of 660cm^2/Vsec for n-channel TFT and 220cm^2/Vsec for p-channel TFT. These processes are rather complex and have a drawback in yield.

ZMR process had been studied in early days to fabricate SOI wafers[4]. It has been known to be a relatively simple process with reasonable crystalline quality.

In this work, a fabrication of SOQ (Silicon-on-Quartz) structured single crystalline silicon thin film by conventional ZMR (Zone Melting Recrystallization) process has been proposed. It is a very simple process and expected to be applicable to high performance display panel, especially a projection panel. The electrical properties of the x-Si TFT's fabricated on a recrystallized film on a quartz have been characterized. To enhance the electrical properties of the film, additional heat treatment process after ZMR has been proposed.

EXPERIMENT

Preparation of SOQ specimen

Figure 1 shows the structure of SOQ specimen. 0.5μm thick a-Si film was deposited on a 6-inch quartz wafer by LPCVD (Low Pressure Chemical Vapor Deposition) using Si$_2$H$_6$ at 485°C, followed by 1.0μm thick LTO (Low Temperature Oxide) deposition for a capping material. Borons were ion-implanted with a dose of 1×10^{13}/cm^2 and accelerating voltage of 80keV to form a p-type substrate doping for n-channel device. The doping concentration is 2×10^{17}/cm^3 assuming that the implanted dopants are distributed uniformly along the film after ZMR. The SOQ wafers were annealed at 900°C, N$_2$ ambient for 30 minutes to improve the mechanical property of LTO capping. It is because the deposited LTO film is too weak to protect the silicon film during the thermal cycles.

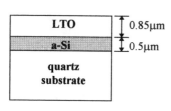

Figure 1. Structure of SOQ specimen.

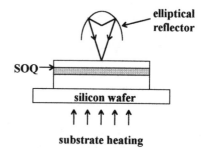

Figure 2. Schematic diagram of SOQ-ZMR. Dummy wafer is a heat susceptor.

ZMR process

The experimental apparatus used in this work consists of two reflectors using halogen lamps as heat sources. One is upper focused elliptical reflector to form a narrow molten zone. The other is lower planar reflector using lamp array for uniform substrate heating. The SOQ sample was heated on a dummy silicon wafer because it cannot absorb the radiation energy efficiently due to its semi-transparency (figure 2). The dummy wafer also prevents the warpage of an SOQ wafer during ZMR. The main parameters were restricted to upper focused lamp power and scanning speed. The solidifying interface morphologies during ZMR can be observed by interface demarcation method. Post-annealing was performed at 1000°C in N_2 ambient for 2~3 hours and cooled down slowly in tube to reduce the stress after ZMR. The textures and the residual stresses were analyzed by XRD (X-Ray Diffraction) and Raman Spectroscopy.

Fabrication of SOQ TFT

N-channel TFT's were fabricated on a recrystallized film to estimate the crystalline quality. 0.5 μm thick recrystallized film is thinned to about 137nm thickness, then the active regions were patterned by island isolation method. 40nm thick gate oxide was formed by dry oxidation, followed by undoped LPCVD p-Si gate formation. Source, drain and gate were doped by solid diffusion source(P_2O_5 glass) in a diffusion tube. Devices were fabricated also on a non-annealed films to compare the effect of the annealing on device characteristics.

RESULTS

SOQ-ZMR

Typical solidifying interface morphologies are shown in figure 3. In cellular interface (a), cells are running parallel to the scanning direction with defect trails formed after the interior corners of the interface[5]. The observed defects are mainly SGB's (Sub Grain Boundary) and isolated threading dislocations. In dendritic interface(b), dendrites with secondary arms are observed. The defect bands, which consist of SGB's with side branches formed by impingement of neighboring secondary arms, are the main defect shape. In this case the hill-rocks are observed, which are originated from the expansion of the liquid silicon droplets captured in solid silicon during ZMR. These are not desirable for device operation because they can cause the reliability problem especially at the gate oxide. These dendritic interfaces are observed frequently at low

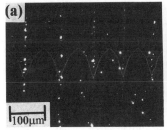 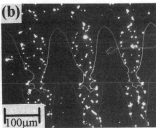

Figure 3. Observed solidifying interface morphologies, (a) cellular interface and (b) dendritic interface.

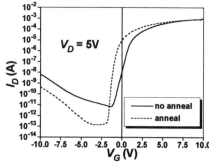

Figure 4. I_D-V_G transfer curves of n-channel TFT fabricated on non-annealed(solid) and annealed(dotted) silicon film. Film thickness, gate length and width are 137nm, 15μm and 15μm, respectively.

power density because of high supercooling at solidifying interface. It is desirable to fabricate silicon film with cellular interface for its low defect density.

The recrystallized film shows perfect (100) texture as single crystalline wafer, while fresh film shows a random crystal orientation. It is advantageous for MOS device to have (100) texture because of high surface mobility and low density of interface state. So, the development of (100) texture even without seeding is desirable for the FPD application.

The estimation of residual tensile stress is important because it can cause the film cracking during the following device fabrication steps. The stress is originated from the difference of thermal expansion coefficient between silicon (2.5×10^{-6}) and quartz($(2.5 \sim 5) \times 10^{-7}$). It should be relaxed as possible for the FPD application. The residual tensile stress was measured from the Raman peak shift. Large tensile stresses were measured for both SOI and SOQ specimen Before ZMR. However, the stress becomes very small for SOI film while it is nearly unchanged in SOQ film after ZMR, which is due to the difference of specimen structure. In case of post-annealed film, the stress is significantly reduced to half of its original value and the crack formation is greatly suppressed. So the post-annealing is essential for ZMR of SOQ structure to get a device quality silicon film.

SOQ-TFT

Figure 4 shows I_D-V_G curve for n-channel TFT fabricated on a recrystallized film for both ZMR only(solid) and ZMR with additional post-annealing(dotted). The gate length and width are 15μm and 15μm, respectively. In annealed film, the device properties are greatly enhanced compared with the non-annealed film : subthreshold swing from 317mV/dec to 101mV/dec, off-state leakage current from 7pA to 0.15pA for V_D = 5V. The electron mobility is about 420cm^2/Vsec for both case regardless of annealing. It is expected that the post-annealing heals the trap sites at silicon/quartz interface as well as relaxes the tensile stress.

Current stress effect

To analyze the effect of post-annealing, back channel current stress was forced and the electrical properties were measured again. V_G and V_D were fixed to -10V and 8V, respectively, The forcing time was 20 seconds and during the stress leakage current flows mainly along the

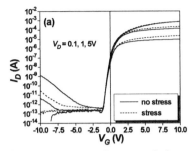 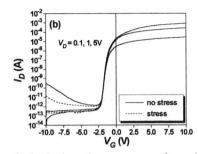

Figure 5. I_D-V_G transfer curves before and after the back channel current stress for n-channel TFT fabricated on (a) non-annealed and (b) annealed silicon film.

back channel with current level of $10^{-9} \sim 10^{-8}$A. Figure 5 shows I_D-V_G curves before and after the current stress for both non-annealed and annealed film. In case of non-annealed film(a), off-state leakage current increases as V_G goes to negative even at $V_D = 1$V, which is expected to be related with the back channel traps. After current stress, however, the leakage current for $V_D = 1$V is completely disappeared and it is reduced by 2 orders of magnitude in case of $V_D = 5$V. In post-annealed film, the subthreshold swing is improved and the leakage current about 0.2pA is observed even at $V_D=1$V for both stressed and non-stressed TFT. The off-state leakage current seems to be due to the electron-hole pair generation at defects or traps at unstable interface like silicon/quartz interface. In single crystalline silicon there are no defects, so the silicon/quartz interface seems to be the dominant source of leakage. It turns out that the post-annealing can stabilize the silicon/quartz interface, reduce the leakage currents as well as reduce the residual tensile stress.

Defects in channel

The defects in channel region can alter the electrical properties such as mobility or leakage current. In ZMR process, the defect formation is unavoidable because the defects are formed after the cusps in non-planar interface. So, it is important to examine the effects of defects on TFT characteristics. The main defects observed in ZMR are SGB's and isolated threading dislocations. Figure 6(a) shows I_D-V_G curves for the devices with and without SGB's running parallel to channel. No noticeable difference is found in transfer characteristics for 15μm and even for 5μm gate length, which shows that the SGB's in channel region can be electrically negligible. However, it turns out that the twin boundaries in channel degrade the leakage current characteristics(b). Fortunately, twin boundaries are not observed at normal ZMR conditions, but they may be generated .under large supercooling.

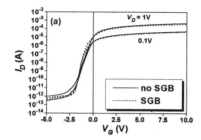 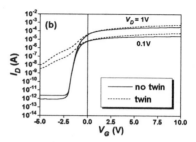

Figure 6. I_D-V_G transfer curves for n-channel TFT depending on the defects running parallel to channel, (a) subgrainboundaries and (b) twin boundaries.

CONCLUSIONS

ZMR process of SOQ structure has been proposed to fabricate single crystalline TFT's. The single crystalline silicon thin films are successfully recrystallized with (100) texture at cellular interface regime. Large residual tensile stress is observed after ZMR, which can be reduced by post-annealing. N-channel TFT fabricated on a recrystallized film shows excellent device characteristics compared with a-Si and p-Si TFT : $\mu n \sim 420 cm^2/Vsec$, S $\sim$ 100mV/dec, leakage current $\sim$ 0.2pA and on/off ratio $> 10^9$. Post-annealing can improve the overall device performance especially a leakage current characteristics. And SGB's in channel region hardly affect the electrical performance of TFT's. ZMR is expected to be a potential process to fabricate the single crystalline TFT for high performance flat panel displays, especially a small geometry panel.

ACKNOWLEDGMENTS

This work has been supported by KOSEF through RETCAM in Seoul National University. Authors would like to thank Samsung Electronics Co. Ltd. and LG Electronics Research Center for their helpful supports.

REFERENCES

[1] F.Okumura and K.Sera, AMLCD Digest of Technical Papers, **94**, 24(1994).
[2] J.P.Salerno, SID Int. Display Res. Conf., **94**, 39(1994).
[3] K.R.Sarma, C.Rogers and C.Chanley, SID Digest of Technical Papers, **XXV**, 419(1994).
[4] J.C.C.Fan, B.Y.Tsaur and M.W.Geis, J. Cryst. Growth, **63**, 453(1983).
[5] R.A.Lemons, M.A.Bosch and D.Herbst in <u>Laser-Solid Interactions and Transient Thermal Processing of Materials</u>, edited by J.Narayan, W.L.Brown and R.A.Lemons (Mater. Res. Soc. Proc. 13, Boston , MA, 1982) pp. 581-592

POLYSILICON THIN FILM TRANSISTORS WITH COBALT AND NICKEL SILICIDE SOURCE AND DRAIN CONTACTS

G.T. SARCONA and M.K. HATALIS
Display Research Laboratory, EECS Department, Lehigh University, Bethlehem, PA 18015

ABSTRACT

Polysilicon n-type thin film transistors have been fabricated with self-aligned cobalt and nickel silicide contacts to the source and drain regions. The sheet resistance of the source and drain without silicide is above 200 $\Omega/\square$. The cobalt and nickel silicide films have sheet resistance below 30 $\Omega/\square$. The contact resistance of the silicided devices is also much lower. The reduced extrinsic resistance is shown to improve the current in the "on" state, without increasing the leakage current. This study includes examination of cobalt and nickel silicidation on thin polysilicon films at temperatures compatible with polysilicon TFT-LCD processes.

INTRODUCTION

Polysilicon thin film transistors are necessary for integrated driver circuitry for active matrix displays. The thin island causes the sheet resistance of the source and drain regions to be high, which will slow the circuit performance [1]. A polysilicon gate TFT with self-aligned silicided source and drain regions would reduce the device series resistance and improve device and circuit speed and drive capability. Previously, sputtered silicides have been investigated for reducing the scan line resistance [2]. But there has been no work on silicided TFTs for integrated drivers.

The thermal limit of low-cost glasses prohibits the use of titanium and tungsten for silicides in polysilicon TFTs. We have shown recently that cobalt and nickel silicidation processes are compatible with TFT fabrication temperatures ($\leq 700°C$) [3]. In this paper, we report the characteristics of polysilicon TFTs with cobalt and nickel silicided source and drain regions.

EXPERIMENT

For sheet resistance study, silicides were prepared on single-crystal silicon wafers, and polysilicon layers on SiO_2. Polysilicon films were deposited by LPCVD, with film thickness of 1000Å. The fabrication process was RCA cleaning, followed by 25:1 H_2O:HF dip for 30 sec after the silicon surface became hydrophobic, DI water rinsing, IPA immersion for 1 min, then nitrogen blow-drying immediately prior to loading into the load-lock of the sputtering system. The sample was pumped down via a cryopump to less than 1μtorr. Then, RF Ar plasma would be ignited, and the target pre-sputtered at 250W for 5 min at 9mtorr. Cobalt was sputtered to thicknesses of 30 and 60Å. Nickel was sputtered to thicknesses of 43 and 86Å. After sputtering, the plasma was extinguished, and the Ar evacuated. When the pressure was below 1μtorr, the sample was in-situ lamp-annealed for 10 to 30 min. Cobalt was annealed over temperatures ranging from 600 to 700°C. Nickel was annealed over temperatures ranging from 250 to 600°C After cooling in N_2 in the load lock, the samples were removed and were metal-etched. Cobalt silicides were etched in 3:1 H_2O_2:HCl; nickel silicides were etched in 3:1 H_2O_2:H_2SO_4. The sheet resistance was measured using a four-point probe setup.

159

Mat. Res. Soc. Symp. Proc. Vol. 424 © 1997 Materials Research Society

For TFTs, the island silicon was deposited to a thickness of 1100Å by SiH$_4$ LPCVD on silicon wafers with 1μm SiO$_2$. The films were crystallized in N$_2$ at 600°C. After island patterning, 900Å gate oxide was grown in dry oxygen at 1000°C for 90 min. Assuming 45% of the gate oxide is silicon consumptive, the island thickness after oxidation was about 695Å. Gate silicon was deposited to a thickness of 1200Å. After patterning the silicon gate stripes, and source/drain oxide etch, the gate silicon was crystallized and doped, and the source and drain regions formed, by POCl$_3$ diffusion at 900°C with a 5 min deposition and 15 min drive-in. The source and drain region sheet resistance for the thick poly islands, and gate stripes on all devices, was above 200 Ω/□. Silicides were grown using the method described above. Cobalt was sputtered to thickness of 60Å only. Nickel was sputtered to 43Å and 86Å. After silicidation, the cobalt devices required annealing in H$_2$/N$_2$ for 2h at 400°C to anneal silicidation process-induced damage. No plasma hydrogenation treatments were performed. The device structure is shown in Figure 1.

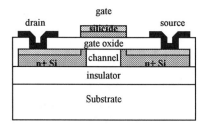

Figure 1. Device Structure

RESULTS

<u>Sheet Resistance</u>

The sheet resistance of the cobalt silicide on single-crystal and polycrystalline silicon as a function of temperature, time, and substrate is shown in Figure 2. The minimum value of resistance is about 13 Ω/□ for the crystalline and polycrystalline samples at 650 and 700 °C when annealed 10 min. The difference of the sheet resistance due to substrate was within the scatter. Annealing at 600°C required 30 min. to achieve low (13 Ω/□) resistance. A cross-sectional electron micrograph of the cobalt silicide is shown in Figure 3.

The nickel silicide sheet resistance on single-crystal and polycrystalline silicon is shown in Figure 4. The monosilicide phase is evident between 400 and 500°C. The minimum sheet resistance on all samples was 8 Ω/□ for 86Å sputter, and 20 - 24 Ω/□ for 43Å sputter.

For all of the silicidations, the measured sheet resistance was unchanged by the metal etch, indicating that the consumption of metal was complete.

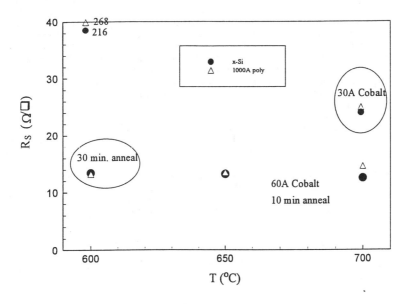

Figure 2. Sheet Resistance of Cobalt Silicide on Crystalline and Polycrystalline Silicon

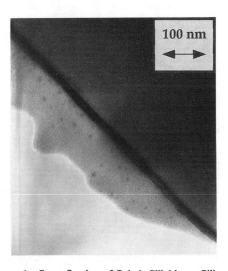

Figure 3. Cross Section of Cobalt Silicide on Silicon

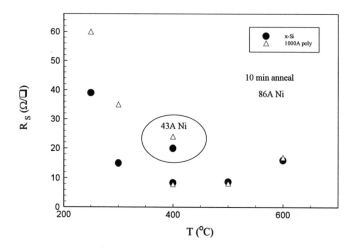

Figure 4. Sheet Resistance of Nickel Silicide on Crystalline and Polycrystalline Silicon

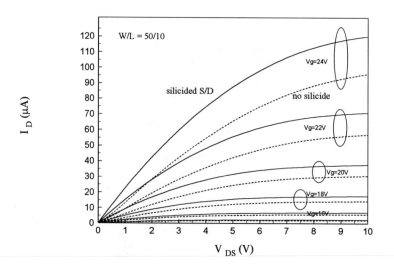

Figure 5. Effect of Cobalt Silicide on TFT Performance

Thin Film Transistors

The subthreshold swing, threshold voltage and effective mobility were measured in the linear mode, at $V_D = 1V$, on devices with W/L = 50/10. The subthreshold swing was determined from the log $I_D(V_{GS})$ characteristic. The threshold voltage was determined by extrapolation of a linear fit of the $I_D(V_{GS})$ characteristic. The mobility was calculated from the transconductance:

$$\mu_{eff} = \frac{g_m}{\frac{W}{L} C_{ox} V_{DS}} \tag{1}$$

The effective mobility incorporates series resistance through the transconductance, g_m, as it relates to the device current. Devices with higher series resistance show a lower effective mobility. A summary of parameters for non-silicided devices and devices with cobalt and nickel silicides is shown in Table 1. The mobility was increased by the silicide, through reduction of the contact and source and drain series resistance. Figure 5 compares the transfer characteristics of a cobalt silicided device to those of a non-silicided device. In the low-V_{DS} region, the non-silicided device exhibits current limitation due to contact resistance. The silicided device does not suffer from contact resistance.

Figure 6 shows the effect of the initial nickel thickness on the device on-off, or log $I_D(V_{GS})$ characteristic. Shown are curves for a 50/10 device, at $V_{DS} = 1V$, for no nickel silicide, and for silicides formed by 43Å and 86Å of nickel sputter. The sheet resistance is halved in the thicker nickel silicide, as shown in Figure 4. The reduction in series sheet resistance, from the non-silicided case, to the thick nickel silicide case, clearly improves the device "on" current. The non-silicided sheet resistance was 260 $\Omega/\square$. The thin nickel silicide had a sheet resistance of 23 $\Omega/\square$. The thick nickel silicide had a sheet resistance of 8 $\Omega/\square$. The mobility increased from 17 to 31 cm^2/V-sec. Also note that the subthreshold swing decreased as the sheet resistance decreased. None of the nickel devices received hydrogen passivation treatments. The decreased swing is due to the decreased series resistance, which causes a voltage drop between the source and drain contacts and the channel ends. As the resistance decreases, the voltage drop in the source and drain is less, hence more of the terminal voltage is applied across the channel, inducing more current. The decreased swing will cause the device to switch between the on and off states more rapidly, in terms of the gate input voltage, which will enhance circuit response.

The subthreshold swing for the nickel silicided devices was the same as for the cobalt silicided devices. However, the nickel devices did not require PMA annealing. Thus, the temperature of silicidation has a negative effect on device performance. The cobalt device performs worse than the nickel device even after receiving PMA; the damage may be fully removed only by plasma hydrogen. The threshold voltage for all of the devices was similar, hence, plasma radiation did not appear to cause damage to the devices. Note however, that exposure was short (1 or 2 min).

silicide	μ eff (cm^2/V-sec)	V_T (V)	S (V/dec-A)	R_{SD} ($\Omega/\square$)
		cobalt		
no	20	15	1.3	200
CoSi$_2$	23	16	1.3	20
		nickel		
no	17	16	1.7	260
NiSi	31	14	1.3	20

Table 1. Parameters for Non-Silicided and Silicided TFTs

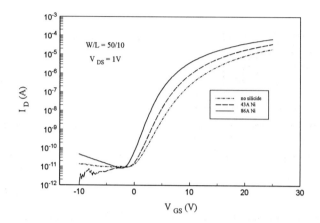

Figure 6. Effect of Nickel Silicide and its Thickness on Device Performance

CONCLUSIONS

Cobalt and nickel silicides were fabricated on thin polysilicon films for TFTs. Cobalt formed low-resistance silicides at 600°C. Nickel formed silicides at 400°C. For low-cost glasses, nickel appears to be the better choice. The sheet resistance of the silicides was shown to have only a minor dependence on the metal film thickness, and underlying silicon structure and thickness.

Silicides were shown to increase the effective mobility of polysilicon TFTs by reducing the series and contact resistance. The subthreshold swing was decreased by incorporation of silicides. The improvement of device characteristics is proportional to the thickness of the silicide, which is inverse to the sheet resistance. Silicides do not degrade device characteristics, though the cobalt silicidation temperature created damage which required PMA. Nickel silicidation did not cause such damage nor require PMA. The impact of silicides on TFT performance. Thus, it was demonstrated that silicides can be incorporated into TFT-LCD processes, and be used to improve integrated driver performance where series resistance is a significant fabrication and design issue.

REFERENCES

1. A.G. Lewis, I.-W. Wu, T.Y. Huang, A. Chiang, R.H. Bruce. *IEDM Dig. Tech. Papers 1990*, pg. 843.
2. M. Kobayashi, T. Nakazono, K. Mori, H. Nakamura, H. Sato, M. Nakagawa. *Soc. Info. Display Int'l Symp. Digest of Technical Papers, 1994*, pg 75, (1994).
3. G.T. Sarcona, M.K. Hatalis. *Mat. Res. Soc. Symp. Proc.* **402**, 191, (1996).

ACKNOWLEDGMENTS

The authors wish to thank Xia Zhou for cross-sectional electron microscopy. This work was supported under DARPA contract # MDA 972-92-J-1037.

HELICAL RESONATOR PLASMA OXIDATION OF AMORPHOUS AND POLYCRYSTALLINE SILICON FOR FLAT PANEL DISPLAYS

Sita R. Kaluri *, Robert S. Howell **, Dennis W. Hess*, Miltiadis K. Hatalis **
* Department of Chemical Engineering, Lehigh University, Bethlehem, Pa 18105
** Department of Electrical Engineering, Lehigh University, Bethlehem, Pa 18015

ABSTRACT

A low temperature process for the oxidation of amorphous and polycrystalline silicon was studied using a helical resonator plasma source. Constant current anodization of amorphous silicon was performed at 350°C and 30 mtorr in an oxygen plasma. Oxidation rate data were analyzed by both parabolic and power law expressions: parabolic rate coefficient B = 23.45 nm^2/min and power law rate coefficients α = 0.3807 and C^* = 7.3 nm/min^{α}. Oxidation at substrate floating conditions yielded thin oxides; the growth rate saturated after 20 mins, at a thickness of about 10 nm. For TFT fabrication, a two layer oxide was used. The first layer was grown by plasma oxidation to a thickness of about 10 nm to ensure a good interface, and the second layer deposited by PECVD to the final thickness of 35 nm. Electrical properties of control oxides grown on c-Si at the same conditions were promising, with a fixed oxide charge of 1.98×10^{11} charges/cm^2 and a mean breakdown field of 5.3 MV/cm after post metallization anneal. The TFTs had an average effective electron mobility of 31.1 V/cm^2 and an average threshold voltage of 4.6 V.

INTRODUCTION

Thin film transistor (TFT) technology with thin polysilicon layers is an area of intense interest for applications in active matrix liquid crystal displays (AMLCDs). The quality of the gate dielectric and the substrate affect the performance of TFTs [1,2]. Since glass substrates are used for these applications, conventional high temperature oxidation is not possible. The electrical properties of deposited oxides and their interfaces with silicon are in general inferior to those of thermally grown oxides. Oxidation of polycrystalline silicon using a helical resonator plasma source offers an attractive alternative, since oxidation can be performed at low temperatures (< 450°C).

A helical resonator is a medium to high density plasma source and it consists of a helical coil surrounded by a grounded co-axial cylinder. The composite structure becomes resonant when an integral number of quarter or half wavelengths of a rf field fit between the two ends of the helix, forming a standing wave. When this condition is satisfied, the intense electromagnetic fields within the helix can sustain a plasma with negligible matching loss at low gas pressure (0.1 - 1 mtorr) [3,4,5].

In this work, the use of the helical resonator for plasma oxidation of amorphous and polycrystalline silicon was explored. Upon optimization of the process, TFTs were constructed employing the plasma oxidation method and compared to those with deposited oxides.

EXPERIMENT

Oxidation

Figure 1 shows a schematic of the helical resonator reactor system. In order to define current densities, the substrate holder was enclosed in quartz except for a 5 cm diameter circular area which was exposed to the plasma for processing [6]. Oxidation was performed at temperatures ranging from 300-400 °C, pressures ranging from 10 mtorr to 50 mtorr, and flow rates ranging from 5-20 sccm, at a radio frequency of 13.25 MHz and an rf power of 300 W. Typical plasma characteristics at these conditions were $n_e = 3\text{-}4 \times 10^{10}/cm^3$ and $T_e = 3\text{-}4$ eV. Oxidation was performed on (100) p-type Si with a resisitivity of 15-30 ohm-cm, and on amorphous silicon deposited on these p-type wafers. Wafers were cleaned by using the RCA clean and given a final HF (1%) dip followed by N_2 dry prior to oxidation. Oxide thicknesses were measured by ellipsometry using a refractive index of 1.465. The oxidations were performed both by using a substrate bias to maintain a constant current as well as under a floating substrate condition.

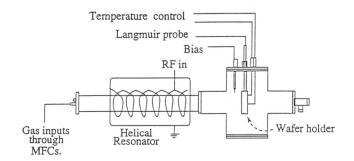

Figure 1. Schematic of Helical Resonator Plasma System

MOS Capacitors

Fixed oxide charge and mean breakdown fields were determined from MOS capacitors fabricated from oxides grown on c-Si. Oxides were grown to 20 nm at 350°C, 30 mtorr, 10 sccm and 3.8 mA/cm^2.

TFT Fabrication

The TFT's were fabricated on 3 inch silicon wafers which had 1.1 μm of silicon dioxide to serve as a barrier layer grown on the wafers with a wet oxidation at 1100°C. A layer of amorphous silicon 105 nm was then deposited at 550°C by LPCVD. This amorphous silicon was subsequently crystallized through furnace annealing in an atmosphere of nitrogen at 600°C. After the polysilicon layer was defined into islands, the source and drain regions were doped by ion implantation of phosphorous at 20 keV to a dose of 2×10^{15} cm^{-2}, while the channel region remained undoped. Activation of dopants was accomplished through another furnace anneal in nitrogen at 600°C. The device wafers were then plasma oxidized to a thickness of 10 nm. A second oxide layer was deposited on the wafers in a PECVD process at 350°C to give a final thickness of 25 nm.

A thin oxide was deposited in order to take advantage of the plasma process, which formed an oxide mostly in the center of the device wafers. Thus devices on the outer edge of the wafer had a dielectric composed almost exclusively of the PECVD oxide, while those in the center had a dielectric of both plasma grown and PECVD oxide. By depositing an oxide in the PECVD process of comparable thickness to the oxide grown on the device wafers, comparisons between devices can readily be made on the same wafer, thus limiting possible fluctuations in process variables that might otherwise affect the outcome.

After depositing the PECVD oxide, contact windows were opened on the oxide. The wafers then received aluminum metallization in order to form the gate, source and drain electrodes of the devices. Metallization was performed by DC Magnetron sputtering at 1 kV in an Ar atmosphere of 4 mTorr. A post metallization anneal was performed at 400°C for 30 minutes in an atmosphere of 20% H_2 and 80% N_2.

RESULTS

Oxidation

Figure 2 shows the kinetic data obtained from constant current anodization in an oxygen plasma at 3.8 mA/cm^2 at 300 °C, 30 mtorr and 10 sccm on two kinds of substrates, single crystal silicon and a-Si/SiO$_2$/Si, with the a-Si layer sputter deposited on thermally grown SiO$_2$. In general, oxide thickness values were higher on a-Si. The thicknesses were higher at short times when compared to c-Si, with the growth rate dropping below that for c-Si at longer times(> 30min). The kinetic data were modeled using both the parabolic law and power law [7,8] rate expressions:

| Parabolic Law | $x^2 = Bt$ | (1) |
| Power Law | $x = C^* t^\alpha$ | (2) |

where x is the oxide thickness, t is the time, B is the parabolic rate constant, C^* is analogous to a rate constant (driving force × ionic conductivity) and α is the exponent of time. Deviation of α from a value of 0.5 indicates the influence of space charge in the oxide on oxide growth rate [7,8]. A value of $\alpha = 0.5$ reduces the power law expression to a parabolic law. As can be seen from the figure, both models describe the oxide growth behavior on c-Si adequately, although the power law model better describes the oxide growth data on a-Si. Table I shows the values of B, C^* and α. On a-Si, the value of the parabolic law constant B and the power law constant C^* are both higher while the exponent α is lower than on c-Si.

Application of an external bias during growth is not feasible with a glass substrate. Therefore, oxidation of a-Si was also performed under 'substrate floating' conditions, with temperature, time, flow rate and pressure as variables. Figure 3 (a) shows oxide thickness values obtained as a function of time at 300°C, 30 mtorr and 10 sccm. At 20 minutes oxidation time, two temperatures were studied, 300 and 400°C. The oxide thickness increased to 12 nm in 20 minutes and then, increased very slowly - the thickness after 40 minutes was 13 nm. No significant change in oxide thickness as a function of temperature was observed. Figure 3 (b) shows the oxide thickness values obtained as a function of flow rate at 400°C, 30 mtorr and 20 minutes oxidation time. At 10 sccm flow rate, two different pressures were investigated, 30 mtorr and 75 mtorr. A flow rate of 10 sccm was optimal; flow rates of 20 sccm and 5 sccm yielded oxides of lower thicknesses, probably due to an inadequate supply of atomic oxygen at the substrate surface. No significant change in oxide thickness was observed by changing the pressure from 30 mtorr to 75 mtorr. As a result of these studies, oxidation conditions of 10 sccm, 30 mtorr, 400°C and a time of 20 minutes were used for TFT applications.

Table I. Parabolic and Power law constants obtained from anodization kinetic data of c-Si and a-Si

Substrate type	B (nm^2/min)	C* (nm/min$^\alpha$)	α
(100) Si	10.5	3.57	0.475
a-Si	23.45	7.3	0.3807

MOS Capacitors

Table II shows the values of fixed oxide charge and mean breakdown fields obtained for as-grown oxides and for oxides after a post metallization anneal in 10% H_2 and 90% N_2 at 400°C for 30 minutes. Considerable improvement in both the fixed oxide charge and the mean breakdown field was obtained after the anneal.

Table II. Values of fixed oxide charge and mean breakdown fields obtained from MOS characteristics of a 20 nm oxide grown on c-Si.

	Fixed oxide charge (charges/cm^2)	Mean breakdown field (MV/cm)
As-grown	8.3 x 10^{11}	3.9
After PMA	2 x 10^{11}	5.3

TFT Characteristics

The effective electron mobility was extracted from the slope of the I_{DS} vs. V_{GS} curve in the linear mode (for V_{DS} = 0.1 V). Table III shows the average measured effective mobilty for devices with only a PECVD deposited oxide and those devices with both a plasma grown oxide and PECVD deposited oxide. The effective mobility is essentially the same for both types of devices. Table III also lists the average threshold voltage for the two types of devices, demonstrating a slight difference between the two. The larger threshold voltage for the dual oxide devices indicates a larger fixed oxide charge in the dielectric. This could be the due to the increased overall oxide thickness or the presence of two interfaces within the dielelectric, or due to exposure to plasma radiation.

The on/off characteristics of two typical devices are shown in Figure 4. The channel width for both devices is 50 μm and the channel length is 5 μm. The voltage across the drain and source was held at 1 V. The device with only a PECVD deposited oxide has a thinner dielectric, which increases its capacitance and is resposible for the higher on current of that device. Figure 4 demonstrates that in the absence of device passivation with a hydrogen plasma, devices with two layers of oxide, one grown in an oxygen plasma and the other deposited in a PECVD process, are analogous to the devices with only a deposited PECVD oxide.

The similarity of TFT characteristics between the two types of gate oxide layers is probably due to the fact that a thin ($\approx$ 3 nm) oxide grows in the silicon regions outside the 5 cm opening in the quartz substrate cover. This oxide growth is due solely to the supply of oxygen atoms/molecules to the silicon surface by diffusion under the quartz cover, and results in thin oxide growth analogous to that observed previously by intentional plasma oxidation [9]. Such results suggest that a thin oxide layer with good interfacial and "bulk" properties may be sufficient to establish the necessary TFT properties.

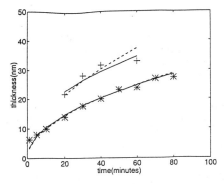

Figure 2. Experimental data and fitted curves for anodization in oxygen plasma at 300 W, 30 mTorr, 350 °C and 3.8 mA/cm². '*' represents (100) Si and '+' represents the parabolic fit and '-' represents the power law fit.

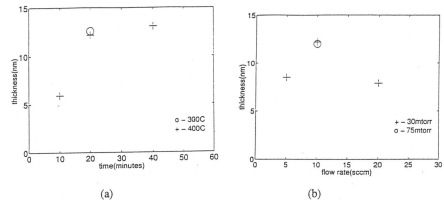

(a)   (b)

Figure 3. (a) Oxide thickness vs. oxidation time in 'floating' plasma, at reactor conditions of 300 W, 30 mTorr, 10 sccm. (b) Oxide thickness vs. flow rate in 'floating' plasma at reactor conditions of 300 W, 400 °C for 20 minutes.

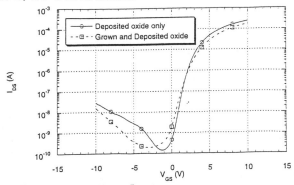

Figure 4. I_D vs V_{GS}, $V_{DS} = 1V$, W = 50 μm, L = 5 μm

Table III. Average values for threshold voltage and effective electron mobility for devices

	Average Threshold Voltage (V)	Average Effective Mobility (V/cm^2)
Devices with only PECVD deposited oxide	3.7	31.1
Devices with PECVD deposited oxide and Oxygen Plasma grown oxide	4.6	31.3

CONCLUSIONS

A method for growing oxides in a helical resonator plasma compatible with low temperature processing has been successfully demonstrated. TFTs with a dielectric of the plasma grown oxide and a PECVD deposited oxide were fabricated and compared to TFTs with primarily a PECVD deposited oxide. The operating characteristics of the two types of devices were very similar. The presence of an ultra-thin plasma oxide on the outer edge of the wafer is likely to have resulted in the similarity. Therefore, the two step process of plasma grown and PECVD deposited oxide is a promising approach for TFT applications.

ACKNOWLEDGEMENTS

The authors gratefully acknowledge support from NSF grant CTS-9214138 and DARPA MDA 972-92-J-1037.

REFERENCES

1) M. Hatalis, J. Kung, J. Kanicki, and A. Bright, Mater. Res. Soc. Symp. Proc., **182**, p.357 (1990).

2) M. Itsumi and K. Kishi, J. Electrochem Soc., **140**, p.3210 (1993).

3) D. L. Flamm, D. E. Ibbotson and W. L. Johnson, U.S. Patent No. 4,918,031, April 17, 1990.

4) J. M. Cook, D. E. Ibbotson, P. D. Foo and D. L. Flamm, J. Vac. Sci. Tech., **A8**, p.1820 (1990).

5) J. M. Cook, D. E. Ibbotson and D. L. Flamm, J. Vac. Sci. Tech., **B8**, p.1 (1990).

6) D. A. Carl, D. W. Hess, M. A. Lieberman, T. D. Nguyen, and R. Gronsky, J. Appl. Phys. **70**, p. 3301 (1991).

7) D. R. Wolters and A. T. A. Zegers-van Duynhoven, J. Appl. Phys. **65**, p. 5126 (1989).

8) D. R. Wolters and A. T. A. Zegers-van Duynhoven, J. Appl. Phys. **65**, p. 5134 (1989).

9) A. A. Bright, J. Batey and E. Tierney, Appl. Phys. Lett. **58**, p. 619 (1991).

PERFORMANCE OF POLY-Si TFTS
WITH DOUBLE GATE OXIDE LAYERS

Byung-Hyuk Min, Cheol-Min Park, Jae-Hong Jun, Byung-Sung Bae[*], and Min-Koo Han
Department of Electrical Engineering., Seoul National Univ., Seoul, 151-742, Korea
[*]Samsung Electronics Co., Kiheung, YongIn-Goon, KyungKi-Do, 449-900, Korea

ABSTRACT

We have fabricated a poly-Si TFT with double gate insulator composed of ECR oxide and APCVD oxide to improved the performance of poly-Si TFTs. The poly-Si TFT with double gate oxide exhibits the remarkable enhancement of the electrical parameters compared with the conventional poly-Si TFTs which has APCVD gate oxide, such as improvement of the subthreshold swing and the low threshold voltage. The proposed poly-Si TFT has a higher oxide breakdown electrical field and the device characteristics are not degraded significantly after an electrical stress. It is found that the ECR oxide plays a key role to improve the device performances and prevent the poly-Si TFTs from degradation due to the electrical stress.

I. INTRODUCTION

The oxides for poly-Si TFTs are required to have low leakage current, high dielectric strength, and good interface with the poly-Si layers. Poly-Si TFTs with ECR(Electron Cyclotron Resonance) plasma thermal oxide as gate insulator exhibit good electrical characteristics for AMLCDs[1]. However, the oxidation rate is very low such as below about 4 Å/minute, so that it may not be suitable to fabricate the poly-Si TFTs only with ECR plasma gate oxide due to a poor throughput. It is also known that the characteristics of APCVD oxide which is usually employed to fabricate gate insulators in low temperature processed poly-Si TFT devices have high leakage current, low dielectric strength, and bad interface with the poly-Si layers. Therefore APCVD Oxide is not suitable for a high performance poly-Si TFTs[2][3]. Therefore the high quality oxide film which has good interface with the poly-Si active layer and stable electrical performances are desired.

The purpose of our work is to investigate the performance and the stability of low-temperature poly-Si TFTs with double gate oxide layers which consist of ECR plasma oxide and APCVD oxide. We have investigated the transfer characteristics of fabricated poly-Si TFTs in order to comprehend the role of the ECR plasma oxide located at interface of poly-Si. We have also examine the effects of electrical stress due to gate bias and the oxide breakdown properties.

II. DEVICE FABRICATION

The main processing steps for the proposed poly-Si TFT device are as follows. Silicon wafers with 5000 $Å$ thermally grown oxide are used as starting substrates. Undoped 1000 $Å$ thick amorphous silicon (a-Si) films are deposited by low-pressure chemical vapor deposition (LPCVD) on substrates at 550 $℃$. The a-Si films were crystallized by solid phase crystallization at 600 $℃$ for 24 *hours* in N_2 ambient. The silicon islands for active layer are formed with a active mask. A 300Å ECR plasma gate oxide was grown at 400 ℃, followed by a 700Å APCVD gate oxide deposition at 370 ℃. The 1000Å thick poly-Si was deposited to form the gate electrode and subsequently RIE etched.   Source/drain and gate electrodes are doped by self-aligned As implantation with $5 \times 10^{15}/cm^2$ at 40 KeV. After a self-aligned implantation, the isolation oxide is deposited. The dopants are activated during a 600 ℃ anneal for 12 hours. Contact lithography and contact etch are then carried out. Finally, aluminum is deposited and defined, followed by a forming gas anneal at 450 ℃ for 30 minutes. Hydrogen plasma passivation was performed at 300 ℃ for 45 minutes to enhance the device performances.

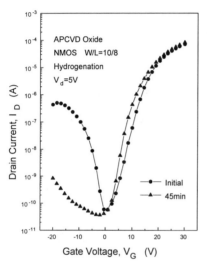

Fig 1. The I_D-V_G characteristics of the conventional poly-Si TFTs with APCVD gate oxide before and after hydrogenation

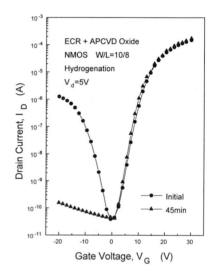

Fig 2. The I_D-V_G characteristics of the poly-Si TFTs with double gate oxide before and after hydrogenation

III. RESULTS AND DISCUSSION

We have found that poly-Si TFTs with double gate oxide has significantly improved a subthreshold swing and field effect mobility in comparison with the poly-Si TFTs with APCVD gate oxide. Fig. 3 shows that the new poly-Si TFTs with double oxide layer have better electrical characteristics than the conventional poly-Si TFTs using APCVD oxide as gate insulator. For the poly-Si TFTs with APCVD gate oxide, the threshold voltage is 5.69V and the subthreshold swing is 1.88V/decade. For the new poly-Si TFTs with double gate oxide layer, the threshold voltages are 4.02V and the subthreshold swings are 1.58V/decade. The breakdown electric field of the gate oxide in the new poly-Si TFTs is 9.32MV/cm and that of the conventional poly-Si TFTs is 3.85MV/cm in Fig. 4. These result implies that the ECR oxide located at the oxide/poly-Si interface may improve the surface roughness of the poly-Si TFTs so the local electric field at interface is reduced that oxide breakdown characteristic was improved.

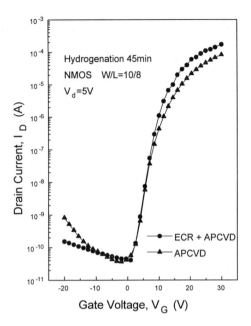

Fig 3. The I_D-V_G characteristics of the poly-Si TFTs with APCVD gate oxide and double gate oxide layer after 45 *minute* hydrogenation.

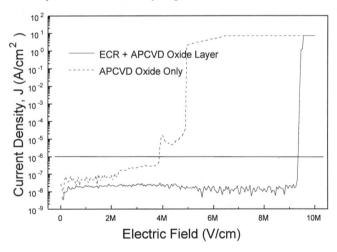

Fig 4. The breakdown electric field of the gate oxide in the new poly-Si TFTs with double gate oxide and that of the conventional poly-Si TFTs.

We also investigated the variation of I_D-V_G characteristics for the conventional and proposed poly-Si TFTs before and after an electrical stress with $V_G = 25V$ and $V_D = 1V$ for 90 *minutes*. The characteristics of the poly-Si TFTs with double gate intermediate oxide layer have a little

173

difference between before and after the electrical stress. However, In the conventional poly-Si TFTs with APCVD oxide, it is seen that the characteristic curve after the stress degraded in ON-current compare with the original curve before the stress. It is found that the value of the maximum ON-current at $V_G = 25V$ degrade from $2.86 \times 10^{-5}A$ to $7.04 \times 10^{-6}A$ and the subthreshold swing is changed from 1.88 *V/decade* to 2.35 *V/decade*. It is also found that the ECR oxide plays a key role in preventing the poly-Si TFTs from degradation due to the electrical stress.

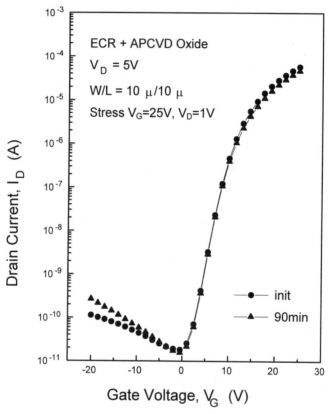

Fig. 5 The I_D-V_G characteristics for proposed poly-Si TFTs with double gate oxide layer before and after an electrical stress (V_G=25V, V_D=1V). The stress time is 90 *minute*. The ECR oxide thickness of the proposed poly-Si TFTs is 300 Å.

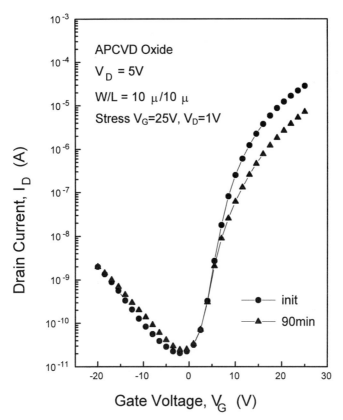

Fig. 6 The I_D-V_G characteristics for conventional poly-Si TFTs with APCVD gate oxide layer before and after an electrical stress (V_G=25 *V*, V_D=1 *V*). The stress time is 90 *minute*. The characteristic curve after the stress degraded in ON-current compare with the original curve before the stress.

IV. CONCLUSION

We have successfully utilized ECR plasma thermal oxide which has passivation effect and smooth interface and APCVD oxide of which deposition rate is high. Our experimental results show that the ECR oxide located at the oxide/poly-Si interface improve the surface roughness of the poly-Si TFTs so the local electric field at interface is reduced that oxide breakdown characteristic was improved. It is also found that the ECR oxide plays a key role in preventing the poly-Si TFTs from degradation due to the electrical stress.

V. ACKNOWLEDGEMENT

This work was performed with the financial support of Samsung Electronics Company.

REFERENCE

1. Mimura, T. Suzuki, N. Konishi, T. Suzuki, and K. Miyata, IEEE Electron Device Letters, **9**, 290 (1988)

2. K. Hatalis, J.-H. Kung, J. Kanicki, and A. Bright, Mater. Res. Symp. Proc., **182**, 357 (1990)

3. Lustig and J. Kanicki, J. Appl. Phys., **65**, 3951 (1989)

Novel N$_2$O Plasma Passivation on Polycrystalline Silicon Thin-Film Transistors

Fang-Shing Wang, Chun-Yao Huang, and Huang-Chung Cheng

Department of Electronics Engineering and Institute of Electronics, Semiconductor Research Center, National Chiao Tung University, National Nano Device Laboratory, Hsinchu, Taiwan, R.O.C.

ABSTRACT

A novel defect passivation process of polycrystalline silicon thin-film transistors (poly-Si TFT's) utilizing nitrous oxide (N$_2$O) plasma was investigated. In terms of the gas flow rate, chamber pressure, and plasma exposure time, the optimum plasma condition has been found to significantly improve the electrical characteristics of poly-Si TFTs. The performance is even better than those passivated with conventional hydrogen plasma. It is believed that the nitrogen radicals from the N$_2$O gas as well as the hydrogen ones from the residual H$_2$O both can diffuse into the gate-oxide/poly-Si interface and the channel poly-Si layer to passivate the defect-states. Furthermore, the gate-oxide leakage current significantly decreases and the oxide breakdown voltage slightly increases after applying N$_2$O-plasma treatment. This novel process is promising for the applications of TFT/liquid crystal displays and TFT/static random access memories.

INTRODUCTION

Hydrogen plasma passivation of polysilicon films to reduce the defect states has been widely investigated for improving the electrical characteristics of poly-Si TFT's.[1-3] In the past, it has been reported that an O$_2$ or a combination of H$_2$ and O$_2$ plasma treatments of poly-Si TFT's significantly promotes the device performance.[4,5] It is suggested that the observed performance enhancement is due to the oxygen atoms and existing hydrogen.[5] Nickel *et al.* found that the performance improvement of TFT's after ECR O$_2$ plasma treatment originated from the residual hydrogen into poly-Si films.[6] Recently, Tsai *et al.* showed that the nitrogen-containing hydrogen (H$_2$/N$_2$) plasma treatments exhibited better passivation effects on the electrical characteristics of the poly-Si TFTs than the pure H$_2$ plasma treatments.[7] In this work, the N$_2$O plasma conditions such as the gas flow rate, chamber pressure, radio-frequency (RF) power density, as well as plasma exposure time on the passivation efficiency of the poly-Si TFTs were studied. By means of typical current-voltage (I-V) characteristics and secondary ion mass spectrometry (SIMS) measurements, the passivation roles of the hydrogen and nitrogen atoms in poly-Si films and gate dielectrics are observed.

EXPERIMENT

The poly-Si TFT's were fabricated on thermally oxidized silicon wafers. A 110 nm-thick amorphous silicon (α-Si) was initially deposited at 550 °C by low-pressure-chemical-vapor-deposition (LPCVD), and then thermally annealed at 600 °C to recrystallize the silicon films. After defining the active islands, a 70 nm-thick SiO$_2$ was grown by dry oxidation. Another 300 nm polysilicon film was deposited at 620 °C by LPCVD and then patterned. A self-aligned POCl$_3$ doping was performed to form the drain, source, and gate electrodes. Then, the samples were subjected to the N$_2$O and H$_2$ plasma treatments, respectively, in a parallel-plate plasma reactor at 300 °C. After a plasma-enhanced CVD (PECVD) SiO$_2$ with a thickness of about 500 nm was deposited, the contact holes were opened and the Al films were deposited and then defined.

RESULTS AND DISCUSSION

The effects of N_2O plasma treatments on the n-channel poly-Si TFT's were investigated in terms of the gas flow rate, chamber pressure, RF power density, and plasma exposure times. Figure 1 shows the changes of the drain current I_d and the threshold voltage V_t of the n-channel poly-Si TFT's treated with different RF power densities. The parameters of the gas flow rate, chamber pressure, treating time and substrate temperature were fixed at 200 sccm, 100 mTorr, 30 min., and 300 °C, respectively. The reason that the device performance is improved with increasing RF power density can be explained by the avalanche multiplication process of the plasma. When the RF power density increases to be above 0.88 W/cm^2, a large amount of plasma damages would be produced and limit the further performance improvement. Figures 2 and 3 show the I_d and V_t as a function of the gas flow rate and chamber pressure, accordingly. It is found that appropriate N_2O gas flow rate (200 sccm) and chamber pressure (100 mTorr) make better passivation effects on those poly-Si TFT's. Figure 4 shows the variations of the I_d and I_{min} (minimum drain current) of the TFT's as a function of plasma exposure times. The I_d monotonically increases with time, but I_{min} degrades after a 60-min plasma treatment. The increase of I_{min} is due to the damages induced by long-time plasma treatments.

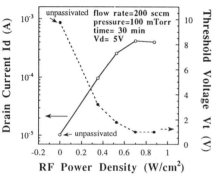

Fig. 1 The variations of the I_d and V_t for different RF power densities.

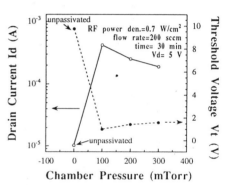

Fig. 3 The variations of the I_d and V_t for different chamber pressures.

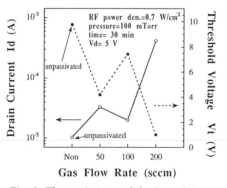

Fig. 2 The variations of the I_d and I_{min} for different gas flow rates.

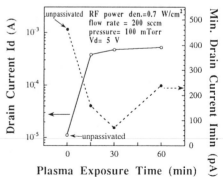

Fig. 4 The variations of the I_d and I_{min} for different plasma exposure times.

The I_d-V_g characteristics of the n-channel poly-Si TFT's with optimum N_2O plasma condition are shown in Fig. 5. The transfer curves of the devices with a 30-min H_2 plasma passivation and without a plasma treatment are also shown for comparison. It is seen that a 30-min N_2O plasma treatment has dramatically improved the performance of the n-channel poly-Si TFT's. The on/off current ratio has increased from 2.40×10^4 to 6.58×10^6, the field effect mobility has increased from 11.4 to 48.4 cm^2/Vsec, the threshold voltage has decreased from 7.18 to 0.55 V, the inverse subthreshold slope has decreased from 1.81 to 0.429 V/decade, and the trap-state density [9,10] has decreased from 6.39×10^{12} to 1.18×10^{12} cm^{-2}. Furthermore, a 60-min N_2O plasma treatment can more effectively improve the device performance except the minimum drain current increases because of plasma induced damages. This significant improvement in the performance of poly-Si TFT's implies that the defect states at the poly-Si/dielectric interface as well as in the active poly-Si layer can be passivated during the N_2O plasma treatment process.

Initially, considering the dissociation of N_2O, one might suggest that the nitrogen or oxygen species were generated in an N_2O plasma. They are responsible for the effective passivation of defect states and gate oxide fixed charges by forming Si-N or Si-O bonds. A SIMS analysis was therefore performed on the specimens exposed to an N_2O and H_2 plasma, respectively, for 1 hr and on the control samples (without plasma treatment). The H and SiN depth profiles are plotted in Figs. 6 (a) and (b), respectively. From the SIMS results, it indicates that not only the nitrogen but also mainly hydrogen incorporate into the gate dielectric and the channel poly-Si films after the N_2O plasma treatment. In Fig. 6(a), as expected, the H_2 plasma treatment causes a pronounced increase of the H concentration in the channel poly-Si layer. On the other hand, the N_2O plasma treatment also cause a markedly increase of the H content with respect to the control samples. The passivating atomic H species seem to be generated by dissociation of residual H_2O

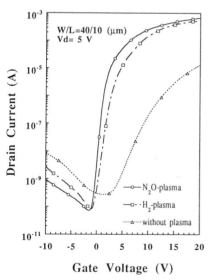

Fig. 5 The Id-Vg characteristics of the n-channel poly-Si TFT's (W/L=40/10) for N_2O-plasma, H_2-plasma, and without plasma treated conditions for V_{ds}=5 V.

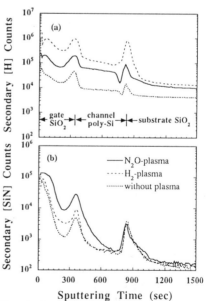

Fig. 6 Depth profiles of (a) H and (b) SiN in the gate oxides/channel poly-Si/oxide substrate structure.

in the PECVD chamber.[6] Although the N_2O and H_2 plasma treatments both have an obviously increase of the H concentration, the increment of H for H_2-plasma treated specimens is much higher than that for N_2O plasma treated ones. However, the performance of poly-Si TFT's treated with N_2O plasma is even better than that treated with H_2 plasma. Hence, the improvement of the electrical properties of TFT's treated with N_2O plasma may come from other reasons. From the SIMS results shown in Fig. 6(b), it can be seen that the Si-N content in the gate oxide and the channel poly-Si films for the N_2O plasma treated specimens is much higher than those for the H_2 plasma and without plasma ones. It is believed that the nitrogen radicals can form the Si-N bonds to reduce the interface states as piling up at the SiO_2/poly-Si interface and to passivate the defect states as diffusing into the poly-Si films.[7,8,11]

Aforementioned in Fig. 6, the hydrogen and nitrogen radicals can diffuse into and accumulate in gate dielectric, so that the effects of the plasma treatments on the electrical properties of gate dielectrics of the TFT's are worthy to be investigated. The characteristics of leakage current density versus gate voltage (J_g-V_g) for the gate dielectric after N_2O-plasma, H_2-plasma, and without plasma treatments are shown in Fig. 7. Obviously, the gate dielectrics for N_2O- and H_2-plasma treatments have lower leakage current than those without plasma ones when the gate is positively biased and the source/drain electrodes are commonly grounded. Moreover, it is found that the N_2O-plasma treated devices exhibit even lower leakage current than the H_2-plasma treated ones. That is, the quality of the gate dielectric for the N_2O-plasma treatments are better than those treated with H_2-plasma.

Furthermore, Fig. 8 shows the cumulative distributions of the time-zero-dielectric-breakdown (TZDB) characteristics of the gate dielectrics with N_2O-plasma, H_2-plasma, and without plasma treatments. The gate dielectric with N_2O-plasma treatments exhibits better TZDB characteristics than the other two cases. This improved V_{bd} of the gate dielectric is also attributed to the nitrogen and hydrogen participation. Originally, due to the high stress in the polyoxide, it is expected that there are many strained bonds within the oxide and near the interface, especially at the local regions nearby the grain boundaries. These strained bonds are easy to be broken by high-energy electrons to cause trap-state generation when a field is applied to the oxide. After N_2O-plasma treatment, the nitrogen and hydrogen in the oxide may strengthen the strained bonds and passivate the interface states. Hence, the oxide breakdown strength is improved.

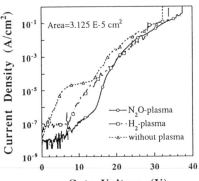

Fig. 7 The Jg-Vg characteristics of the gate oxides after N_2O-plasma, H_2-plasma, and without plasma treatments.

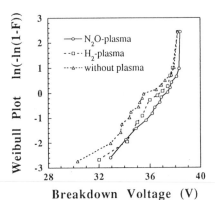

Fig. 8 The cumulative distributions of time-zero-dielectric-breakdown (TZDB) characteristics of the gate oxides with N_2O-plasma, H_2-plasma, and without plasma treatments.

CONCLUSIONS

In conclusion, the N_2O plasma treatment on the poly-Si TFT's has been demonstrated to very effectively passivate the poly-Si TFT's. The on/off current ratio 6.58×10^6, threshold voltage 0.55 V, and field effect mobility 48.2 cm^2/Vsec can be achieved after a 30-min N_2O plasma treatment with respect to the 2.40×10^4, 7.18 V, and 11.4 cm^2/Vsec, correspondingly, for the unpassivated devices. The performance improvement is attributed to the nitrogen radicals dissociated from the N_2O gas as well as the hydrogen radicals dissociated from the residual H_2O, which diffuse into the SiO_2/poly-Si interface and the channel poly-Si layer to passivate the defect states. Furthermore, the characteristics of the gate dielectric after the N_2O-plasma treatment has been also promoted.

ACKNOWLEDGMENTS

This research was supported in part by the Republic of China (ROC) National Science Council under the Contract No. NSC-85-2215-E009-035. I would like to thank Ms. P. F. Chou for her helps in SIMS analysis. In additions, the technical supports from the National Nano Device Laboratory and the Semiconductor Research Center in the National Chiao Tung University were also acknowledged.

REFERENCES

1. T. I. Kamins and P. J. Marcoux, IEEE Electron Device Lett., **EDL-1**, 159 (1980).
2. P. Migliorato, and D. B. Meakin, Applied Surface Sci., **30**, 353 (1987).
3. I-W. Wu, T-Y. Huang, W. B. Jackson, A. G. Lewis, and A. C. Chiang, IEEE Electron Device Lett., **EDL-12**, 181 (1991).
4. S. Ikeda, S. Hashiba, I. Kuramoto, H. Katoh, S. Ariga, T. Yamanaka, T. Hashimoto, N. Hashimoto and S. Meguro, IEDM Tech. Dig., 459 (1990).
5. H. N. Chern, C. L. Lee, and T. F. Lei, IEEE Trans. Electron Devices, **ED-40**, 2301 (1993).
6. N. H. Nickel, A. Yin and S. J. Fonash, Appl. Phys. Lett., **65**, 3099 (1994).
7. M. J. Tsai, F. S. Wang, K. L. Cheng, S. Y. Wang, M. S. Feng, and H. C. Cheng, Solid State Electronics, **38**, 1233 (1995).
8. F. S. Wang, M. J. Tsai, and H. C. Cheng, IEEE Electron Device Lett., **EDL-16**, p.503.
9. J. Levinson, F. R. Shepherd, P. J. Scanlon, W. D. Westwood, G. Este, and M. Rider, J. Appl. Phys., **53**, 1193 (1982).
10. R. E. Proano, R. S. Misage, and D. G. Ast, IEEE Trans. Electron Devices, **ED-36**, 1915 (1989).
11. H. S. Momose, T. Morimoto, Y. Ozawa, K. Yamabe, and H. Iwai, IEEE Trans. Electron Devices, **ED-41**, 546 (1994).

THE POLY-Si TFTS FABRICATED BY NOVEL OXIDATION METHOD WITH INTERMEDIATE OXIDE

C-M Park, J-S Yoo, B-H Min, and M-K Han

Department of Electrical Engineering, Seoul National University, Seoul 151-742, Korea

ABSTRACT

We have fabricated a poly-Si TFT using a novel oxidation method, which improves the surface roughness at the interface between the poly-Si layer and the gate oxide layer. Compared with the poly-Si TFTs fabricated by the conventional oxidation method, the proposed poly-Si TFT exhibits the remarkable enhancement of the electrical parameters, such as the subthreshold swing and the threshold voltage. It is observed that the proposed poly-Si TFT has a higher dielectric strength and the device characteristics are not degraded significantly after an electrical stress. The improvement of the surface roughness at oxide/poly-Si interface is found to be critical to enhance the device performance.

INTRODUCTION

Poly-Si TFTs (Polycrystalline Silicon Thin Film Transistors) is a promising device for active matrix liquid crystal displays (AMLCDs). The thermal oxidation of poly-Si is usually employed to fabricate gate insulators in TFT devices. The thermal oxides grown from poly-Si, referred to as poly-oxides, are required to have low leakage current and high dielectric strength. However, it is well known that the thermally grown poly-oxides have higher leakage current and lower dielectric strength than thermal oxides grown on a single crystalline silicon [1][2]. The inferior properties of the poly-oxides may be attributed to a rough poly-oxide/poly-Si interface. Several studies have been reported regarding to the effects of the surface roughness on the electrical properties of poly-oxides grown on poly-Si film [3][4]. Recently, we have reported the reduction of the surface roughness on the poly-Si film by a novel oxidation method [5]. However, previous investigations are limited to the properties of poly-oxide/poly-Si films, and the effects of the surface roughness on the characteristics of poly-Si TFTs hardly have been reported.

We report a new poly-Si TFT by employing a novel oxidation method to improve the surface roughness at the poly-oxide/poly-Si interface. In order to investigate the effect of the surface roughness on the device performance, we have studied the transfer characteristics and the oxide breakdown properties. We have also investigated the degradation phenomena of the new device due to an electrical stress.

DEVICE FABRICATION

The 1200Å thick a-Si film was deposited on the thermally oxidized silicon substrates at 550°C by LPCVD. The a-Si film was crystallized by solid phase crystallization at 600°C for 24 hours (lower poly-Si). After the formation of poly-Si islands, thin oxide layers (intermediate oxides), of which thickness were 100Å and 200Å, were formed from poly-Si films by thermal oxidation in a dry oxygen ambient. The a-Si film of 400Å thickness was deposited on the thin poly-oxide at 550°C by LPCVD. Next, the a-Si film on the thin oxide was crystallized and

183

changed into poly-Si (upper poly-Si) during the oxidation at 950°C for 6 hours. The grain size of the upper poly-Si is small and differed from that of lower poly-Si due to higher annealing temperature, and its grain boundaries did not coincide with that of the lower poly-Si due to the intermediate oxide, as shown in Fig.1(a). Because the oxidation along the upper poly-Si grain boundaries was faster than that along the center of the upper poly-Si grains, the peak of the initial poly-oxide/lower poly-Si interface under the upper poly-Si grain boundaries was lowered during that oxidation. Finally, the 1500Å thick poly-oxide layer was formed and the surface of the lower poly-Si was smoothened when the oxidation was finished, as shown in Fig. 1(b). For comparison, the poly-Si TFT was fabricated by the conventional oxidation method, in which the 1500Å thick poly-oxide layer was directly grown from poly-Si film by dry thermal oxidation. The 1000Å thick poly-Si was patterned to form the gate electrode and the subsequent device processes such as source/drain implantation and metallization, in the standard manner, were performed to complete the poly-Si TFTs. Hydrogen plasma treatment at 300°C for 30 minute was performed in order to enhance the device performance.

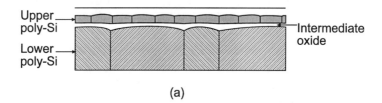

(a)

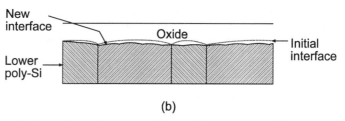

(b)

Fig. 1. The schematic representations of the novel oxidation method

(a) Intermediate oxide (100Å, 200Å) and a-Si film (400Å) are formed on the poly-Si, and the a-Si film is crystallized during the oxidation.

(b) The new poly-oxide/poly-Si interface is formed after the thermal oxidation.

RESULTS AND DISCUSSION

Fig. 2 shows I_D-V_G characteristics of the proposed poly-Si TFTs and the conventional poly-Si TFTs. The proposed poly-Si TFTs with the intermediate oxide thickness of 100Å and 200Å were fabricated by the novel oxidation method. The conventional poly-Si TFT was also fabricated by conventional oxidation method without the intermediate oxide. It is found that the proposed poly-Si TFTs have better electrical characteristics than the conventional poly-Si TFTs. In the conventional poly-Si TFTs, the threshold voltage and the subthreshold swing are 3.4V and

1.75 V/decade, respectively. In the proposed poly-Si TFTs with 100Å and 200Å thick intermediate oxide, the threshold voltages are 1.5V and 2.4V, and the subthreshold swings are 0.82 V/decade and 1.24 V/decade, respectively. The electrical characteristics between two proposed poly-Si TFTs with 100Å and 200Å thick intermediate oxide is varied due to the difference of the surface roughness at the oxide/poly-Si interface. Fig. 3 shows the variation of I_D-V_G characteristics for the conventional and proposed poly-Si TFTs before and after an electrical stress with V_G = 25V and V_D = 1V for 90 min. The characteristics of the proposed poly-Si TFTs with the intermediate oxide thickness of 100Å have a little difference between before and after the electrical stress. However, In the conventional poly-Si TFTs, it is seen that the characteristic curve after the stress shifts to the right of the original curve before the stress. It is found that the value of the threshold voltage increase from 3.4V to 6.5V and the subthreshold swing is changed from 1.75 V/decade to 1.89 V/decade. It is also found that the smooth surface is an important factor to control the degradation of the poly-Si TFTs due to the electrical stress.

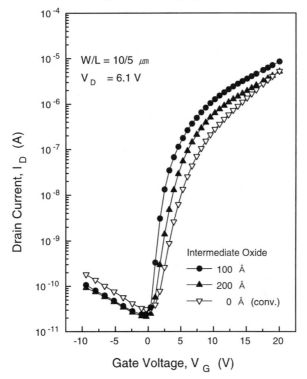

Fig.2. I_D-V_G characteristics of the conventional and proposed poly-Si TFTs. The intermediate oxide thickness of the proposed poly-Si TFTs are 100Å and 200Å, respectively. Hydrogen passivation was performed for 30 minute.

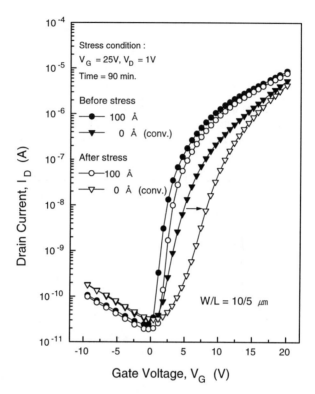

Fig.3. The comparison of I_D-V_G characteristics for the conventional and proposed poly-Si TFTs before and after an electrical stress (V_G=25V, V_D=1V). The stress time is 90 minute. The intermediate oxide thickness of the proposed poly-Si TFTs is 100Å.

The breakdown electric field of the gate oxide is higher in the proposed poly-Si TFTs than in the conventional poly-Si TFTs, as shown in Fig. 4. This result implies that the local electric field at the oxide/poly-Si interface is reduced because the surface roughness of the proposed poly-Si TFTs is smoother than that of the conventional poly-Si TFTs.

CONCLUSION

We have fabricated the poly-Si TFTs by employing the novel oxidation method to improve the surface roughness at the poly-oxide/poly-Si interface. Compared with the conventional poly-Si TFTs, the device performance of the proposed poly-Si TFTs, such as threshold voltage, subthreshold swing and on current, was improved remarkably. It was also found that the breakdown electric field of the gate oxide was increased more than 8 MV/cm and the device characteristics were not degraded due to the electrical stress.

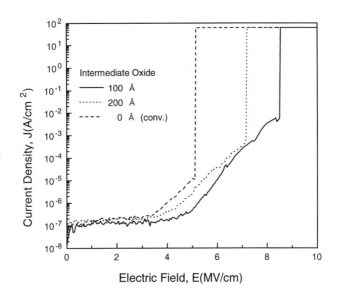

Fig.4. Current density versus electric field (J-E) characteristics of the gate oxide for the conventional and proposed poly-Si TFTs. The intermediate oxide thickness of the proposed poly-Si TFTs are 100Å and 200Å, respectively.

ACKNOWLEDGEMENT

This work was performed with the financial support of Ministry of Education Reaserch Fund for Advanced Materials in 1995

REFERENCES

1. J. DiMaria and D. R. Kerr, Appl. Phys. Lett., **27**, 505 (1975)

2. R .M. Anderson and D. R. Kerr, J. Electrochem. Soc., **48**, 4834 (1977)

3. Takechi, H. Uchida, and S. Kaneko, Proc. Mat. Res. Soc. Symp., **258**, 955 (1992)

4. Moazzami and C. Hu, IEEE Electron Device Lett., **14**, (1993)

5. C. Jun, Y. S. Kim, M. K. Han, J. W. Kim, and K. B. Kim, Appl. Phys. Lett., **66**, 2206 (1995)

ECR PLASMA OXIDATION OF AMORPHOUS SILICON FOR IMPROVEMENT OF THE INTERFACE STATE IN A POLY SILICON THIN FILM TRANSISTOR

TAE-HYUNG IHN, SEOK-WOON LEE, BYUNG-IL LEE, YOO-CHAN JEON* AND
SEUNG-KI JOO
DEPT. OF METALL. ENG. SEOUL NAT'L UNIV., KOREA
LG SEMICON CO. LTD.*

ABSTRACT

A new oxidation process of the poly Silicon thin films has been demonstrated, where the amorphous silicon thin film was oxidized by ECR plasma at room temperature and then crystallized at 600℃ for 30hrs(pre-oxidation). For comparison, amorphous silicon thin film was oxidized after crystallization(post-oxidation). The interface roughness turned out to be only 3Å in case of the pre-oxidation process, while the post-oxidation showed about 8Å of the interface roughness. The pre-oxidized TFT showed the constant mobility at high gate voltages, the post-oxidized TFT showed serious degradation of the mobility at high gate voltages.

INTRODUCTION

For active matrix liquid crystal display(AMLCD's), polycrystalline silicon thin film transistors(poly-Si TFT's) are preferable to amorphous silicon(a-Si) TFT's because of the fast response and integrability of the driving circuits with pixel transistors on the same substrate[1-5]. Since the gate oxide of the poly silicon TFT's has to be formed on the poly crystalline silicon substrate rather than the single crystalline silicon substrate, the interface state between the gate oxide and the substrate may seriously affect the operation characteristics of the TFT's. It has been reported that the interface roughness between the oxide and the poly-Si substrate causes deterioration of the breakdown characteristics[6,7,8] and the field effect mobility decrease with increase of the interface roughness[9,10]. Recently, it has been reported that the SiO_2 formed by ECR plasma oxidation of the single crystalline silicon shows dielectric qualities as good as thermally oxidized SiO_2[11]. According to Lee et al., the gate oxide formed by ECR plasma oxidation of the poly crystalline silicon showed the interface roughness of 8Å, which is much smoother than that formed by thermal oxidation of polycrystalline silicon[12].

In this work, a-Si was oxidized by ECR plasma at room temperature before crystallization of the a-Si substrate. Relation between the interface roughness and poly-Si TFT's characteristics was examined. Difference between ECR oxidation of a-Si and ECR oxidation of poly-Si in terms of device performance was carefully investigated

EXPERIMENT

1000Å amorphous silicon thin films were deposited on thermally grown SiO_2 substrates by LPCVD at 550℃ using SiH_4 gas. ECR plasma oxidation was carried out at the microwave power of 600W. Oxygen was introduced into the chamber during oxidation at the flow rate of 10 SCCM. Crystallization of a-Si was done by SPC(solid phase crystallization) at 600°C for 30hrs. The interface between the poly-Si substrate and the oxide was examined with AFM after the oxide was removed with BHF.

For poly-Si TFT's, thickness of the gate oxide was fixed to 1000 Å. 300 Å of gate oxide was formed by ECR oxidation and the rest by ECR CVD, where SiH_4 and O_2 gas were simultaneously flowed into the chamber. Pre-oxidized TFT means the gate oxide was formed by ECR oxidation of a-Si thin film and then the substrate was crystallized into poly-Si, while SPC was carried out prior to ECR oxidation for the gate oxide in post-oxidized TFT in this work. The hydrogen plasma passivation was done in ECR H_2 plasma.

RESULTS AND DISCUSSIONS

Average roughness of a-Si thin film deposited by LPCVD was 2.5 Å, and it was almost unchanged after the solid phase crystallization (SPC) at 600°C for 30hrs (figure 1 (a)).

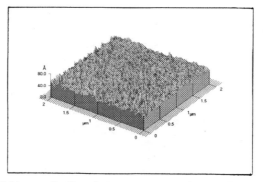

(a) AFM image of a-Si surface, average rms surface roughness : 2.5 Å

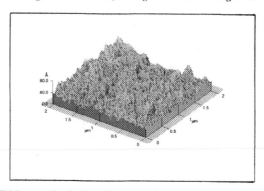

(b) AFM image of poly-Si surface, average rms surface roughness : 8 Å

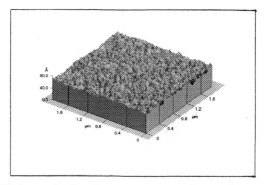

(c) AFM image of a-Si surface, average rms surface roughness : 2.9 Å

Figure 1. AFM images of the interface of the oxide(400 Å) and the substrate.

    (a) a-Si surface before oxidation

    (b) polycrystalline surface where the oxide was formed after the SPC

    (c) polycrystalline surface where the oxide was formed prior to the SPC

When 400 Å of the oxide was formed by ECR plasma oxidation, average roughness of the poly silicon substrate became 8 Å as shown in figure 1 (b). However, when a-Si was oxidized to 400 Å by ECR plasma oxidation prior to crystallization, average roughness measured after the crystallization was only 2.9 Å, which is essentially same to that of a-Si before oxidation(figure 1 (c)).

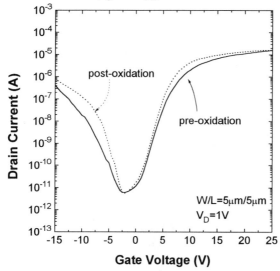

Figure 2 Comparison of the transfer characteristics of the pre-oxidized
and the post-oxidized TFT's

Since there is no grain boundaries in a-Si, oxidation of a-Si at room temperature by ECR oxidation might be expected to produce smoother interface than oxidation of poly-Si. The crystallinity of the substrate after SPC was 65% in both cases measured by UV reflectance spectroscopy. Figure 2 showed that the transfer characteristics of the pre-oxidized TFT and the post-oxidized TFT. Both cases show almost same properties except that the leakage current of the pre-oxidized TFT is slightly lower than the post-oxidized TFT. The field effect mobility was calculated in the linear region at $V_D=0.1V$. Figure 3 shows that the calculated mobility variation with the gate voltages. It was reported that the variation of field effect mobility with increasing gate voltage is mainly due to the interface roughness between gate oxide and poly-Si substrate[9,10]. In figure 3, the maximum field effect mobility of the pre-oxidized TFT is maintained even at high gate voltages but in case of the post-oxidized TFT, the field effect mobility is drastically decreased at the high gate voltages. This drastic degradation of the field effect mobility with increase of the gate voltage might be related to the rough interface between the gate oxide and the poly-Si substrate. After ECR H_2 plasma passivation, the pre-oxidized TFT showed decrease of subthreshold slope(V/dec), threshold voltage(V_T) and leakage current (figure 4) but in case of the post-oxidized TFT, no appreciable affect of the hydrogen plasma passivation could be observed except the leakage current (figure 5). The values of device parameters are summarized in Table I .

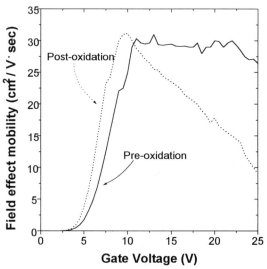

Figure 3 Field effect mobilities with increase of the gate voltages.
The W/L is 5μm/5μm and the mobility was calculated at
the linear regions.

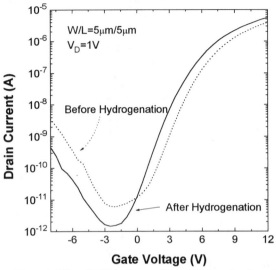

Figure 4 **Effect of ECR hydrogenation on the transfer characteristics in pre-oxidized.**

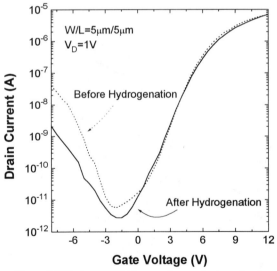

Figure 5 **Effect of ECR hydrogenation on the transfer characteristics in post-oxidized.**

Table I Device parameters of pre-oxidized and post-oxidized TFT's

Parameter	Pre-oxidized TFT		Post-oxidized TFT	
	Before Hydrogenation	After Hydrogenation	Before Hydrogenation	After Hydrogenation
Field effect mobility(cm^2/Vsec)	31	42	32	41
Subthreshold slope(V/dec)	1.2	1.0	1.2	1.2
Threshold voltage(V)	6.5	5	5.5	5.5
On/Off ratio(V$_G$=-15~25V)	8×10^6	1.2×10^7	8×10^6	1×10^7
Leakage current(pA/μ m)	1.2	0.3	1.2	0.5

CONCLUSIONS

According to AFM observation, when the gate oxide was formed by ECR plasma oxidation of a-Si thin films(pre-oxidation), the interface roughness between the oxide and the substrate turned out to be more smoother than the oxidation of poly-Si thin film(post oxidation), and in case of the pre-oxidation, interface roughness was not changed by SPC of a-Si at 600℃ for 30hrs.

The current-voltage characteristics of the pre-oxidized TFT were similar to the post-oxidized TFT, but in case of pre-oxidized TFT, no appreciable degradation of the mobility with increase of gate voltage could be observed in comparison with the post-oxidized TFT's. After ECR H$_2$ plasma passivation, the efficiency of the passivation effect was apparent in the pre-oxidized TFT's.

ACKNOWLEDGEMENTS

This work has been supported by KOSEF through RETCAM in Seoul National University.

REFERENCES

1. K. Nakazawa, J. Appl. Phys., **69(3)**, p.1703 (1991).
2. T. J. King and K. C. Saraswat, IEEE Electron Device Lett, **13(6)**, p.309 (1992).
3. T. Aoyama, G. Kawachi, N. Konishi, T. Suzuki, Y. Okajima, and K. Miyata, J. Electrochem. Soc., **136(4)**, p.1169 (1989).
4. T. W. Little, K. Takahara, H. Koike, T. Nakazawa, I. Yudasaka, and H. Ohshima, Jpn. J. Appl. Phys. **30(12B)**, p.3724 (1991).
5. H. Kuriyama, S. Kiyama, S. Noguchi, T. Kuwahara. S. Ishida, T. Nohda, K. Sano, H. Iwata, S. Tsuda, and S. Nakano, IEDM 1991 Tech. Dig., p.563 (1991).
6. E.A Irene, E, Tierndy and D.W. Dong, J. Electrochem. Soc., **127**, p.705 (1980).
7. L. Faramle et al, J. Electrochem. Soc., **133**, p.1410 (1986).
8. P. A. Heimann et al, J. Appl. Phys., **53**, p.6240 (1982).
9. K. Takechi et al, MRS Proc., **258**, p.955 (1992).
10. Ji-Ho Kung, Miltiadis K. Hatalis and Jerzy Kanicki, Thin Solid Films, **216**, p.137 (1992).
11. D. A. Carl, D. W. Hess and M. A. Liberman, J. Vac. Sci. Technol., **A8(3)**, p.2924 (1990).
12. J. Y. Lee, C. H. Han and C. K. Kim, IEEE Electron Device letters, **15(8)**, p.301(1994).

DEVICE CHARACTERISTICS OF A POLY-SILICON THIN FILM TRANSISTOR FABRICATED BY MILC AT LOW TEMPERATURE

SEOK-WOON LEE*, BYUNG-IL LEE**, TAE-HYUNG IHN**, TAE-KYUNG KIM**, YOUNG-TAE KANG**, and SEUNG-KI JOO**
*Samsung Electronics Co., Ltd., Nongseo-ri, Kiheung-eup, Yongin-city, Kyungki-do, Korea
**Dept. of Metall. Eng., Seoul Nat'l Univ., Shillim-dong, Kwanak-ku, Seoul, Korea

ABSTRACT

High performance poly-Si thin film transistors were fabricated by using a new crystallization method, Metal-Induced Lateral Crystallization (MILC). The process temperature was kept below 500 ℃ throughout the fabrication. After the gate definition, thin nickel films were deposited on top of the TFT's without an additional mask, and with a one-step annealing at 500 ℃, the activation of the dopants in source/drain/gate a-Si films was achieved simultaneously with the crystallization of the a-Si films in the channel area. Even without a post-hydrogenation passivation, mobilities of the MILC TFT's were measured to be as high as 120cm^2/Vs and 90cm^2/Vs for n-channel and p-channel, respectively. These values are much higher than those of the poly-Si TFT's fabricated by conventional solid-phase crystallization at around 600 ℃.

INTRODUCTION

Poly-Si TFT's are attracting considerable attention for the replacement of amorphous TFT's for the AMLCD (Active Matrix Liquid Crystal Devices) requiring high resolution and high response speed. Conventional amorphous Si TFT's can switch the LCD pixel, but cannot drive the LCD peripheral circuits, so LCD should be connected to discrete IC's fabricated on a single crystal Si wafer. Since poly-Si TFT's show high speed, they enable the integration of LCD pixel and driving circuit on the same substrate, which result in high performance and high productivity. Currently, however, a fused quartz plate is being used as a substrate because of its high processing temperature, which prevents cost-effectiveness and development of a large size LCD. Among the other methods, Poly-Si formation needs the highest temperature. The most widely used poly-Si formation methods are SPC (Solid Phase Crystallization) [1-4] or ELA (Excimer Laser Annealing)[5]. The SPC can be adapted easily to industrial application by its simple process and equipments, but relative high temperature above 600 ℃ and a long process time of 20~30 hours are the main barriers to the application of glass substrates. The ELA enables low temperature process, but a high cost machine as well as non-uniformity still remains as a problem.

Among the attempts to lower the crystallization temperature of a-Si, the most remarkable one is metal induced crystallization (MIC). It has been reported that some metals such as Ag[6,7], Au[8], Al[7,9,10], Sb[10], and In[11], lower the crystallization temperature below 500 ℃. In spite of successful low temperature crystallization, MIC can not be used in electronic fields due to its original metal contamination drawbacks which degrade crystalline qualities. Recently, a new low temperature crystallization phenomenon has been reported[12] and it was shown that MILC (Metal Induced Lateral Crystallization) can crystallize the a-Si films in metal-free area.

In this work, a new poly-Si TFT fabrication method is proposed and actual devices were fabricated with whole processing temperature below 500 ℃. Through only one annealing step, source/drain and gate areas were crystallized via MIC and channel area via MILC from the seeds in the source/drain area. Device characteristics of N- and P-channel TFT's are investigated.

EXPERIMENT

Fabrication Method of MILC poly-Si TFT

Schematic diagrams showing the fabrication steps of MILC poly-Si TFT's are shown in Fig. 1. A thermally oxidized Si (100) wafer was used as a substrate. A 1000 Å-thick a-Si film was deposited by LPCVD at 480 ℃ using Si$_2$H$_6$, and active island was defined. For gate dielectric, 300 Å plasma oxidation and 700 Å CVD oxide was formed in ECR chamber[13].

195

The gate and gate dielectric was patterned and defined by conventional SF_6 plasma etching and wet chemical etching(diluted BHF solution), respectively. Self aligned ion implantation was carried out to form p-type gate at P-channel TFT's and n-type at N-channel TFT's. A thin (5 Å) Ni film was then deposited without an additional mask as in Fig. 1(b). Through the annealing at 500 ℃, the a-Si film of source/drain and gate areas was crystallized by Ni-MIC, while the a-Si in the channel area was laterally crystallized from the source/drain area by MILC. After the crystallization, unreacted Ni films on oxide area was removed by wet etching. Finally, Al metallization was performed for electrical measurement as shown in Fig. 1(c). Summarized process conditions are listed in table 1.

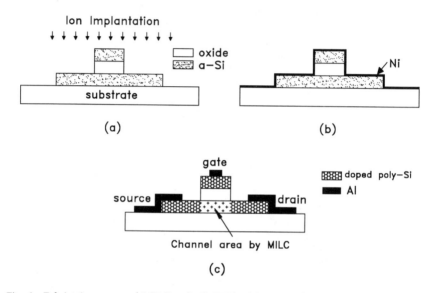

Fig. 1 Fabrication steps of MILC poly-Si TFT's. (a) ion implantation for source/drain and gate formation (b) deposition of a Ni layer (20 Å) (c) crystallization annealing and electrode formation.

Table 1 Key processes for the fabrication of MILC poly-Si TFT's.

Process	Method
a-Si deposition	LPCVD (1000 Å, 460 ℃)
gate oxide formation	ECR Oxidation (300 Å) + ECR CVD (700 Å)
gate (a-Si)	LPCVD (2000 Å, 460 ℃)
source/drain doping	Ion Implantation (P, 5E15cm^{-2}, 40keV or BF$_2$, 3E15cm^{-2}, 40keV)
Ni deposition	Sputtering (20 Å)
Al deposition	Sputtering (2000 Å)

Low temperature dopant activation method by MIC

A 5 Å-thick Ni film was deposited on a-Si and then P was doped into the a-Si using PH_3 (5% diluted in H_2) plasma in an ECR chamber[14]. To investigate the change of sheet resistance and crystallization behavior of P doped a-Si, RTA (Rapid Thermal Annealing) was performed at 500 ℃. Sheet resistance was continually monitored by 4-point probe method and the crystallinity of the Si film was calculated by UV/VIS reflectance measurement[15,16].

RESULTS AND DISCUSSIONS
Crystallization of channel area by MILC

Fig. 2 is Nomarski optical micrographs showing the progress of MILC in the channel area of a TFT. The cross-sectional structure without gate Si is shown in Fig. 2(a) for observing MILC in the channel region. Channel width and length are both 40μm, and a 20Å-thick Ni layer was selectively deposited only on the source/drain area by a lift-off technique where any dopants were not implanted. In 5 hours at 500℃, 8μm-long lateral crystallization from the source/drain area is observed as in Fig. 2(b). And in 15hours, channel area is seen to be fully crystallized as in Fig. 2(c). Crystallization rate is much higher than that of conventional SPC, which is attributed to the formation of $NiSi_2$ for which the lattice mismatch with Si is only 0.4%. Ni deposited on a-Si reacts with Si to form a thin $NiSi_2$ in the first stage, and the $NiSi_2$ precipitates propagate into the a-Si region leaving the crystallized Si behind continuously[17,18].

In MILC structured TFT, time for full crystallization depends only on channel length. Crystallization length with respect to time at 500℃ is shown in Fig. 3. The thickness of a Ni layer on a-Si was 20Å. Up to 30 hours, Ni-MILC growth rate remained constant, 1.6μm/hour.

Low temperature dopant activation by MIC

For the low temperature fabrication of poly-Si TFT's, dopants implanted on the source/drain area should be activated at lower or equal temperature to that of crystallization one. In case of single crystalline or poly-crystalline Si, a relative high temperature (up to 600℃) annealing is required to activate dopants. But when the crystalline-Si is amophized during implantation, dopant activation temperature is reported to be slightly lowered through recrystallization of Si[19]. This suggests that lowering the crystallization temperature of a-Si can also pull down the dopant activation temperature. Fig. 4 shows the relationship between crystallization rate of the a-Si and dopant activation in MIC or MILC process. In 40 min, a-Si started crystallization and was fully crystallized in 120min. At the beginning of crystallization, sheet resistance dropped sharply, but no more drop occurred after 50% level of crystallization of Si. Sheet resistance drop with crystallization of a-Si at 500℃ enables the whole process temperature to be kept to the crystallization temperature.

Fig. 5 shows sheet resistances of Si film which was doped with P or BF_2 and then activated at 500℃. The implantation energies were 40KeV, and doses were varied from $1E14/cm^2$ to $3E15/cm^2$. For comparison, Si films annealed at 600℃ without Ni films were prepared (SPC). Annealing time was 5 hours for MIC and 30 hours for SPC, and further annealing did not decrease the sheet resistance. 500℃-annealed Ni-MIC sheet resistance showed a lower value than that of 600℃ annealed SPC, in spite of the lower annealling temperature. Resistivity of poly-Si depends on dopant concentration as well as microstructure. Crystal defects such as Dislocations, microtwins and grain boundaries, degrade the mobility of electrons or holes. According to the TEM analysis[28], it was observed that the grains formed by MILC grew up and merged to several μm, and contained no microtwins which can always be observed in poly-Si grains grown by SPC. From this, the low resistivity of low temperature crystallized Ni-MIC can be explained.

Characteristics of MILC poly-Si TFT

I-V characteristics of N and P-channel poly-Si TFT's by MILC are shown in Fig. 6, and key device parameters are summarized in table 2. The field effect mobilities of N- and P-channel TFT were $120cm^2/Vs$ and $90cm^2/Vs$, respectively. Compared with SPC a mobility of which ranged $40~60cm^2/Vs$[20], MILC TFT gives about 3 times larger value. It is known that the mobility largely depends both on grain microstructure and grain size at the channel region. The excimer laser crystallized poly-Si TFT's of which the grain size is smaller than that of SPC, show a high electron mobility due to its micro defect free grains. Thus, Ni-MILC poly-Si TFT's having larger grain size and micro defect free grains give much higher field effect mobility than that of SPC.

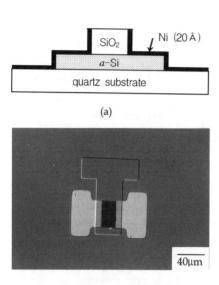

(a)

(b)

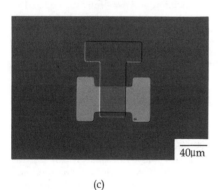

(c)

Fig. 2 Nomarski optical micrographs showing the progress of MILC in the channel area of a TFT. A 20Å-thick nickel layer was deposited after source/drain definition and the width/length of the TFT was 40μm/40μm. (a) cross-sectional sample structure before annealing (b) 500℃ 10h (c) 500℃ 15h

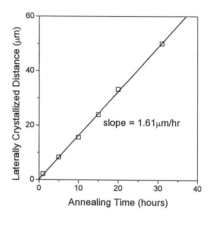

Fig. 3 Laterally crystallized distance as a function of annealing time at 500℃. The distance was measured from the interface between the source/drain and the channel area.

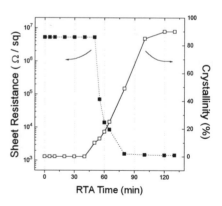

Fig. 4 Variation of crystallinity and sheet resistance with annealing time at 500℃. A 5Å-thick nickel layer was deposited on top of a-Si films and then P doping was performed before rapid thermal annealing.

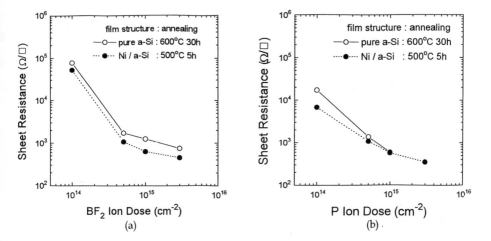

Fig. 5 Sheet resistance vs. ion dose after crystallization. For Ni-MIC, a 5Å-thick Ni layer was deposited after ion implantation and then the sample was annealed at 500℃ for 5 hours. For SPC, the annealing was done at 600℃ after the ion implantation.   (a) P, 40keV   (b) BF$_2$, 35keV.

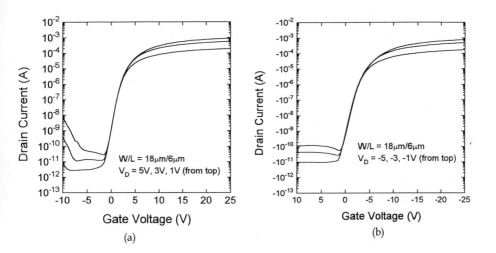

Fig. 6 I$_D$-V$_G$ characteristics of MILC poly-Si TFT's. The width/length of TFT's were 18μm/6μm.   (a) N-channel (V$_D$ = 1, 3, 5V)   (b) P-channel (V$_D$ = -1, -3, -5V).

Table 2 Device parameters of N-channel and P-channel poly-Si TFT's. (W/L=18/6)

Parameter	N-ch	P-ch
threshold voltage (V)	1.2	1.7
subthreshold slope (V/dec)	0.6	0.7
field-effect mobility $(cm^2/V \cdot s)$	120	90
maximum on/off current ratio	6×10^7	2×10^7
minimum leakage current/channel width $(pA/\mu m)$	0.15	0.45

For an N-channel (P-channel) TFT, the threshold voltage was defined at a normalized drain current $(I_D \times W/L)$ of $0.1 \mu A$ $(-0.1 \mu A)$ at $V_D = 1V$ (-1V). The field-effect mobility was calculated in the linear region at $V_D = 0.1V$ (-0.1V). The maximum on/off current ratio was determined at $V_D = 1V$ (-1V) and $V_G = -10 \sim 25V$ $(10 \sim -25V)$.

CONCLUSIONS

High performance poly-Si TFT's were fabricated at 500℃ by a newly developed crystallization method ; Metal Induced Lateral Crystallization (MILC). The dopants implanted into a-Si in the source/drain area were activated simultaneously with the crystallization of a-Si, and the crystallized Si film showed lower resistiviyies than that activated at 600℃ by SPC.

Field effect mobility of the N- and P-channel poly-Si TFT's by MILC at 500℃ was as high as $120cm^2/Vs$ and $90cm^2/Vs$, respectively, which is a much larger value than that of conventional TFT's fabricated by solid phase crystallization at 600℃ for several tens of hours. This novel low temperature crystallization method, MILC, is thought to enable the fabrication of LCD's having high resolution and high reliability on large area glass substrates.

REFERENCES

1 K. Nakazawa, J. Appl. Phys. 69(3), 1703, 1991.
2 T. J. King and K. C. Saraswat, IEEE Electron Device Lett. 13(6), 309, 1992.
3 T. Aoyama, G. Kawachi, N. Konishi, T. Suzuki, Y. Okajima, and K. Miyata, J. Electrochem. Soc. 136(4), 1169, 1989.
4 T. W. Little, K. Takahara, H. Koike, T. Nakazawa, I. Yudasaka, and H. Ohshima, Jpn. J. Appl. Phys. 30(12B), 3724, 1991.
5 H. Kuriyama, S. Kiyama, S. Noguchi, T. Kuwahara, S. Ishida, T. Nohda, K. Sano, H. Iwata, S. Tsuda, and S. Nakano, IEDM 1991 Tech. Dig., 563, 1991.
6 G. Ottaviani, D. Sigurd, V. Marrello, H. W. Mayer, and J. O. McCaldin, J. Appl. Phys. 45(4), 1730, 1974.
7 T. J. Konno and R. Sinclair, Mat. Sci. Eng. A179/A180, 426, 1994.
8 L. Hultman, A. Robertsson, H. T. G. Hentzell, and I. Engstrom, J. Appl. Phys. 62, 3647, 1987.
9 G. Radnoczi, A. Robertsson, H. T. G. Hentzell, S. F. Gong, and M. -A. Hasan, J. Appl. Phys. 69(9), 6394, 1991.
10 S. F. Gong, H. T. G. Hentzell, A. E. Robertsson, L. Hyltman, S. -E. Hornstrom, and G. Radnoczi, H. Appl. Phys. 62, 3726, 1987.
11 E. Nygren, A. P. Pogany, K. T. Short, H. S. Williams, R. G. Elliman, and H. M. Posate, Appl. Phys. Lett. 52, 439, 1988.
12 S. W. Lee, Y. C. Jeon, and S. -K. Joo, Appl. Phys. Lett., 66(13), 1671, 1995.
13 T. H. Ihn and S. -K. Joo, 2nd Pacific Rim Inter. Conf. on Adv. Mat. and Proc., 1333, 1995.
14 S. W. Lee and S. -K. Joo, Digest of AMLCD '95, 113, 1995.
15 G. Harbeke and L. Jasrzebski, J. Electrochem. Soc., 137, 696, 1990.
16 S. W. Lee, Y. C. Jeon, and S. -K. Joo, Mater. Res. Soc. Proc. 321, 707, 1993.
17 C. Hayzelden and J. L. Batstone, J. Appl. Phys. 73(12), 8279, 1993.
18 S. W. Lee and S. -K. Joo, IEEE Electron Dev. Lett., 17(4), 160, 1996.
19 B. L. Crowder and F. F. Morehead, Appl. Phys. Lett., 14, 313, 1969.
20 J. Y. Lee, C. H. Han, and C. K. Kim, IEDM 1994 Tech. Dig., 523, 1994.

IMPROVEMENT OF POLY-SILICON THIN FILMS
AND THIN FILM TRANSISTORS USING ULTRASOUND TREATMENT

S. OSTAPENKO*, L. JASTRZEBSKI*, J.LAGOWSKI* and R.K.SMELTZER**
* Center for Microelectronics Research, University of South Florida, Tampa, FL 33620
** David Sarnoff Research Center, Princeton, NJ 08543

ABSTRACT

Ultrasound treatment (UST) was applied to improve electronic properties of polycrystalline silicon films on silica-based substrates. A strong decrease of sheet resistance by a factor of two orders of magnitude was observed in hydrogenated films at UST temperatures lower than 100°C. This is accompanied by improvement of a film homogeneity as confirmed by spatially resolved photoluminescence study. The UST effect on sheet resistance demonstrates both stable and metastable behavior. A stable UST effect can be accomplished using consecutive cycles of UST and relaxation. An enhanced passivation of grain boundary defects after UST was directly measured by nano-scale contact potential difference with atomic force microscope. Two specific UST processes based on interaction between ultrasound and atomic hydrogen are suggested: enhanced passivation of grain boundary states and UST induced metastability of hydrogen related defects. We also demonstrate a utility of UST for improvement of leakage current and threshold voltage in hydrogenated poly-Si thin film transistors.

INTRODUCTION

Polycrystalline silicon (poly-Si) thin films on glass or fused silica substrates are promising for thin film transistor (TFT) technology in active matrix liquid crystal displays. Compared with transistors using amorphous silicon films, poly-Si TFTs have improved operational parameters due to a substantially higher electron mobility. However, grain boundary and interface defects in poly-Si lead to high off-state current and affect threshold voltage. A conventional approach to passivate these defect states and to reduce inter-grain barriers for electron transport is hydrogenation. Plasma hydrogenation using radio frequency or electron cyclotron resonance technique were proven to be effective to improve parameters of poly-Si TFT's [1]. The hydrogen defect passivation occurs in two steps: plasma penetration and a subsequent hydrogen diffusion. The diffusion of hydrogen in poly-Si is slow compared with single crystal silicon due to a trap limiting mechanism at grain boundaries [2], typically resulting in a long hydrogenation time and electrical inhomogeneity within passivated regions of poly-Si. Previously we reported the improvement of recombination properties in bulk and thin film poly-Si by using Ultrasound Treatment (UST), which is a new technological approach in defect engineering [3,4]. UST was proven to be extremely beneficial in polycrystalline material where ultrasound vibrations enhance an interaction between extended lattice defects, like grain boundaries and dislocations, with mobile point defects. In particularly, the UST enhanced hydrogenation was suggested in poly-Si thin films [4]. In this paper we present additional facts, which provide an evidence of ultrasound stimulated effect on the atomic hydrogen in poly-Si. The effect of UST on threshold voltage and leakage current in commercial hydrogenated TFTs is documented.

EXPERIMENTAL DETAILS

Semi-insulating silicon films with thicknesses ranging from 0.35 to 0.55μm were deposited at 625°C or 550°C on Corning 7059 glass or fused silica substrates by low pressure chemical vapor deposition. The 625°C films had a crystallite structure. The films deposited at 550°C were amorphous, and their annealing under nitrogen flow for 8 to 75hr at 550°C caused a gradual crystallization. Poly-Si films were selectively plasma hydrogenated using an opening in the Al mask. This enabled us to study UST effects by comparing hydrogenated with non-hydrogenated areas on the same film. The hydrogenation was performed at 300°C in a parallel plate RF plasma system operating at a 100ccm H_2 flow with 0.3Torr pressure and 200W radio frequency

Mat. Res. Soc. Symp. Proc. Vol. 424 © 1997 Materials Research Society

power or using a low pressure electron cyclotron resonance (ECR) plasma system. For UST experiments, ultrasound vibrations were applied to poly-Si films through a glass or fused silica substrates from external circular piezoceramic transducers of 70mm diameter and 3mm thickness. Transducers were driven at there resonance frequency. Each UST is specified by a set of parameters: ultrasound resonance frequency (25KHz for radial mode and 650KHz for thickness mode); amplitude of acoustic strain generated by transducer, typically, on the order of 10^{-5}; temperature from 20 to 100°C; and UST time of 5 to 120min. The temperature of a sample during UST was measured in-situ by a remote infrared pyrometer, while the amplitude of the acoustic vibrations was controlled by a miniature piezoelectic detector. Details of UST set-up were published elsewhere [4].

The UST effect was monitored by measurements of sheet resistance at room temperature using the four-point-probe method. Concurrently, spatially resolved photoluminescence (PL) and nano-scale contact potential difference (CPD) mapping were performed. PL spectra at temperatures 4.2 to 300K were analyzed using the single SPEX 500M spectrometer coupled with cooled Ge or PbS detector. The 514 nm Ar+-laser line with power from 50mW to 150mW was used as the PL excitation source. The light absorption coefficient of our films at 514nm according to room temperature transmission is $3 \times 10^4 \text{cm}^{-1}$. The CPD mapping with spatial resolution of 20nm was performed in a two-step procedure [5] using the atomic force microscope "Nanoscope III". In the first step, a regular topographic map of a poly-Si surface was obtained by a standard AFM scan in the "tapping" mode. In the second step, the AFM tip was lifted over a sample surface with an elevation of 20nm. This constant elevation was accurately held by a computer control system during the second map, and therefore, the tip-to-sample capacitance was maintained constant. When an external AC voltage is applied to the AFM tip, the amplitude of the first harmonic of tip oscillations is proportional to the tip-to-sample potential difference. Using a compensating DC voltage, we obtained the spatial distribution of contact potential difference in the poly-Si surface in a form of CPD map. Details of the CPD technique have been published elsewhere.

RESULTS

We have found that plasma hydrogenation applied to poly-Si films reduces resistance by a factor of one order of magnitude which is consistent with previously published plasma hydrogenation data. We notice that the decreased resistance is saturated to a minimum value of the order of 10^9 Ω/sq after 3 to 5hr RF plasma processing of 0.35μm films. In the films where the plasma hydrogenation process was not completed, we observed the additional dramatic reduction of sheet resistance by a factor of *two orders* of magnitude after UST (Figure 1). Resistance in non-hydrogenated region was practically not changed after the UST. Our control 100°C annealing of the same hydrogenated films without UST did not reveal any change in resistance with accuracy of 10%. We notice that the UST-induced reduction of sheet

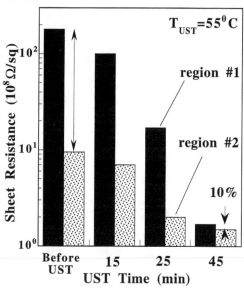

Fig.1 UST effect on sheet resistance in two regions of the hydrogenated poly-Si film. UST parameters: T=55°C, f=25kHz (radial vibrations)

resistance occurs at temperatures lower than 100°C, which is specifically beneficial for large scale poly-Si TFT applications requiring low-cost glass substrates with low thermal stability. By varying the UST time, we found that the kinetics of UST effect is characterized by a time constant of about 25min at 55°C. Another feature of the UST effect is an improvement of resistance homogeneity within the same poly-Si film. In Figure 1 we present the UST induced change of resistance in two different regions (#1 and #2) of the same film. Initially, a more than one order of magnitude variation in resistance was reduced to approximately 10% after a few consecutive UST runs. Based on these findings, we strongly suggest that ultrasound vibrations applied to hydrogenated films can promote the process of defect passivation with atomic hydrogen. This statement is justified by the following spatially resolved PL and nano-scale CPD study.

It was reported recently that PL spectroscopy allows defect monitoring in polycrystalline Si films, and is specifically sensitive to the state of film hydrogenation [6]. At 4.2K the 0.9eV PL band corresponding to band-tail recombination dominates in the luminescence spectrum of films deposited at 625°C.

With increasing temperature, the 0.9eV band is strongly quenched and an oxygen-related PL bands with maxima at 0.65eV and 0.72eV are retained and persist at room temperature [7,8]. We used these "oxygen" PL bands intensity to monitor the result of hydrogenation and UST on poly-Si film properties. In Figure 2 (insert), the increase of the 0.72eV PL band intensity versus ECR plasma hydrogenation time is shown. The PL increasing occurs due to hydrogen passivation of non-radiative recombination centers at grain boundaries [6]. We used room-T PL mapping technique with a resolution of 100μm to monitor a distribution of recombination centers before and after UST. The results are presented in Figure 2 as two histograms of exactly the same hydrogenated film area prior to and immediately after UST. The average PL intensity after UST exhibits an additional 25-30% increase and also narrowing by a factor of two of the half-width of the PL

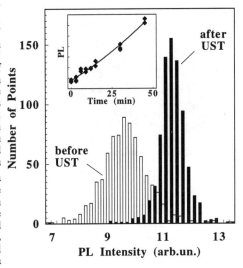

Fig.2 UST change in distribution of room-T PL intensity (0.75eV) in ECR hydrogenated films. Insert: increase of PL with ECR time.

histogram. This result is consistent with the UST-induced improvement of resistance homogeneity in Figure 1.

Contact potential difference (CPD) mapping at 5μ-by-5μ film area using the atomic force microscope with a resolution of 20nm is presented in Figure 3a for the film subjected to 30min ECR hydrogenation. CPD contrast originates from the charge trapped at grain boundary defects in poly-Si. The contrast of a CPD map is gradually reduced with hydrogenation time due to passivation of the grain boundary defects [9]. We performed the UST on the hydrogenated film and compared CPD contrast before and after treatment (Figures 3a and 3b). We precisely maintained the measurement condition of CPD mapping in both cases, in particularly, the elevation of a tip above sample surface. The reduction of the CPD contrast after UST is seen and quantified by a narrowing of the contact potential histogram in Figures 3c and 3d. We notice, that at samples exposed with ECR plasma treatment more than 45min the UST effect on CPD contrast was negligible. The UST reduction of nano-scale CPD mapping contrast provides a direct proof of ultrasound enhanced hydrogenation of grain boundary defects in poly-Si films where plasma hydrogenation was not completed.

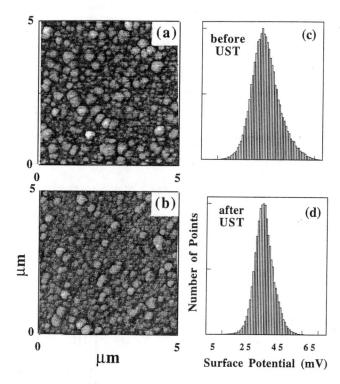

Fig.3. Contact potential difference mapping

(5μmx5μm) of poly-Si films before UST (a) and after UST (b). Hystograms of CPD distribution (c) and (d) across the area shown in (a) and (b), respectively

The stability of the UST effect on sheet resistance was examined. We found that the UST-induced reduction of sheet resistance in a hydrogenated region partially recovers after storage in the dark at room temperature for 10-12hr (see Table I). We noticed, however, that the value of resistance after relaxation was not exactly the same as the initial one, but usually lower. Using consecutive cycles of UST and relaxation, we were able to stabilize a reduced sheet resistance on the level about one order of magnitude lower compared with the initial value (Table I). This final value of resistance did not show any significant change for a few months at room temperature. This relaxation study suggests that two competing UST processes occur in hydrogenated films. One of them produces a stable reduction of sheet resistance, while the second promotes a metastable phenomenon. We notice that the change of CPD contrast presented in Figure 3 corresponds to the stable UST effect.

The UST was applied to poly-Si TFTs designed at CMR and David Sarnoff Research Center. We have found that 3 to 8 times reduction of

Table 1. Cyclic UST and Relaxation at Hydrogenated Poly-Si Thin Films

Film processing	Cycle number	Resistance Normalized to Initial Value
Initial	--	1.0
UST	1	0.21
Relaxation		0.70
UST	2	0.07
Relaxation		0.12
UST	3	0.02
Relaxation		0.11

Notice: each cycle consisted of 30min @ 60°C UST, followed by 12hr @ 20°C relaxation

TFT's leakage current and a shift of threshold voltage by as much as 0.5V could be achieved with representing results shown in Figure 4. This observations are encouraging to explore the UST mechanism in poly-Si TFTs and to optimize UST parameters for improvement of large scale flat panels.

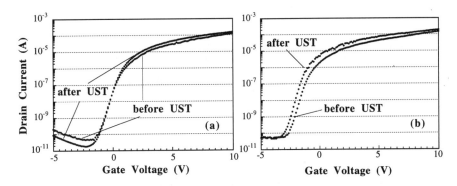

Fig.4 UST reduction of leakage current (a) and shift of threshold voltage (b) in two poly-Si TFT. UST parameters: Δt=15min, T_{ust}=60°C.

DISCUSSION AND CONCLUSIONS

Based on our findings, we suggest that two parallel processes can be activated by UST in hydrogenated poly-Si thin films: (a) stable passivation with atomic hydrogen of grain boundary defects (e.g. dangling bonds), which is responsible for a permanent change of sheet resistance and CPD contrast, and (b) UST generation of metastable hydrogen-related defects responsible for relaxation behavior. It is known that after plasma hydrogenation the total hydrogen concentration in poly-Si films can exceed the number of non-passivated dangling bonds by as much as two orders of magnitude [10]. Therefore, a significant reservoir of electrically non-active hydrogen is available in poly-Si films after hydrogenation. It had been observed that the hydrogen in this reservoir is weakly bound and can be released using a low temperature annealing at 460K followed by a rapid quenching [11]. A thermal liberation of hydrogen from the reservoir requires an activation energy of about 1eV. After fast diffusion to grain boundary regions, this atomic hydrogen can be captured by dangling bonds and form stable at room temperature centers or, alternatively, form electrically active metastable hydrogen-related complexes. The Si-Si bonds strongly distorted by tensile stress at grain boundaries are known to be a place to accommodate bond-centered hydrogen atoms [12]. These defects are metastable donors responsible for the relaxation of resistivity change after annealing and quenching of poly-Si films [11].

We postulate that low-temperature UST can promote a hydrogen liberation from the reservoir, presumably as a result of decreasing the binding energy of the atomic hydrogen in reservoir. Moreover, the UST can also increase the diffusivity of hydrogen by reducing the activation energy for hydrogen diffusion. Due to a competition of the stable (a) and the metastable (b) processes, cyclic UST and relaxation are necessary to bring together released hydrogen and dangling bonds at grain boundaries, which results in formation of stable Si-H centers. It is recognized that extended defects such as grain boundaries and dislocations can effectively absorb the ultrasound and transfer its energy to mobile point defects. In this respect, the atomic hydrogen introduced by plasma treatment into polycrystalline silicon is a favorable application for UST processing. The reported effects of stable UST passivation of grain boundary defects may be a cause of UST improvement of poly-Si TFT's leakage current and threshold voltage (Fig.4). We believe that practical benefits offered by UST defect engineering (simplicity, compatibility with in-line processing, and large-scale device treatment) are promising for integration with polycrystalline silicon technology for flat panel display applications.

ACKNOWLEDGMENTS

We acknowledge expert assistance of G.Nowak in the atomic force microscopy and a help of A.U.Savtchouk in luminescence measurements. We also thank W.Henley, A.M.Hoff, B.Moradi, and S.Karimpannakkel for deposition and hydrogenation of poly-Si films. This work was supported by Advanced Research Projects Agency, under NASA cooperative agreement #NCC8-31. Matching funds were provided by the State of Florida, Inc. and Semiconductor Diagnostics, Inc.

REFERENCES

1. I.W.Wu, T.Y.Huang, W.B.Jackson, A.G.Lewis, and A.Chiang, IEEE Elec.Dev.Lett., **12**, 181 (1991).

2. W.B.Jackson, N.M.Johnson, C.C.Tsai, I.-W.Wu, A.Chiang, and D.Smith. Appl.Phys.Lett., **61**, 1670 (1992).

3. S.Ostapenko, J.Lagowski, L.Jastrzebski, and B.Sopori. Appl.Phys.Lett., **65**, 1555 (1994).

4. S.Ostapenko, W.Henley, S.Karimpannakkel, L.Jastrzebski and J.Lagowski, in Defects and Impurity Engineered Semiconductors and Devices, ed. S.Ashok, J.Chevallier, I.Akasaki, N.M.Johnson, B.L.Sopori. (Pittsburg, PA), **378**, p.405.

5. H.Yokoyama and T.Inoue, Thin Solid Films **242**, 33 (1994).

6. A.U.Savchouk, S.Ostapenko, G.Nowak, J.Lagowski and L.Jastrzebski. Appl.Phys. Lett., **67**, 82 (1995).

7. M.Tajima, J.Cryst.Growth, **103**, 1 (1990).

8. O.King and D.G.Hall, Phys.Rev.B, **50**, 10661 (1994).

9. G.Nowak and J.Lagowski, private communication.

10. 11. N.H.Nickel, N.M.Johnson, and W.B.Jackson, Appl.Phys.Lett., **62**, 3285 (1993).

11. N.H.Nickel, N.M.Johnson, and C.Van der Walle, Phys.Rev.Lett, **72**, 3393 (1994).

12. C. Van der Walle and N.H.Nickel, Phys.Rev.B, **51**, 2636 (1995).

A NEW ELECTRICALLY REVERSIBLE DEPASSIVATION/PASSIVATION MECHANISM IN POLYCRYSTALLINE SILICON

Vyshnavi Suntharalingam* and Stephen J. Fonash
Electronic Materials and Processing Research Laboratory, The Pennsylvania State University, University Park, PA 16802, *vyshi@lion.emprl.psu.edu

ABSTRACT

An electrically reversible depassivation/passivation phenomenon, recently found for hydrogen passivated polysilicon [1] is further explored in this report. This reversible effect is seen in both ECR and RF h, 'rogen passivated n-channel thin film transistors (TFTs) but is not seen in the corresponding hydrogen passivated p-channel TFTs, nor is it seen in either n- or p-channel TFTs before hydrogenation. This phenomenon has been observed when room temperature bias stressing TFTs fabricated on solid phase or laser crystallized polysilicon films on quartz substrates [1]. A model involving hydrogen release or capture at defects, positively charged hydrogen motion in device electric fields, and subsequent hydrogen capture at other defects has been proposed. This phenomenon has significant implications for polycrystalline silicon TFT design and operation. By extension, it also offers significant insight into the stability problems of hydrogenated amorphous silicon.

INTRODUCTION

Degradation of polycrystalline silicon (poly-Si) thin film transistors (TFTs) has been investigated by many groups over the last decade. Several degradation mechanisms in these devices have been reported including carrier trapping in the gate oxide [2,3], creation of states in the poly-Si film by channel carriers [4], hot carrier induced effects [5,6,7], as well as mobile ion and water related effects [8]. However, to the best of our knowledge, degradation mechanisms tied to hydrogen in the poly-Si film have not been reported. In single crystal transistors, hydrogen has been thought to create instabilities in thermal silicon dioxide and at the oxide/silicon interface [9] and in amorphous silicon, it has long been implicated in the metastability of the material [10]. However, in the case of poly-Si TFTs, the role of hydrogen has primarily been considered to be beneficial in the poly-Si itself. There has been sparse recognition that hydrogen must affect silicon dioxide and silicon dioxide/silicon interfaces with the same behavior exhibited in single crystal FETs [11] and no consideration of its stability in the poly-Si film itself.

In this paper, we show that certain bias stresses cause a decrease in n-channel poly-Si TFT OFF current. This phenomenon is seen to be electrically reversible and the data indicates that the reduction in OFF current is due to defect passivation near the source of the stress, rather than field reconfiguration by state creation or charge trapping. To explain this behavior we propose a model involving hydrogen release or capture at defects and positively charged hydrogen motion in device electric fields.

EXPERIMENTAL DETAILS

Poly-Si top gate n- and p-channel TFTs were fabricated on quartz substrates in undoped solid phase (2μm grain size) or laser crystallized (1μm grain size) polycrystalline silicon films of 1000Å thickness. The devices had implanted source and drain regions, a 1000Å silicon dioxide gate dielectric, and were fabricated using a high temperature (950°C) process. More detailed information on the high-temperature device fabrication process has been reported elsewhere [12]. Hydrogenation of these devices occurred in the last step of fabrication and was performed with either an electron cyclotron (ECR) or RF plasma source. Devices of 8, 10, and 15μm channel length and 5 to 50μm channel width were examined.

The n- and p-channel TFTs used in this study were analyzed by taking current-voltage measurements before and after stress. Additional measurements were taken several days or months later to monitor the stability of the phenomenon. Measurements were taken with a computer controlled Hewlett-Packard 4142B semiconductor parameter analyzer and samples were measured in a light-tight, grounded, shielding enclosure with a model 7000-LTE Micromanipulator probe system. Quasi-static transfer characteristics were taken with drain voltages of +0.1, +5, and +10V, gate voltages between -10V and +20V (opposite polarities for p-channel TFTs), and measurements were taken in forward mode and reversed mode (roles of source and drain reversed). Current-Voltage (I_d-V_g) measurements on n-channel devices were made by sweeping the gate voltage from negative voltages to positive voltages to avoid electron trapping effects, similarly, on p-channel devices I-V measurements were made by sweeping from positive voltages to negative.

The threshold voltage (V_{th}) was defined as the gate voltage which yields a normalized drain current [I_d/(W/L)] of 10nA at V_{ds}=0.1V and 100nA at V_{ds}=10V. The maximum and minimum drain currents at V_{ds}=10V are noted as the ON and OFF currents , respectively.

RESULTS AND DISCUSSION

After room temperature bias stressing RF-hydrogenated n-channel TFTs with V_{ds}=V_{gs}=+20V for 2.5 hours, the forward mode transfer characteristics (Fig. 1) show an increase in the threshold voltage and OFF current which is consistent with other published results [4]. The fact that the threshold voltage shift is accompanied by little degradation in the subthreshold swing for n-channel devices subjected to this stress indicates this threshold voltage shift could be due to either electron trapping in the oxide or to the formation of acceptor-like interface states below the equilibrium Fermi energy in the poly-Si film and interface. The most noteworthy feature however, is that the stress has caused a reduction in the OFF current when measured in reversed mode at V_{ds}=+10V (cf. Fig 1b), as first reported in Ref. 1. As seen in Fig. 2, ECR-hydrogenated devices of the same channel length exhibit similar behavior but appear to be susceptible to a larger shift in the threshold voltage and, unlike RF hydrogenated devices, do show a degradation in the subthreshold swing with this stress. The latter suggests that the high density plasma (i.e. ECR) passivation has made the interface more vulnerable to the stress.

Another perspective on the stress induced reversed mode n-TFT OFF current reduction is given in the output characteristics of Fig. 3 which were measured by biasing the terminal which was the source during the stress. After stress, the current

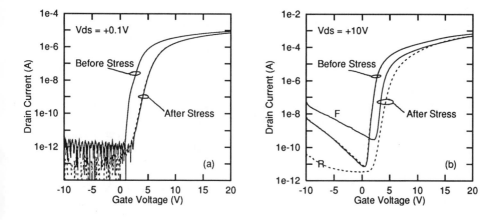

Fig. 1: Transfer characteristics of a RF-Hydrogenated n-TFT measured at (a) Vds = 0.1V and (b) 10V before and after a 2.5 hour Vds=Vgs=+20V stress. Both Forward (–) and Reversed (- -) modes are shown, though in Fig. 1a they are barely distinguishable. In Fig. 1b, Forward and Reversed measurements after stress are indicated by "F" and "R", respectively. W/L=20μm/10μm.

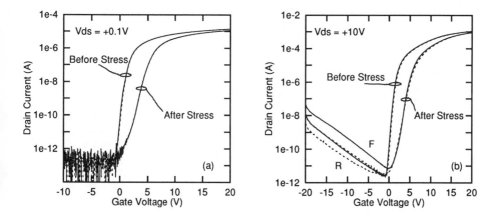

Fig. 2: Transfer characteristics of an ECR-Hydrogenated n-TFT measured at (a) Vds = 0.1V and (b) 10V before and after a 2.5 hour Vds=Vgs=+20V stress. Both Forward (–) and Reversed (- -) modes are shown, though in Fig. 2a they are barely distinguishable. In Fig. 2b, Forward and Reversed meausurements after stress are indicated by "F" and "R", respectively. W/L=30μm/10μm.

from the source terminal no longer exhibits a strong field dependence in the gate voltage regime shown. This OFF current reduction has been observed on solid phase and laser crystallized devices, indicating that the mechanism responsible is not dependent on crystallization process. It has also been observed on devices with different channel lengths, after stresses of shorter duration, and with certain other biasing voltages.

This observed reduction in the reversed mode n-channel OFF curent after stress must be due to one of, or a combination of, the following taking place near the source contact during the stress: (1) a reduction in the electric field in the channel region for a given negative gate bias, or (2) a reduction in the number of states generating the current. For the case of field reconfiguration near the source contact during the stress, either positive charge has to be injected into the oxide/interface or donor-like interface states have to be created below the equilibrium Fermi level to produce the required field reduction in the channel region near the source. However, hole injection is inconsistent with the transverse field direction in the oxide near the source contact during the stress. Further, if hole injection were present, one would expect some recovery with time. However, we reiterate that this enhanced OFF-state behavior has been seen to be stable at room temperature for at least nine months beyond the stress.

Interface and channel region donor-like state creation between the equilibrium and OFF state Fermi energy positions of the OFF state would again reduce the OFF current in n-channel devices due to field reduction, but it would also have the effect of significantly deteriorating the subthreshold swing in the forward state which, as noted, is not seen.

As noted, the second potential mechanism for the observed reduction in the n-channel reversed mode OFF current is a decrease in the number of generation sites near the source contact of the stress. In Ref. 1 we propose that such a decrease occurs due to hydrogen passivation of defects near the source during stress. We proposed that the supply of this hydrogen is from the stress-induced breakage of Si-H bonds and the resulting release of H+ near the drain and along the channel/dielectric interface (manifested as an increase in the forward mode OFF current). The positive potential on the drain and gate during the stress, in the n-channel case, tends to repel H+ from the drain and the interface and, as a result, mobile H+ species drift in the channel electric field of the stress toward the source where they subsequently passivate defects (manifested as a decrease in the reversed mode OFF current). Since this phenomenon has not been observed after stresses in the saturation and "kink" regimes (e.g. Vds=20V & Vgs=10V or Vds=20V & Vgs=5V), the appearance of this phenomenon is dependent on the particular bias stress condition. It is only under stress conditions in the linear regime or near pinchoff that significant amounts of mobile hydrogen will drift toward the source, in other cases the transverse field orientation will drift H+ toward the drain-dielectric interface.

The phenomenon is electrically reversible, as illustrated in Fig. 4. In this example a second stress has been applied to the device of Fig. 2. This second stress is identical to the first except that the roles of the source and drain during the stress were reversed. After this reversed stress, the region around what was the drain for the stress of Fig. 2 is now passivated while the region around what was the source for the stress of Fig. 2 is now depassivated. Devices which were not hydrogenated do not exhibit this reversible depassivation/passivation phenomenon. The

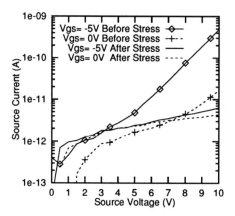

Fig. 3: Output characteristics obtained by biasing the Source and Gate of a RF hydrogenated n-TFT before and after a 2.5 hour Vds=Vgs=+20V stress. W/L=20μm/10μm.

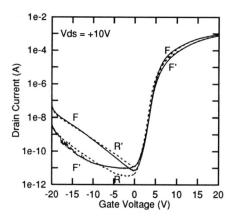

Fig. 4: Transfer characteristics at Vds = +10V of ECR hydrogenated n-TFT after a 2.5 hour Vds=Vgs=20V stress, and after a sequential stress in which the roles of source and drain during the stress were reversed. The Forward and Reversed mode characteristics after the first stress are indicated by "F" and "R", respectively. The Forward and Reversed mode charactersitics after the second (reversed) stress are indicated by " F' " and " R' ", respectively. W/L =30μm/10μm.

improvement in OFF current and the electrically reversibility is also not observed in p-channel TFTs because the direction of the electrical field of the stress forces the positively charged hydrogen released during the stress to be attracted toward the channel/dielectric interface.

CONCLUSIONS

We have demonstrated an electrically reversibly passivation/depassivation mechanism involving hydrogen motion in poly-Si TFTs. The mechanism is stable for several months at room temperature and the behavior observed is consistent with a model of H+ moving in stress electric field configurations.

ACKNOWLEDGMENTS

The authors thank Tsu-Jae King for helpful discussions. Xerox PARC generously provided the samples for this study. This work was supported by the Advanced Research Projects Agency. V.S. gratefully acknowledges a NSF Graduate Engineering Education fellowship.

REFERENCES

[1] V. Suntharalingam and S. J. Fonash, *Appl. Phys. Lett.* **68**, 1400 (1996).
[2] C.A. Dimitriadis and P.A. Coxon, *Appl. Phys. Lett.* **54**, 620 (1989).
[3] N.D. Young and A. Gill, *Semicond. Sci. Technol.* **5**, 72 (1990).
[4] I.-W. Wu, W.B. Jackson, W.B. Huang, T.-Y Hyang, A.G. Lewis, and A. Chiang, *IEEE Electron Device Lett.* **11**, 167 (1990).
[5] M.S. Rodder and D.A. Antoniadis, *IEEE Trans. Electron Devices* **ED-34**, 1079 (1987).
[6] S. Banerjee, R. Sundarasen, H. Shichijo, and S. Malhi, *IEEE Trans. Electron Devices* **ED-35**, 152 (1988).
[7] J.R. Ayres and N.D. Young, *IEE Proc. Circuits Devices Syst.* **141**, 38 (1994).
[8] N.D. Young and A. Gill, *Semicond. Sci. Technol.* **7**, 1103 (1992).
[9] O.O. Awadelkarim, S. J. Fonash, P. I. Mikulan, and Y.D. Chan, *J. Appl. Phys.* **79**, 517 (1996).
[10] R.A. Street, *Physica B* **170**, 69 (1990).
[11] Y.S. Kim, K.Y. Choi, M.C. Jun, and M.K. Han, *Mat. Res. Soc. Symp. Proc.* , **345**, 155 (1994).
[12] A.G. Lewis, I.-W. Wu, T.Y. Huang, A. Chiang, and R.H. Bruce, *IEDM Tech. Dig.*, 843 (1990).

AC AND DC CHARACTERIZATION AND SPICE MODELING
OF SHORT CHANNEL POLYSILICON TFTs

M.D. JACUNSKI, M.S. SHUR, T. YTTERDAL, A.A. OWUSU*, and M. HACK**
*Department of Electrical Engineering, University of Virginia, Charlottesville, VA 22903
**dpiX, 3406 Hillview Ave., Palo Alto, CA 94302

ABSTRACT

We present an analytical SPICE model for the AC and DC characteristics of n and p channel polysilicon TFTs which scales fully with channel length and width in all regimes of operation (leakage, subthreshold, above threshold, and kink) and accounts for the frequency dispersion of the capacitance. Once physically based parameters have been extracted from long channel TFTs, which include the gate length and drain bias dependencies of the device parameters, our model accurately reproduces short channel device characteristics. The AC model includes the input channel resistance in series with the gate oxide capacitance. As a result, our model is able to fit the frequency dispersion of the device capacitances. The model has been implemented in the AIM-Spice simulator and good agreement is observed between measured and modeled results for gate lengths down to 4 µm.

INTRODUCTION

Polysilicon thin film transistors (TFTs) have become increasingly important for applications such as AMLCDs [1-3], projection displays, and SRAMs [4, 5]. As the technology matures and gate lengths decrease, more accurate CAD models must be developed.

In previous papers [6, 7], we presented a physically based analytical SPICE model for long channel TFTs. It accurately reproduced DC device characteristics in all four regimes of operation (leakage, subthreshold, above threshold, and kink) for n and p channel TFTs with channel lengths down to 15 µm. Because the model is physically based, the number of parameters is kept to a minimum, and those parameters can be readily related to the device structure. Therefore, in addition to meeting the needs of circuit designers, the parameters can be used in QC and QA programs which are critical for large area, high yield applications.

In the present work, our long channel DC SPICE model is extended to include short channel effects for n and p channel TFTs with gate lengths down to 4 µm. This is accomplished by describing the threshold voltage of the device as a function of both channel length and drain bias. In addition, the expression for the kink regime current now includes a multiplier which takes into account the channel length dependent feedback of the impact ionization mechanism. The result is a complete DC SPICE model for polysilicon TFTs which accurately reproduces the device characteristics for a wide range of gate geometries using a single parameter set.

An AC model of the TFT is also presented and implemented in SPICE. Our new AC model includes bias dependent access resistors in series with the source and drain capacitances. The access resistance values are calculated as a fraction of the channel resistance, similar to what is done in BSIM3v3 [8]. The model accurately reproduces the frequency dispersion of the gate capacitances and properly scales with channel length.

213

SHORT CHANNEL DC SPICE MODEL

Our long channel, analytical DC SPICE model was presented in [6] and [7], but will be briefly reviewed here. Fig. 1 shows measured and modeled data for an n channel polysilicon TFT with $W/L = 50\mu m/50\mu m$. The transfer characteristic in Fig. 1a reveals a large subthreshold leakage current which is the result of thermionic field emission of carriers through the grain boundary trap states. The leakage current is a function of V_{DS}, V_{GS}, and the flatband voltage, but is independent of channel length and V_t. Because no short channel effects are observed, this regime will not be discussed except to say that the "long channel" model results in a good fit for all channel lengths examined.

In the subthreshold regime, the drain current, I_{sub}, is given by

$$I_{sub} = \mu_s C_{ox} \frac{W}{L} (\eta_i V_{th})^2 \exp\left(\frac{V_{GS} - V_t}{\eta_i V_{th}}\right)\left[1 - \exp\left(\frac{-V_{DS}}{\eta_i V_{th}}\right)\right] \tag{1}$$

where μ_s and η_i are extracted constants and all other symbols have their usual meanings. Above threshold, the drain current, I_a, is given by [6, 7]

$$I_a = \begin{cases} \mu_{FET} C_{ox} \dfrac{W}{L} \left[(V_{GS} - V_t) V_{DS} - \dfrac{V_{DS}^2}{2\alpha_{sat}}\right] & \text{for } V_{DS} < \alpha_{sat}(V_{GS} - V_t) \\[4mm] \mu_{FET} C_{ox} \dfrac{W}{L} \dfrac{(V_{GS} - V_t)^2 \alpha_{sat}}{2} & \text{for } V_{DS} \geq \alpha_{sat}(V_{GS} - V_t) \end{cases} \tag{2}$$

α_{sat} is an extracted constant which accounts for the variation of depletion charge across the channel, and μ_{FET} is the field effect mobility:

$$\frac{1}{\mu_{FET}} = \frac{1}{\mu_0} + \frac{1}{K_\mu \left|V_{GS} - V_t\right|^m} . \tag{3}$$

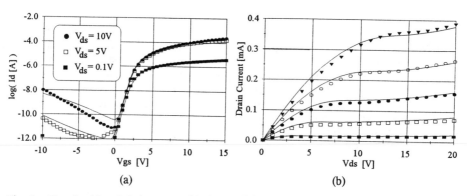

Fig. 1. Transfer (a) and drain current (b) characteristics of n channel polysilicon TFT with $W/L = 50\mu m/50\mu m$. Markers represent measured data and curves indicate model results. In (b), $V_{GS} = 5V$, 8V, 11V, 14V, and 17V.

μ_0, K_μ, and m are constants. The expressions for the above and below threshold currents are combined to give a continuous total drain current, I_{DI} (excluding the kink effect):

$$\frac{1}{I_{DI}} = \frac{1}{I_{sub}} + \frac{1}{I_a} . \tag{4}$$

The drain current and gate transfer characteristics of a short n channel TFT are shown in Fig. 2. Here, $W/L = 50\mu m/6\mu m$. Note in Fig. 2a that the $V_{DS} = 10V$ transfer curve is shifted from the 5V curve, indicating that V_t is a function of drain bias. By comparing Figs. 1a and 2a, it is also evident that V_t is changing with channel length.

Threshold voltage has been extracted from gate to channel capacitance measurements as detailed in [9] and [10] for several different L and V_{DS}. Note that this voltage is different from the ON voltage, V_{ON}, extrapolated from the I_D - V_{GS} curve. For the n channel TFTs examined here, the V_t at 0.1V drain bias occurs when $I_D \approx 2\cdot10^{-12}$A per micrometer of channel width. In contrast, the drain current is about three orders of magnitude higher for $V_{GS} = V_{ON}$. The difference is even more severe for p channel devices. This definition of threshold voltage is important in order to assure that (i) the extracted transport parameters are independent of gate geometry and (ii) the same V_t will also correctly describe the gate capacitance at low frequencies. (The second point will be demonstrated in the next section.)

The short channel threshold voltage shift for n channel TFTs is plotted versus $1/L$ in Fig. 3 for two different drain biases. In both cases, a linear dependence is observed. In Fig. 4, the short channel V_t shift is found to be a function of $(V_{DS})^2$ for the 6 µm long device. Similar results were observed for other gate lengths and for p channel devices. The threshold voltage can therefore be described by

$$V_t = V_{t0} \mp \frac{A_t V_{DS}^2 + B_t}{L} \tag{5}$$

where the "minus" is used for n channel TFTs and the "plus" is used for p channel. V_{t0} is the long channel threshold voltage at zero drain bias, while A_t and B_t are empirical constants.

The kink effect is visible in Figs. 1b and 2b for large V_{DS}. We have extended our long

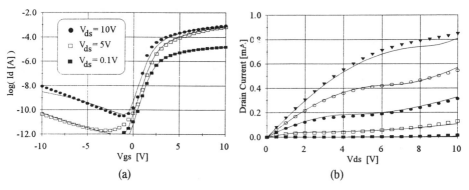

Fig. 2. Transfer (a) and drain current (b) characteristics of n channel polysilicon TFT with $W/L = 50\mu m/6\mu m$. Markers represent measured data and curves indicate model results. Model parameters are the same as those used in Fig. 1. In (b), $V_{GS} = 2V$, 4V, 6V, 8V, and 10V.

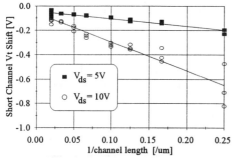

Fig. 3. Threshold voltage shift of short n channel polysilicon TFTs as function of reciprocal gate length. A linear relationship is observed.

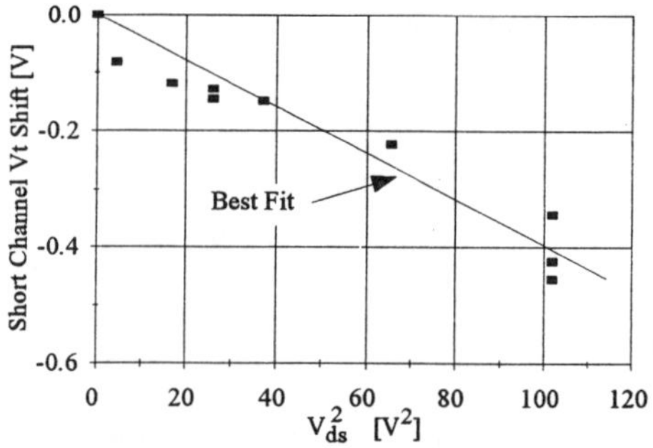

Fig. 4. Threshold voltage shift of 6μm long n channel polysilicon TFT as a function of drain bias. A parabolic relationship is observed.

channel impact ionization model to include a gate length dependent feedback multiplier. The details of this model will be presented elsewhere [11]. Here we state only the result:

$$\Delta I_{kink} = I_{DI} \frac{A_{kink}}{L^{mkink} V_{kink}} [V_{DS} - \alpha_{sat}(V_{GS} - V_t)] \exp\left[\frac{-V_{kink}}{V_{DS} - \alpha_{sat}(V_{GS} - V_t)}\right]$$ (6)

for $V_{GS} > V_t$ and $V_{DS} > \alpha_{sat}(V_{GS} - V_t)$. A_{kink}, $mkink$, and V_{kink} are constants. The total drain current is $I_D = I_{DI} + \Delta I_{kink}$.

For the model calculations shown in Figs. 1 and 2, the expression for V_t given in (5) has been used in (1) - (4) and (6). A very good fit is observed. Similar results are observed for p channel TFTs.

AC SPICE MODEL

Because polysilicon TFTs have such a low field effect mobility (as compared to crystalline silicon devices), the time required for a carrier to traverse the channel is on the order of the period for even "low frequency" signals (< 1MHz). This results in the frequency dispersion of the gate to channel capacitance shown in Fig. 5. Our AC model accounts for this phenomena by placing fractions of the channel resistance in series with the gate to source and gate to drain capacitances as shown in the equivalent circuit of Fig. 6.

In Fig. 6, the channel resistance, R_{ch}, is defined by

$$\frac{1}{R_{ch}} \equiv \frac{\partial I_D}{\partial V_{DS}}.$$ (7)

The access resistors, r_{sx} and r_{dx}, are given by

$$r_{sx} = \left.\frac{R_{ch}}{K_{ss}}\right|_{V_{DS} = 0V} \qquad r_{dx} = \left.\frac{R_{ch}}{K_{ss}}\right|_{V_{DS}}.$$ (8)

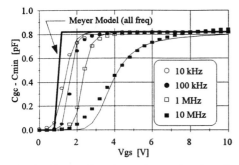

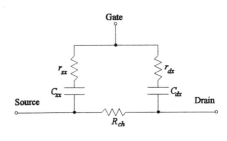

Fig. 5. Gate to channel capacitance of n channel TFT with $W/L = 50\mu m/50\mu m$. Thin lines indicate the model presented here.

Fig. 6. Equivalent small signal circuit model for polysilicon TFT. The small signal parameters r_{sx} and r_{dx} are fractions of the channel resistance, R_{ch}.

K_{ss} is a small signal constant similar to the Elmore constant used in BSIM3v3 non-quasi static equivalent circuit model [8]. Note that because r_{sx} physically originates near the source, it is calculated using $V_{DS} = 0V$ (i.e., independent of V_{DS}).

The expressions for the capacitors C_{sx} and C_{dx} are

$$C_{sx} = \frac{2}{3} C_{gc} \left[1 - \frac{\left(V_{GD} - V_t\right)^2}{\left(V_{GS} + V_{GD} - 2V_t\right)^2} \right], \qquad C_{dx} = \frac{2}{3} C_{gc} \left[1 - \frac{\left(V_{GS} - V_t\right)^2}{\left(V_{GS} + V_{GD} - 2V_t\right)^2} \right]. \quad (9)$$

The expressions in (9) are very similar to those of the Meyer model [12], except that the gate to channel capacitance, C_{gc}, is taken from the Universal Charge Control Model [13]. The UCCM expression for C_{gc} is valid both above and below threshold and allows the effects of the grain boundary trap states to be included:

$$C_{gc} = \frac{C_{ox} L W}{1 + 2 \exp\left[\frac{-\left(V_{GS} - V_t\right)}{\eta_c V_{th}} \right]} . \quad (10)$$

η_c is the capacitance ideality factor which, because of the trap states, is bias dependent near the drain. As for r_{sx}, C_{sx} is calculated as though $V_{DS} = 0V$. Empirically we have found that $\eta_c = \eta_0 + \eta_{00}V_{DS}$. η_0 and η_{00} are extracted constants.

SPICE CIRCUIT APPLICATION

The TFT model described in the previous sections has been implemented in the AIM-Spice [13] circuit simulator. The simulator was then used in the analysis of a TFT operational amplifier fabricated at Xerox PARC. Fig. 7 shows the DC transfer characteristic of the opamp, and Fig. 8 shows the frequency response for a 25 pF load. Reasonable agreement between measured and modeled results is observed in both cases.

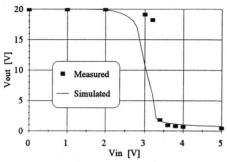

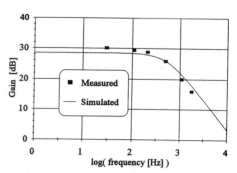

Fig. 7. DC transfer characteristic of a CMOS TFT opamp. Data points are measured and curves are AIM-Spice simulation results.

Fig. 8. Frequency response of a CMOS TFT op amp with 25 pF load. Data points are measured and curves are AIM-Spice simulation results.

CONCLUSIONS

A complete short channel analytical SPICE model has been presented for both n and p channel polysilicon TFTs. All four DC regimes of operation are accurately described by a single parameter set for devices with gate lengths down to 4 μm. In addition, the frequency dispersion of the gate capacitance is modeled using bias dependent access resistors in series with the gate to source and gate to drain capacitors in the small signal equivalent circuit. The model has been implemented in the AIM-Spice circuit simulator, and a sample circuit analysis has been demonstrated.

ACKNOWLEDGEMENT

The authors gratefully acknowledge the support of the Advanced Research Projects Agency for its support under contract N61331-94-K-0034.

REFERENCES

[1]   M. Clark, *IEE Proc. Circuits Dev. Syst.*, vol. 141, p. 3, 1994.
[2]   Y. Morimoto et al., in *IEDM Tech. Dig.*, p. 837, 1995.
[3]   T. Morita et al., in *IEDM Tech. Dig.*, p. 841, 1995.
[4]   S. Batra, in *ECS Proc.*, vol. 94-35, p. 291, 1994.
[5]   C. Liu, P. Diodato, K. Lee, and H. Cong, in *IEDM Tech. Dig.*, p. 919, 1995.
[6]   M. Shur et al., *J. Soc. Info. Display*, vol. 3, no. 4, p. 223, 1995.
[7]   M. Jacunski et al., in *AMLCDs '95 Workshop Proc.*, p. 134, 1995.
[8]   *BSIM3v3 Manual*, University of California, p. 5-3, 1995.
[9]   M. Jacunski, M. Shur, and M. Hack, in *53rd Annual DRC Dig.*, p. 158, 1995.
[10]   M. Jacunski, M. Shur, and M. Hack, to be published in *IEEE Tran. Elec. Dev.*
[11]   A. Owusu et al., submitted to *54th Annual DRC*, 1996.
[12]   J. Meyer, *RCA Review*, vol. 32, p. 42, 1971.
[13]   K. Lee et al., *Semiconductor Device Modeling for VLSI*, Prentice Hall, 1993, p. 219.

SOLUTION GROWN POLYSILICON FOR FLAT PANEL DISPLAYS

R. L. Wallace and W. A. Anderson
State University of New York at Buffalo, Center For Electronic and Electro-optic Materials,
Department of Electrical and Computer Engineering, 217C Bonner Hall, Amherst, NY 14260

ABSTRACT

Thin-film poly-Si on low-cost substrates is useful for thin-film transistors in flat panel displays or for photovoltaics. The films reported herein were formed at 600°C by d.c. magnetron sputtering from a Si-target onto a SiO_2/Mo substrate, pre-coated with Sn or In/Ti to give a liquid phase growth. Poly-Si films have given a preferred (111) orientation, grain size up to 20µm, a carrier mobility exceeding 100cm^2/Vs and carrier lifetime of 8µs when using the Sn pre-layer. The Sn pre-layer typically gave smaller grain size than did the In/Ti solvent but the latter gave lower carrier lifetime from Ti incorporation in the film. The film from the Sn pre-layer gave carrier mobility of 140cm^2/V-s after hydrogenation by microwave electron cyclotron resonance. Properties of the Si thin-film can be controlled by type of pre-layer, doping of the target, substrate temperature and deposition environment.

INTRODUCTION

Thin-film poly-Si silicon on a low cost substrate is important for both photovoltaic and thin-film transistor applications. Many methods have been used to form such thin films. These techniques have included d.c. magnetron sputtering [1], chemical vapor deposition (CVD) [2], hot wall - low pressure CVD [3], microwave-assisted CVD [4] and radio frequency diode glow discharge [5]. Some have utilized an anneal following low pressure CVD of a-Si:H [6,7]. Several issues limit the desired maximum processing temperature to about 600°C, such as, diffusion of substrate impurities into the growing film, residual thermal stresses, and the need for reduced cost of substrates. One popular substrate has been Corning 7059 glass [1,3,4,7]. Grain size has ranged from 32 nm to 2500 nm. Values of carrier mobility have ranged from 1.5 to 25 cm^2/V-s [5].

In this work, a polished Mo substrate is utilized (with an SiO_2 buffer layer) which is first coated with a eutectic-forming solvent metal. The Mo was chosen since it withstands later high temperature steps, if needed for grain growth, and resists inter-diffusion. SiO_2 prevents silicide formation with the Mo. This solvent metal promotes formation of the polycrystalline Si film as will be explained below. We have utilized an In/Ti or Sn solvent metal whereas others have utilized Al-Ga-Si [8] or Pb-Bi-Si [9], for example. These lower temperature deposits have generally resulted in poor electrical properties whereas deposits near 900°C have yielded much better electrical properties [10]. This work describes a new low temperature process for producing thin-film Si on a low-cost substrate where the electrical properties are quite good.

EXPERIMENTAL

Depositions were done in a diffusion-pumped high vacuum system [11]. A liquid nitrogen cold finger assists in achieving a base pressure in the low 10^{-7} Torr range. A molecular sieve foreline trap and a chilled baffle over the diffusion pump reduce backstreaming. A filament evaporation source deposits the solvent metal prior to sputtering of the Si by d.c. magnetron from a 2" target. Typically, a polished Mo substrate, coated with SiO_2, is first coated with a 50-500nm layer of Sn while heated to about 400°C, or at room temperature for the In/Ti pre-layer, Ti followed by In to give In/Ti/SiO₂/Mo (hereafter called In/Ti). Ti is used as a wetting agent to allow full wetting of the substrate surface by the liquid In layer. This is followed by the Si deposition in 2 mTorr of 5% H₂ in Ar at a rate of 1 micron per hour with the substrate at about 600°C. The H₂ reduces defects in the resulting films, especially at grain boundaries. During deposition, a metal-Si melt is formed which is kept saturated in Si by the sputter gun. This liquid layer remains on the surface and leaves behind a poly-Si film. Chemical etching later removes the metal-Si residue and exposes the poly-Si layer. Films have been grown with thickness ranging from less than one micron to about 6 microns.

RESULTS

The structure, shown in Figure 1, may be evaluated by depositing Schottky contacts on top. Vias in the SiO₂ buffer layer form silicide channels during deposition, establishing a good ohmic contact to the film. As an alternative, the vias may be omitted and thin-film field effect transistors fabricated on the top surface. The poly-Si can also be removed from the substrate by chemical etching and mounted upon a different substrate in an inverted fashion. A previous report [12] focussed upon structural analysis utilizing X-ray diffraction, Raman spectroscopy, and transmission electron microscopy. The films are strongly (111) oriented with grains of up to 20 microns. These

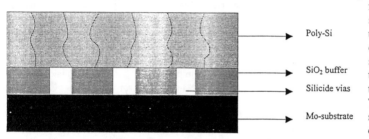

Poly-Si

SiO₂ buffer

Silicide vias

Mo-substrate

Fig. 1 Cross-section of thin-film poly-Si and substrate.

properties and the electrical behavior can be controlled by solvent (or prelayer) thickness, solvent composi-tion, doping of the target or substrate temperature.

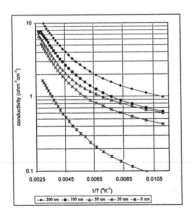

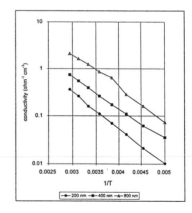

Fig. 2 (a) Effect of prelayer thickness on dark conductivity for the Sn prelayer films.

Fig. 2 (b) Effect of prelayer thickness on dark conductivity for the In/Ti pre-layer films.

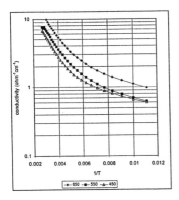

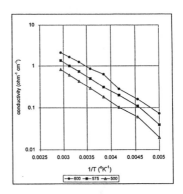

Fig. 3 (a) Effect of substrate temperature on dark conductivity for the Sn prelayer films.

Fig. 3 (b) Effect of substrate temperature on dark conductivity for the In/Ti prelayer films.

The effect of prelayer thickness on dark conductivity (σ), shown in Figure 2, indicates increased thickness to increase conductivity for both Sn and In/Ti solvents and that the Sn solvent gives higher values of σ. Figure 3 indicates that conductivity increases with substrate temperature which may be related to the increased grain size or increased doping from the prelayer at these higher temperatures. The In/Ti prelayer material shows a single conduction mechanism (linear 1/T plot) whereas the Sn prelayer data seems to show two distinct mechanisms. This may be explained by the potential fluctuation model proposed by Taniguchi [13], where similar behavior was found for CVD polysilicon. The higher activation energy was attributed to band conduction, while the lower activation energy was thought to result from hopping through a continuum of trapping or donor states at the grain boundaries. Data on Hall effect carrier mobility, shown in Figure 4, indicates mobility to increase with In prelayer thickness, to exceed 100 cm^2/Vs. A lesser value is seen for the Sn prelayer material due to grain boundary effects in the finer grain material. However, the mobility of Sn-prelayer material increases significantly upon

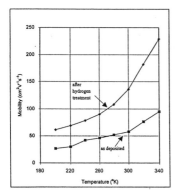

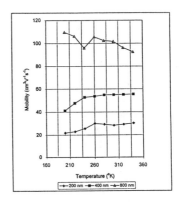

Fig. 4 (a) Hall effect carrier mobility for the Sn prelayer films.

Fig. 4 (b) Hall effect carrier mobility for the In/Ti prelayer films.

hydrogenation by microwave electron cyclotron resonance. Photoconductive decay data of Figure 5 indicate a carrier lifetime of 0.3 μs for the In prelayer material and this was confirmed by R.K. Ahrenkiel at National Renewable Energy Laboratory. A value of 6 μs was seen for the Sn prelayer material since the contamination by Ti, seen in the In/Ti material, is avoided. Table 1 gives a summary of the electrical data.

Table 1. Properties of Sputtered Thin Film Silicon

	Sn Prelayer	In/Ti Prelayer
Conductivity type	N	P
Grain size (μm)	≤ 1	20 or more
Carrier concentration (cm^{-3})	10^{18}	10^{16}
Dark conductivity (Ω-cm)$^{-1}$	1 - 10	0.1 - 2
Carrier mobility (cm^2/V-s)	60 @300K	100 @300K
Carrier mobility (cm^2/V-s) after hydrogenation	140 @300K	100 @300K
Lifetime (μs)	~ 8	~ 0.3
Photoconductivity (Ω-cm)$^{-1}$ @35 mW/cm^2, white light	10^{-3}	10^{-5}

CONCLUSIONS

This process provides for control of carrier type, grain size, carrier concentration, and carrier mobility. The respectable values of carrier mobility and carrier lifetime make this a suitable material for electronic applications. The technique may be scaled for larger area depositions and is potentially a low-cost process. Other solvent metals must be studied as the process is not yet optimized.

ACKNOWLEDGMENT

This research was supported by the National Renewable Energy Laboratory (John Benner) and New York State Energy Research and Development Authority (Jennifer Harvey).

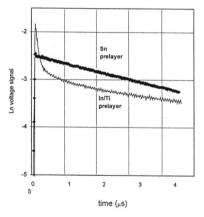

Fig. 5 Lifetime measurement from photoconductive decay for In/Ti and Sn prelayer films.

REFERENCES

1.   Y. H. Yang and J.R. Abelson, Appl.Phys. Lett.,67, 3623 (1995).
2.   J. R. Heath, S. M. Gates and C. A. Chess, Appl. Phys. Lett.,3569 (1994).
3.   W. Czubatyz, D. Beglau, R. Himmler, G. Wicker, D. Jablonski and S. Guha, IEEE Elec. Dev. Lett., 10, 349 (1989).
4.   D. He, N. Okada, C. M. Fortmann, and I. Shimizu, J. Appl. Phys., 76, 4728 (1994).
5.   F. Finger, P. Hapke, M. Luysberg, R. Carius, H. Wagner and M. Scheib, Appl. Phys. Lett., 65, 2588 (1994).
6.   S.Hasegawa, S. Watanabe, T. Inokuma and Y. Kurata, J. Appl. Phys., 77, 1938 (1995).

7.   M. Sarret, A. Liba, F. Le Bihan, P. Joubert and B. Fortin, J. Appl. Phys., 76, 5492 (1994).

8.   B. Girault, F. Chevrier, A. Joullie and G. Bougnot, J. Cryst. Growth, 37, 169 (1977).
9.   S. Lee and M. Green, J. Elec. Mtls., 20, 635 (1991).
10.   A. Barnett, R. Hall, D. Fardig and J. Culik, Proc 18th IEEE Photovoltaic Specialists Conf., IEEE Piscataway, 1094 (1985).
11.   R. L. Wallace and W. A. Anderson,First World Conference on Photovoltaic Energy Conversion, Hawaii, Dec. 5-9, 1994.
12   R. L. Wallace and W. A. Anderson, Materials Research Society Meeting, Boston, Nov. 28-Dec.2, 1994.
13.   M.Taniguchi, M. Hirose, Y. Osaka, S. Hasegawa, Jap. J. Appl. Phys., 19, 4 (1991).

COPPER-ENHANCED SOLID PHASE CRYSTALLIZATION
OF AMORPHOUS SILICON FILMS

Dong Kyun Sohn, Dae Gyu Moon* and Byung Tae Ahn
Department of Materials Science and Engineering, Korea Advanced Institute of Science and Technology, 373-1 Koosung-dong, Yusung-gu, Taejon 305-701, Korea
*Current address : LG Electronics Inc., LCD R&D Center, Kyungki-do, Korea

ABSTRACT

Low-temperature crystallization of amorphous Si (a-Si) films was investigated by adsorbing copper ions on the surface of the films. The copper ions were adsorbed by spin-coating of Cu solution. This new process lowered the crystallization temperature and reduced crystallization time of a-Si films. For 1000 ppm solution, the a-Si film was partly crystallized down to 500°C in 20 h and almost completely crystallized at 530°C in 20 h. The adsorbed Cu on the surface acted as a seed of crystalline and caused fractal growth. The fractal size was varied from 10 to 200 µm, depending on the Cu concentration in solution. But the grain size of the films was about 400 nm, which was similar to that of intrinsic films crystallized at 600°C.

INTRODUCTION

Polycrystalline silicon (poly-Si) thin-film transistors (TFTs) have attracted much attention because of their wide range of applicability in large-area electronics, such as in the active matrix liquid crystal display (AMLCD) with peripheral driving circuits [1]. In recent years, many studies on poly-Si TFTs have focused either on reducing the process time or on lowering the process temperature [2]. The latter is important since it enables the use of inexpensive glass substrates. One of the common methods of preparing poly-Si films is solid-phase crystallization (SPC) of a-Si films at 600°C for 20-48 hours [3]. However, this temperature is too high to use glass substrates for large-area TFT-LCDs. By using metal-Si eutectic temperature, a-Si/metal layer can crystallized bellow 500°C [4]. But in that case the films form silicides or alloys, which are not applicable to channel layer of TFTs.

To avoid silicide formation, we proposed a metal adsorption method instead of metal deposition [5]. From the method, we found Cu and Au were effective metals to enhance crystallization of a-Si films. In this work we report the crystallization behavior of the Cu-adsorbed a-Si films with various concentrations.

EXPERIMENTAL

Amorphous Si films of 150 nm thickness were deposited on thermally oxidized Si substrates by low-pressure chemical vapor deposition (LPCVD) at 540°C using 100% silane (SiH_4) gas. The thickness of the thermal oxide layer on Si was 750 nm. Cu was adsorbed by spin-coating of Cu solutions with various concentrations after cleaning the Si with a mixture of H_2SO_4 and H_2O_2 and 50:1 HF. The Cu solution was common standard solution, namely, $CuCl_2$ dissolved in 1N HCl.

In this study, the annealing time, temperature and Cu concentration in the solution were varied, and the resulting crystallization behavior was investigated. The concentration of Cu ions in solution was controlled by adding deionized water. Secondary ion mass spectroscopy (SIMS) and energy-dispersive spectroscopy (EDS) showed that Cl was completely evaporated from the surface of the a-Si films within 30 min at 530°C. The speed of the spin-coater used for coating and drying the solutions was 3000-5000 rpm. Subsequently the Cu-adsorbed a-Si films were annealed in Ar atmosphere. The microstructure was examined using a scanning electron microscope (SEM) and the proportion that had crystallized was evaluated from thin film X-ray diffraction (XRD) patterns and Raman spectroscopy. The final grain size of the crystallized films were measured by transmission electron microscopy (TEM).

RESULTS AND DISCUSSION

Figure 1 shows the annealing temperature dependence of (111) XRD intensity of the Cu-adsorbed Si film from 1000 ppm Cu solution and intrinsic Si film after 20 h annealing. The intensity curve of the Cu-adsorbed film is shifted to a lower temperature than that of the intrinsic Si film, suggesting that crystallization can occur below 530°C. The Cu-adsorbed film started to

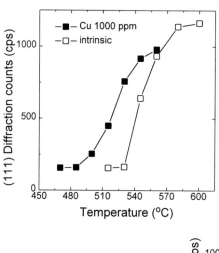

Fig. 1. Annealing temperature dependence of (111) XRD intensity of intrinsic and Cu-adsorbed Si films after 20 h annealing.

Fig. 2. Annealing time dependence of (111) XRD intensity of intrinsic and Cu-adsorbed Si films annealed at 530°C.

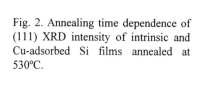

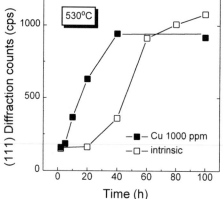

crystallize at 500°C which is about 50°C lower than the crystallization temperature of intrinsic a-Si. Figure 2 shows the annealing time dependence of (111) XRD intensity of the intrinsic and Cu-adsorbed Si films after annealing at 530°C. The Cu-adsorbed film from 1000 ppm solution started to crystallize immediately, while the intrinsic a-Si film does not crystallize even after 20 h. Figures 1 and 2 indicate that Cu adsorbed on the Si surface enhances crystallization of a-Si at a lower temperature and in a shorter time.

Figure 3 shows the SEM surface micrographs of Si films, spin-coated with 500 ppm Cu solution and annealed at 530°C for 1, 5, 10 and 20 h. The nucleation of crystallites is almost completed within 1 h. The size of the crystallite fractals is about 50 μm within 10 h and the growth rate is 5 μm/h. The film is not fully crystallized after 10 h annealing. The fractal size of the film adsorbed from the 10 ppm solution in Fig. 4, is almost 200 μm, which is much larger than that of 500 ppm Cu solution.

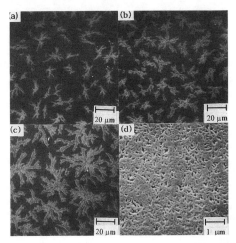

Fig. 3. SEM surface micrographs of the crystallized film spin-coated from 500 ppm Cu solution and annealed at 530°C for (a)1 (b)5 (c)10 and (d)20 h.

Fig. 4. SEM surface micrograph of the crystallized film spin-coated from 10 ppm Cu solution and annealed at 530°C for 10 h.

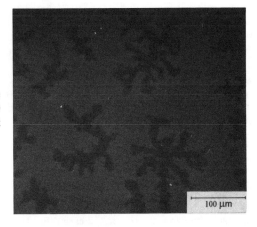

Figure 5 shows the average center-to-center distance and final size of fractal as a function of Cu concentration in solution, for the Si films annealed at 530°C. The center-to-center distance of fractal is about 200 μm for the Cu-adsorbed Si film from 10 ppm and decreases to 20 μm with increasing the Cu concentration in solution. The reason for larger fractal size is that the adsorbed Cu on a-Si surface forms a crystal seed in the initial stage of annealing. The number of crystal seeds increases with increasing the Cu concentration of in solution. As a result, the final fractal size is saturated by the impingement of each fractals and the center-to-center distance of fractals decreased.

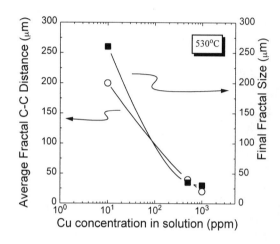

Fig. 5. Cu concentration in solution dependence of average center to center (C-C) distance and final size of fractal of the crystallized films annealed at 530°C.

Figure 6 shows the annealing time dependence of crystalline fraction of the Si films, spin-coated from 1000 ppm Cu solution and annealed with various temperatures. The crystalline fraction is calculated from Raman spectrum. As the annealing temperature increases, the incubation time for crystallization decreased and the crystallization time is shortened. It takes 20 h to crystallize 90% of a-Si films at 530 °C, but reduces to 3 h at 600°C. Figure 7 is the Arrhenius plot of the characteristic crystallization time as a function of annealing temperature for 10 and 1000 ppm. The characteristic time is defined as the required time for crystalline fraction from 5% to 95% in Fig. 6. The activation energies calculated from the slope of Fig. 7 are almost the same regardless of the Cu concentration in solution.

Figure 8 shows the TEM dark field images of the annealed Si films spin-coated from 1000 ppm Cu solution and annealed at 530°C, 560°C, 580°C and 600°C for 20 h. The grain size of the crystallized film increased with increasing the annealing temperature. The film annealed at 600°C has grain size of 400 nm.

CONCLUSION

The adsorption of Cu ions by spin-coating Cu solutions on the surface of a-Si films lowered the crystallization temperature and reduced the crystallization time The adsorbed Cu acted as a seed of crystalline and lowered the crystallization temperature down to 500°C. The films showed fast fractal growth with size of 20-200 μm. But low-temperature crystallization did not improve grain size.

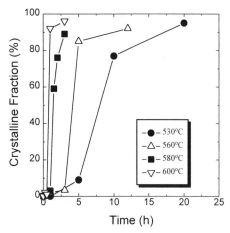

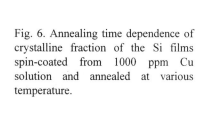

Fig. 6. Annealing time dependence of crystalline fraction of the Si films spin-coated from 1000 ppm Cu solution and annealed at various temperature.

Fig. 7. Annealing temperature dependence of characteristic crystallization time of the Si films spin-coated from 10 and 1000 ppm Cu solution.

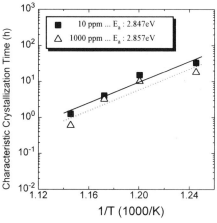

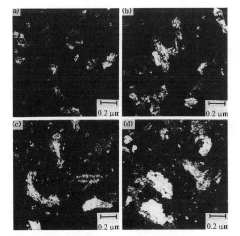

Fig. 8. TEM dark field images of the annealed Si films spin-coated from 1000 ppm Cu solution, annealed at (a) 530°C (b) 560°C (c) 580°C and (d) 600°C for 20 h.

229

ACKNOWLEDGMENT

This work was performed with the financial support of the Korea Ministry of Information and Communication.

REFERENCE

1. H. Oshima and S. Morozumi, Ext. Abstr. Int. Conf. Solid State Devices and Materials, Yokohama, 1991, p 577 (1991).
2. A. Kohno, T. Sameshima, N. Sano, M. Sekiya and M. Hara, IEEE Trans. Electron Devices **42,** 251 (1995).
3. K. Nakazawa, J. Appl. Phys. **69**, 1703 (1991).
4. S. W. Russel, J. Li and J. W. Mayer, J. Appl. Phys. **70,** 5153 (1991).
5. D. K. Sohn, J. N. Lee, S. W. Kang and B. T. Ahn, Jpn. J. Appl. Phys. **35**, 1005 (1996).

SOLID-PHASE CRYSTALLIZATION OF AMORPHOUS $Si_{0.7}Ge_{0.3}$/Si AND Si/$Si_{0.7}Ge_{0.3}$ BILAYER FILMS ON SiO_2

Tae-Hoon Kim, Myung-Kwan Ryu, Jin-Won Kim, Chang-Soo Kim* and Ki-Bum Kim
Division of Material Science & Engineering, Seoul National University, Seoul, Korea 151-742
*Materials Evaluation Center, Korea Research Institute of Standards Science, Chungnam, Korea, 305-606

ABSTRACT

We have investigated the solid phase crystallization of a-(Si/$Si_{0.7}Ge_{0.3}$) and a-($Si_{0.7}Ge_{0.3}$/Si) bilayer films deposited on SiO_2 for an annealing temperature of 550 °C. It was found that, in case of a-($Si_{0.7}Ge_{0.3}$/Si), nucleation of crystalline phases occurred at the free surface, while in a-(Si/$Si_{0.7}Ge_{0.3}$) crystalline phase nucleated at $Si_{0.7}Ge_{0.3}$/SiO_2 interface. The crystallization rate of an a-($Si_{0.7}Ge_{0.3}$/Si) is much slower than that of an a-(Si/$Si_{0.7}Ge_{0.3}$) films. After full crystallization, poly-($Si_{0.7}Ge_{0.3}$/Si) has many equiaxed grains and the defect density of the upper $Si_{0.7}Ge_{0.3}$ was much lower than that of lower $Si_{0.7}Ge_{0.3}$ in a poly-(Si/$Si_{0.7}Ge_{0.3}$) film whose grain morphology was elliptical. The average grain size of poly-($Si_{0.7}Ge_{0.3}$/Si) was ~7 μm and this film had strong (111) preferential orientation, while poly-(Si/$Si_{0.7}Ge_{0.3}$) had weak (311) or random oriented grains with the average size of ~0.3 μm.

INTRODUCTION

In recent years, many researches have been worked on utilizing $Si_{1-x}Ge_x$ as an active layer of a thin film transistor (TFT) due to its low thermal burdget.[1,2] However, some device characteristics of poly-$Si_{1-x}Ge_x$ TFT are not still optimized and thus there were some challenges to improve the characteristics of poly-$Si_{1-x}Ge_x$ TFT by using Si/$Si_{1-x}Ge_x$ bilayer structure.[3,4] In view of the material engineering, a systematic study of the crystallization mechanism of a-(Si/$Si_{1-x}Ge_x$) bilayer and the resulting microstructure has not been performed yet. In this paper, the SPC behavior of a-(Si/$Si_{0.7}Ge_{0.3}$) bilayer on SiO_2 and the microstructure after crystallization are comparatively investigated with the case of a-($Si_{0.7}Ge_{0.3}$/Si) bilayer on SiO_2.

EXPERIMENT

Amorphous Si/$Si_{0.7}Ge_{0.3}$ and $Si_{0.7}Ge_{0.3}$/Si bilayer films were deposited on thermally oxidized 4-inch Si wafer by using Si_2H_6 and GeH_4 source gases in a low pressure chemical vapor deposition (LPCVD) chamber. The deposition pressure was 1 Torr and the deposition temperature was 425 °C and 450 °C. The deposition rate of Si and $Si_{0.7}Ge_{0.3}$ was 4 Å/min and 20 Å/min, respectively.[5] and the thickness of each layer was 1000 Å. As-deposited samples were annealed at 550 °C in a tube furnace with nitrogen ambient. The crystallization behavior with annealing time and the preferred orientation of a crystallized film were examined by using x-ray diffractometry (XRD). Finally, the microstructure evolution and nucleation behavior during the crystallization were observed by using transmission electron microscopy (TEM).

EXPERIMENTAL RESULTS

As-deposited films

Figure 1(a) and (b) shows that XRD spectra of a-(Si/$Si_{0.7}Ge_{0.3}$) and a-($Si_{0.7}Ge_{0.3}$/Si) deposited at 425 °C and 450 °C, repectively. For the deposition temperature of 425 °C, both of these bilayers were amorphous since no XRD peak was appeared. However, if the deposition temperature was increased upto 450 °C, (220) diffraction peak of $Si_{0.7}Ge_{0.3}$ appeared, indicating that the lower $Si_{0.7}Ge_{0.3}$ layer of Si/$Si_{0.7}Ge_{0.3}$ is polycrystalline while upper Si layer is amorphous. In the case of 450 °C-deposited $Si_{0.7}Ge_{0.3}$/Si, upper $Si_{0.7}Ge_{0.3}$ as well as lower Si

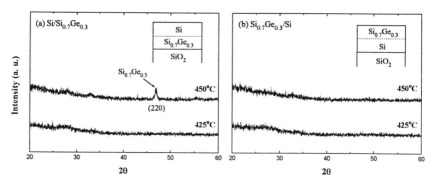

Fig. 1. XRD spectra of (a) Si/Si$_{0.7}$Ge$_{0.3}$ and (b) Si$_{0.7}$Ge$_{0.3}$/Si bilayers deposited on SiO$_2$ at 425°C and 450 °C.

was still amorphous state.

Figure 2(a) and (b) show the plan-view TEM images of as-deposited a-(Si/Si$_{0.7}$Ge$_{0.3}$) for the deposition temperature of 425 °C and 450 °C, respectively. In 425 °C-deposited film, no crystalline phase was observed and the selected area diffraction (SAD) pattern of this sample was halo-type, which demonstrate that the Si/Si$_{0.7}$Ge$_{0.3}$ deposited at 425 °C is amorphous state. In the plan-view TEM image of 450 °C-deposited Si/Si$_{0.7}$Ge$_{0.3}$, plenty of crystallites (d=100 Å~500 Å) are observed in amorphous matrix, i.e., mixed-phase. The plan-view TEM image and SAD pattern of 425 °C- and 450 °C-deposited Si$_{0.7}$Ge$_{0.3}$/Si are shown in Fig. 3(a) and (b). Unlike the Si/Si$_{0.7}$Ge$_{0.3}$, the Si$_{0.7}$Ge$_{0.3}$/Si is amorphous at both deposition temperature.

As a next step, the crystallization behaviors of a-(Si/Si$_{0.7}$Ge$_{0.3}$) and a-(Si$_{0.7}$Ge$_{0.3}$/Si) were investigated for the deposition temperature of 425 °C in order to compare the SPC behavior of *amorphous* films.

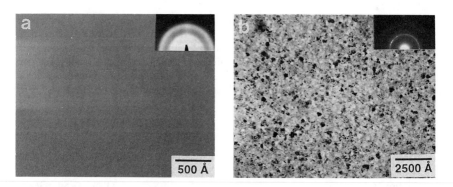

Fig. 2. Plan-view TEM images of as-deposited Si/Si$_{0.7}$Ge$_{0.3}$ for the deposition temperature of (a) 425 °C and (b) 450 °C.

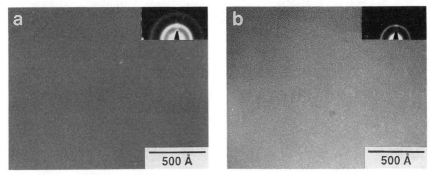

Fig. 3. Plan-view TEM images of as-deposited $Si_{0.7}Ge_{0.3}/Si$ for the deposition temperature of (a) 425 °C and (b) 450 °C.

Crystallized films

The crystallization behaviors of a-($Si/Si_{0.7}Ge_{0.3}$) and a-($Si_{0.7}Ge_{0.3}/Si$) at 550 °C for the deposition temperature 425 °C are compared in Fig. 4. The crystalline fraction was obtained by normalizing (111) peak intensity of each annealed bilayer to that of fully crystallized one.[4] As is shown in this results, the crystallization rate of a-($Si/Si_{0.7}Ge_{0.3}$) is much faster than that of a-($Si_{0.7}Ge_{0.3}/Si$). Figure 5 shows XRD spectra of fully crystallized a-($Si/Si_{0.7}Ge_{0.3}$) and a-($Si_{0.7}Ge_{0.3}/Si$) at 550 °C. After calibrating the XRD peak intensity of (111), (220), and (311) with correction factor[2], it was found that the preferred orientation of poly-($Si/Si_{0.7}Ge_{0.3}$) was weak (311) or random, while that of poly-($Si_{0.7}Ge_{0.3}/Si$) was strong (111).

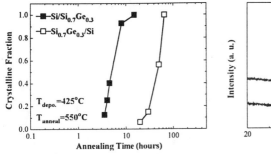
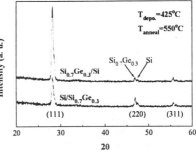

Fig. 4. SPC behavior of an a-($Si/Si_{0.7}Ge_{0.3}$) and an a-($Si_{0.7}Ge_{0.3}/Si$) at 550 °C.(T_d=425°C)

Fig. 5. XRD spectra of a poly-($Si/Si_{0.7}Ge_{0.3}$) and a poly-($Si_{0.7}Ge_{0.3}/Si$) crystallized at 550 °C.

Figure 6(a) and (b) are plan-view TEM images showing the sequence of crystallization process of a-($Si/Si_{0.7}Ge_{0.3}$) at 550 °C. Many elliptical grains are observed in amorphous $Si/Si_{0.7}Ge_{0.3}$ matrix and after full crystallization, ~0.3 μm of elliptical grains with microtwins along its long axis of each grain were formed. The sequence of the crystallization process of a-($Si_{0.7}Ge_{0.3}/Si$) for the same experimental conditions are shown in Fig. 7(a) and (b). Unlike $Si/Si_{0.7}Ge_{0.3}$ bilayer, 6~9 μm of large equiaxed grains were observed in poly-($Si_{0.7}Ge_{0.3}/Si$).

Fig. 6. Plan-view TEM images of Si/Si$_{0.7}$Ge$_{0.3}$ bilayer annealed at 550 °C with annealing times of (a) 3.5 hours and (b) 20 hours for the deposition temperature of 425 °C.

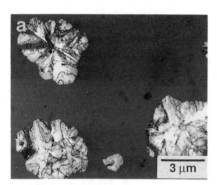

Fig. 7. Plan-view TEM images of Si$_{0.7}$Ge$_{0.3}$/Si bilayer annealed at 550 °C with annealing times of (a) 20 hours and (b) 65 hours for the deposition temperature of 425 °C.

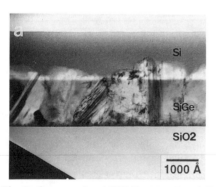
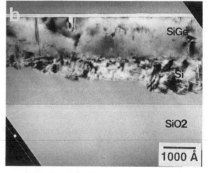

Fig. 8. Cross-sectional TEM image of partially crystallized (a) Si/Si$_{0.7}$Ge$_{0.3}$ (3.5 hours) and (b) Si$_{0.7}$Ge$_{0.3}$/Si (20 hours) at 550 °C for the deposition temperature of 425 °C.

Cross-sectional TEM image of partially crystallized a-(Si/Si$_{0.7}$Ge$_{0.3}$) and a-(Si$_{0.7}$Ge$_{0.3}$/Si) at 550 °C in Fig. 8(a) and (b) demonstrate the nucleation and initial grain-growth stages of these two bilayers. In a-(Si/Si$_{0.7}$Ge$_{0.3}$) [Fig. 8(a)], grains were nucleated at the interface between Si$_{0.7}$Ge$_{0.3}$ and SiO$_2$, i.e.,"*interface nucleation*", followed by the growth proceeding upto the surface of upper Si layer. In the case of a-(Si$_{0.7}$Ge$_{0.3}$/Si) [Fig. 8(b)], the nucleation of crystalline phase occurred at the surface of upper Si$_{0.7}$Ge$_{0.3}$ layer, i.e.,"*surface nucleation*". After a nucleation event, grains grow down toward lower Si layer. Particularly, the defect density of upper Si$_{0.7}$Ge$_{0.3}$ in Si$_{0.7}$Ge$_{0.3}$/Si is much lower than that of lower Si$_{0.7}$Ge$_{0.3}$ in Si/Si$_{0.7}$Ge$_{0.3}$. Figure 8(b) also shows that as the surface nucleated grains grow toward lower Si layer, many crystalline defects begin to appear. The solid-phase crystallization behavior and the resulting microstructure after annealing of a-(Si/Si$_{0.7}$Ge$_{0.3}$) and a-(Si$_{0.7}$Ge$_{0.3}$/Si) bilayer are summarized in Table. I.

Table I. Summary of the experimental results of solid-phase crystallization of a-(Si/Si$_{0.7}$Ge$_{0.3}$) and a-(Si$_{0.7}$Ge$_{0.3}$/Si) bilayer films at 550 °C for the film deposition temperature of 425 °C.

	Crystallization behavior				Microstructure		
	τ_0	t_{full}	nucleation mode	P.O.	D_g (μm)	defect density	grain shape
Si/Si$_{0.7}$Ge$_{0.3}$	~2 hours	~20 hours	interface nucleation	weak (311) or random	0.3	high	elliptical
Si$_{0.7}$Ge$_{0.3}$/Si	~10 hours	~65 hours	surface nucleation	strong (111)	7	low	equiaxed

τ_0 : incubation time. t_{full} : full crystallization time. P.O. : preferred orientation. D_g : average grain size.

DISCUSSION

From the XRD results of Fig 1, it was found that the transition temperature of Si$_{0.7}$Ge$_{0.3}$, at which the phase of an as-deposited film change from amorphous state to crystalline one, is higher on a-Si than on SiO$_2$. Extending the results to amorphous films, it can be expected that a-Si$_{0.7}$Ge$_{0.3}$ layer will have higher "*structural orderness*" on a-SiO$_2$ than on a-Si. The higher structural orderness of the lower a-Si$_{0.7}$Ge$_{0.3}$ of a-(Si/Si$_{0.7}$Ge$_{0.3}$) suggests that the lower a-Si$_{0.7}$Ge$_{0.3}$ should have more embryos than the upper a-Si$_{0.7}$Ge$_{0.3}$ of a-(Si$_{0.7}$Ge$_{0.3}$/Si). Thus, nuclei created from many embryos as well as random nucleation in an amorphous matrix can reduce the incubation time of a-(Si/Si$_{0.7}$Ge$_{0.3}$) compared with a-(Si$_{0.7}$Ge$_{0.3}$/Si). In these bilayer structure, there exist three types of nucleation site: free surface, interface between a-Si and a-Si$_{0.7}$Ge$_{0.3}$, and interface between lower amorphous layer and SiO$_2$. Since the interface between two amorphous layer is thermodynamically stable relative to the other two sites, one has to consider a competetion of the interface- and the surface-nucleation. However, Fig. 8(a) and (b) clearly show that the nucleation sites is determined by the location of a-Si$_{0.7}$Ge$_{0.3}$ layer.

The elliptical grains observed in poly a-(Si/Si$_{0.7}$Ge$_{0.3}$) can be explained that the grains already have defects such as microstwins even at the nucleation step on SiO$_2$ and then following grain growth proceeds along the microtwins.[6-8] This twin-assisted process make the growth more efficient when the crystallization temperature is low (< 600 °C) and thus thermal vibration energy is not sufficient for the grain growth. Since the crystallization of upper Si layer is performed by the continuing growth of grains nucleated in the lower Si$_{0.7}$Ge$_{0.3}$ layer across the

Si/Si$_{0.7}$Ge$_{0.3}$ interface, the preferred orientation of poly-(Si/Si$_{0.7}$Ge$_{0.3}$) as well as the microstructure has been affected by the lower Si$_{0.7}$Ge$_{0.3}$ layer. It has been reported that the preferred orientation of poly-Si is (111) while that of poly-Si$_{1-x}$Ge$_x$ is (311).[2] Based on the results, the weak (311) or random orientation of poly-(Si/Si$_{0.7}$Ge$_{0.3}$) can be explained as the seed effect of the lower Si$_{0.7}$Ge$_{0.3}$ layer.[4]

As for the equiaxed grains and the strong (111) preferred orientation of poly-(Si$_{0.7}$Ge$_{0.3}$/Si), some more considerations are needed. We believe that the equiaxed grains resulted from the rearrangement of surface atoms of the upper a-Si$_{0.7}$Ge$_{0.3}$ layer when the surface nucleation occurs. In this case, the surface atoms have high degree of freedom for the rearrangement although native oxide exists at the film surface. Therefore, crystalline phases propagate equivalently toward all growth direction to form equiaxed grains with lowest energy plane (111) pararell to the film surface. Additionally, the defect density of the grains can be much lower than that of interface-nucleated grains shown in Fig. 8(a), since the atomic rearrangement is more favorable at the surface of upper Si$_{0.7}$Ge$_{0.3}$ layer of a-(Si$_{0.7}$Ge$_{0.3}$/Si) than at the interface a-Si$_{0.7}$Ge$_{0.3}$/SiO$_2$ of a-(Si/Si$_{0.7}$Ge$_{0.3}$) on SiO$_2$. Howerver, as grains in the upper Si$_{0.7}$Ge$_{0.3}$ grow toward the lower Si layer, many crystalline defects begin to appear due to lower thermal vibration energy in a-Si than that in a-Si$_{0.7}$Ge$_{0.3}$, as mentioned above. The results of Si$_{0.7}$Ge$_{0.3}$/Si bilyer are very similar to the report of Sakai *et al.*[9], in which the crystallization of an a-Si with clean surface performed by *in-situ* ultra high vacuum (UHV)-annealing was investigated. They reported that plenty of round-shaped grains formed at the surface of a-Si having no native oxide after the *in-situ* UHV-annealing and mentioned that the grains are the result of the migration of surface atoms with high mobility.

CONCLUSION

In conclusion, we have studied the solid-phase crystallization behavior of a-(Si/Si$_{0.7}$Ge$_{0.3}$) and a-(Si$_{0.7}$Ge$_{0.3}$/Si) bilayer on SiO$_2$. The tranisition termperature of as-deposited Si$_{0.7}$Ge$_{0.3}$ layer from amorphous phase to crystalline one was higher on a-Si than on SiO$_2$. The nucleation site of these two bilayer was the film surface for the a-(Si$_{0.7}$Ge$_{0.3}$/Si) and the interface (a-Si$_{0.7}$Ge$_{0.3}$/SiO$_2$) for the a-(Si/Si$_{0.7}$Ge$_{0.3}$), repectively. The crystallization rate is much faster in a-(Si/Si$_{0.7}$Ge$_{0.3}$) than in a-(Si$_{0.7}$Ge$_{0.3}$/Si). After full crystallization at 550 °C, ~0.3 μm of elliptical grains were observed in poly-(Si/Si$_{0.7}$Ge$_{0.3}$) with weak (311) or random orientaion, while ~7 μm of equiaxed grains existed in poly-(Si$_{0.7}$Ge$_{0.3}$/Si) with strong (111) orientation. The defect density of the upper Si$_{0.7}$Ge$_{0.3}$ layer of poly-(Si$_{0.7}$Ge$_{0.3}$/Si) is much lower than that of the lower Si$_{0.7}$Ge$_{0.3}$ layer of poly-(Si/Si$_{0.7}$Ge$_{0.3}$).

REFERENCES
1. T.-J. King, J. R. Pfiester, and K. C. Saraswat, IEEE Electron Dev. Lett. **EDL-12**, 584 (1991).
2. C. W. Hwang, M. K. Ryu, K. B. Kim, S. C. Lee, and C. S. Kim, J. Appl. Phys. 77, 3042 (1995).
3. A. J. Tang, J. A. Tsai, R. Reif, and T.-J. King, **IEDM95**, 513 (1995).
4. M. K. Ryu, J. W. Kim, T. H. Kim, K. B. Kim, C. W. Hwang, Jpn. J. Appl. Phys. **34**, L1031 (1995).
5. J. W. Kim, M. K. Ryu, K. B. Kim, S. J. Kim, J. Electrochem. Soc. **143**, 363 (1996).
6. D. R. Hamilton and R. G. Seidensticker, J. Appl. Phys. **31**, 1165 (1960).
7. R. Drosd and J. Washburn, J. Appl. Phys. **53**, 397 (1982).
8. A. Nakamura, F. Emoto, E. Fujii, A. Yamamoto, Y. Uemoto, K. Senda, and G. Kano, J. Appl. Phys. **66**, 4248 (1989).
9. A. Sakai, H. Ono, K. Ishida, T. Nini, and T. Tatsumi, Jpn. J. Appl. Phys. **30**, L941 (1991).

COMPARISON BETWEEN RAPID THERMAL AND FURNACE ANNEALING FOR a-Si SOLID PHASE CRYSTALLIZATION

Reece Kingi, Yaozu Wang, Stephen J. Fonash, Osama Awadelkarim, *John Mehlhaff and *Howard Hovagimian, Electronic Materials Processing and Research Laboratory, Penn State University, University Park, PA16802, and *Intevac RTP Systems, 3845 Atherton Road, Suite 1 Rocklin, CA95765

ABSTRACT

Rapid thermal annealing and furnace annealing for the solid phase crystallization of amorphous silicon thin films deposited using PECVD from argon diluted silane have been compared. Results reveal that the crystallization time, the growth time, and the transient time are temperature activated, and that the resulting polycrystalline silicon grain size is inversely proportional to the annealing temperature, for both furnace annealing and rapid thermal annealing. In addition, rapid thermal annealing was found to result in a lower transient time, a lower growth time, a lower crystallization time, and smaller grain sizes than furnace annealing, for a given annealing temperature. Interestingly, the transient time, growth time, and crystallization time activation energies are much lower for rapid thermal annealing, compared to furnace annealing.

We propose two models to explain the observed differences between rapid thermal annealing and furnace annealing.

INTRODUCTION

Polycrystalline silicon (poly-Si) thin film transistors (TFTs) are currently receiving considerable attention for active matrix flat panel display technology[1-4], as well as for SRAM applications[5]. The crystallization of amorphous silicon (a-Si) thin films is a promising approach to obtain these poly-Si thin films, as opposed to depositing the poly-Si film directly, because larger grains and smoother surfaces can be attained[1,2,4]. Solid phase crystallization (SPC) has several advantages over laser crystallization (a liquid phase crystallization technique) which include smoother surfaces, better uniformity, and higher throughputs[6-8]. However, SPC does result in lower poly-Si TFT mobilities than laser crystallization[6-8]. Nevertheless, the advantages of SPC over laser crystallization make it a promising process for the fabrication of high performance, high throughput, poly-Si TFTs.

In this work we have made a direct comparison between two common solid phase crystallization techniques, namely rapid thermal annealing and furnace annealing. Specifically we have compared these two approaches for the solid phase crystallization of a-Si thin films deposited using plasma enhanced chemical vapor deposition (PECVD) from silane diluted with argon at 210°C.

EXPERIMENTAL PROCEDURE

PECVD was used to deposit 1000Å of a-Si onto Corning code 7059 glass from silane diluted with argon (1:10) at 210°C. Films were then annealed in a conventional furnace and a rapid thermal processing system in a N_2 ambient over a range of temperatures. In each case the temperature was monitored with k-type thermocouples located next to the a-Si samples.

The UV reflectance technique[9,10] was used to determine the degree of crystallinity of all annealed films, and TEM was used to determine the grain size of all fully crystallized films.

Mat. Res. Soc. Symp. Proc. Vol. 424 © 1997 Materials Research Society

EXPERIMENTAL RESULTS AND DISCUSSION

Figure 1 presents the degree of crystallinity versus annealing time for both rapid thermal annealing and furnace annealing, for some of the annealing temperatures investigated. As can be seen, in each case the crystallization process begins with a transient period, during which time no crystallization is observed[3,4,11]. Following this transient period is the growth period, during which time nuclei form, and grains grow outwardly from these nuclei. The crystallization process is complete at the end of this growth period (at which point the degree of crystallinity is unity), which is designated as the total crystallization time. As one would expect, all of these curves exhibit similar characteristics, and as the temperature is increased the overall crystallization time decreases. From these curves (and those not shown), the transient time, the growth time, and the crystallization time have been extracted. These are plotted on a log scale (in seconds) versus 1000/T in Figure 2, where T is the annealing temperature in degrees Kelvin. Error bars corresponding to $\pm 10^\circ C$ have been included in Figure 2.

Figure 2 illustrates that for either furnace annealing or RTA the transient time, growth time and crystallization time are temperature activated processes. Thus, these times are proportional to $\exp(E/kT)$, where E is some activation energy associated with either the transient time, growth time, or the crystallization time, k is Boltzmann's constant, and T is the annealing temperature in degrees Kelvin. The transient time, growth time and crystallization time activation energies for both furnace and rapid thermal annealing have been determined from Figure 2, and are reported in Table I. This clearly illustrates that the activation energies for these times are lower for rapid thermal annealing compared to furnace annealing.

The poly-Si grain size is plotted in Figure 3 versus 1000/T, where again T is the annealing temperature in degrees Kelvin. The grain size was calculated by averaging the largest twenty grains in a given TEM micrograph. Figure 3 shows that the resulting grain size is inversely proportional to the annealing temperature.

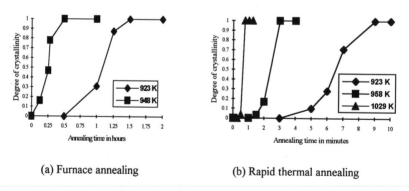

(a) Furnace annealing   (b) Rapid thermal annealing

Figure 1. Degree of crystallinity versus annealing time for (a) furnace annealing, and (b) rapid thermal annealing. The temperatures are reported in degrees Kelvin.

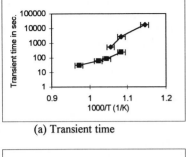

(a) Transient time

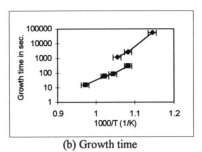

(b) Growth time

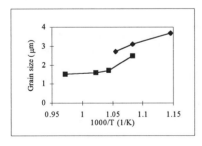

(c) Crystallization time

Figure 2. (a) Transient time in seconds (log scale), (b) growth time in seconds (log scale), and (c) crystallization time in seconds (log scale), versus 1000/T, for both rapid thermal annealing (■) and furnace annealing (◆). Here T is the annealing temperature in degrees Kelvin

Table I. Transient time and crystallization time activation energies for rapid thermal annealing and furnace annealing.

Crystallization technique	E_t (transient time activation energy)	E_G (growth time activation energy)	E_c (crystallization activation energy)
Furnace Annealing	3.57 eV	3.40 eV	3.45 eV
RTA	1.68 eV	2.20 eV	1.94 eV

Figure 3. Grain size in μm versus 1000/T, where T is the annealing temperature in degrees Kelvin, for furnace annealing (◆) and a rapid thermal annealing (■).

Figures 2 and 3 illustrate that for a given annealing temperature, RTA results in lower transient, growth, and crystallization times, with smaller grain sizes than furnace annealing. We note that with the error bars in place in Figure 2 it may be possible that the differences in activation energies between rapid thermal annealing and furnace annealing result from experimental error. However, it should be noted that even with these error bars, the transient time, growth time, and crystallization time (and grain size) are unambiguously lower for rapid thermal annealing, compared to furnace annealing. This indicates that there are differences in crystallization kinetics resulting from these two annealing techniques, which provides a basis for the difference in activation energies.

Two explanatory models for this phenomena are hereforth proposed. The first model is based on the fact that the light spectra present during annealing in the two systems are different. The RTP system uses tungsten-halogen lamps to heat the a-Si film, whereas the conventional furnace uses a resistive coil to heat the environment. The light spectrum present during rapid thermal annealing is much more energetic and intense than that present during furnace annealing. Because of this, it is proposed that the energetic photons present during rapid thermal annealing create photocarriers which subsequently recombine and thereby break weak and strained bonds in the a-Si:H thin film. These sites then serve as recombination centers for further photocarriers producing localized heating. The locations at which this localized heating occurs then become favorable sites for nucleation. This results in the ability for nuclei to form more readily which results in a higher nucleation rate and hence a lower crystallization time. A direct result of this increased nucleation rate is the reduction in the grain size that is observed.

The second model proposed to explain the reduction in thermal budget and resulting grain size for rapid thermal annealing compared to furnace annealing is based upon the sudden release of stored energy within the a-Si:H thin film. Experimental observations made by Takamori et al[12,13] illustrated that stored energy in amorphous films could be released resulting in "explosive" crystallization. Specifically, they deposited relatively thick (>10μm) Ge films by sputtering which could then be crystallized by simply "pricking" the surface of the film with a sharp object. Based on their experimental observations, they concluded that once the crystallization process was triggered (i.e. by "pricking" the surface of the film with a sharp object), the free energy released (resulting from the difference in free energy between the amorphous state and the polycrystalline state) was sufficient to drive the crystallization of the remainder of the film. Based upon these results, it is proposed that the sudden heating of the a-Si:H thin film that occurs during rapid thermal annealing is the "pricking" that results in the sudden release of stored energy, in the form of heat. This excess heat results in a higher "effective" temperature which drives the nucleation and grain growth processes. In the same way that the grain size decreases with increasing annealing temperature, compared to furnace annealing the grain size is reduced for rapid thermal annealing simply because of the higher "effective" temperature, resulting from this sudden release of stored energy within the film. Although this energy must also be released during furnace annealing, because it is released over such a long period of time (because the heating rate is much slower), it does not have the same impact that occurs during rapid thermal annealing.

The lower activation energies associated with rapid thermal annealing reported in Table I indicate that additional energy is supplied to the a-Si film during crystallization. We propose that the difference in activation energies for rapid thermal annealing and furnace annealing is related to the amount of additional energy supplied to the a-Si film by either one of the two processes described above.

CONCLUSIONS

Rapid thermal annealing results in smaller grain sizes than furnace annealing, with lower transient, growth, and crystallization times, for a given annealing temperature. Two models to explain this phenomena have been proposed. The transient time, growth time, and crystallization time were found to be temperature activated for both rapid thermal and furnace annealing, with rapid thermal annealing having lower activation energies than furnace annealing. We propose that this difference in activation energies results from additional energy being supplied during crystallization by rapid thermal annealing as a result of one of two proposed models.

REFERENCES

1   I-Wei Wu, Solid State Phenomena Vols 37-38 (1994) pp. 553-564
2   K. Nakazawa and K. Tanaka, J. Appl. Phys. **68** (3) 1990 (1029)
3   T. Kretz, R. Stroh, P. Leganeux, O. Huet, M. Magis and D. Pribat, Solid State Phenomena Vols 37-38 (1994) pp. 311-316
4   J. Guillemet, B. Pieraggi, B. de Mauduit and A. Claverie, Solid State Phenomena, Vols 37-38 (1994) pp. 293-298
5   C. T. Liu, P. W. Diodato, K. H. Lee, H. I. Cong, Tech. Dig. of 1995 IEDM, p.919
6   T. E. Dyer et al, Solid State Phenomena, Vols. 37-38 (1994) p. 329
7   T. Sameshima et al, J. Appl. Phys. **76** (11) 1994 (7377)
8   T. King, Trends in Polycrystalline Silicon Thin Film Transistor Technologies for AMLCDs, Second International Workshop on AMLCDs, September 1995
9   A. Yin, S. Fonash, D. Reber, T. Li and M. Bennett, MRS Symposium Proceedings **345**, 1994, pp. 81-86
10   T. Kammins, Polycrystalline Silicon for IC Applications, (Klumer Academic Publishers, 1988)
11   I-Wei Wu, A. Chiang, M. Ovecoglu and T. Huang, J. Appl. Phys. **65** (10) 1989 (4036)
12   T. Takamori, R. Messier, and R. Roy, Appl. Phys. Lett. Vol. 20, No. 5, 1972, p. 201
13   T. Takamori, R. Messier, and R. Roy, J. Mats. Sci. **8** (1973) p. 1809

POLYCRYSTALLINE SILICON FILMS FORMED BY SOLID-PHASE CRYSTALLIZATION OF AMORPHOUS SILICON: THE SUBSTRATE EFFECTS ON CRYSTALLIZATION KINETICS AND MECHANISM

Y. -H. SONG,* S. -Y. KANG,* K. I. CHO,* H. J. YOO,* J. H. KIM,** and J. Y. LEE**
*Semiconductor Division, ETRI, Taejon 305-600, KOREA
**Department of Materials Science and Engineering, KAIST, Taejon 305-701, KOREA

ABSTRACT

The substrate effects on the solid-phase crystallization of amorphous silicon (a-Si) have been extensively investigated. The a-Si films were prepared on two kinds of substrates, a thermally oxidized Si wafer (SiO_2/Si) and a quartz, by low-pressure chemical vapor deposition (LPCVD) using Si_2H_6 gas at 470 °C and annealed at 600 °C in an N_2 ambient for crystallization. The analysis using XRD and Raman scattering shows that crystalline nuclei are faster formed on the SiO_2/Si than on the quartz, and the time needed for the complete crystallization of a-Si films on the SiO_2/Si is greatly reduced to 8 h from ~15 h on the quartz. In this study, it was first observed that crystallization in the a-Si deposited on the SiO_2/Si starts from the interface between the a-Si film and the thermal oxide of the substrate, called interface-induced crystallization, while random nucleation process dominates on the quartz. The very smooth surface of the SiO_2/Si substrate is responsible for the observed interface-induced crystallization of a-Si films.

INTRODUCTION

Polycrystalline silicon (poly-Si) films have been a key material of thin-film transistors (TFTs) which can serve as a basic transistor of peripheral driver circuits as well as a pixel switch of active matrix liquid crystal display (AMLCD). Also, it has been widely accepted that poly-Si TFTs with staggered structure are necessary in high-density static random-access memories (SRAM) as a load transistor of the memory cells [1]. The performance of poly-Si TFTs is strongly influenced by grain boundaries and intragranular traps of the poly-Si active layer. Solid-phase crystallization of amorphous silicon (a-Si) films at low temperature is a useful method for enlarging the grain size of the finally obtained poly-Si films, leading to a decrease in the effective defect density per unit area of the films [1-3].

When a-Si film precursor was deposited by low-pressure chemical vapor deposition (LPCVD), the deposition parameters such as temperature, pressure, and source gas were found to directly affect the crystallization behaviors. The use of Si_2H_6 gas instead of conventional SiH_4 enables to deposit a-Si films at low temperature as low as 430 °C, resulting in poly-Si films with large grains due to a low nucleation rate [2]. However, the a-Si films deposited by using Si_2H_6 requires a prolonged annealing time (> 20 h below 600 °C) for the complete crystallization, which is a critical problem in device applications. Furthermore, it has been reported that the a-Si films deposited at low temperature using Si_2H_6 gas are crystallized throughout the films without preferred nucleation sites, so called bulk-induced crystallization process [3], while crystallization in the a-Si films deposited at relatively high temperature (> 550 °C) using SiH_4 gas starts from the interface between the substrate and the a-Si films, called interface-induced crystallization [4].

In this paper, we report the substrate effects on the solid-phase crystallization of the a-Si films deposited by LPCVD using Si_2H_6 gas. The crystallization behavior is found to be predominantly changed with the substrate, and the surface roughness of the substrate plays an important role in

243

crystallization of a-Si films.

EXPERIMENTS

In the experiments two types of substrates, a thermally oxidized Si wafer (SiO_2/Si) and a quartz, were mainly used. The oxidation process was performed at 920 °C in a wet ambient, resulting in growth of 3000-Å-thick oxide on Si wafer. Amorphous silicon films with a thickness of 1200 Å were deposited by LPCVD using Si_2H_6 gas at 470 °C. The pressure was 0.25 Torr during the deposition process. These films were annealed at 600 °C in a dry N_2 ambient for solid-phase crystallization. In order to investigate the crystallization kinetics, x-ray diffraction (XRD) and Raman scattering studies were carried out for the a-Si samples with annealing time. Planar and cross-sectional views of grains in the Si films were also examined by transmission electron microscope (TEM). The surface state of the substrate has been studied by using atomic force microscope (AFM).

RESULTS AND DISCUSSION

Figure 1 shows the relative intensity of the (111) peak of XRD for Si films on the SiO_2/Si and the quartz substrates as a function of annealing time. The main crystalline plane of the crystallized film is (111) while (220) and (311) peaks are observed for fully annealed samples, which agrees well with the results already reported [2, 5]. From the XRD studies, we have found that a-Si film is faster crystallized on the SiO_2/Si than on the quartz. The same result was also observed by Raman scattering as shown in Fig. 2. Figure 2 exhibits the crystalline fraction that was evaluated from the Raman spectra for the Si films on the SiO_2/Si and the quartz as a function of annealing time. The crystalline fraction was calculated by the integrated Raman scattering intensity of the crystalline phase relative to that of the amorphous part, in which the scattering cross-section for crystalline and amorphous phases was taken into account [6]. The time needed for the complete crystallization of a-Si films on the SiO_2/Si is greatly reduced to 8 h from ~15 h on the quartz. Crystalline nuclei are faster formed on the SiO_2/Si than on the quartz, and also grain-growth rate is higher on the SiO_2/Si substrate than on the quartz. The crystallization kinetics of the a-Si films

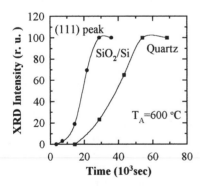

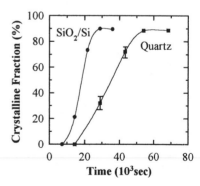

Fig. 1. XRD intensity of the (111) texture for a-Si films on a quartz and a thermally oxidized Si wafer as a function of annealing time.

Fig. 2. Crystalline fraction for a-Si films on the two substrates as a function of annealing time.

deposited on a quartz by LPCVD using Si_2H_6 gas has been widely studied by others [2-3, 5], and our results agree well with their data. However, the different crystallization behavior of the a-Si films on between the SiO_2/Si and the quartz was first reported in this paper.

The plan-view TEM photographs for poly-Si films formed by annealing of the a-Si films on the SiO_2/Si and quartz substrates are shown in Fig. 3. The a-Si films on the thermal oxide and on the quartz were annealed for 10 h and 19 h, respectively, at 600 °C. These annealing processes lead to fully crystallized poly-Si films on the two substrates. The grain size of the fully crystallized poly-Si films on the SiO_2/Si is about 2 μm, which is comparable to or slightly smaller than that on the quartz (2~3 μm). Although the time needed for the complete crystallization of the a-Si films on the SiO_2/Si is greatly reduced to 8 h, the decrease in grain size of the finally obtained poly-Si films is relatively small. Also, the poly-Si films on the SiO_2/Si substrate have grains of dendritic structure that is generally observed in the solid-phase crystallized poly-Si films [7-8]. The electrical properties of poly-Si films are governed by the density of electrically active defects that are strongly influenced by the crystallinity (grain size and its structure) of the films. From the view point of crystallinity, the electrical properties of the poly-Si films formed on the SiO_2/Si may be identical to those on the quartz. These results imply that poly-Si films with large grains can be formed by a short annealing of the a-Si deposited at low temperature by LPCVD using Si_2H_6 gas. Therefore, the disadvantage of a prolonged annealing for complete crystallizing a-Si films

Fig. 3. Plan-view TEM photographs of poly-Si films formed by solid-phase crystallization of a-Si deposited cn the SiO_2/Si and quartz substrates.

deposited at low temperature can be settled if one selects a proper substrate.

We have studied cross-sectional TEMs in order to explain the difference of crystallization kinetics of the a-Si films on between the SiO_2/Si and the quartz substrates. Figure 4 shows the cross-sectional TEM photographs of the a-Si film that was deposited on the quartz and then annealed at 600 °C for 8 h. The crystalline fraction of the film is around 30 %, which is measured from the Raman spectrum. We could observe the grains that were generated over all regions of the a-Si film, center and upper (free surface) and lower (interface between the a-Si and the quartz) parts of the film. The result indicates that random or homogeneous nucleation process dominates in the a-Si films deposited on the quartz. Our observation agrees well with the result of Hasegawa, et al. [3]. They deduced the bulk-induced crystallization through the homogeneous nucleation for the a-Si films on quartz from the fact that the signals from a crystalline phase start to be appear at the same time in both XRD and Raman scattering measurements. However, we directly found the homogeneous nucleation process in the a-Si films using cross-sectional TEM

⊢——⊣ 120 nm

Fig. 4. Cross-sectional TEM photographs of the a-Si deposited on the quartz and then annealed at 600 °C for 8 h, showing grains generated at center and upper (free surface) and lower (interface with the substrate) parts of the a-Si film.

⊢——⊣ 120 nm

Fig. 5. Cross-sectional TEM photograph of the a-Si film deposited on the SiO$_2$/Si and then annealed at 600 °C for 4 h. The arrows indicate the grains generated at the interface between the a-Si and the thermal oxide of the substrate.

observations.

In contrast to the quartz substrate, crystalline nuclei in the a-Si films on the SiO$_2$/Si were observed at the interface between the a-Si and the substrate as shown in Fig. 5. Figure 5 shows the cross-sectional TEM photograph of the Si film deposited on the SiO$_2$/Si substrate. The film was annealed at 600 °C for 4 h, which results in a crystalline fraction of ~20 %. We could find the grains generated at the interface between the a-Si and the thermal oxide of the substrate and then grown. TEM studies show that most of nuclei in the a-Si films on the SiO$_2$/Si substrate are generated at the interface and then grow until the film is completely crystallized. It should be noted here that the a-Si film, consisting of a composite double layer with the mixed-phase film (crystallites embedded in an amorphous matrix), deposited on SiO$_2$ using SiH$_4$ gas is crystallized from the interface between the a-Si and the mixed-phase film [9]. The mixed-phase film acts as a seeding layer for crystallization of the amorphous regions. In our study, however, we observed interface-induced crystallization in the a-Si films deposited at a low temperature using Si$_2$H$_6$ gas without any seeding layer.

It has been confirmed from the cross-sectional TEM observations that the a-Si deposited on the SiO$_2$/Si starts to be crystallized from the interface between the SiO$_2$ of substrate and the a-Si film, called interface-induced crystallization, while random nucleation process dominates in the a-Si on the quartz. The different kinetics and mechanism of solid-phase crystallization is considered to be strongly correlated to the surface state of the substrates upon which a-Si film is deposited.

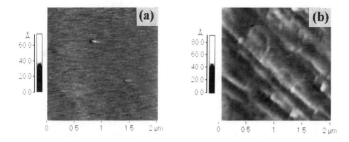

Fig. 6. Surface structure and its roughness of the thermal oxide grown on Si wafer (a), and the quartz (b) observed by AFM.

The surface structure of the substrate was investigated by AFM before a-Si deposition, and its data are shown in Fig. 6. The surface of the thermal oxide grown upon Si wafers is very smooth with a root mean square (RMS) roughness of 4.2 Å. However, the quartz has a ditch-like surface with a large RMS roughness of 13.0 Å as shown in Fig. 6 (b), which we think results from the polish process of the substrate. The very rough surface of the quartz substrate might severely decrease the surface diffusivity of Si atom, specifically, at the initial stage of the deposition process, resulting in the a-Si film with a large structural disorder. Therefore, a longer annealing time is required for the crystallization of a-Si films deposited on the quartz than the SiO_2/Si substrate, and the interface between the a-Si film and the quartz substrate can be no longer a preferred site for nucleation due to the large disorder at the initial stage of the deposition process.

In general, nucleation and grain growth in the solid-phase crystallization of a-Si films have been found to be directly dependent on the structural disorder of Si network. As the structural disorder is increased, the nucleation and grain growth rates are decreased. It was also reported by Hasegawas, et al. that crystallization of the a-Si deposited on quartz by LPCVD using SiH_4 gas is controlled by the interface-induced nucleation process [4], while the a-Si deposited on the same substrate at low temperature using Si_2H_6 is randomly crystallized throughout the film [3]. They have not given a clear explanation for the difference. Based on our results, we suggest that the structural disorder of Si network nearby on the substrate plays a crucial role in the interface-induced nucleation. Our results show that the interface-induced crystallization is continued to the a-Si film deposited at low temperature using Si_2H_6 gas when the thermal oxide with a very smooth surface is used, as shown in Fig. 4. The very smooth surface of the substrate can remarkably relieve the increase of the structural disorder due to lowering the deposition temperature from 580 °C (SiH_4) to 470 °C (Si_2H_6). However, the use of quartz as a substrate causes the structural disorder of a-Si film, at least, nearby on the substrate to further increase due to the large surface roughness shown in Fig. 6 (b). As a result, the a-Si films deposited on the quartz by LPCVD using Si_2H_4 gas has the largest structural disorder because of the surface roughness of the substrate and the low deposition temperature. Furthermore, the interface between the a-Si film and the quartz substrate has a structural disorder comparable to or larger than the bulk has, because the surface diffusivity of Si atom is very low at the initial stage of the deposition process due to the rough surface of the substrate. Therefore, the interface between the a-Si film and the quartz can be no longer a preferred nucleation site, and crystallization proceeds through the homogeneous bulk process rather than the interface-induced one.

CONCLUSIONS

In this study, the crystallization kinetics and mechanism of a-Si films deposited by LPCVD using Si_2H_6 gas has been extensively investigated. The a-Si films are faster crystallized on a thermally grown oxide than on a quartz through interface-induced nucleation process. The structural disorder of silicon network, which is strongly affected by the surface roughness of the substrates, is found to govern the nucleation mechanism in solid-phase crystallization of a-Si films. The very smooth surface of the thermally grown oxide could considerably relieve the increase of the structural disorder due to lowering the deposition temperature. However, the quartz with a rough surface might largely decrease the diffusivity of Si atoms at the initial stage of a-Si deposition, resulting in an additional disorder nearby on the substrate. Therefore, the interface with thermal oxide maintains preferred sites for nucleation, while that with quartz can be no longer preferred sites when the deposition temperature of a-Si films is decreased using Si_2H_6 instead of SiH_4 gas.

ACKNOWLEDGMENT

The authors would like to thank Dr. S. -D. Jung for AFM observations.

REFERENCES

1. N. Yamauchi and R. Reif, J. Appl. Phys. **75**, 3235 (1995).

2. K. Nakazawa, J. Appl. Phys. **69**, 1703 (1991).

3. S. Hasegawa, S. Watanabe, T. Inokura and Y. Kurata, J. Appl. Phys. **77**, 1938 (1995).

4. S. Hasegawa, T. Nakamura and Y. Kurata, Jpn. J. Appl. Phys. **31**, 161 (1992).

5. C. H. Hong, C. Y. Park and H.- J. Kim, J. Appl. Phys. **71**, 5427 (1992).

6. R. Tsu, J. G.- Hernandez, S. S. Chao, S. C. Lee and K. Tanaka, Appl. Phys. Lett. **40**, 534 (1982).

7. A. Nakamura, F. Emoto, E. Fujii, A. Yamamoto, Y. Uemoto, K. Senda and G. Kano, J. Appl. Phys. **66**, 4248 (1989).

8. J. H. Kim, J. Y. Lee and K. S. Nam, J. Appl. Phys. **77**, 95 (1995).

9. A. T. Voutsas and M. K. Hatalis, J. Elec. Mat. **23**, 319 (1994).

APPROACHES TO MODIFYING SOLID PHASE
CRYSTALLIZATION KINETICS FOR a-Si FILMS

Reece Kingi, Yaozu Wang, Stephen Fonash, Osama Awadelkarim, and Yuan-Min Li*, Electronic Materials and Processing Research Laboratory, The Pennsylvania State University, University Park, PA16802, and *Solarex Corporation, 826 Newtown-Yardley Road, PA18940.

ABSTRACT

Three approaches to modifying the solid phase crystallization kinetics of amorphous silicon thin films are examined with the goal of reducing the thermal budget and improving the poly-Si quality for thin film transistor applications. The three approaches consist of (1) variations in the PECVD a-Si deposition parameters; (2) the application of pre-furnace-anneal surface treatments; and (3) using both rapid thermal annealing and furnace annealing at different temperatures. We also examine the synergism among these approaches.

Results reveal that (1) film deposition dilution and dilution/temperature changes do not strongly affect crystallization time, but do affect grain size; (2) pre-anneal surface treatments can dramatically reduce the solid phase crystallization thermal budget for diluted films and act synergistically with deposition dilution or dilution/temperature effects; and (3) rapid thermal annealing leads to different crystallization kinetics from that seen for furnace annealing.

INTRODUCTION

Polycrystalline silicon (poly-Si) thin film transistors are currently receiving considerable attention for active matrix flat panel display technology[1-4]. The solid phase crystallization (SPC) of plasma enhanced chemical vapor deposited (PECVD) amorphous silicon (a-Si) thin films is a promising approach to obtain these poly-Si thin films, as opposed to depositing them directly, because larger grains and smoother surfaces can be attained with a low thermal budget[1,2,4].

In this work we have studied three approaches to modifying the SPC kinetics of PECVD deposited a-Si thin films, with the goal of reducing the thermal budget and improving the poly-Si quality for thin film transistor applications. These three approaches consist of (1) varying the PECVD deposition parameters, (2) using pre-anneal surface treatments, and (3) using both furnace annealing and rapid thermal annealing.

EXPERIMENTAL PROCEDURE

Eleven 1000Å a-Si thin films were deposited onto Corning code 7059 glass substrates from pure SiH_4, SiH_4 diluted with Ar (SiH_4:Ar 1:10) and SiH_4 diluted with H_2 (SiH_4:H_2 1:10), using a capacitively coupled rf PECVD system. The deposition temperature was varied between 290°C and 130°C. The pressure was 0.5 Torr for pure SiH_4 and 1 Torr for SiH_4:Ar and SiH_4:H_2 mixtures. The deposition rates for pure SiH_4, SiH_4 diluted with H_2 and SiH_4 diluted with Ar were 2.4 Å/sec, 0.87 Å/sec and 3.3 Å/sec, respectively. All other parameters were kept constant.

Films were then annealed in a conventional furnace and a rapid thermal processing system in an N_2 ambient over a range of temperatures. In each case the temperature was monitored with k-type thermocouples located next to the a-Si samples. Some films were subjected to pre-furnace-anneal surface treatments, which consisted of the deposition of 15 Å of either Ni or Pd onto the film surface, or the exposure of the film surface to high density N_2 or He plasmas. The thin metal layers were deposited using a high vacuum thermal evaporation system, and the high density

Mat. Res. Soc. Symp. Proc. Vol. 424 © 1997 Materials Research Society

plasma source was an ASTeX electron cyclotron resonance system, the later of which is described elsewhere[5].

Spectroscopic ellipsometry (SE) was performed on a selection of samples prior to crystallization to determine the relative microvoid density as a function of PECVD deposition parameters. The UV reflectance technique[6,7] was used to determine the degree of crystallinity of all annealed films, and TEM was used to determine the grain size of all fully crystallized films.

EXPERIMENTAL RESULTS AND DISCUSSION

Effects of PECVD Deposition Parameters

The degree of crystallinity versus annealing time for a-Si thin films on substrates begins with a transient (or incubation) period, during which time no crystallization is observed[3,4,8]. Nucleation then begins, and grains grow laterally outward from these nucleation sites until they impinge upon adjacent grains, at which point crystallization is complete. The period between the end of the transient period and the total crystallization time shall hereafter be referred to as the growth period (or growth time). Figure 1 presents the transient time, growth time, and the total crystallization time as a function of the PECVD deposition temperature for the films deposited with various types of dilution, i.e. pure SiH_4, SiH_4:Ar, and SiH_4:H_2 source gas mixtures.

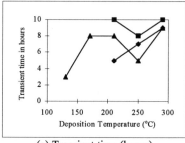

(a) Transient time (hours)

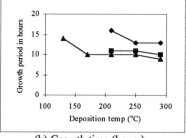

(b) Growth time (hours)

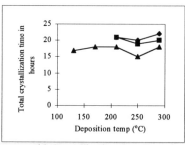

(c) Crystallization time (hours)

Figure 1. (a) transient time, (b) growth time and (c) crystallization time in hours versus deposition temperature (°C) for PECVD a-Si films deposited from SiH_4 (■), SiH_4:Ar (♦) and SiH_4:H_2 (▲) source gas mixtures, and annealed in a conventional furnace at 600°C in flowing N_2 ambient.

The total crystallization time is lower for films deposited from a SiH_4:H_2 source gas mixture, but does not show a specific trend as a function of temperature. The growth period, during which time the nucleation and grain growth processes occur, seems to increase slightly with decreasing deposition temperature, and is longest for a SiH_4:Ar source gas mixture, and lowest for a SiH_4:H_2

source gas mixture. Figure 2 presents the grain size for fully crystallized films as a function of the PECVD deposition temperature for pure SiH_4, SiH_4:Ar, and SiH_4:H_2 source gas mixtures. The grain size was calculated by averaging the largest twenty grains in a given TEM micrograph. Figure 2 shows that the grain size increases as the deposition temperature is reduced, and that the grain size is largest for SiH_4:Ar source gas mixtures, and lowest for SiH_4:H_2 source gas mixtures.

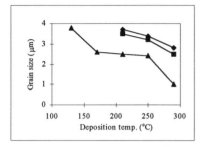

Figure 2. Grain size (μm) versus deposition temperature (°C) for PECVD a-Si films deposited from SiH_4 (■), SiH_4:Ar (◆) and SiH_4:H_2 (▲) source gas mixtures, and crystallized in a conventional furnace at 600°C in flowing N_2 ambient.

It can be seen that, as the deposition temperature is reduced and as the source gas mixture is changed from SiH_4:H_2 to pure SiH_4, or from pure SiH_4 to SiH_4:Ar, the results presented in Figures 1(a) and 2 are consistent with a decreasing nucleation rate. For example, as the deposition temperature is reduced the growth period and the average grain size increase. Because the growth period increases nucleation occurs over a longer period of time, and because the average grain size increases, it follows that the number of total nuclei that form per unit area decreases. Thus, the nucleation rate decreases. We note these correlations between grain size, growth time, etc, and deposition parameters were not observed for the transient period, suggesting the state of the film/glass interface dominates in determining this time.

The relative microvoid density, obtained from SE, for the various films of Figures 1 and 2 are presented in Table I. These results show that as the deposition temperature is reduced, or as the source gas mixture is changed from SiH_4:H_2 to pure SiH_4, or from pure SiH_4 to SiH_4:Ar, the relative microvoid density increases. Furthermore, it is well known that as the deposition temperature is reduced, or as the deposition rate is increased (which in this case corresponds to changing the source gas mixture from SiH_4:H_2 to pure SiH_4, or from pure SiH_4 to SiH_4:Ar), the structural disorder in the resulting a-Si film increases[2,9]. Thus, these results coupled with those presented above illustrate that, as the structural disorder in the a-Si film increases, then the resulting nucleation rate decreases. This impact on nucleation has previously been observed using deposition temperature variations to affect disorder[2]. This is the first report of using both temperature and dilution to affect disorder and subsequent crystallization.

Table I. Relative microvoid density for various PECVD deposition conditions determined from spectroscopic ellipsometry.

Dep. temp. (°C)	Source gas mix.	Relative microvoid density
290	SiH_4:H_2	0.02
250	SiH_4:H_2	0.04
130	SiH_4:H_2	0.08
250	SiH_4	0.05
250	SiH_4:Ar	0.11

Effects of Pre-Anneal Surface Treatments

Pre-anneal surface treatments have been compared for a-Si films deposited from both hydrogen and argon diluted silane. Table II presents the grain size and the total crystallization time for a-Si films subjected to the various pre-anneal surface treatments described above, and then annealed in a conventional furnace at 600°C in flowing N_2 ambient. Table II shows that metal-catalyzed crystallization treatments of these various diluted films has a very dramatic impact on crystallization time. These reductions in crystallization time are much larger than any reductions we had seen previously on non-diluted films[12]. This result suggests a synergism between these metal treatments and dilution. Table II also shows that, in general, the greater the reduction in the thermal budget, the greater the reduction in the grain size, for both metal and plasma pre-anneal surface treatments.

Table II. Grain sizes and crystallization times for a-Si films subjected to various pre-anneal surface treatments. Annealing was carried out in a conventional furnace at 600°C in flowing N_2 ambient.

Dep. Temp (°C)	Source gas mix.	Pre-Anneal Treatment Applied	Cryst. time	Grain Size (μm)
210	$SiH_4:H_2$	none	18 hours	2.5
		He plasma	14 hours	2.5
		N_2 plasma	8 hours	0.9
		Ni metal	~ 5 min.	0.16
		Pd metal	~ 15 min.	~ 0.05
130	$SiH_4:H_2$	none	17 hours	3.8
		Ni metal	< 30 min.	0.15
		Pd metal	< 30 min.	~ 0.07
210	$SiH_4:Ar$	none	18 hours	3.7
		Pd metal	~ 5 min.	~ 0.08

The crystallization of a-Si thin films with Ni and Pd overlayers have been studied by Kawazu et al[10] and Liu et al[11,12], respectively. Both parties concluded from their experimental observations that the Ni and Pd overlayers result in the formation of a silicide at a low temperature (~350°C), and that this silicide induces nucleation. Liu et al[12] found that this Pd layer could also be effectively placed between the substrate and the a-Si layer. The reduction in thermal budget resulting from pre-anneal plasma surface treatments has been studied by Yin et al[6], who postulated that plasma treatments increase the stored energy in the surface of the a-Si film, in the form of strained and broken Si-Si bonds. This energy then aids the nucleation process upon annealing. It is interesting to note that Table II shows that the plasma treatments on diluted films produce almost the same reduction in crystallization time reported earlier for plasma treatments, also for diluted films[6]. We reiterate that the metal treatments dramatically lower the crystallization times - much more so than for this treatment on non-diluted films.

Effects of Annealing Technique and Temperature

Figure 3 presents the grain size and crystallization time for a-Si films deposited at 210°C from a $SiH_4:Ar$ source gas mixture annealed in a conventional furnace and a RTP system, versus 1000/T, where T is the annealing temperature in degrees Kelvin. Figure 3 illustrates that for either furnace annealing or RTA the transient and crystallization times are temperature activated processes, and the resulting poly-Si grain size is inversely proportional to the annealing temperature. Both phenomena have previously been observed for furnace annealing only[14].

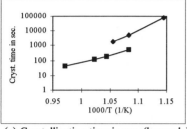
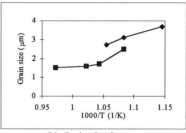

(a) Crystallization time in sec. (log scale)   (b) Grain size in μm

Figure 3. (a) crystallization time in seconds (log scale), and (b) grain size in μm versus 1000/T where T is the annealing temperature in degrees Kelvin, for a-Si films annealed in a conventional furnace (■) and a rapid thermal processing system (◆) in flowing N_2 ambient.

Figure 3 also illustrates that for a given annealing temperature, RTA results in lower crystallization times with smaller grain sizes than furnace annealing. Two explanatory models for this phenomena are hereforth proposed. The first model is based on the fact that the light spectra present during annealing in the two systems are different. The RTP system uses tungsten-halogen lamps to heat the a-Si film, whereas the conventional furnace uses a resistive coil to heat the environment. The light spectrum present during rapid thermal annealing is much more energetic and intense than that present during furnace annealing. Because of this, it is proposed that the energetic photons present during rapid thermal annealing create photocarriers which subsequently recombine and thereby break weak and strained bonds in the a-Si:H thin film. These sites then serve as recombination centers for further photocarriers producing localized heating. The locations at which this localized heating occurs then become favorable sites for nucleation. This results in the ability for nuclei to form more readily which results in a higher nucleation rate and hence a lower crystallization time. A direct result of this increased nucleation rate is the reduction in the grain size that is observed.

The second model proposed to explain the reduction in thermal budget and resulting grain size for rapid thermal annealing compared to furnace annealing is based upon the sudden release of stored energy within the a-Si:H thin film. Experimental observations made by Takamori et al[15,16] illustrated that stored energy in amorphous films could be released resulting in "explosive" crystallization. Specifically, they deposited relatively thick (>10μm) Ge films by sputtering which could then be crystallized by simply "pricking" the surface of the film with a sharp object. Based on their experimental observations, they concluded that once the crystallization process was triggered (i.e. by "pricking" the surface of the film with a sharp object), the free energy released (resulting from the difference in free energy between the amorphous state and the polycrystalline state) was sufficient to drive the crystallization of the remainder of the film. Based upon these results, it is proposed that the sudden heating of the a-Si:H thin film that occurs during rapid thermal annealing is the "pricking" that results in the sudden release of stored energy, in the form of heat. This excess heat results in a higher "effective" temperature which drives the nucleation and grain growth processes. In the same way that the grain size decreases with increasing annealing temperature, compared to furnace annealing the grain size is reduced for rapid thermal annealing simply because of the higher "effective" temperature, resulting from this sudden release of stored energy within the film. Although this energy must also be released during furnace annealing, because it is released over such a long period of time (because the heating rate is much slower), it does not have the same impact that occurs during rapid thermal annealing.

CONCLUSIONS

The PECVD deposition conditions affect the SPC kinetics of a-Si thin films. As the deposition rate is increased, or as the source gas mixture is changed from $SiH_4:H_2$ to pure SiH_4, or from pure SiH_4 to $SiH_4:Ar$, the nucleation rate decreases resulting in an increased grain size. This decreasing nucleation rate was correlated with increasing structural disorder in the a-Si film.

RTA was found to modify the kinetics of SPC. Two models were propose for this. RTA was found to produce smaller grains at a lower thermal budget than furnace annealing, for a given temperature.

Substantial thermal budget reduction is offered by pre-anneal surface treatments when applied to diluted films. These treatments do result in smaller poly-Si grain sizes. The thermal budget reduction occurs as a result of an enhanced nucleation rate, and hence the greater the enhancement, the smaller the resulting poly-Si grain size.

ACKNOWLEDGMENT

The authors would like to thank Dr. Y. Lu for the spectroscopic ellipsometry measurements.

REFERENCES

1   I-Wei Wu, Solid State Phenomena Vols 37-38 (1994) pp. 553-564
2   K. Nakazawa and K. Tanaka, J. Appl. Phys. **68** (3) 1990 (1029)
3   T. Kretz, R. Stroh, P. Leganeux, O. Huet, M. Magis and D. Pribat, Solid State Phenomena Vols 37-38 (1994) pp. 311-316
4   J. Guillemet, B. Pieraggi, B. de Mauduit and A. Claverie, Solid State Phenomena, Vols 37-38 (1994) pp. 293-298
5   R. A. Ditizio, G. Liu, S. Fonash, B. Hseih and D. Greve, Appl. Phys. Lett. **56** (12) 1990 (1140)
6   A. Yin, S. Fonash, D. Reber, T. Li and M. Bennett, MRS Symposium Proceedings **345**, 1994, pp. 81-86
7   T. Kammins, Polycrystalline Silicon for IC Applications, (Klumer Academic Publishers, 1998)
8   I. Wu, A. Chiang, M. Ovecoglu and T. Huang, J. Appl. Phys. **65** (10) 1989 (4036)
9   S. Nag, PhD Thesis, The Pennsylvania State University, 1992
10   Y. Kawazu, H. Kudo, S. Onari and T. Arai, Jpn. J. Appl. Phys. Vol. 29, No. 12, 1990, pp. 2698-2704
11   G. Liu and S. Fonash, Appl. Phys. Lett. **55** (7) 1989 (660)
12   G. Liu and S. Fonash, Appl. Phys. Lett. **62** (20) 1993 (2554)
13   S. Herd, P. Chaudhari, and M. Brodsky, J. Non-Cryst. Solids Vol 7, 1972, pp.309-327
14   R. Iverson and R. Reif, J. Appl. Phys. **62** (5) 1987 (1675)
15   T. Takamori, R. Messier, and R. Roy, Appl. Phys. Lett. Vol. 20, No. 5, 1972, p. 201
16   T. Takamori, R. Messier, and R. Roy, J. Mats. Sci. **8** (1973) p. 1809

EFFECTS OF GERMANIUM ON GRAIN SIZE AND SURFACE ROUGHNESS OF THE SOLID PHASE CRYSTALLIZED POLYCRYSTALLINE $Si_{1-x}Ge_x$ FILMS

Jin-Won Kim*, Myung-Kwan Ryu, Tae-Hoon Kim, Ki-Bum Kim, and Sang-Joo Kim
Department of Metallurgical Engineering, Seoul National University, Seoul, Korea
*Present Address : Semiconductor R&D Center, Samsung Electronics Co., Korea

ABSTRACT

$Si_{1-x}Ge_x$ ($x \leq 0.5$) films were deposited by using Si_2H_6 and GeH_4 source gases in a low pressure chemical vapor deposition (LPCVD). The deposition temperature was varied from 375 °C for $Si_{0.5}Ge_{0.5}$ to 450 °C for Si film in order to deposit amorphous $Si_{1-x}Ge_x$ films and the deposition pressure was about 1 Torr. The grain size of polycrystalline $Si_{1-x}Ge_x$ films made by solid phase crystallization (SPC) decreases with germanium content in the films, and the activation energy obtained from the dependence of grain size on annealing temperature was 0.4 eV for Si and 0.45 eV for $Si_{0.69}Ge_{0.31}$. From obtaining a similar activation energy irrespective of Ge content in the film, the decrease of grain size with germanium content is attributed to the difference of the as-deposited film conditions. The surface roughness of $Si_{1-x}Ge_x$ films investigated by atomic force microscope (AFM) increases with germanium content in the film before and after SPC. For instance, the root-mean square (rms) values of the surface roughness of the as-deposited Si and $Si_{0.5}Ge_{0.5}$ films were 2.3 and 17 Å, while those values were increased to 2.6 and 41 Å, respectively, after SPC. In order to reduce the surface roughness of $Si_{0.5}Ge_{0.5}$ film, we have deposited a thin Si capping layer on top of the $Si_{1-x}Ge_x$ layer and identified that this capping layer effectively reduces the increase of surface roughness after SPC.

INTRODUCTION

In poly-Si thin film transistor (TFT) technology, the solid phase crystallization (SPC) of amorphous Si (α-Si) film has become an important method for obtaining a high quality active layer since a polycrystalline film with both large grain size and low surface roughness can be obtained compared with as-deposited poly-Si film.[1,2,3] However, one of the major drawbacks of SPC method is a large thermal budget, which makes it incompatible with low cost glass substrates. This has motivated many researchers to study $Si_{1-x}Ge_x$ alloy film as an alternative to Si film since one can decrease the thermal budget of SPC with germanium content in the film.[4,5,6,7]

There have been several reports on the SPC behavior of α-$Si_{1-x}Ge_x$ films.[4,6,7,8] For instance, Hwang et al.[4] suggested the crystallization mechanism of α-$Si_{1-x}Ge_x$ films deposited by molecular beam epitaxy (MBE), Tsai et al.[6] reported the effects of germanium on the material and electrical properties of poly-$Si_{1-x}Ge_x$ films grown from SiH_4 and GeH_4 in a very low pressure chemical vapor deposition (VLPCVD) or plasma-enhanced VLPCVD (PEVLPCVD), and Mäenpää et al.[8] investigated the growth kinetics of α-$Si_{1-x}Ge_x$ films made by e-beam evaporation and subsequent ion beam mixing. All of these results asserted that the thermal budget of SPC reduces with germanium content. However, the effects of germanium on the grain size of the solid phase crystallized $Si_{1-x}Ge_x$ films have not been well understood; Hwang et al.[4] reported that the grain size of the crystallized films decreases with germanium content, while Tsai et al.[6] reported the opposite trend. In addition, as far as we know, the surface roughness of the crystallized $Si_{1-x}Ge_x$ films, which is known to be a important parameter affecting the field-effect mobility of a TFT,[9] has not been systematically investigated yet.

In this work, we have investigated the effects of germanium on grain size and surface roughness of poly-$Si_{1-x}Ge_x$ films made by SPC. We believe that this investigation is important in optimizing the processing condition and thereby obtaining the optimum device characteristics of poly-$Si_{1-x}Ge_x$ TFT.

EXPERIMENTAL

The $Si_{1-x}Ge_x$ films of which germanium concentration x=0, 0.16, 0.25, 0.31, 0.4, and 0.5 were deposited using Si_2H_6 and GeH_4 source gases in an LPCVD and H_2 was used as carrier gas. The substrates were Si (100) wafers having thermally grown SiO_2. The working pressure was about 1 Torr and the deposition temperature was varied from 375 to 450 °C in order to deposit α-$Si_{1-x}Ge_x$ films; Si, $Si_{0.84}Ge_{0.16}$ and $Si_{0.75}Ge_{0.25}$ films were deposited at 450 °C, $Si_{0.69}Ge_{0.31}$ and $Si_{0.6}Ge_{0.4}$ films were deposited at 425 °C, and $Si_{0.5}Ge_{0.5}$ films were deposited at 375 °C. The nominal thickness of all the films was either 500 or 1000 Å and the deposition rate was in the 12 ~ 30 Å/min. range. The detailed deposition behaviors of the above films were presented in the previous papers.[10] The as-deposited films were annealed at 600 °C in a tube furnace using a high purity nitrogen gas as an ambient. For the study of the origin of surface roughness, both $Si_{0.75}Ge_{0.25}$ and $Si_{0.5}Ge_{0.5}$ films having Si capping layer were also prepared. The Si capping layers were *in situ* deposited on the $Si_{1-x}Ge_x$ alloy films in the reaction chamber. In the case of the Si/$Si_{0.75}Ge_{0.25}$ film, there was no time interval between the deposition of $Si_{0.75}Ge_{0.25}$ and that of Si film because the deposition temperature of each film was the same. However, in the deposition of Si/$Si_{0.5}Ge_{0.5}$, it takes about 5 min. to raise the deposition temperature from 375 °C for $Si_{0.5}Ge_{0.5}$ to 450 °C for Si, so that the $Si_{0.5}Ge_{0.5}$ film was exposed to H_2 ambient during this time interval.

The composition of the films was identified by Rutherford backscattering spectrometry (RBS) and the annealed samples were investigated by using transmission electron microscopy (TEM) and atomic force microscopy (AFM).

RESULTS AND DISCUSSION

<u>TEM Results of As-deposited $Si_{0.6}Ge_{0.4}$ and $Si_{0.5}Ge_{0.5}$ Alloy Films</u>

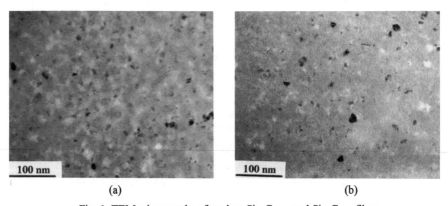

(a)   (b)

Fig. 1. TEM micrographs of as-dep. $Si_{0.4}Ge_{0.6}$ and $Si_{0.5}Ge_{0.5}$ films.

Fig. 1(a) and 1(b) are the TEM micrographs of the $Si_{0.6}Ge_{0.4}$ and $Si_{0.5}Ge_{0.5}$ alloy films deposited at 425 and 375 °C, respectively. It is observed that a number of small crystallites whose average size is about 50 Å are embedded in an amorphous matrix. These TEM results clearly demonstrate that the respective transition temperature from amorphous to polycrystalline phase of $Si_{0.6}Ge_{0.4}$ and $Si_{0.5}Ge_{0.5}$ alloy films during deposition is lower than 425 and 375 °C, which are even lower than those previously reported by King et al..[11] They noted that the transition temperature of $Si_{0.6}Ge_{0.4}$ and $Si_{0.5}Ge_{0.5}$ alloy films was about 450 and 425 °C, respectively, which was determined by the XRD analysis of the $Si_{1-x}Ge_x$ alloy films deposited using SiH_4 and GeH_4 source gases in an LPCVD.

<u>Grain Size of the SPC Processed Poly-$Si_{1-x}Ge_x$ Films</u>

The average grain size of the poly-$Si_{1-x}Ge_x$ as a function of the germanium content is presented in the Fig. 2. The data show that the average grain size reduces with germanium content in the film.

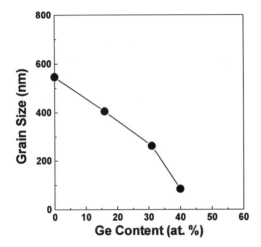

Fig. 2. Average grain size of the poly-$Si_{1-x}Ge_x$ films as a function of germanium content

Hatalis et al.[3] theoretically calculated a final average grain size of the crystallized films assuming that both the nucleation rate and growth rate are independent of time and are expressed as Arrhenius equation, and that a two-dimensional grain growth occurs. According to their calculation, the final average grain size can be written as

$$D \propto \left(\frac{r_G}{r_N}\right)^{1/3} \propto \exp\left(\frac{(E_N - E_G)}{3kT}\right)$$

where the E_N and E_G are the activation energies for nucleation and grain growth, respectively. According to this equation, the final average grain size is determined by the interrelation of the nucleation rate and the growth rate, which means that the variation of the grain size with germanium content in the film can not be just interpreted by the lowering of the melting temperature in $Si_{1-x}Ge_x$ alloy films since the decrease of the melting temperature results in enhancing both the nucleation rate and the grain growth rate.

In order to study the difference of the activation energy between the nucleation and the grain growth ($E_N - E_G$), we have investigated the dependence of grain size on the crystallization temperature in both Si and $Si_{0.69}Ge_{0.31}$ films (Fig.3). The data show that the average grain size decreases as the annealing temperature increases and that the ($E_N - E_G$)s of the two films are close each other; 0.4 eV for Si and 0.45 eV for $Si_{0.69}Ge_{0.31}$. In Si films, similar studies have been performed by other researchers.[3,12] For instance, Hatalis et al.[3] reported that the ($E_N - E_G$) was about 0.25 eV by studying of the SPC of α-Si films deposited by SiH_4 in an LPCVD and subsequently crystallized. In particular, they demonstrated that the ($E_N - E_G$) is independent of the deposition temperature although the final grain size depends on the deposition temperature and asserted that the grain size is mainly determined by the degree of the structural ordering of as-deposited amorphous films. This assertion was supported by Nakazawa et al.[13], who reported that the nucleation rate strongly decreases with the increase of the structural disorder, which is determined by Raman spectroscopy, in the Si network.

In this respect, it is considered that the dependence of the grain size of poly-$Si_{1-x}Ge_x$ films on germanium content also arises from the difference of the structural ordering in the as-deposited films and that the structural ordering in the as-deposited $Si_{1-x}Ge_x$ films increases with germanium content because the difference between the transition temperature from amorphous to polycrystalline during deposition and the deposition temperature decreases with germanium content; for example, the difference is about 125 °C in Si film,[14] while it is about 50 °C in $Si_{0.69}Ge_{0.31}$ film.[10] Consequently, the nucleation rate increases with germanium content in the film during SPC, which results in decreasing the final average grain size of the poly-$Si_{1-x}Ge_x$.

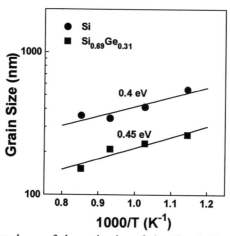

Fig. 3. Dependence of the grain size of the Si and $Si_{0.69}Ge_{0.31}$ films on the annealing temperature. (Annealing condition : 600 °C for 20 hr., 700 °C for 5 hr., 800 °C for 1 hr., and 900 °C for 1hr.)

Surface Roughness

In Fig. 4, the root mean square (rms) surface roughness values of both as-deposited and annealed $Si_{1-x}Ge_x$ films [x=0, 0.25, 0.4 and 0.5] are plotted as a function of the germanium content in the films. The rms surface roughness values of $Si_{0.75}Ge_{0.25}$ and $Si_{0.5}Ge_{0.5}$ films with Si capping layer are also presented in that figure. From the data, it is observed that the germanium incorporation in the film enhanced the increase of the surface roughness of poly-$Si_{1-x}Ge_x$ films made by SPC. In Si films, it is well known that the smooth surface roughness is preserved after SPC when the poly-Si film is formed from the amorphous by SPC.[2,15,16,17] The mechanism of the preservation of the surface roughness was explained by Kamins,[16] who proposed that the presence of a thin native SiO_2, which forms when the film is exposed to the air, preserves the smoothness of an amorphous films even after the films are crystallized by annealing. This hypothesis was supported by Ibok et al.,[17] who reported that the polycrystalline film deposited in the amorphous and annealed after the exposure of the film to air exhibits a smooth surface morphology, while the polycrystalline film deposited in amorphous but *in situ* annealed at 610 °C exhibits rough surface morphology. Thus, we can explain the increase of the surface roughness of poly-$Si_{1-x}Ge_x$ films by adopting the native oxide hypothesis. In the case of $Si_{1-x}Ge_x$, it has been reported that when $Si_{1-x}Ge_x$ film is oxidized the germanium atoms are rejected from the oxide and pile up at the oxide/$Si_{1-x}Ge_x$ interface.[18,19,20] In addition, Liou et al.[20] noted that when the native oxide forms on the $Si_{1-x}Ge_x$ alloy films the same phenomena also occur. Considering this, one possible assumption which can be derived from the above results is that the thickness of native oxide decreases with germanium content in the film. Consequently, it is considered that the germanium incorporation in the film renders the native oxide less stiff, thus increasing the surface roughness of the poly-$Si_{1-x}Ge_x$ made by SPC.

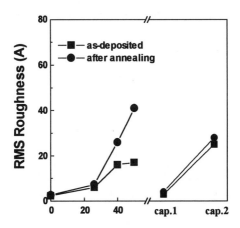

Fig. 4. rms surface roughness values of both as-deposited and annealed $Si_{1-x}Ge_x$ films with or without Si capping layer. The nominal film thickness of $Si_{1-x}Ge_x$ films and Si capping layer were about 500 and 80 Å, respectively. [cap.1:$Si/Si_{0.75}Ge_{0.25}$, cap.2: $Si/Si_{0.5}Ge_{0.5}$]

The another thing can be noted that the Si capping layer effectively reduces the increase of surface roughness resulted from SPC. The use of Si capping layer is considered to have two advantages over $Si_{1-x}Ge_x$ single layer; one is to reduce the surface roughness of the films made by SPC, the other, as is suggested by Tsai et al.,[6] is to form a better quality Si/SiO_2 interface compared with $Si_{1-x}Ge_x/SiO_2$ interface since the presence of Ge atom at the interface increases the interface trap density.[7]

CONCLUSION

In this study, we have investigated the effects of germanium on grain size and surface roughness of the solid phase crystallized polycrystalline $Si_{1-x}Ge_x$ films. It is found that the increase of germanium content in the film decreases the average grain size of poly-$Si_{1-x}Ge_x$ films made by SPC and increases the surface roughness of those films. We have explained above two results qualitatively; the former can be explained by using the concept of the structural ordering in an amorphous film, the latter can be explained by the thickness variation of the native oxide on $Si_{1-x}Ge_x$ films with germanium content in the film.

REFERENCE

1. M. K. Hatalis and D. W. Greve, IEEE Electron Device Lett. **8**, 361 (1987)
2. G. Harbeke, L. Krausbauer, E. F. Steigmeier, A. E. Widmer, H. F. Kappert, and G.Neugebauer, Appl. Phys. Lett. **42**, 249 (1983)
3. M. K. Hatalis and D. W. Greve, J. Appl. Phys., **63**, 2260 (1988)
4. C. W. Hwang, M. K. Ryu, K. B. Kim, S. C. Lee, and C. S. Kim, J. Appl. Phys. 77, 3042 (1995)
5. J. W. Kim, M. K. Ryu, K. B. Kim, M. K. Han, C. S. Kim, and S. J. Kim, Proceedings of the Second Pacific Rim International Conference on Advanced Materials and Processing, Vol. 2, p1273 (1995)
6. J. A. Tsai, A. J. Tang, T. Noguchi, and R. Reif, J. Electrochem. Soc. **142**, 3220 (1995)
7. T.-J. King and K. C. Saraswat, IEEE Trans. Electron Devices **41**, 1581 (1994)
8. M. Mäenpää and S. S. Lau, Thin Solid Films **82**, 343 (1981)
9. J. H. Kung, M. K. Hatalis, and J. Kanicki, Thin Solid Films **216**, 137 (1992)
10. J. W. Kim, M. K. Ryu, K. B. Kim, and S. J. Kim, J. Electrochem. Soc. **143**, 363 (1996)
11. T.-J. King, and K. C. Saraswat, J. Electochem. Soc. **141**, 2235 (1994)
12. R. B. Iverson, and R. Reif, J. Appl. Phys. **62**, 1675 (1987)
13. K. Nakazawa, J. Appl. Phys. **69**, 1703 (1991)
14. C. H. Hong, C. Y. Park, and H. J. Kim, J. Appl. Phys. **71**, 5427 (1992)
15. G. Harbeke, L. Krausbauer, E. F. Steigmeier, A. E. Widmer, H. F. Kappert, and G. Neugebauer, J. Electochem. Soc. **131**, 675 (1984)
16. T. Kamins, Polycrystalilne Silicon For Integrated Circuit Application, pp71, Kluwer Academic Publishers, USA, 1988
17. E. Ibok and S. Garg, J. Electrochem. Soc. **140**, 2927 (1993)
18. D. Nayak, K. Kamjoo, J. C. S. Woo, J. S. Park, and K. L. Wang, Appl. Phys. Lett. **56**, 66 (1990)
19. K. Prabhakaran, T. Nishioka, K. Sumitomo, Y. Kobayashi, and T. Ogino, Appl. Phys. Lett. **62**, 864 (1993)
20. H. K. Liou, P. Mei, U. Gennser, and E. S. Yang, Appl. Phys. Lett. **59**, 1200 (1991)

CHARACTERIZATION OF POLYSILICON FILMS USING ATOMIC FORCE MICROSCOPY

Francis P. Fehlner and Chad B. Moore
Corning Incorporated, Science and Technology Division, Corning, NY 14831
J. Greg Couillard
Cornell University, MS&E, Bard Hall, Ithaca, NY 14853

ABSTRACT

A simplified technique for characterizing crystallites in polysilicon films has been demonstrated based on use of the atomic force microscope (AFM). The crystallization of films deposited by two different techniques was examined.

INTRODUCTION

Polysilicon films on glass provide the basis for fabricating higher performance thin film transistors (TFTs) for active matrix liquid crystal displays (AMLCDs). The advantage of polycrystalline over amorphous silicon films is a higher mobility for electrons and thus improved performance in terms of larger displays, faster response, and more gray scale (better color). In addition, peripheral circuitry can be fabricated directly on the glass substrate.

The size of the crystallites in the polysilicon films, relative to the gate dimensions of the TFT, determines the uniformity of display performance. For a $10\,\mu$m gate length, a crystallite size of approximately $1\,\mu$m is desired to maintain a similar number of grain boundaries in each transistor of the TFT array [1]. Size determinations thus require scanning or transmission electron microscope (SEM or TEM) measurements which can be difficult and time consuming.

A simpler method for determining grain size is desired. In the present work, it is shown that atomic force microscopy can provide that simplified method. Comparisons between AFM, X-ray diffraction (XRD), TEM, and SEM show that AFM gives reliable information on the size and morphology of the crystallites.

EXPERIMENTAL PROCEDURE

Two sets of experiments were run. The first was designed to test the applicability of AFM to characterize individual crystallites in a polysilicon film. The second applied the technique to films being used in TFT fabrication at the Cornell University Nanofabrication Facility (CNF).

The first experiments were run at Corning using Code 1735 glass and a thin noncrystalline silicon film deposited using plasma enhanced chemical vapor deposition (PECVD). The glass was an alkali earth boroaluminosilicate with a strain point of 665 °C. Samples were cleaned using a detergent followed by 20 min of oxygen plasma to remove organics. Texwipe^{TM} X-100 Glass and Plastic Cleaner was then used to remove sodium and any remaining dust. A silica/silicon/silica film stack was deposited using PECVD. Settings for the amorphous silicon were: RF power 400 W, pressure 550 mT, temperature 300 °C, and gas flow rate 2000 sccm of 5% silane in Ar. These deposition conditions resulted in a film growth rate of 1.1 nm/sec for silicon. The silicon films were noncrystalline by XRD and had an orange brown color in transmission. The samples were heat treated at 645 °C for 70 hrs in lab atmosphere (approximately 20% relative humidity). The transmission color of the heat treated samples was a light lemon yellow indicating crystallization of the film. Detailed measurements were carried out on a sample having a 100 nm thick barrier layer of silica, a 200 nm thick silicon film, and a 100 nm thick capping layer (Sample No. 35-4).

Examination of the silicon layer required removal of the oxide capping layer. This was accomplished using a 150 sec dip in buffered HF (Transene Siloxide^{TM}). It was found that delineation of the crystallite structure by AFM required an additional 5 sec dip in Wright-Jenkins etch [2] which preferentially attacks defective regions in silicon crystals. Etch rate at room temperature for as-deposited films was 20 nm/sec while for heat treated films it was 8 nm/sec. Sample preparation for the TEM required etching of the silicon film into a wedge shaped cross section. The film on the substrate was then scribed into 1 mm squares and the whole substrate immersed in dilute (20%) HF. As a result, the silicon film was undercut and the squares floated to the surface of the HF where they were picked up on TEM grids. The films on the grids were rinsed in DI water before examination in the microscope.

The second set of experiments was carried out using films prepared at the CNF. Fused silica wafers cut into 3" dia. circles were used as substrates. They were RCA cleaned to remove organic and metallic residues. No HF dip was used. The DI water rinse was continued until the bath resistivity exceeded 16 M Ω-cm. A fused silica hot wall LPCVD reactor was utilized to deposit noncrystalline films of silicon at a rate of 3-5 nm/min. They were 100 nm thick, formed at 220 mTorr using a silane flow rate of 80 sccm. Deposition temperatures were 560°, 570°, and 580 °C to bracket the film growth region in which some nuclei initially grow at the film-substrate interface [3]. The film coated silica was cut into sections. One was kept as a control sample. The remaining sections were annealed in dry nitrogen at 580 °or 600 °C for 2 to 79 hrs. For AFM measurements, half the thickness of the silicon films was removed using Wright-Jenkins etch diluted 5:1 with DI water to slow the etch rate, thus making it more controllable.

RESULTS

XRD was applied to the unetched PECVD films to determine their state of crystallinity. As-deposited PECVD films were noncrystalline. Sample 35-4 which had been heat treated at 645 °C for 70 hrs showed diffraction peaks from the (111), (220), and (311) planes of silicon. The relative size of the peaks indicated that crystal orientation was almost completely random. An analysis of the full width at half maximum value for the (111) peak gave an approximate crystallite size of 100 nm perpendicular to the plane of the film. A similar analysis of the films deposited by LPCVD at various temperatures and then heat treated at 600 °C for 8 hrs gave a value of 28 nm irrespective of the deposition temperature.

Examination of the PECVD films in the AFM showed 50 μm hillocks on the surface of the silicon films after removal of the silica over-layer using BHF. The as-deposited and heat treated films were similar. However, subsequent exposure to full strength Wright-Jenkins etch revealed larger 200 nm hillocks only on the heat treated film, as shown in Fig. 1. TEM results for the same sample are presented in Fig. 2. The dark field TEM (holes in the film are black) highlights single grains of silicon having the same orientation. These vary in lateral dimensions from 200 to 500 nm. The irregular shapes are similar to those shown by Hatalis and Greve [3].

The study of LPCVD films was done to test the use of AFM in distinguishing the extent of crystallite nucleation and growth in heat treated silicon films. There was no need for the BHF etch since no silica under or overlayer was used. Removal of half the film thickness by dilute Wright-Jenkins etch exposed the crystallites for examination. Representative results are shown in Figs. 3A, 3B, and 3C for the films deposited at 580 °C and heat treated at

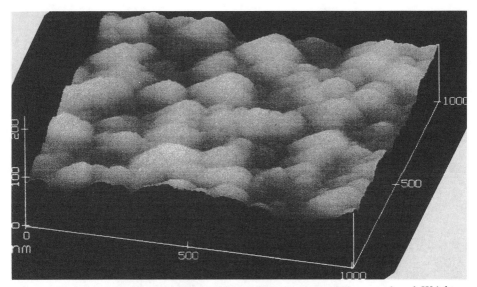

Figure 1: AFM of a heat treated PECVD film after capping oxide removal and Wright-Jenkins etch.

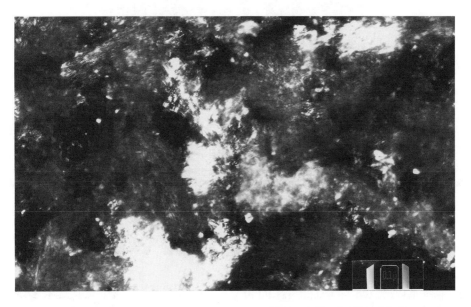

Figure 2: Dark field TEM of heat treated PECVD film (100 nm label).

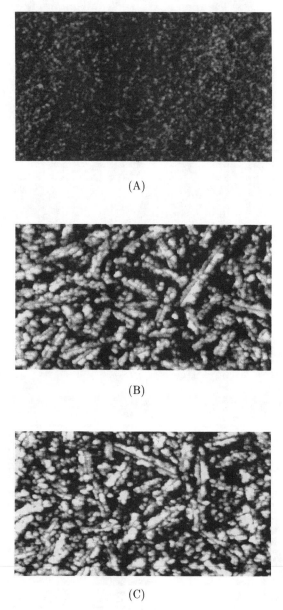

(A)

(B)

(C)

Figure 3: AFM of Wright-Jenkins etched LPCVD films deposited at 580 °C and heat treated at 600 °C for (A) 0 hrs. (B) 8 hrs. and (C) 79 hrs. The area shown is 1.0 μm x 2.0 μm .

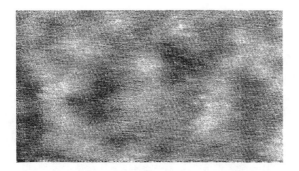

Figure 4: AFM of unetched LPCVD film deposited at 580 °C and heat treated at 600 °C for 8 hrs. The area shown is 1.0 µm x 2.0 µm .

600 °C for 0, 8, and 79 hrs respectively. It is apparent that crystallite growth was almost complete after 8 hrs. Note that the growth habit is dendritic with detailed structure of the crystallites shown.

It was also found using AFM that: (1) the lower deposition temperatures produced a smaller density of nuclei and (2) heat treatment at 580 °C caused less rapid crystal growth in the films. Unetched, heat treated films were atomically smooth as shown in Fig 4.

AFM Fig. 3B can be compared with the SEM micrograph of the same sample as shown in Fig. 5. It is apparent that the same morphology is found.

CONCLUSIONS

The first experiment on PECVD films of silicon shows that both TEM and AFM give the same general size for crystallites in the heat treated film. The original 50 nm hillock size seen on as-deposited films before Wright-Jenkins etch is taken to be an artifact of the deposition process. This did not change during the transition from an amorphous to crystallized film where heat treatment was carried to the point of maximum crystal growth [4]. Use of Wright-Jenkins etch exposed the crystalline structure as shown by the increase in hillock size, giving a crystallite size similar to that found with TEM.

The second experiment shows the value of the AFM in delineating the growth morphology of the crystallites. The dendritic form is revealed and the progression of nucleation and growth defined. The higher deposition temperatures gave rise to a greater density of nuclei. It was also apparent that longer times and higher temperatures during heat treatment resulted in a higher degree of crystallization.

The reason for the difference in crystallite morphology between Experiments 1 and 2 is not apparent. It is probably due to the method of silicon film deposition, but may also relate to the use of diluted Wright-Jenkins etch in the second experiment.

Our overall conclusion is that the AFM offers a simple, low cost method for characterizing polysilicon silicon films. The technique can be extended to other situations where crystals are growing in a noncrystalline matrix. It should also be noted that, while the presentation of AFMs in this text is in black and white, a color presentation greatly enhances discrimination of the crystallites.

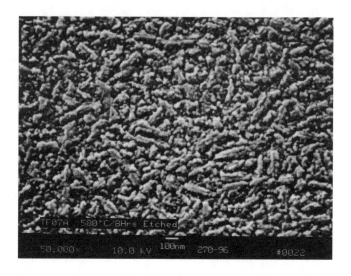

Figure 5: SEM of Wright-Jenkins etched LPCVD film deposited at 580 °C and heat treated at 600 °C for 8 hrs.

ACKNOWLEDGMENTS

The authors would like to thank Paul Sachenik for depositing the film stacks by PECVD, Hans Holland for carrying out the XRD, and Dave Pickles and Bryan Wheaton for performing the AFM measurements. We would also like to thank Bob Schaeffler for Wright-Jenkins etching the samples and Dieter Ast for his encouragement.

REFERENCES

[1] T. Kamins, *Polycrystalline Silicon for Integrated Circuit Applications*, page 239. Kluwer Academic Publishers, 1988.

[2] M. W. Jenkins, "A New Preferential Etch for Defects in Silicon Crystals", *J. Electrochem. Soc.*, **124**:757, 1977.

[3] M. K. Hatalis and D. W. Greve, "Large Grain Polycrystalline Silicon by Low-temperature Annealing of Low-pressure Chemical Vapor Deposited Amorphous Silicon Films", *J. Appl. Phys.*, **63**:2260, 1988.

[4] I. W. Wu, S. Stuber, C. C. Tsai, W. Yao, A. Lewis, R. Fulks, A. Chiang, and M. Thompson, "Processing and Device Performance of Low-Temperature CMOS Poly-TFTs on 18.4" Diagonal Substrates for AMLCD Application", *SID 92 Digest*, :615, 1992.

A NOVEL TECHNIQUE FOR IN-SITU MONITORING OF CRYSTALLINITY AND TEMPERATURE DURING RAPID THERMAL ANNEALING OF THIN SI/SI-GE FILMS ON QUARTZ/GLASS

V. SUBRAMANIAN, F. L. DEGERTEKIN, P. DANKOSKI, B.T. KHURI-YAKUB, K.C. SARASWAT
Electrical Engineering Department, Stanford University, Stanford, CA 94305

ABSTRACT

A novel technique is presented to simultaneously measure temperature and crystallinity in-situ during the rapid thermal annealing of thin Si / SiGe films on transparent substrates for active matrix liquid crystal display applications. The technique uses acoustic waves to monitor temperature, by measuring changes in velocity with temperature. The technique enables accurate tracking of crystalline phase transitions along with temperature, since it is independent of emissivity. This provides a methodology for closed-loop control and end-point detection. The experiments on thin amorphous Si on Quartz demonstrate temperature repeatability of 2%. Also, the technique proved sensitive enough to detect the onset of nucleation, as evidenced by TEM.

INTRODUCTION

Thin film transistors (TFTs) find extensive applications in active matrix liquid crystal displays (AMLCDs). The ability to fabricate low temperature polysilicon TFTs will enable the incorporation of driver circuitry onto glass display substrates[1], improving performance at a reduced cost. To prevent glass warpage, it is necessary to minimize thermal budget and peak process temperatures. Rapid thermal annealing (RTA) to crystallize amorphous films to form the channel is a promising technique for the processing of polycrystalline TFTs. In-situ endpoint detection will allow the minimization of thermal budgets. The ability to measure temperature accurately is critical to ensuring glass compatibility and enabling process control and modeling.

Pyrometers, which track sample emissivity to determine temperature, are commonly used for temperature monitoring during RTA. Emissivity of transparent substrates is highly dependent on surface film optical parameters. This results in inaccurate measurements, particularly in thin film crystallization processes, where surface emissivity changes rapidly during crystalline phase transitions. An *in-situ* temperature sensor[2] using acoustic waves for temperature measurement in rapid thermal processing overcomes this difficulty by sensing the variation of elastic constants with temperature. This technique is independent of emissivity, enabling isolation of emissivity variations from temperature measurement. Hence, phase transitions can be tracked due to changes in light absorption.

In this paper, we present the methodology for incorporating the acoustic temperature sensor in *in-situ* monitoring of crystallinity and temperature during RTA of thin films on transparent substrates. Using this technique, we demonstrate accurate temperature measurement and the ability to track crystalline phase transitions on thin Si/SiGe films on quartz substrates.

MEASUREMENT PRINCIPLES

The acoustic sensor uses the variation in the elastic properties of the sample with temperature as the principle for measurement. This results in temperature measurements which are independent of such factors as doping, thin surface films, and emissivity. The variation in elastic properties is monitored using the phase velocity of the lowest order antisymmetric (A_0) mode Lamb wave applied to the sample. The schematics of the acoustic sensor in the RTA is shown in figure 1. The quartz support pins are modified by bonding PZT acoustic transducers at one end and shaping the tip at the other end for reliable contact with the sample.

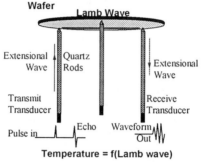

Figure 1: Schematic of acoustic sensor

To make a temperature measurement, a high voltage pulse is applied to the PZT transducer to excite extensional waves in the transmitter quartz pin. At the quartz pin-wafer interface, some of the acoustic energy is coupled to the A_0 mode in the sample. The reflected wave is detected again in the transmitter to start a high resolution time interval counter. The A_0 mode Lamb waves travel in the sample and are detected at the receiver. The time interval value is the time of flight (TOF) of the A_0 mode Lamb wave in the wafer and is inversely proportional to the velocity of the acoustic wave. Since there is a linear relationship between the temperature of the sample and the TOF, it is possible to measure the average temperature of the sample *in-situ* during RTA.

For acoustic temperature measurements, the frequency of operation is chosen around 250KHz. At this frequency, the temperature sensitivity is mainly due to the substrate since the acoustic power flow is uniform along the thickness of the plate, and the substrate is much thicker than the surface film. By using 3 or more quartz pins, the sensor has also been used to map the temperature distribution on silicon wafers during rapid thermal processing.

The isolation of emissivity variations from temperature measurement provides a unique means of tracking crystallinity in-situ. Thin amorphous Si films are significantly more opaque than polycrystalline films of the same thickness. Therefore, the absorption of light from the heating lamp by the film drops substantially during the amorphous-polycrystalline phase transition. Quartz is essentially transparent to the optical frequencies used for RTA. Therefore, heating of the quartz occurs through conduction from the surface film. Thus, the decrease in heat absorption by the surface film results in a decrease in temperature. This can be detected accurately using the acoustic sensors. The rapidly changing surface optical characteristics make the use of emissivity for this simultaneous detection impractical. Therefore, this technique is suitable for monitoring crystallinity and temperature for a large variety of films.

EXPERIMENTAL METHODOLOGY

The sample used for this experiment were prepared on quartz substrates. 100nm LTO SiO_2 was deposited by LPCVD on the substrates. This provided a clean and smooth surface for film deposition, and mimics the conventional process used for both quartz and glass substrate AMLCDs. 150nm amorphous Si / $Si_{0.7}Ge_{0.3}$ was deposited on the films by LPCVD using SiH_4 and GeH_4. The samples were annealed using a 3-zone tungsten halogen lamp system[3]. The lamp powers were sequenced open-loop. After an 8 second preheat (to avoid the cold inrush current problem in tungsten halogen bulbs) and a 30 second ramp up, the steady state powers were held constant in the three zones. The powers were chosen to ensure crystallization within 15 minutes, providing good temporal resolution. The TOF data from the acoustic sensors was monitored in real time at 10Hz.

The elastic coefficients of quartz required to convert the TOF data to temperature accurately were measured in an oven. The TOF coefficient was calculated to be -3.11nsec/°C, with an estimated error of 4%. The error sources are the temperature-dependent behavior of the transducers, and the oven temperature and TOF path measurement uncertainties. The temperature of the quartz wafer was inferred from TOF data using the calculated coefficients. The temperature represents the average over the path length of the wafer. The onset of crystallization was indicated by a drop in the temperature. The end point of the phase change was indicated by a re-stabilization of the temperature. The recipe was initially run until the sample was fully crystallized. For subsequent samples, the recipe was aborted at various points along the temperature-time curve, and the samples were withdrawn for materials analysis, using plan view TEM micrographs to compare grain size, and diffraction patterns to compare relative crystallinity. The correspondence of crystallinity to position along the temperature-time curve was analyzed.

RESULTS AND DISCUSSION

The temperature vs. time plot for the recipe used is shown below, in figure 2. The peak temperature of the quartz wafers is measured to be 588°C with a standard deviation of 12°C, determined over several runs, providing a repeatability of approximately 2%. Clearly, the temperature drops during the amorphous-polycrystalline transition. Films which were initially polycrystalline, on the other hand, do not exhibit this drop in temperature, and also do not heat up as much initially, verifying the higher absorbance of the amorphous films.

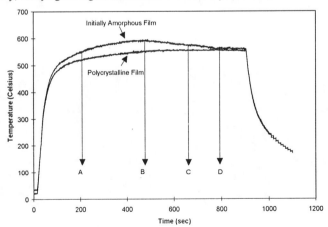

Figure 2: Temperature vs. Time plot during RTA of amorphous and polycrystalline samples.

TEM micrographs and diffraction patterns for samples annealed to the points indicated are shown below. The sample annealed to point 'A' along the temperature-time curve is still completely amorphous. This is in agreement with substantial work done on crystallization which suggests that crystallization begins only after a specific incubation time[4].

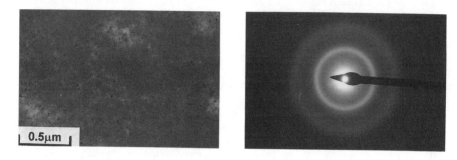

Figure 3: TEM micrograph and diffraction pattern for sample annealed to point 'A'

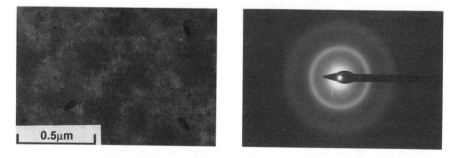

Figure 4: TEM micrograph and diffraction pattern for sample annealed to point 'B'

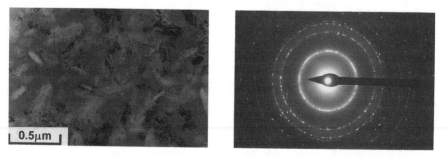

Figure 5: TEM micrograph and diffraction pattern for sample annealed to point 'C'

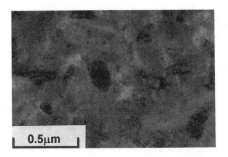
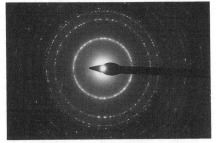

0.5μm

Figure 6: TEM micrograph and diffraction pattern for sample annealed to point 'D'

The sensitivity of the technique is apparent upon examination of the micrograph for samples annealed to point 'B', shown in figure 4. A few small nuclei are present within the sample. The diffraction rings exhibit no significant tightening to the naked eye. However, the temperature curve shows a drop almost immediately afterwards. Further into the anneal, at point 'C', samples have large grains surrounded by patches of amorphous material, indicative of incomplete crystallization, shown in figure 5. Upon re-stabilization of the temperature, at point 'D', the samples are fully crystallized and have no large amorphous regions visible. Additional annealing does not increase grain size any further. The diffraction rings are well defined. This is shown in figure 6.

A study of the intensification of the lines in the diffraction rings using a normalized quantization image enhancement technique illustrates the crystallization progress clearly. This is shown in figure 7. The tightening of lines is clearly visible. The correspondence to temperature data is excellent, suggesting that tracking is reliable. Repeatability of the experiments was also verified using multiple anneals. Deviations were found to be minimal.

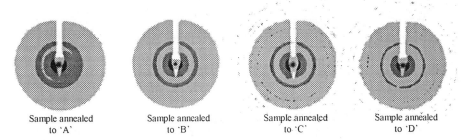

| Sample annealed to 'A' | Sample annealed to 'B' | Sample annealed to 'C' | Sample annealed to 'D' |

Figure 7: Progressive tightening of diffraction rings.

The results obtained from this experiment show that the acoustic sensor is an excellent means of monitoring temperature and crystallinity *in-situ*. Similar results were obtained for SiGe films, which actually exhibit enhanced contrast due to the larger change in absorption. The system has high sensitivity to phase changes, and tracks temperature, independent of surface film conditions. The substrate temperature, as determined using the sensor, is glass compatible. There is substantial evidence in the literature suggesting that crystallization of Si films at such low temperatures takes several hours. Therefore, it is suspected that the film temperature is higher than the average substrate temperature. This is explained by the low thermal conductivity of quartz. The quartz exhibits a strong temperature gradient, hottest near the film, and coolest in the bulk of the quartz. Since substrate temperature is the critical issue for warpage considerations, this is beneficial and reinforces the promise of RTA as a technique for the production of high throughput, high quality polycrystalline Si/SiGe TFTs for AMLCD applications.

The sensitivity of the sensor to crystallinity can be used to provide input for a closed loop control system for multi-step anneals. For example, the technique could be extended to provide an initial high temperature step to reduce the incubation time, followed by a low temperature quench upon nucleation, to suppress further nucleation and potentially result in the growth of extremely large grains. The temperature sensors have been extended to provide full-wafer tomography of temperature and crystallinity by using multiple sensors over different paths across the wafer. This could also be used for closed loop control to reduce non-uniformities across large area display substrates. Since the sensor measures substrate temperature independent of film temperature, it could be used to develop techniques to minimize warpage while maximizing film heating.

CONCLUSIONS

A novel technique for simultaneously monitoring temperature and crystallinity during the rapid thermal annealing of thin Si/SiGe films on quartz/glass substrates has been presented. Acoustics sensors are used to measure the variation of sound velocity with temperature. This technique for measuring temperature is independent of surface emissivity. This enables tracking of film crystallinity based on variations in optical absorbance. The technique has been demonstrated to be highly repeatable and sensitive. Phase changes can be tracked from the onset of nucleation to full crystallization. This should result in optimized techniques for the production of high performance high throughout glass compatible TFTs for AMLCD applications.

ACKNOWLEDGEMENTS

We gratefully acknowledge Mr. L. Booth for his assistance in the operation and use of the Stanford Rapid Thermal Multiprocessor, and Dr. A. F. Marshall for her assistance in TEM usage. Support for Vivek Subramanian and Paul Dankoski is provided by the Advanced Research Projects Agency.

REFERENCES

1. T. Morita, S. Tsuchimoto and N. Hashizume in Flat Panel Display Materials, edited: J. Batey, A. Chiang, and P. H. Holloway (Mat. Res. Symp. Proc. 345, Pittsburgh, PA 1994), p. 71-80.

2. F.L. Degertekin, J. Pei, B.T. Khuri-Yakub, K. Saraswat, Appl. Phys. Lett. **64**, p. 1338-1340 (1994).

3. K. C. Saraswat et al, IEEE Trans. on Semiconductor Manufacturing, 7, p. 159-175 (1994).

4. T. Kretz, D. Pribat, P. Legagneux, F. Plais, O. Huet and M. Magis in Flat Panel Display Materials, edited by J. Batey, A. Chiang, and P. H. Holloway (Mat. Res. Symp. Proc. 345, Pittsburgh, PA 1994), p. 123-128.

ATOMIC-RATIO EFFECT OF TRANSITION LAYER
FOR LINEAR MOBILITY IN POLY-SI TFTs

M.Y. JUNG[*], Y.H. JUNG, S.S. BAE, S.M. SEO, D.G. MOON, G.H. LEE, H.S. SOH
LG Elelectronic Inc, LCD R&D center, Anyang-Shi, Kyungki-Do 430-080, KOREA

ABSTRACT

Poly-Si TFTs with high field effect mobility are fabricated by using PECVD SiO_2 layer deposited with a new method: two-step (graded) oxide deposition. To adjust stoichiometry of the poly-Si/oxide interface and the bulk oxide layer, the double layer oxide films were deposited. The oxide films near the interface were deposited with high N_2O/SiH_4 gas ratio to obtain the stoichiometric layer for good matching between poly-Si and SiO_2. The remaining bulk oxide films were deposited with low N_2O/SiH_4 gas ratio. The composition of the bulk oxide film was measured by using ESCA and the interface layer was analized with ESR. The poly-Si TFT with the double layer gate oxide resulted to the better performance than conventional TFT wth single layer gate oxde.

INTRODUCTION

Hydrogenated a-Si TFTs have been widely used in AMLCD panels such as notebook PC display, personal information display. Poly-Si TFTs provide two or three orders of magnitude higher carrier mobility than that of amorphous silicon. This characteristic of poly-Si TFTs enables the formation of high density pixel array and the construction of peripheral drive circuits on the display panel.

In poly-Si TFTs, the understanding of Si dangling-bond structure at poly-Si/SiO$_2$ interface is an important since the Si dangling-bond plays a role to electrically active interface traps. In thermal oxide, the thickness of the interface transition layer is about 70 Å (1). The atomic density of the interface transition layer is lower than that of the bulk oxide layer. Lower atomic density is due to improper atomic composition which can not be controlled in thermal oxide.

In this paper, to avoid this problem, we used a new method with two-step deposition using PECVD. Since we can get any compositions between $SiO_{x=0}$ and $SiO_{x\approx2}$, we can study the characteristics of dangling-bonds in suboxide states on various compositions. Initial poly-Si surface consists of only ionic Si dangling-bonds. To segregate Si dangling-bond, excess oxgens need to excessively consume at initial plasma condition than bulk deposition. Using this technique we deposited high N_2O/SiH_4 ratio at transition layer to obtain O/Si~2, and low N_2O ratio at bulk oxide to obtain O/Si~2 as shown in ESCA spectroscopic. So graded method allows a freedom to alter film quality that is impossible with thermally-grown oxides. We applied a Electron Spin Resonance(ESR) technique to examine transition layer before fabrication of TFT devices, and then we found that Si dangling-bond energy is correlated closely with electronegativity.

EXPERIMENTAL

PECVD a-Si films were deposited   on the glass substrate coated with buffer SiO₂ film. After the evolution of hydrogen in the a-Si film by furnace, the films were crystallized by XeCl excimer laser. 700 Å thick gate dielectric SiO₂ films were deposited by PECVD with using He, SiH₄, N₂O gases. The flow ratio of N₂O to SiH₄ was varied from 270 to 33 to alter the atomic composition of the oxde layer. The flow rate of He gas was fixed. After depositing and patterning of the gate metal, the phosphorous ions were doped for source and drain ohmic layer. The interlayer oxide depostion, contact hole opening, Al sputtering and sintering were performed.

The O/Si atomic ratio was analized by ESCA. ESR was used to measure Si dangling-bond. MOS structure was prepared to measure leakage of oxide film.

RESULTS AND DISCUSSION

Fig.1 shows the O/Si atomic ratio as a function of the N₂O/SiH₄ gas flow ratio. The O/Si ratio was measured by ESCA.   O/Si ratio decreases with increasing SiH₄ flow.   Above 0.9 of gas flow ratio, the O/Si ratio is about 2, which indicates the stochiometric oxide film.

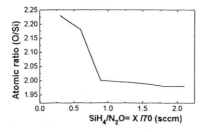

Fig.1.Atomic ratio and gas ratio charateristics for a-SiO₂ films at fixed He flow. For values of SiH₄>0.9, atomic ratios of PECVD oxide were consisted of stoichiometric or quasi-stoichoimetric compositions.

Fig.2 shows the   current density (J) - electric field (E) relations for gas ratios 270 (atomic ratio 2.25), 270+33 (2.25+1.97) and 33 (1.97) conditions (t = 700 Å).   In Fig.2 (A), 2.25+1.97 represents that the interface layer and bulk oxide were deposited with gas ratio of N₂O/SiH₄ = 270 and 33, respectively.   According to ESCA, (a) O/Si (1.97) film has stoichiometry only in bulk. And then, from ESR data, excess Si layer forms on transition layer since initial gas contains lower N₂O.   (b) O/Si (2.25) has stoichiometry in only transition layer since   initial gas contains high N₂O, then really execess O for bulk. (c) O/Si (2.25+1.97) has stoichiometric layers on both interface and bulk since initial high N₂O, later low N₂O flow condition.   Fig.2 (A) indicates that leakage property is no   strong dependence on film compositions.   In this case of thick film (700 Å),   chemical structure like atomic ratio can not be standard evaluation about interface trapping (2).   However, as shown in Fig.2(B), in a thinner a-SiO₂ film (200 Å), leakage current is a strong dependence on oxide stoichiometry near poly-Si.   In case of a thinner films, leakage current is   a strong dependence on oxide stoichiometry near poly-Si, and it is known as carrier conduction is contact limited current (2). Dielectric   breakdown endurance becomes   higher at the graded-gas ratio 270+30.   For   these samples, Fig.3 (A), (B), (C) show the dependence   on gate bias polarity. (A) O/Si (1.97) film, in case of substrate injection(+Vg), shows a lower

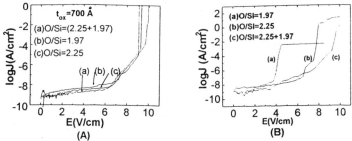

Fig.2. (A) The leakage current charateristics of atomic ratios on a-SiO₂ films, t_{ox}= 700Å. Bulk limited leakage current (Poole-Frankel current)does not depend on atomic ratios. (B) Leakage current dependence for oxide films(200Å) on atomic ratios. Contact limited current depend on interface stoichiometry.

breakdown strengh because of barrier lowering at non-stoichiometric initial layer. This polarity asymmetry is opposite to that of the (B) O/Si (2.25) with stiochiomeric initial layer. (C) O/Si (2.25+1.97) shows high breakdown quality at both substrate injection(+V_g) and gate injection(-V_g) since bulk and initial layer togather gained stichiometric composition, and then, in case of substrate injection film strength endures a stronger electricfield, 8MV/cm. It shows that stoichiometric initial layer has stronger interface barrier-height than Al/SiO₂ interface. In conclusion, these facts indicate that stoichiometry at transition layer has a very importance about device characteristics than composistion of bulk oxide. Elelectron spin resonnance for the samples at 300 K is shown in Fig.4 (A), (B). Approximately 50 Å thick films were deposited on poly-Si for surfacial dangling-bond segregation by Si-O bonds or grain boundary segregation by atom O. The spectra (A) observed for bare poly-Si and poly-Si coated thin oxide is significant difference in the spectra. The differency is due to passivation of grain boundary or grain surface by oxygens. The spectrum observed for bare poly-Si shows a broad line at g=2.0057 known as Si dangling-bond, .Si≡Si₃. On the other hand , poly-Si coated thin oxide shows the smaller broadline than bare poly-Si. It indicates the reduction of random orientation and site -to-site variation in structure (3). At gas ratio 270, .Si≡Si₃ centers contains similarly lower intensity than 33.

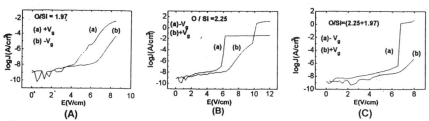

Fig.3. Contact limited Leakage current dependence of bias polarity on atomic ratios O/Si=1.97(A), 2.25(B), 2.25+1.97(C), t_{ox}=200 Å a-SiO₂.

This is assumed that high N_2O plasma contains excess oxygens, segregating more $.S \equiv Si_3$ comparison to low N_2O. Here, ESR samples have smaller grain size than real devices from SEM data. Because grain size was smaller on a-Si/Si-wafer than on a-Si/glass and ESR intensity was sum of 3 times scanning , quantitative intensity on samples is not coincidence with real device. In addition, these samples were not sintered. Fig.4 (B) shows difference of intensity between graded-deposition oxide and conventional. Since samples were not sintered and intensity was adding value with 9 times , spin intensities have very higher values. Therefore, spin values of ESR samples is not quantitatively equal to it of real TFT devices. The spectra greatly changes in multiple peak line, in case of composition 1.97. ESR centers, $.Si \equiv Si_3$ (g=2.0057) and $.Si \equiv SiO_2$(or $.Si \equiv Si_2O$) (g=2.0027) are shown in curve (b). Graded composition oxide (curve (a)) has only $.Si \equiv SiO_2$(or $.Si \equiv Si_2O$) (g=2.0029) dangling-bond center. g-values of $.Si \equiv SiO_2$ or $.Si \equiv Si_2O$ is not clear between 2.007 and 2.009. Conventional oxide (b) contain not only $.S \equiv Si_3$ center at interface, but also $.Si \equiv SiO_2$(or $.Si \equiv Si_2O$) center at bulk oxide. Low N_2O condition induces $.Si=Si_3$ dangling-bond centers at transition layer since oxygen deficiency occur initial layer. Table 1 shows the correlation between dangling-bond clusters, dangling-bond energy relative to vacuum level (E_{db}), percentage local probablity of unpaired electron (ρ) at the Si dangling-bonded atom ($\equiv Si.$) and additional partial charge (q) due to electronegativitic difference (4). In $.Si \equiv SiO_2$(or $.Si \equiv Si_2O$) cluster, because oxygen has higher electronegativity than Si, electron crowding-probability (ρ (%)) at unpaired Si atom seems to decreased.

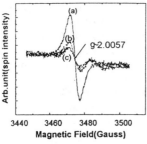

Fig.4(A). First derivative ESR signals of (a)bare poly-Si and (b)poly-Si coated thin oxide(50 Å), N_2O/SiH_4=33, (c) poly-Si coated oxide(50Å), N_2O/SiH_4=270.

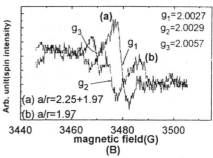

Fig.4(B). First derivative ESR spectra dependence on atomic ratios(a/r) of oxides, (a) a/r= 2.27+1.97, (b)1.97. (a) g- value 2.0027 correlates with $.Si=Si_2O$ (or $.Si=SiO_2$) ,and (b) g-values 2.0057, 2.0029 correlate with $.Si=Si_3$, $.Si=SiO_2$ (or $.Si=Si_2O$), respectively.

Additional partial charge of $.Si \equiv Si_3$ cluster means ratio x initial charge(0.04 x initial charge e)that transfered into unpaired Si atom from paired Si atom. In high N_2O condition, $.Si \equiv Si_2O$(or $.Si \equiv SiO_2$) with g=2.0027~2.0029, reduce partial charge (+0.31e, +0.57e) because electron transforts into oxyen from unpaired Si atom.

Table 1. various electric characteristics related silicon cluster, silicon dangling-bond energies(E_{db}), percentage local probability(ρ) at ionic Si dangling-bond($.Si\equiv$) and additional partial charge(p) at ionic Si dangling-bond due to $\equiv Si_2O$ or $\equiv SiO_2$.

cluster	E_{db}(eV)	ρ (%)	q (e)
$.Si\equiv Si_3$	-7.9 eV	78	-0.04 e
$.Si\equiv Si_2O$	-7.3 eV	69	+0.31 e
$.Si\equiv SiO_2$	-6.7 eV	61	+0.57 e

Therefore, the path of electron running for trap-site ($.Si\equiv$) is modified by electronegativitic attraction due to $\equiv Si_2O$(or $\equiv SiO_2$). Oxygen atom near $.Si\equiv$ perturbes free electron trap-path, weaking potential near trap-point($.Si\equiv$). On the other hand, in case of trapped electron, $\equiv Si_2O$ (or $\equiv SiO_2$) with higher electronegativity decreases trap potential of $.Si\equiv$. And then, reduction of Edb increases the rate of emision state from trap state by lateral drain voltage. In low N_2O conditions, $.Si\equiv Si_3$(g=2.0057) existing at interface and $.Si\equiv Si_2O$ (or $.Si\equiv SiO_2$) (g=2.0027~2.0029) existing bulk were shown in ESR curves(B). $.Si\equiv Si_3$ get 0.04 times more partial charge (-0.04e) than unpaired unit atom state (.Si, charging e). In $.Si\equiv Si_3$ cluster, paired silicon atoms wrriten as $\equiv Si_3$ atom have neutral states, and unpaired silicon ion wrriten as $.Si\equiv$ is charged more negative than paired Si atom (Electrionegativity of unpaired silicon ion wrriten $.Si\equiv$ is increased relatively higher than paired Si atoms). In case of low N_2O, electron injected from source may be occured larger trapping than in high N_2O. Fig.5 (A), (B). shows Id-V_g(A), and gm-V_g (B) characteristics in the linear region (V_d=0.1V), respectively. The increase of gm (gas ratio 230+33) is more remarkable than that of gas ratio 230 (atomic ratio 2.25) or 33 (1.97) at same Vg. At high drain voltage, carrier conduction is influenced not only by the grain boundary but also by grain resistance.

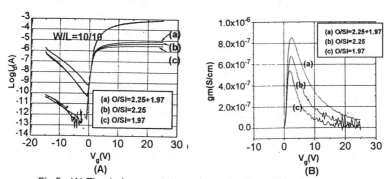

Fig.5. (A) The drain current dependence for Poly -Si / a-SiO$_2$ TFFs on atomic ratios. (B)Transconductance dependence for on atomic ratios. W/L=10/10.

However, at low voltage, drain electric field concentrate grain boundaries with high resistivity (5). In this case, drain current depend on grain boundary dangling-bonds (N_t) and the density n_o of detraped free carrier as shown in fomular (1).

$$I_d \propto n \exp(-qN_t{}^2/8\,\varepsilon_{\,s}n_0kt) \tag{1}$$

Therefore low voltage advantages inspection of grain boundary and interface properties (5). Number of detraped free carrier have closely relation to the trap-depth of dangling-bonds as shown in ESR spectra. At transtion layer, dangling-bond energy decreases in high N_2O condition. This is consistent with experimental results from Fig.5. Difference of drain current at linear region is remarkable. Fig.6 (A) shows drain current dependence on different oxide depositions at Vg=12V, indicating significant modulation at 270+33 condition. Fig.6 (B) shows drain current dependence on various oxide depositions, incrementing interval 0.001 V from Vd=0.001V at Vg=12V. Even if lateral elelctric field is very low, the modulation shows similaly difference. According to these results, trap depths of dangling-bond sites in Fig.4 (B) and table.1 have good relation with linear drain currents, using electronegativity concept. Also, ESR intensities of Fig.4(a) relate the result to (N_t)at fomular(1). (n_o) in fomular(1) correlates to dangling-bond energy in table 1.

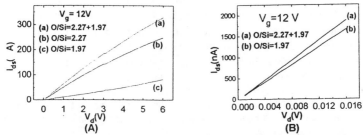

Fig.6. (A)Drain current characteristics at fixed Vg=12V, W/L=10/10 for n-channel TFTs. A graded oxide(a) device is compared to devices (b),(c) with conventional oxides. (B) Drain current chereceteristics vs very low drain voltage. A grade oxide(a) device is shown higher modulation.

Grain boundary carrier heating of poly-Si films is shown in Fig.7. At large voltage (space charge limited current region), oscillating current is obserbed. This osillation effect may be caused by local heating of grain boundary microstructures defects (6). As shown in this figure, carrier heating effect is not shown at graded oxide device that grain boundary passivated by excess oxgion at high N_2O plasma. The fact that slope is <1 shows that the current is not a simple channel or bulk resistance limited current(7). It was because this measurement was characterized after bias stress experiments arising defect state at interface.

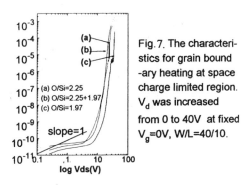

Fig. 7. The characteri-
stics for grain bound
-ary heating at space
charge limited region.
V_d was increased
from 0 to 40V at fixed
V_g=0V, W/L=40/10.

CONCLUSIONS

Our idea about transistion layer is support since these show;
1) a correlation between bias polarity and leakge current
2) a good dependence on ESR signal; electronegativity on $.S \equiv Si_3$, $.Si \equiv Si_2O$(or $.Si \equiv SiO_2$)
3) a correlations at grain boundary heating and drain current modulation at very low drain voltage
4) a linear mobility, transconductance dependences on N_2O flow
We demonstrated the superity of atomic ratio(2.25+1.97) PECVD SiO_2/poly-Si TFTs having both high linear mobility, Graded deposited SiO_2 film products a promising approach to poly-Si TFTs devices.

ACKNOWLEDGEMENTS

We thank Dc. Il-Woo Park, Korea Basic Science Institute for help in researching the samples for electron spin resonnance. And table 1. part reported in this paper is quoted at reference(4).

REFERENCES

(1). Semiconductor international, pp34, March 1995.
(2). E. Suzuki, H. Hiraishi, Appl. Phys. Lett., 40, 681, 1982.
(3). K. Tanaka, Phil. Mag. B, Vol.69, No. 2, 263-275, 1994.
(4). T. Inokuma, L. He, Y. Kurata and S. Hasegawa, J. Electrochem. Soc., Vol.142, No. 7, July 1995.
(5). So Yamada, Shin Yokotana, S.S.D.M. , 1003, 1990.
(6). R. L. Weisfield, J. Appl. Phys. 54(11). November 1983.
(7). J. R. Ayres and N. D. Young, IEE. Proc.-Circuits Devices Syst., Vol.141. No.1, February 1994.

TMCTS FOR GATE DIELECTRIC IN THIN FILM TRANSISTORS

ALBERT W. WANG, NAVAKANTA BHAT, KRISHNA C. SARASWAT
Department of Electrical Engineering, Stanford University, Stanford, CA 94305

ABSTRACT

The use of the liquid source tetramethylcyclotetrasiloxane (TMCTS) for gate dielectric deposition in low-temperature polysilicon thin film transistor (TFT) processes is investigated. TMCTS was reacted with O_2 in an LPCVD furnace at 580°C to form a gate dielectric. For comparison, a low temperature oxide (LTO) was deposited as a gate dielectric using SiH_4-O_2 LPCVD at 450°C. Capacitance and charge pumping measurements indicate fewer interface states for TMCTS gate dielectric. Both NMOS and PMOS TFTs show comparable or superior performance with TMCTS oxide. Post-deposition annealing has less effect on TMCTS gate oxides. Although TMCTS gate dielectrics appear slightly more susceptible to damage in bias-temperature stress tests, TFTs with TMCTS gate oxides still retain better performance after stressing.

INTRODUCTION

Polysilicon thin film transistor (poly-Si TFT) technology is the subject of extensive research because of its application to active-matrix liquid crystal displays (AMLCDs) and static memory elements. However, most poly-Si TFT processes are restricted to low thermal budgets and thus the gate dielectrics are usually deposited. These gate dielectrics need to be of the best possible quality. For silicon dioxide gate dielectrics, chemical vapor deposition (CVD) using SiH_4 and O_2 is most often used.

Organometallic liquid sources for oxide deposition such as tetraethylorthosilicate (TEOS), diethylsilane (DES), and tetramethylcyclotetrasiloxane (TMCTS) are potential alternatives to traditional SiH_4-O_2 deposition technology. TEOS has already been used in AMLCD processes [1] but its low-pressure CVD (LPCVD) temperature (about 700°C) restricts its use to plasma-enhanced CVD for low thermal budget processes. TMCTS and DES offer attractive qualities when compared with traditional SiH_4-O_2 deposition, including reasonable deposition rate, low deposition temperature (< 600°C), and excellent conformality. In addition, silane is stored in gas phase and is both toxic and pyrophoric, requiring more safety precautions for its use [2]. The liquid sources, while still flammable and toxic, are considerably easier to handle.

A previous paper [3] reported the feasibility of gate dielectrics formed by TMCTS, but the study was limited in detail and objectives, and apparently was never followed up. In this paper, the use of TMCTS as a source for gate dielectric deposition in a low-temperature TFT process is investigated in much more detail.

EXPERIMENT

Two experiments were conducted to evaluate the effectiveness of using TMCTS to deposit gate dielectrics in TFT processes. Both experiments used very similar processing.

In both experiments, TMCTS was reacted with O_2 in an LPCVD furnace at 580°C at a deposition pressure of 500 mtorr and an O_2:TMCTS flow ratio of 10:1 to form a gate dielectric. This temperature is typical of TMCTS deposition for interlayer dielectrics, but TMCTS can deposit oxides down to 525°C [4]. The deposition rate was about 35 Å/min. For comparison, a low temperature oxide (LTO) was deposited using SiH_4-O_2 LPCVD at 450°C. The gate material was n-type *in situ* doped polysilicon deposited at 580°C. As the last step, forming gas anneal was done at 400°C.

MOS capacitors on single crystal substrates and polysilicon TFTs were fabricated in both experiments. The MOS capacitors were formed on <100> p-type substrates using LOCOS isolation. The self-aligned NMOS and PMOS TFTs were fabricated on thermally oxidized silicon wafers, with an amorphous Si channel layer film deposited by LPCVD and crystallized at 600°C. The maximum process temperature used to fabricate the TFTs was 600°C.

The first experiment used a gate oxide thickness targeted to a value of 200 Å. The TMCTS gate oxide was given no further oxide anneal while the LTO gate oxide was annealed at 600°C for 3 hr in O_2 followed by 3 hr in Ar. For the TFTs, a 1400 Å-thick channel layer was used. The TFTs were measured and then hydrogenated by ion implantation [5] using a hydrogen dose of 5×10^{15} cm^{-2}. They were then remeasured.

In the second experiment, the gate oxide thickness was targeted to a value of 1000 Å, which is more typical of an AMLCD process. For each type of dielectric, there were three annealing splits: no further oxide anneal; 600°C anneal for 3 hr in O_2 followed by 3 hr in Ar; and 600°C anneal for 6 hr in Ar. The TFTs used a 1000 Å-thick poly-Si channel layer. The TFTs were measured, subjected to plasma hydrogenation for 8 hr at 350°C, and then remeasured.

RESULTS

Table I shows the results of high-frequency and quasi-static CV measurements on both oxides for $t_{ox} \approx 200$ Å. The measurements show that TMCTS oxide has slightly lower absolute flatband and threshold voltage than LTO, and its D_{it} is about half that of LTO.

Figure 1 show the I_D-V_G characteristics of the n- and p-TFTs for $t_{ox} \approx 200$ Å, while Table II summarizes the key TFT parameters. In both cases, TFTs with TMCTS oxide have slightly higher leakage current and comparable on-off ratio. However, the TMCTS TFTs have superior field-effect mobility, subthreshold slope, and NMOS threshold voltage. This is despite the lack of a special oxide anneal for the TMCTS oxide.

Charge pumping measurements [6] were also carried out on n-TFTs using the substrate wafer as the p-type contact using a 100 kHz square wave with an amplitude of 4 V. The results are shown in Figure 2. The TMCTS TFTs have lower charge-pumping current than the LTO TFTs, and since the channel materials are identical, the difference is most likely due to the quality of the oxide and the oxide-channel interface. After bias-temperature stressing (BTS) of ± 8 V at 300°C for 10 min, TMCTS oxide shows considerably more increase in charge pumping current. It is possible that this difference between TMCTS oxide and LTO is due to increased incorporation of hydrogen [7] in the TMCTS oxide. However, it is important to note that the charge pumping current, and therefore the trap densities, of the TMCTS oxide is still lower than that of LTO even after stressing.

Table I. Results of CV measurements of single-crystal MOS capacitors ($t_{ox} \approx 200$ Å).

Split	t_{ox} (Å)	V_{FB} (V)	D_{it} (10^{10}cm^{-2})
LTO/O_2+Ar	205	-1.24	16
TMCTS/no an.	200	-1.17	8.8

Table II. Performance of TFTs ($t_{ox} \approx 200$ Å, W/L = 20 μm/20 μm).
Parameters defined as ($|V_D| = +5$ V unless noted): V_T =V_G at $|I_D| = +100$ nA; μ_{FE} at max. g_m and $|V_D| = +0.1$ V; S = steepest subthreshold slope; I_{off} = min. $|I_D|$; $I_{on} = |I_D|$ at $|V_G| = +10$ V.

Split	V_T (V)	μ_{FE} (cm^2/Vs)	S (V/dec)	I_{off} (pA)	I_{on}/I_{off} (10^6)	V_T (V)	μ_{FE} (cm^2/Vs)	S (V/dec)	I_{off} (pA)	I_{on}/I_{off} (10^6)
NMOS	Before hydrogenation					After hydrogenation				
LTO/O_2+Ar	2.8	22	0.53	83	0.81	2.8	24	0.53	88	0.79
TMCTS/no an.	2.1	32	0.45	137	0.90	1.4	40	0.36	171	1.0
PMOS	Before hydrogenation					After hydrogenation				
LTO/O_2+Ar	-5.1	21	-0.50	-58	0.42	-5.1	24	-0.55	-63	0.39
TMCTS/no an.	-5.1	26	-0.41	-62	0.58	-4.7	32	-0.39	-75	0.79

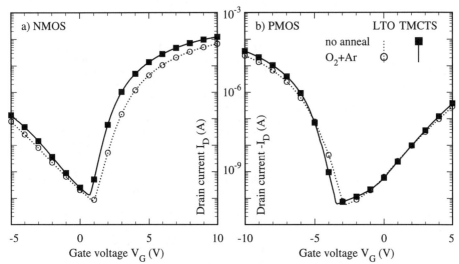

Figure 1. I_D-V_G characteristics of unhydrogenated TFTs ($t_{ox} \approx 200$ Å, W/L = 20 µm/20 µm). $|V_D| = +5$ V, $V_S = 0$ V. Key to symbols and lines for both graphs are shown in figure 1b.

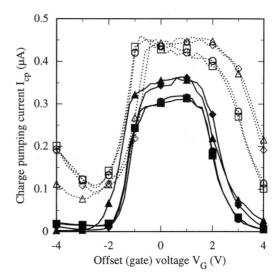

Figure 2. Charge pumping currents of hydrogenated TFTs ($t_{ox} \approx 200$ Å, W/L = 50 µm/20 µm). Pulses were 100 kHz square waves with 4 V amplitude. Charge pumping currents before and after bias-temperature stress (300°C, 10 min, ± 8 V) are shown.

Table III shows the results of high-frequency C-V measurements on the various oxide splits for $t_{ox} \approx 1000$ Å. The measurements show that TMCTS oxide has lower absolute flatband and fixed charge than LTO.

Figure 3 shows the I_D-V_G characteristics of the *n*- and *p*-TFTs after hydrogenation for $t_{ox} \approx 1000$ Å, and Table IV summarizes the key TFTs parameters. For both NMOS and PMOS, TFTs with TMCTS oxide have better mobility and on-off ratio (about 33% improvement for both parameters in NMOS). The TMCTS *n*-TFTs also have superior subthreshold slope and comparable leakage. The TMCTS *p*-TFTs have worse subthreshold slope than LTO *p*-TFTs; however, measurements of annular (radial conduction) *p*-TFTs show the opposite result. This suggests the presence of parasitic sidewall transistor conduction in the LTO *p*-TFTs.

The performance of the TMCTS *n*-TFTs were not apparently affected by the different annealing conditions, as can be seen in Figure 3. In contrast, the LTO *n*-TFTs improved after either type of annealing, as expected. The O_2-Ar annealed *p*-TFTs with both LTO and TMCTS dielectrics show improvement in threshold before hydrogenation. After hydrogenation, however, the difference between O_2-Ar annealed *p*-TFTs and *p*-TFTs with other annealing conditions is markedly reduced. The overall results suggest that post-deposition annealing could be omitted in TFTs using TMCTS gate oxides, reducing the processing thermal budget.

Table III. Results of CV measurements of single-crystal MOS capacitors ($t_{ox} \approx 1000$ Å).

Split	t_{ox} (Å)	V_{FB} (V)	Q_f ($10^{11} cm^{-2}$)
LTO/no anneal	945	-5.3	9.9
LTO/O_2+Ar	950	-6.0	11
LTO/Ar	911	-5.4	10
TMCTS/no an.	948	-4.2	7.4
TMCTS/O_2+Ar	894	-2.9	4.7
TMCTS/Ar	867	-3.6	6.5

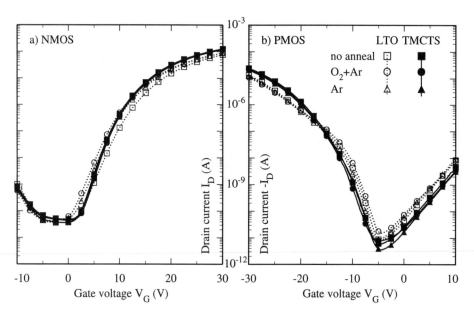

Figure 3. I_D-V_G characteristics of hydrogenated TFTs ($t_{ox} \approx 1000$ Å, W/L = 20 µm/20 µm). $|V_D| = +10$ V, $V_S = 0$ V. Key to symbols and lines for both graphs are shown in figure 3b.

Table IV. Performance of TFTs ($t_{ox} \approx 1000$ Å, W/L = 20 μm/20 μm).
Parameters defined as ($|V_D| = +10$ V unless noted): $V_T = V_G$ at $|I_D| = +100$ nA; μ_{FE} at max. g_m and $|V_D| = +0.1$ V; S = steepest subthreshold slope; I_{off} = min. $|I_D|$; $I_{on} = |I_D|$ at $|V_G| = +30$ V.

Split	V_T (V)	μ_{FE} (cm²/Vs)	S (V/dec)	I_{off} (pA)	I_{on}/I_{off} (10^6)	V_T (V)	μ_{FE} (cm²/Vs)	S (V/dec)	I_{off} (pA)	I_{on}/I_{off} (10^6)
NMOS	Before hydrogenation					After hydrogenation				
LTO/no anneal	11.8	19	2.4	43	1.0	9.7	30	2.2	36	2.1
LTO/O$_2$+Ar	10.2	19	2.4	66	0.8	7.9	28	2.1	43	2.1
LTO/Ar	10.6	21	2.3	55	0.9	8.4	28	2.0	49	1.8
TMCTS/no an.	11.0	27	2.1	56	1.2	8.3	41	1.6	48	2.6
TMCTS/O$_2$+Ar	11.1	26	2.1	56	1.1	8.5	39	1.7	45	2.6
TMCTS/Ar	11.2	28	2.2	48	1.3	8.3	38	1.7	37	3.3
PMOS	Before hydrogenation					After hydrogenation				
LTO/no anneal	-18.8	11	-1.9	-36	0.2	-15.1	14	-1.3	-8.3	1.6
LTO/O$_2$+Ar	-17.5	8	-1.9	-71	0.08	-14.6	12	-1.5	-16	0.7
LTO/Ar	-18.9	11	-2.0	-49	0.1	-15.6	13	-1.3	-9.3	1.3
TMCTS/no an.	-20.0	11	-2.5	-41	0.2	-15.3	16	-1.3	-5.9	4.0
TMCTS/O$_2$+Ar	-17.8	10	-2.0	-40	0.3	-15.6	14	-1.4	-7.9	2.8
TMCTS/Ar	-20.0	9	-2.5	-36	0.2	-15.6	15	-1.2	-3.6	5.8

Both negative BTS [8] and positive BTS [7] measurements of ±30 V at 200°C for 10 min were also carried out on both NMOS and PMOS TFTs. The results are shown in Table V. In all cases, the subthreshold slope degraded and the threshold voltage became more negative. The shift for negative BTS was greater than for positive BTS, as expected. Overall, the TMCTS TFTs retained superior performance after BTS. They were more prone to subthreshold slope degradation, as might be expected from the charge pumping measurements, but they also tended to exhibit less threshold degradation than their LTO counterparts. Annealing improved the BTS degradation of LTO, but only O$_2$-Ar annealing retarded threshold degradation in TMCTS TFTs.

Oxide breakdown measurements on TFTs shown in Figure 4 indicate that TMCTS oxide is much more resistant to breakdown than LTO. This is likely due to both better step coverage and the inherent properties of each oxide.

Table V. Performance change after bias-temperature stress of hydrogenated TFTs ($t_{ox} \approx 1000$ Å, W/L = 20 μm/20 μm). Stress was 200°C for 10 min with V_G as shown below. Parameters defined as ($|V_D| = +10$ V): $V_T = V_G$ at $|I_D| = +100$ nA; S = steepest subthreshold slope.

Split	ΔV_T (V)	ΔS (V/dec)	ΔV_T (V)	ΔS (V/dec)
NMOS	$V_G = -30$ V		$V_G = +30$ V	
LTO/no anneal	-3.9	0.2	-1.3	0.1
LTO/O$_2$+Ar	-3.3	0.2	-0.9	0.1
LTO/Ar	-2.2	0.2	-1.2	0.1
TMCTS/no an.	-2.4	0.3	-0.9	0.1
TMCTS/O$_2$+Ar	-1.3	0.3	-1.3	0.2
TMCTS/Ar	-2.2	0.3	-1.1	0.1
PMOS	$V_G = -30$ V		$V_G = +30$ V	
LTO/no anneal	-5.6	-0.6	-2.8	-0.6
LTO/O$_2$+Ar	-4.5	-0.7	-3.1	0.0
LTO/Ar	-5.4	-0.6	-2.2	-0.3
TMCTS/no an.	-6.0	-0.3	-1.3	-0.3
TMCTS/O$_2$+Ar	-3.1	-0.5	-1.3	-0.3
TMCTS/Ar	-5.4	-0.7	-1.7	-0.4

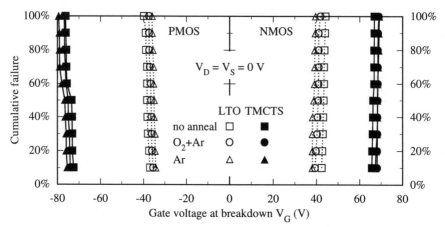

Figure 4. Gate oxide breakdown of hydrogenated TFTs ($t_{ox} \approx 1000$ Å, W/L = 20 µm/10 µm).

CONCLUSIONS

TMCTS has been used as a deposition source for gate dielectric in a low-temperature polysilicon TFT process. The experimental results show TMCTS gate oxide has lower interface trap density and produces TFTs with comparable or better characteristics than similar devices fabricated with conventional LTO dielectric. Furthermore, TMCTS oxides do not need post-deposition annealing, in contrast to LTO. Therefore, oxide anneals could be omitted in processes using TMCTS, thus reducing the thermal budget. Some of the TMCTS TFTs were more susceptible to damage in bias-temperature stress tests but still retained better performance than LTO TFTs. This suggests that TMCTS gate oxide could be used with reasonable reliability in TFT processes.

ACKNOWLEDGMENTS

This research has been funded by Advanced Research Projects Agency grant number MDA972-95-1-0004. The authors would also like to thank Frank Yaghmaie and Schumacher for supplying the TMCTS for this experiment, Dr. Tsu-Jae King of the Xerox Palo Alto Research Center for the plasma hydrogenation, and the technical staff of the Center for Integrated Systems.

REFERENCES

1. T. Morita, Y. Yamamoto, M. Itoh, H. Yoneda, Y. Yamane, S. Tsuchimoto, F. Funada, and K. Awane, *International Electron Devices Meeting 1995 Technical Digest*, pp. 841-3.
2. L.G. Britton and P. Taylor, Semiconductor International, **14** (5), 88-92 (1991).
3. H. Pattyn, K. Baert, P. Debenest, M. Heyns, M. Schaekers, J. Nijs, and R. Mertens, *Extended Abstracts of the 1990 International Conference on Solid State Devices and Materials*, **22**, pp. 959-62.
4. B. Gelernt, Semiconductor International, **13** (3), 82-85 (1990).
5. M. Cao, T. Zhao, K.C. Saraswat, and J.D. Plummer, IEEE Trans. Electron Devices **42** (6), 1134-40 (1995).
6. M. Koyanagi, Y. Baba, K. Hata, I.-W. Wu, A.G. Lewis, M. Fuse, and R. Bruce, IEEE Electron Device Lett. **13** (3), 152-4. (1992).
7. N. Bhat, M. Cao, and K.C. Saraswat, *Society for Information Display International Symposium Digest of Technical Papers*, **26**, pp. 393-5 (1995).
8. N.D. Young and J.R. Ayres, IEEE Trans. Electron Devices **42** (9), 1623-7 (1995).

EFFECT OF ANNEALING AMBIENT ON PERFORMANCE AND RELIABILITY OF LOW PRESSURE CHEMICAL VAPOR DEPOSITED OXIDES FOR THIN FILM TRANSISTORS

N. BHAT, A. WANG, and K. C. SARASWAT
Electrical Engineering Department, CIS, Stanford University, Stanford, CA-94305.

ABSTRACT

The performance and reliability of low pressure chemical vapor deposited (LPCVD) oxides subjected to oxidizing, inert and nitriding annealing ambients is characterized both at low temperature (600°C) and high temperature (950°C). The oxidizing ambient results in worse initial interface state density and charge to break down. We attribute this to the interfacial stress developed during the oxidation, due to the volume mismatch between Si and SiO_2. The C-V measurements on poly-Si substrate capacitors and the charge pumping measurements on poly-Si thin film transistors (TFTs) indicate lower trap density for inert and nitriding ambients. The TFTs with inert anneal exhibit lower bias temperature instability compared to oxidizing ambient.

INTRODUCTION

Poly-Si TFTs are used in active matrix liquid crystal displays (AMLCDs) and high density SRAMs. The LPCVD oxide gate dielectric for poly-Si TFTs is typically annealed in oxidizing ambient to densify and enhance the quality of the oxide. However interfacial oxidation takes place during this step which creates stress at the Si-SiO$_2$ interface due to the volume mismatch of Si and SiO$_2$. In this paper we have studied the effect of annealing in oxidizing, inert and nitriding ambient. We demonstrate that oxidizing ambient is worse both with respect to initial characteristic and reliability. We attribute this to the increased interfacial stress. We have also used the rapid thermal annealing (RTA) of LPCVD oxides to improve the quality of the oxide, which is important due to the thermal budget limit imposed by glass substrate in AMLCDs and source/drain junction depth of bulk transistors in SRAM applications.

EXPERIMENT

We have used both MOS capacitors made on single crystal silicon and poly-Si thin film transistors in this study. MOS capacitors were formed on $<100>$ p-type substrates using standard LOCOS isolation. The LPCVD gate oxide, about 180 $\overset{\circ}{A}$ thick, was deposited using SiH$_4$ and O$_2$ at 450°C. The oxides were annealed either at 600°C in a furnace or at 950°C using rapid thermal processing as described in Table 1. The gate electrode was in-situ doped n^+ poly-Si deposited at 580°C. For the TFTs, Si wafers with thermally grown oxide were used to fabricate self aligned n^+ poly-Si gate n- and p-channel MOSFETs. A 1400 $\overset{\circ}{A}$ thick Si channel layer was deposited in amorphous form by LPCVD and crystallized at 600°C. The gate oxide deposition and annealing was identical to that of the MOS capacitors. The TFTs were hydrogenated by H$^+$ ion implantation at 50KeV with a dose of 5×10^{15} cm^{-2}, followed by forming gas anneal at 375°C for 45 minutes.

The interface-state density in MOS capacitors was measured using high frequency and quasistatic C-V measurement, while the charge pumping technique was used to measure grain boundary traps in TFTs. The charge to breakdown for 100×100 μm capacitors was measured

Mat. Res. Soc. Symp. Proc. Vol. 424 © 1997 Materials Research Society

Annealing ambient	600°C, Furnace				950°C, RTA		
	NH$_3$	O$_2$	Ar	N$_2$O	O$_2$	Ar	N$_2$O
O$_2$,Ar		180	180				
Ar			360				
N$_2$O-LP				360			
N$_2$O				120			
NH$_3$,O$_2$,Ar	120	120	120				
O$_2$,Ar					1	1	
Ar						2	
N$_2$O							2

Table 1. Annealing condition. The time is in minutes. LP:Low pressure at 250mT. All other anneals at 1 atm.

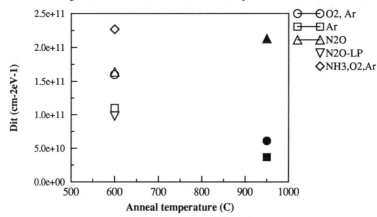

Fig. 1 Effect of annealing ambient on midgap D$_{it}$. Filled symbols are RTA.

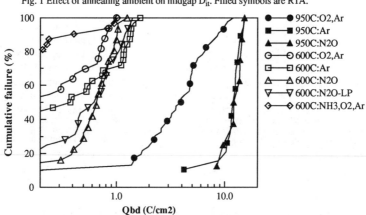

Fig.2 Q$_{bd}$ distribution. J=-10mA/cm^2, with substrate in accumulation.

by gate injection at a current density of 10mA/cm². The bias temperature stress was done on TFTs at 300°C with oxide field of 4MV/cm.

RESULTS AND DISCUSSION

Fig. 1 shows the midgap interface-state density (D_{it}) extracted from C-V measurements. The (Ar) oxides show lower D_{it} compared to (O_2,Ar) oxides. The RTA annealed (N_2O) oxide has higher D_{it} indicating that the nitridation condition has to be optimized to get lower initial D_{it}. However among the 600°C annealed oxides (N_2O-LP) shows the lowest D_{it}.

The distribution of charge to breakdown (Q_{bd}) under gate injection (Fig. 2) indicates, (O_2,Ar) oxide has lower Q_{bd} compared to (Ar) and (N_2O) oxides. The Q_{bd} is a very good measure of oxide degradation. The degradation takes place by physical bond breaking in the oxide by the energetic electrons [1]. In the case of (O_2,Ar) oxide, the Si-O bonds at the substrate interface are strained because of the significant stress at the substrate interface. This interfacial stress develops during oxidation of Si substrate, due to the difference in volume between Si and SiO_2. It is easier for energetic electrons to break the strained bonds and hence we see lower Q_{bd}. Fig. 3 shows the oxide thickness as measured from high frequency capacitance. The small increase in thickness after high temperature Ar anneal is due to volume relaxation [2]. There is further increase in oxide thickness, in an oxidizing ambient, as a result of interfacial oxidation.

High frequency C-V (HFCV) curves on the (O_2,Ar) oxides exhibit significant hysteresis after high field stressing. This is due to the generation of slow interface traps or border traps. Fig. 4 shows the slow traps as a function of injected electron fluence, calculated from the magnitude of the hysteresis at the midgap potential. The (N_2O) oxide exhibits lowest density of slow traps. We speculate that the high field generated slow traps arise due to the stressed SiO_x region in (O_2,Ar) oxides.

We have measured HFCVs on capacitors made on poly-Si substrate in the TFT wafers (Fig. 5). The source and drain contacts act as the sources of carriers. The transition from minimum to maximum capacitance depends on the grain boundary traps in the channel region and the oxide traps. Notice a sharp rise in (Ar) oxide compared to (O_2,Ar) oxide indicating fewer traps. Except for a lateral shift, presumably due to higher fixed oxide charge, (N_2O) oxide also has a sharper rise.

Fig. 6 shows the I-V characteristics for the n-channel TFTs after hydrogenation. The TFTs with (Ar) anneal give the best results both at 600°C and 950°C, with a field effect mobility of 25 cm²/V-s and 37 cm²/V-s respectively. At 950°C the (N_2O) oxide is much better than (O_2,Ar) oxide probably due to the trap passivation by nitrogen as seen by charge pumping measurement. However, at 600°C, (N_2O) oxide is inferior to (O_2,Ar) oxide, although it gives better performance before hydrogenation [3].

The charge pumping measurement [4] was done on n-channel TFTs with W=50 μm and L=20 μm. The substrate contact was obtained by depositing the poly-Si channel layer directly on to the p- type Si substrate. The source and drain terminals were connected to substrate through the current meter. The gate pulse was a square wave with frequency 100KHz and amplitude 4V. The charge pumping current is directly proportional to the grain boundary and interface trap density:

$$I_{cp} = q.f.A.N_t.T_{Si} \qquad (1)$$

where I_{cp} is the peak charge pumping current, q is the electron charge, f is frequency of gate pulse, A is the area of TFT, T_{Si} is the thickness of channel region and N_t is the total volume

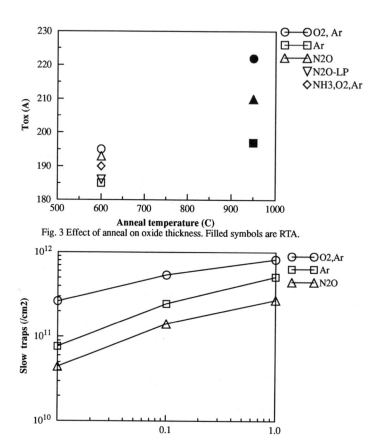

Fig. 3 Effect of anneal on oxide thickness. Filled symbols are RTA.

Fig. 4 Slow interface traps extracted from HFCV hysteresis at midgap potential.

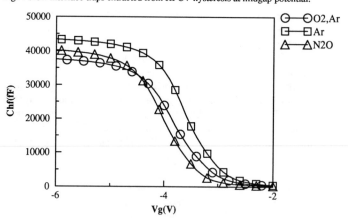

Fig. 5 HFCVs on poly-Si capacitors. The slope of capacitance rise is a measure of grain boundary traps.

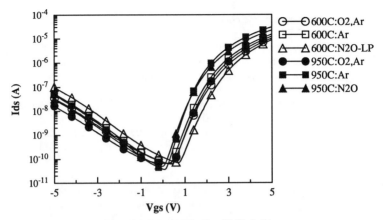

Fig. 6 I-V characteristics of *n*-channel TFTs. V_{ds}=5V. W=L=20µm.

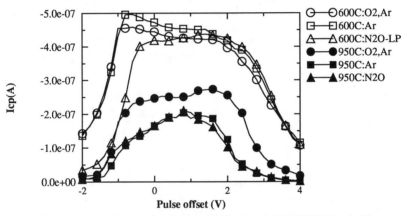

Fig. 7 Charge pumping characteristics of *n*-channel TFTs. Pulse height=4V. W=50µm, L=20µm.

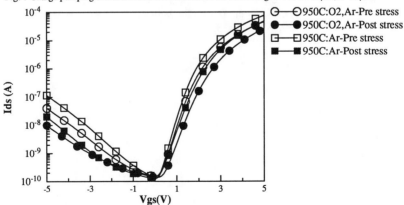

Fig. 8 Bias temperature instability. Stressed at 300°C, for 10 minutes with Vg=-8V.

trap density which includes the contribution from interface-states, grain boundary traps and intra grain defects.

The difference between 600°C and 950°C anneal is very apparent from Fig. 7. Also, at 950°C, it can be seen that (N_2O) anneal results in lower grain boundary traps compared to (O_2,Ar) anneal, which is consistent with the I-V measurements. The 950°C (Ar) anneal results in the lowest volume trap density of 9×10^{16} cm^{-3}.

Fig. 8 shows the comparison of (Ar) and (O_2,Ar) oxide TFTs with respect to bias temperature instability. The stress was done at 300°C for 10 minutes with an applied gate voltage of -8V. The (Ar) TFTs retain the better performance even after the stress. Similar results were obtained under positive stress also, with the a slightly lower degradation as would be expected [5].

CONCLUSION

We have demonstrated that the LPCVD oxides annealed in oxidizing ambient result in poor performance and reliability compared to inert and nitriding ambients. This is probably due to stressed Si-SiO$_2$ interface during interfacial oxidation. The reliability is significantly affected under high field stressing by the generation of slow interface traps in the strained SiO$_x$ region. Annealing of the LPCVD oxides in inert or nitriding ambient significantly improves the quality of TFTs by reducing interface and grain boundary traps. The (Ar) anneal gives the best results with respect to initial performance and bias temperature instability. For high temperature TFTs, RTA for a short time is very attractive, when constrained by thermal budget.

ACKNOWLEDGMENTS

We wish to acknowledge ARPA grant number MDA972-95-1-0004 for supporting this project. We also thank Len Booth for rapid thermal processing and Gladys Sarmeinto for hydrogen implantation.

REFERENCES

1. P.P.Apte and K.C.Saraswat, *IEEE. Trans. Electron Devices*, vol. 41, no. 9, p. 1595, 1994.
2. C.Rafferty, *Ph.D. dissertation, Stanford University*, p.33, 1989.
3. N.Bhat, A.Wang and K.C.Saraswat, *Society for Information Display International Symposium Digest of Technical Papers*, vol. 27, p. 785, 1996.
4. M.Koyanagi, I.Wu, A.G.Lewis, M.Fuse and R.Bruce, *IEDM Tech. Dig.*, p.863, 1990.
5. N.Bhat, M.Cao and K.C.Saraswat, *Society for Information Display International Symposium Digest of Technical Papers*, vol. 26, p. 393, 1995.

Part III

Liquid Crystal Display Materials and Technology

LIQUID-CRYSTALS FOR MULTIMEDIA DISPLAY USE: MATERIALS AND THEIR PHYSICAL PROPERTIES

S. NAEMURA
Liquid Crystal Technical Center, Merck Japan Limited,
Aikawa-machi, Aiko-gun, Kanagawa 243-03, Japan

ABSTRACT

Display devices are one of the key elements of the coming multimedia infrastructure, and Thin-Film-Transistor addressed Liquid Crystal displays are most promising for this application.   Physical properties such as dielectric permitivities and electric resistivity are essentially important for TFT-addressing use LC materials.   Both theoretical and experimental studies on these properties are reviewed and discussions are made on these in relation to the chemical structure of the liquid crystal substances, providing semiquantitative explanations on the superiority of fluorinated LC materials in respect to the coexistence of the high polarity and the high resistivity.   Further studies to improve LC material properties are finally suggested.

INTRODUCTION

It is often said that the 21st century is an era of multimedia.   According to a definition recently presented [1], multimedia is an infrastructure which enables the providing of an integrated and harmonized environment for digital AV (audio and video) and CC (computer and communications) technologies.   As can seen in this definition, display devices are one of the key elements of the multimedia infrastructure.   And, therefore, tremendous efforts are being made to develop advanced display devices to cope with the trend of the society moving towards the multimedia era.   Among various kinds of flat panel displays, liquid crystal (LC) displays are the leading devices in this field.

Although remarkable developments have been made on LC display related technologies, there still remain several performances to be improved for LC displays for multimedia use.   Those are larger screen size (above 13 inches in diagonal), higher resolution (XGA and SXGA, corresponding 1024×768 and 1280×1024 picture elements respectively), higher gray-scale capability (260 thousands to 167 million colors) and so on, and quite naturally it is also required to overcome the weaknesses of the conventional LC displays such as narrow viewing cone, insufficient picture-switching speed, poor brightness and so on.   The present LC displays can be broadly divided into two categories from a driving scheme point of view: passive addressing and active addressing. The typical ones in these two categories are STN (Super Twisted Nematic) and TFT (Thin Film Transistor) LC displays, and it is generally supported that TFT LC displays are most promising in regard to an applicability to multimedia displays [2].

Then, what are the requirements for LC materials for TFT-addressing display uses? Independent from an addressing scheme, whether it is an active (TFT, for example) addressing or a passive addressing, most of the requirements for physical properties of LC materials depend on the electro-optical mode, applied to the display.   Moreover, physics teaches us some universal truths such that larger dielectric anisotropy of an LC phase makes the threshold and also the saturation voltages smaller to induce an optical change in a nematic LC phase for all electro-optical modes using electric field effects. Another example is a viscous property of a LC phase, which is required to be small for a fast switching of the optical performances.   Thus, large anisotropy of the dielectric permitivity and low viscosity are requested for a nematic LC phase in general for display applications and some other physical properties, such as an elastic property and optical birefringence, are requested to be optimized for each electro-optical mode.

For TFT-addressing use, in addition, there exists one more important factor of LC materials from a display quality point of view.   That is a voltage (or charge) holding characteristic of a LC phase in display panels, which corresponds to the RC time of the LC phase.   It is because discharge through a LC phase causes problem in maintaining the induced optical change, resulting in fluctuation of optical performance called a flicker in display.   As the voltage holding characteristic is related to the specific resistivity and the dielectric permitivity of a LC phase, both of which are requested to be sufficiently high for LC materials in order to enable TFT LC display to exhibit a high quality picture image.

As can be understood from the above, such electrical properties as the dielectric permitivity and the specific resistitity are essentially important for LC materials especially for use in TFT addressing LC displays, which are most promising display devices in the coming multimedia era.   In the following two sections, studies on the dielectric permitivity and the specific resistivity of a LC phase are reviewed and explanations are made on those properties relating to the chemical structure of the LC substances.   Finally, the status of the studies on those physical properties are summarized and the direction of research and development works on LC materials toward the multimedia era is presented.

DIELECTRIC PERMITIVITY OF LC MATERIALS

<u>Molecular field theory of dielectric permitivity of LC phase</u>

In a nematic phase, LC molecules align with their long axis at an angle θ with the macroscopic symmetrical axis.   Here, the symmetrical axis is in a direction of a unit vector n called a director, corresponding to an optical axis of the nematic LC phase. The order parameter S is given by a formula,

$$S = (3/2)< \cos^2\theta - 1/3 > . \tag{1}$$

LC molecules are modeled to possess a total dipole moment μ at an angle β with the long molecular axis, which corresponds to the axis of maximum polarizability.   The longitudinal and transverse components of μ are represented as μ_l and μ_t.   The longitudinal and transverse elements of the electric polarizability tensor of a LC molecule are also represented as α_l and α_t .   The statistical average of each component of the polarizability, $\alpha_{//}$ parallel to the director and $\alpha_\perp$ perpendicular to the director, is given for

a nematic phase as follows,
$$< \alpha_{//} > = (1/3)\{ \alpha_l (1+2S) + 2\alpha_t (1\text{-}S) \}, \tag{2}$$
$$< \alpha_\perp > = (1/3)\{ \alpha_l (1\text{-}S) + \alpha_t (2+S) \} . \tag{3}$$
Here, a rotational symmetry of the electric polarizability around the long molecular axis is taken into account.

In regard to the statistical average of the components of the permanent dipole moment, $\mu_{//}$ which is parallel to the director and $\mu_\perp$ which is perpendicular, the dipole-dipole association has to be taken into account especially in case of polar molecules presented in this paper.   This case was studied by D.A. Dunmur et $al.$ and the following formulas are given for uniaxial LC materials [3].
$$< \mu_{//}^2 > = (1/3) g_1^{(//)} \{\mu_l^2 (1+2S) + \mu_t^2 (1\text{-}S) \} \tag{4}$$
$$< \mu_\perp^2 > = (1/3) g_1^{(\perp)} \{\mu_l^2 (1\text{-}S) + (1/2)\mu_t^2 (2+S) \} \tag{5}$$
Here, $g_1^{(i)}$ ($i = //, \perp$) are Kirkwood factors, which are a measure of the correlation of molecular dipole components projected onto the symmetry axes of the LC phase.
The dielectric permitivity tensor can be given by
$$\varepsilon^{(i)} = 1 + (NhF/\varepsilon_0) [< \alpha^{(i)} > + (F/k_BT) < \mu^{(i)2} >] \quad (i = //, \perp) , \tag{6}$$
where h and F are cavity field and reaction field factors and N is the number density.
That is, the dielectric anisotropy, $\Delta\varepsilon \equiv \varepsilon_{//} - \varepsilon_\perp$, is
$$\Delta\varepsilon = (NhF/\varepsilon_0)[(\alpha_l - \alpha_t)S + (F/3k_BT)\{(\Delta g_1 + g_1*S)\mu_l^2 + (\Delta g_1 - g_1*S/2)\mu_t^2\}] , \tag{7}$$
where $\Delta g_1 \equiv g_1^{//} - g_1^\perp$ and $g_1* \equiv 2g_1^{//} + g_1^\perp$.   In case when there exists no dipole-dipole association ($g_1^{//} = g_1^\perp = 1$), the formula (7) comes to the Maier-Meier formula [4],
$$\Delta\varepsilon = (NhF/\varepsilon_0)[(\alpha_l - \alpha_t) - (\mu^2/2k_BT)(1 - 3\cos^2\beta)]S . \tag{8}$$
As can easily be foreseen from the formula (8), a larger permanent dipole moment μ in parallel to the long molecular axis ($\beta=0$) is preferable in order to achieve large positive-$\Delta\varepsilon$ values.   In the following, discussions are made on permanent dipole moments, μ and β values, of LC substances.

Permanent dipole moment of LC molecules

Attempts have been made to investigate the applicability of semiempirical calculations of μ and β values to the forecast of the dielectric anisotropy of LC molecules [5].   Although there exists a limitation of semiempirical methods, the calculation of the permanent dipole moment of free molecules is useful for the design of molecular structures.   Measurements of dipole moments are also reported on several LC substances [6], supporting the meaningfulness of the semiempirical calculations by providing semiquantitative agreement with the calculations.   The μ and β values listed in the following Tables 1-5 were those calculated by using a combination of empirical force field (PCMODEL) and semiempirical MO method (MOPAC) by M. Bremer and K. Tarumi [5].

Table 1 shows the μ and β values for 2- and 3-ring structures with -CN end group. Comparing the 2-ring core structures with n-alkyl chain at the terminal, phenylcyclohexane core is inferior to biphenyl in regard to linearity and the angle between the axis of the core and the end -CN group is larger for the structure (II) compared to the structure (I).   A second cyclohexane ring in the structure (IV) makes the β value larger compared to the structure (III).   Regarding the μ value of -CN group, no effects can be seen from the *n*-alkyl chain at the opposite terminal (Compare (II) and (III)) and also almost no effects can be seen from an additional cyclohexane ring in the structure (IV) when compared to (III).   What is seen rather clearly, however, is the effect of the conjugation length on the dipole moment of the -CN end group.   As can be seen in the substances (I) and (V), larger dipole moments were obtained most probably due to the elongated conjugation compared to the substances (II) and (IV).

Table 1   Calculation of LC molecular dipole moment [5]
(effects of conjugation length)

	β	μ (Debye)
(I) C_5H_{11} ⟨○⟩⟨○⟩ CN	7.5○	4.1₅
(II) C_5H_{11} ⟨⟩⟨○⟩ CN	13.3○	3.9
(III) C_3H_7 ⟨⟩⟨○⟩ CN	8.4○	3.9
(IV) C_3H_7 ⟨⟩⟨⟩⟨○⟩ CN	9.9○	3.9
(V) C_3H_7 ⟨⟩⟨○⟩⟨○⟩ CN	6.0○	4.2

    In Table 2, comparison is made on μ and β values of LC substances with cyclohexyl-cyclohexyl-phenyl or biphenyl-cyclohexane core and *n*-pentyl and different polar end groups.   In case when the terminal substitution is made by perfluoromethoxy group with an ether linkage (substances (VI) and (IX)), an interesting result can be found in the calculated β values.   The energy minimum configuration of the dipole moment (-OCF3) and the molecular long axis seems to be preferable with respect to the parallelism for the substance (IX) with biphenyl-cyclohexane core compared to the substance (VI) with cyclohexyl-cyclohexyl-phenyl core.   However, as is also pointed out by the authors of the Ref.[5], the rotation around the C_{Ar} - O bond requires only 2.6 kcal/mol according to their AM1 calculation, which is similar to the intermolecular interactions in a packed LC bulk.   Thus, it is not clear whether the actual configuration of (VI) and (IX) molecules in a nematic phase can be like that as calculated.   The similarity of calculated β values for the terminal substitution by -CN, -F and -Cl (substances (IV), (VII) and (VIII)) looks rather reasonable, and the differences in μ values of these three polar terminals are also presented properly.   The relatively large μ values obtained for the -OCF3 substitution (substances (VI) and (IX)) compared to a simple halide substitution reflect the effect of "super fluorinated" structure well.

Table 2   Calculation of LC molecular dipole moment [5]
(effects of terminal polar group)

	β	μ (Debye)
(IV) C_3H_7 —◯—◯—◯— CN	9.9°	3.9
(VI) C_3H_7 —◯—◯—◯—O– CF_3	10.0°	3.0
(VII) C_3H_7 —◯—◯—◯— F	10.8°	2.0
(VIII) C_3H_7 —◯—◯—◯— Cl	9.4°	1.7
(V) C_3H_7 —◯—◯—◯—CN	6.0°	4.2
(IX) C_3H_7 —◯—◯—◯—O– CF_3	2.7°	2.8

Table 3 shows effects of the fluoro substitution at the ortho positions in respect to the terminal -CN or -F group.   The substitution at one ortho position (substances (X) and (XII)) naturally makes the total dipole moment off axis, and the symmetrical double substitutions (substances (XI) and (XIII)) make no significant changes in the β value. The calculated total dipole moments show a reasonable tendency to increase in accordance with the number of substitutions corresponding to the number of additive dipoles in parallel.

Table 3   Calculation of LC molecular dipole moment[5]
(Effects of fluoro substitution-1-)

	β	μ (Debye)
(II) C_5H_{11} —◯—◯— CN	13.3°	3.9
(X) C_5H_{11} —◯—◯— CN (F)	24.8°	4.8
(XI) C_5H_{11} —◯—◯— CN (F, F)	12.6°	5.4
(VII) C_3H_7 —◯—◯—◯—F	10.8°	2.0
(XII) C_3H_7 —◯—◯—◯—F (F)	11.4°	3.2
(XIII) C_3H_7 —◯—◯—◯—F (F, F)	10.6°	3.8

Same kind of discussions can be made on some other structures listed in Table 4. Although similar tendencies as the above can be found qualitatively, the substitution at one ortho position of these cases (substances (XV) and (XVIII)) decreases the β value preferably and the double substitution (substances (XVI) and (XIX)) does not make the β value similar to that of substances (XIV) and (XVII) with no fluoro substitutions at the ortho positions unlike the case of the structures in Table 3.   This could be an actual case due to the difference of the core structures, biphenyl-cyclohexane in case of substances in Table 4 and cyclohexyl-cyclohexyl-phenyl in case of those in Table 3, but , at the same time, it can not be ignored that the calculation is less reliable for the substances in Table 4 due to the terminal groups, which are energetically easier to rotate. This uncertainty also exists in the calculated μ values.   According to the calculation [5] using a Boltzmann-type weighting scheme, the μ value can be larger at a magnitude of 2 Debye for difluoro-methoxybenzene.

Table 4   Calculation of LC molecular dipole moment [5]
(effects of fluoro substitution-2-)

	β	μ (Debye)
(XIV) C_3H_7⬡-⬡-⬡-O - CH$_2$F	325o	3.7₅
(XV) C_3H_7⬡-⬡-⬡-O - CH$_2$F	17.7o	4.5
(XVI) C_3H_7⬡-⬡-⬡-O- CH$_2$F	245o	5.5
(XVII)C_3H_7⬡-⬡-⬡- O-CH$_2$ - CF$_3$	41.7o	3.5
(XVIII)C_3H_7⬡-⬡-⬡-O-CH$_2$- CF$_3$	9.5o	3.3₅
(XIX) C_3H_7⬡-⬡-⬡-O-CH$_2$ - CF$_3$	27.9o	4.6₅

The above results give semiqualitative explanations of the relation between the total dipole moment and the chemical structure of LC molecules, providing some idea of LC molecular structures with a large dipole moment, which is more-or-less in parallel to the molecular axis.   Some examples of this kind of substances and their calculated dipole moments are presented in Table 5, showing that large dipole moments, effective for large Δε values, can be expected for fluorinated substances without a highly polar -CN end group.

Table 5 Calculation of LC molecular dipole moment [5]
(highly polar fluorinated substances)

	β	μ (Debye)
(XIII) C$_3$H$_7$—◯—◯—◯—F F F	10.6o	3.8
(XX) C$_3$H$_7$—◯—◯—◯—F F O - CF$_3$	103o	4.6
(XXI) C$_3$H$_7$—◯—◯—◯—F F CF$_3$	9.7o	5.4$_5$

Calculation and Measurement of Dielectric permitivities of LC phase

As for substances with large permanent dipole moments, the first term of electric polarizability in the formula (8) can be neglected, and the dielectric anisotropy $\Delta\varepsilon$ at a fixed reduced temperature is attempted to be a function of ($-\mu^2(1-3\cos^2\beta)$). Calculation results of the value "$\Delta\varepsilon$"$_{calc} \equiv f(-\mu^2(1-3\cos^2\beta))$ by using the μ and β values obtained as above were reported for 80 substances including those in Tables 1-5 [5]. As is seen in Fig.1, reasonable correlation was found with the correlation function R^2 of 0.80.

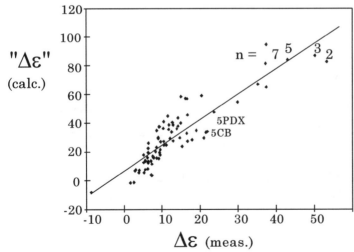

Fig.1 Correlation between calculated "$\Delta\varepsilon$" and measured $\Delta\varepsilon$ [5]

The scattering of the data is partly explained to be due to discrepancies between the dipole moments calculated for one conformer with minimum free energy in the gas phase and those of packed molecules in the nematic phase, as briefly introduced above. Another factor is the association effect on the $\Delta\varepsilon$ value in the nematic phase, because the calculation of "$\Delta\varepsilon$"$_{calc}$ was made on a basis of the formula (8), which does not include this effect.   The association effect is taken into account in the molecular field theory of LC dielectric permitivity, yielding the formula (7), where Kirkwood factor g_1 is used as the parameter.   The g_1 values less than unity correspond to anti-parallel dipole-dipole associations and those above unity to parallel associations.

Existence of association is demonstrated experimentally and g_1 values are reported for some LC substances.   Figure 2 shows the g_1 values of the homologues,

(XXII) C_nH_{2n+1}⟨○⟩$\overset{O}{\overset{\|}{C}}$O⟨○⟩$_F$ C≡N   (n=2,3,5,7)

at temperatures slightly above the nematic-isotropic transition temperature T_{NI}, which were calculated from the isotropic permitivity measurements [6].   In case of these substances, parallel dipole-dipole associations seem to exist and the g_1 value increases as the alkyl chain length becomes shorter, providing an idea that long alkyl chains at the terminal of this LC core hinder the parallel association.   This parallel association can

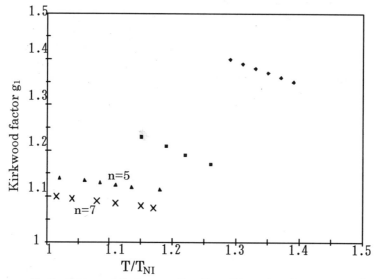

Fig.2   Kirkwood factors as a function of the reduced temperature for substances (XXII) [6]

be a reason of the different $\Delta\varepsilon$ values measured for the above homologues (dots in Fig.1, numbered 2, 3, 5 and 7, corresponding to the number of carbon atoms n in the alkyl chain) despite the result that the calculated μ values, and therefore the calculated "$\Delta\varepsilon$"$_{calc}$ values, are quite similar.

The association effect can sometimes give a qualitative explanation of the scattering in Fig.1, but this is not always the case.   Figure 3 shows the g_1 values obtained for the two substances,

(I)5CB; C_5H_{11} ⟨◯⟩—⟨◯⟩— CN   (XXIII)5PDX C_5H_{11}—⟨◯⟩—⟨◯⟩—CN

for which occurrence of anti-parallel dipole-dipole association is easily predicted [6]. The calculated g_1 values less than unity support the idea of anti-parallel association, and predict that actual $\Delta\varepsilon$ values will be smaller than the calculated "$\Delta\varepsilon$"$_{calc}$ values.   This relation between measured $\Delta\varepsilon$ and calculated "$\Delta\varepsilon$"$_{calc}$, however, is rather difficult to see in the results in Fig.1 for 5CB and 5PDX.

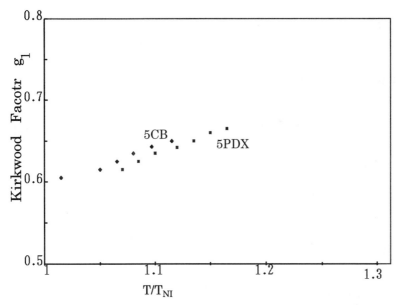

Fig.3   Kirkwood factors as a function of the reduced temperature
for substances (I) and (XXIII) [6]

SPECIFIC RESISTIVITY OF LC MATERIALS

Highly purified isotropic liquids such as paraffin hydrocarbons are known to show the specific resistivity ρ of the order of $10^{18} \sim 10^{19}$ $\Omega \cdot$cm. In case of standard non-polar liquids, however, the ρ values are in the order of $10^{14} \sim 10^{15}$ $\Omega \cdot$cm, where the dominant cause of the electric conduction is said to be residual ions. Typical non-polar LC materials with the mean ε value ($\equiv (\varepsilon_{//} + 2_\perp)/3$) of around 3 also show the ρ value of the order of $10^{14} \sim 10^{15}$ $\Omega \cdot$cm, and, in case of polar LC materials with the mean ε value of 5 $\sim$ 9, the typical ρ values are down to the order of $10^{11} \sim 10^{14}$ $\Omega \cdot$cm. One estimation says that the ion density in a polar LC phase, 5CB, is around 8×10^{14} cm^{-3} [7].

In case when the electric conduction is caused by ions, the specific resistivity ρ is inversely proportional to the conductivity σ, which is given by the formula,

$$\sigma = nq\mu_\pm \tag{9}$$

where n and $\mu_\pm$ are the number and the mobility of ions respectively and q is the quantity of electricity. It is easily predicted that the number of ions in a solvent is dependent on the polarity of the solvent and, therefore, the number of ions, that is the specific resistivity of polar LC substances, will exhibit some dependence on the permanent dipole moment of the LC substance.

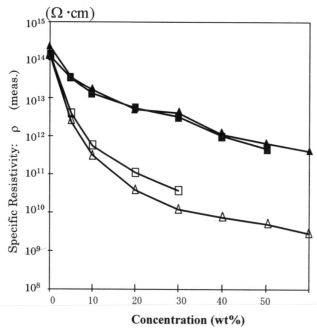

Fig.4 Dependence of specific resistivity on the concentration of LC solute in a non-polar solvent (filled symbols: fluorinated LC substances, open symbols: cyano LC substances)

Figure 4 shows observed dependence of the ρ value on the concentration of LC materials in a non-polar isotropic solvent.   Results for fluorinated substances are presented by filled symbols and those for substances with CN- group are by open symbols   The concentration dependence can qualitatively explained by a predicted dependence of the number of ions on the polarity of the solution, which increases according to the concentration of the LC solute substances.   The difference observed between the fluorinated substances and the cyano substances can be due to the difference of the polarity of these substances.

Figure 5 shows the relation between measured ρ valucs for solutions with a fixed concentration of each LC solute and calculated μ values of the LC substances.   As measurements of this kind of property can easily afford some scattering of the data, any conclusion should be deduced from this figure in a very careful manner.   Nevertheless, it cannot be neglected to see a possibility that the specific resistivity differs according to the chemical structure of LC substances, even though the polarity is same.

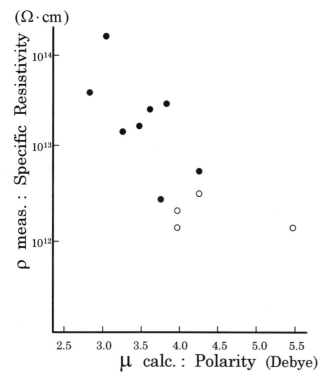

Fig.5   Correlation between specific resistivity and calculated [5] polarity of LC substances (● fluorinated LC substances, O cyano LC substances)

Temperature dependence of ρ values was measured for three classes of LC substances, as is shown in Fig.6.   One is for a neutral LC material (●) and the others are for polar LC substances, among which one is a fluorinated substance (■) and the other two are cyano substances (▲,✖).   Although the temperature dependence looks to be similar in the temperature range sufficiently above the nematic-isotropic transition (→), it looks to be stronger for the cyano substances than the fluorinated substance in the namatic phase. One explanation of this chemical-structure dependence can be that residual ions, which are strongly constrained by the cyano molecules, become mobile at higher temperatures, and thus the difference of number of "mobile" charge carriers becomes significant between cyano and fluoro substances at a high temperature, even though it is similar for both classes of substances at a low temperature.   As the constraining force can be related to associations, the Kirkwood correlation factor is also important for characterization of LC substances in this respect.

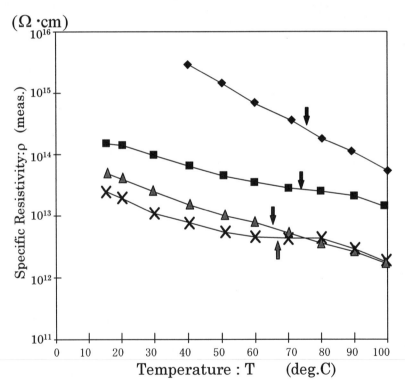

Fig.6   Temperature dependence of specific resistivity
(● neutral LC, ■ fluorinated LC, ▲✖ cyano LC substances)

CONCLUSION

The molecular field theory provides quantitative discussions on the contributions of molecular parameters, such as the dipole moment μ and its angle β, to macroscopic properties, such as the dielectric permitivity, of a LC phase.   On the other hand, experiments have been made to deduce the permanent dipole μ of LC substances, and even more active attempts are being made on the semiempirical calculations of the dipole. As the present calculations are carried out for isolated molecules at 0 K only for one conformer, some deviations from the real state in a LC phase can not be avoided. Nevertheless, the calculations provide qualitatively reasonable results from a viewpoint of the chemical structures.   The proposal to use the Boltzmann-type weighting scheme [5] will make the calculation more informative, and, moreover, a complementary approach together with experimental investigations on LC molecular dipole moment will make these studies more effective for the LC molecular modeling.

In regard to the computational prediction of $\Delta\varepsilon$ value of a nematic LC phase, effects of the dipole-dipole association, which are discussed by using the molecular field theory, can not be neglected, in addition to the effects of the calculation errors in the dipole moments.   Experimental approach to deduce the Kirkwood correlation factor g_1 is being made, but results are only for limited number of LC substances so far. Considerations are desired on a reliable numerical estimation method of the g_1 values by referring the experimental results on variety of substances and also improved calculation results of the dipole moment of LC substances.

As for the specific resistivity ρ of a LC phase, residual ions are considered to be the main factor and their number density seems to make the ρ value of a LC phase correlated to the polarity of LC substances.   Experimental results exhibit a possible relation between the ρ value of a LC phase and the chemical structure of the LC substance.   It is also suggested that the g_1 factor can be the parameter to connect these two.   Further investigation on this subject seems to be a key point to improve the LC performance in regard to the coexistence of a high polarity and a high resistivity.

The present studies provide an idea that the temperature dependence of the ρ value of cyano LC materials is characterized by the behavior of residual ions, which are constrained by the polar cyano molecules at a low temperature and are released as charge carriers at a high temperature.   This temperature dependence can be behind the difference of voltage holding characteristics between cyano and fluorinated LC substances, as is shown in Fig.7.   This figure shows a clear superiority of fluorinated LC materials for TFT-addressing display uses.

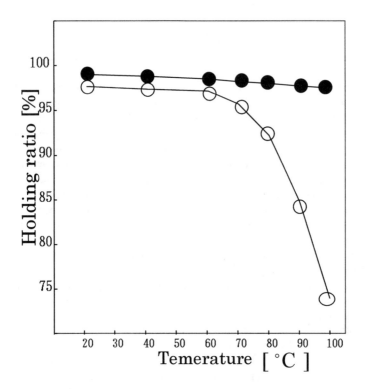

Fig.7   Temperature dependence of voltage holding ratio

To sum up, enormous studies, partly introduced in this paper, have supported the progress of fluorinated LC substances, which are now successfully used for TFT-LC displays[8][9].   It was in 1990 when LC mixtures, containing not cyano substances but only fluorinated substances as the polar components, were introduced from E.Merck (present Merck KGaA) for TFT-addressing display use, demonstrating their superior voltage holding characteristic.   The Δε values of those fluorinated LC mixtures, at that time, were relatively small compared to the conventional cyano LC mixtures, and the electro-optical threshold voltage was not sufficiently low.   Since then, improvements have been made on fluorinated LC mixtures to make the Δε large without sacrificing the superior voltage-holding characteristic [10], as is shown in Fig.8.   Discussions, presented in this paper, hopefully suggest a further improvement of the performance of fluorinated LC substances and also development of further advanced class of LC substances for the coming multimedia era.

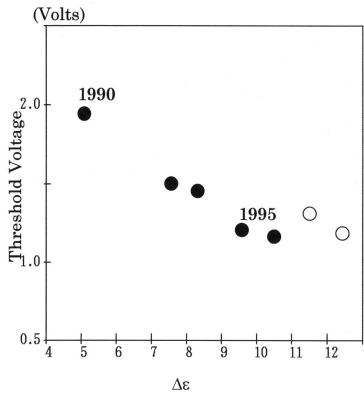

Fig.8   Threshold voltage and Δε values of typical LC materials
(● fluorinated LC, ○ cyano LC)

ACKNOWLEDGMENTS

The author wishes to express his appreciation to Dr.Pauluth and Dr.Bremer from LC Chemical Research, Merck KGaA, and Dr.Geelhaar, Dr.Tarumi and Dr.Hirschmann from LC Physical Research, Merck KGaA, for their helpful discussions.   It should be mentioned with sincere thanks that many valuable data and information on the molecular modeling are introduced in this paper from the referred paper from Dr.Bremer and Dr.Tarumi.   Appreciation is also due to Ms.Kimura from LC R&D, Merck Japan, who performed the electric resistivity measurements.

REFERENCES

1. H.Mizuno, Proc. 15th Int'l. Display Res. Conf. 1995, 3.
2. H.Kawakami, Digest Tech. Papers AM-LCD 95 1995, 63.
3. D.A.Dunmer and K.Toriyama, Mol. Cryst. Liq. Cryst. **198**, 201 (1991).
4. W.Maier and G.Meier, Z. Naturforsch. **A16**,262 (1961).
5. M.Bremer and K.Tarumi, Adv. Mater. **5**, 842 (1993).
6. D.G.McDonnell, E.P.Raynes and R.A.Smith, Liq. Cryst. **6**, 515 (1989).
7. H.Naito, K.Yoshida, M.Okuda and A.Sugimura, J. Appl. Phys. **73**, 1119 (1993).
8. S.Naemura, Digest Tech. Papers AM-LCD 94 1994, 76.
9. M.Bremer, K.Tarumi H.Ichinose and H.Numata, Digest Tech. Papers AM-LCD 95 1995, 105.
10. T.Tarumi, E.Bartmann, T.Geelhaar, B.Schuler, H.Ichinose and H.Numata, Proc. 15th Int'l. Display Res. Conf. 1995, 559.

THE EFFECT OF PHENYL BENZOATE SELF-ASSEMBLED MONOLAYERS ON LIQUID CRYSTAL ALIGNMENT

R. J. ONDRIS-CRAWFORD*, L. M. LANDER**, C. W. FRANK*, AND R. J. TWIEG**
*Department of Chemical Engineering, Stanford University, Stanford, CA 94305-5025, renate@leland.stanford.edu
**IBM Almaden Research Center, 650 Harry Rd., San Jose, CA 95120

ABSTRACT

We have synthesized compounds of the form $M-R-SiCl_3$ where M is a mesogen based on octyloxy phenyl benzoate, and R is an alkyl chain (tether) that connects M to the silane head group. Tethering lengths from the mesogenic unit to the silane moiety involved 3, 6, or 10 methylene units. These compounds were used to form self-assembled monolayers (SAMs) on silicon wafers, fused silica, and SiO_2 coated indium tin oxide (ITO) coated glass. Characterization of these systems includes contact angle measurements, ellipsometry, and atomic force microscopy. Results suggest that the monolayer packing is governed by the lateral interactions between the mesogenic units and not by the tethering length to the solid surface. We have investigated the utility of these chemically-tuned surfaces for liquid crystal alignment. The octyloxy phenyl benzoate self-assembled monolayers induce homeotropic alignment of a nematic liquid crystal.

INTRODUCTION

Self-assembled monolayers (SAMs) can be formed from many different surface-active compounds on a wide variety of surfaces [1]. Two of the most widely studied SAMs are alkanethiol monolayers on gold [1, 2] and alkylsiloxane monolayers formed on either silicon-silicon dioxide, glass, or other oxide-bearing substrates [1, 3, 4, 5]. Silanes of the form $RSiX_3$ where R can be an alkyl chain or functionalized alkyl chain (provided that the functional group does not interact with the surface or the reactive silicon end group), and X is either Cl or OEt, have been found to form stable, robust, ordered monolayer films.

Although thiol monolayers containing large aromatic functional groups have been prepared and extensively studied [1, 6], there has not been an extensive body of work performed on siloxane monolayers containing aromatic functionalities. Sagiv [7] has performed coadsorption experiments using surfactant dye molecules and octadecyltrichlorosilane (OTS). In this system, however, while the OTS was chemically bound to the substrate, the dye molecules were merely physisorbed and easily removed. Typically, the dye molecules employed adsorbed with the chromophore aligned perpendicular to the substrate surface.

Tillman, et al. [8, 9] studied the effects of the incorporation of an aromatic functional group on the quality, ordering, and orientation of chemically-bound siloxane monolayers. Any perturbations caused by the incorporation of the aromatic groups were not sensed at the outer surface, as measured using contact angle methods. The aromatic groups were found to be almost perpendicular to the substrate surface with the alkyl chains tilted more than normal to compensate for the difference in cross-sectional area between the alkyl chains and the bulky functional group. This tilt preserved the close-packed, ordered structure of the monolayer maximizing the packing density of the aromatic groups. Placing a sulfone or ester functionality in the monolayer along with the aromatic ring further increased the tilt of the alkyl chains, while the aromatic groups remained perpendicular [9]. Monolayers formed from molecules containing a benzyl ester functionality were found to be more disordered than those containing a sulfone functionality.

Investigations into the SAM-forming ability of azobenzene-modified molecules have also been undertaken [10, 11]. Monolayers can be formed with azobenzene moieties residing in predominantly all-trans conformations. These can then be photoisomerized to the cis configuration by irradiation with UV light. Transition from cis to trans is accomplished by the application of visible light. These 'command surfaces' are of primary interest as liquid crystal alignment films that are capable of inducing reversible homeotropic to planar alignments.

Much has been published on the diverse methods of obtaining the desired alignment of liquid crystalline materials. Investigations on the use of surface coupling agents [12] and siloxane self-assembled monolayers for the alignment of liquid crystalline compounds are found among the copious literature [13]. It has been established that simple alkylsiloxane monolayers will impart alignment to liquid crystals as will films containing azobenzene moieties and those with other aromatic groups at the monolayer surface [10, 14]. Hayashi and Schimura used liquid crystals to observe intimate molecular interactions [15]. Other than the above mentioned examples, to our knowledge no siloxane self-assembled monolayers incorporating mesogenic groups have been evaluated for use as liquid crystal alignment films.

EXPERIMENT

Synthesis of Phenyl Benzoates

The 4-n-alkyenyloxyphenyl-4'-n-alkyloxybenzoates series was prepared via the Mitsunobu reaction and hydrosilylation [5]. A schematic overview of the synthesis procedure is shown in Figure 1.

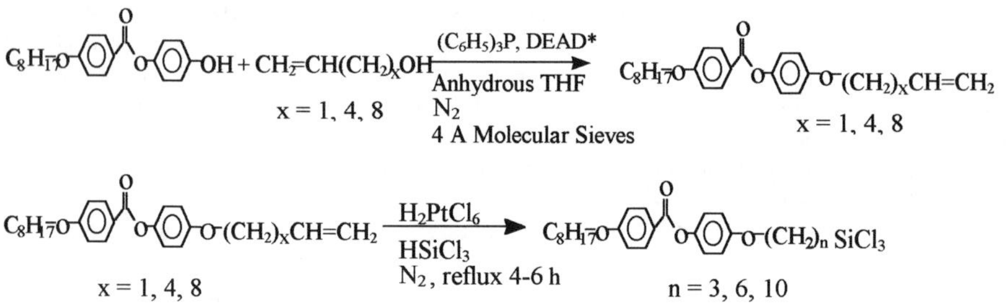

*DEAD = Diethyl azodicarboxylate

Figure 1: Synthesis procedure for tri-chlorosilanes incorporating octyloxy phenyl benzoate mesogenic units

Self-Assembled Monolayer Formation

Several substrates were used: SiO$_2$ (300 A thickness) coated indium tin oxide (ITO, resistance 125 MΩ) coated soda-lime glass, fused silica, and two types of silicon wafers: (1): <100>, undoped, double polished silicon and (2): n-type, phosphorous doped, single side polished silicon. Prior to SAM formation, the substrates were made purely hydrophilic by immersion in a 70/30 v/v sulfuric acid/hydrogen peroxide solution and then rinsed with copious amounts of high purity water. Before placing the substrates in the deposition solution, they were dried in a stream of nitrogen.

The deposition solution was prepared by dissolving 5-10 milligram of the phenyl benzoate in 20 milliliters of a 4/1 (v/v) mixture of chloroform and toluene. The substrates were placed in the reaction jar and the latter was placed in a sealed secondary container with Drierite®. The solutions were allowed to sit undisturbed for 20-24 hours.

After 20-24 hours, the substrates were removed from the deposition solution and immersed in methylene chloride and further rinsed with chloroform and air dried. A schematic representation of the resulting self-assembled monolayer is shown in Figure 2. The procedure followed was that developed by Wasserman [4] and Lander [5].

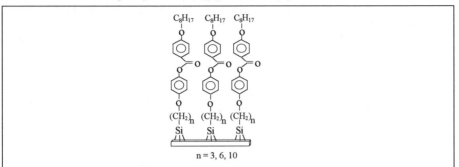

Figure 2: Schematic representation of self-assembled monolayers incorporating octyloxy phenyl benzoate mesogenic units on silicon wafer, fused silica, or SiO₂ coated ITO glass

RESULTS

Self-Assembled Monolayer Characterization

Contact angle measurements of high purity water on the self-assembled monolayers on silicon wafers, fused silica, and SiO₂ coated Indium Tin-Oxide (ITO) glass were taken using a Ramé-hart NRL Contact Angle Goniometer. Two values were taken for the advancing (θ_a) and receding (θ_r) contact angles for each substrate. The average values for the Milli-Q water contact angles on the phenyl benzoate (PB-n) self-assembled monolayers are shown in Table 1.

PB-3 on:	θ_a (deg.)	θ_r (deg.)	PB-6 on:	θ_a (deg.)	θ_r (deg.)
SiO₂ ctd. ITO	98.5 (2.0)	65.0 (6.0)	SiO₂ ctd. ITO	104.0 (2.0)	43.5 (6.0)
Fused silica	98.5 (3.0)	70.0 (8.0)	Fused silica	102.0 (2.5)	67.0 (2.5)
Silicon (1)	103.0 (3.0)	78.5 (3.0)	Silicon (1)	105.5 (2.0)	90.5 (3.0)
Silicon (2)	105.0 (4.0)	81.0 (2.0)	Silicon (2)	104.0 (2.0)	73.0 (8.0)

PB-10 on:	θ_a (deg.)	θ_r (deg.)
SiO₂ ctd. ITO	105.0 (2.0)	72.0 (3.0)
Fused silica	105.0 (2.0)	68.0 (2.0)
Silicon (1)	110.5 (2.5)	91.0 (2.0)
Silicon (2)	99.0 (5.0)	72.0 (5.0)

Table 1: High purity water contact angles on the SAMs incorporating octyloxy phenyl benzoate moieties with a tethering length of n=3 (PB-3), 6 (PB-6), and 10 (PB-10).

The water contact angles for the self-assembled monolayers incorporating the octyloxy phenyl benzoate mesogenic units appear to be independent of tethering length within the range studied (n = 3 - 10), within experimental error. In the initial molecular design phase of the project, the tethering length was varied in order to modify the extent of coupling to the solid surface, anticipating that a crucial tethering length would be necessary to decouple the mesogenic unit from the surface, much like the situation in side-chain liquid crystal polymers. The lack of dependence of the contact angle on the tethering length suggests that the lateral interactions between mesogenic units are strong enough to overcome any steric effects associated with the shorter tether chains. In effect, the phenyl benzoate units may be interacting as a relatively uniform, coherent layer with the underlying tethering chains accommodating the variations in surface roughness and chemistry, which are known to exist for silicon oxide surfaces, and the overlying octyloxy chains providing a relatively uniform surface.

Ellipsometric film thickness measurements were performed using a Gaertner ellipsometer. The average film thicknesses of the SAMs are shown in Table 2. The film thicknesses are given in Angstroms. The index of refraction of the films were estimated as n_f=1.46.

SAM	Film thickness (Å)
PB-3	27 ± 3
PB-6	33 ± 2
PB- 10	37 ± 1

Table 2: Ellipsometric film thickness measurements of the SAMs on silicon wafers.

The measured film thicknesses for the phenyl benzoate system compare well with calculated values for the nearly extended chain. The approximated chain length for a nearly extended PB-3 system is 25 Å, 28 Å for PB-6, and 32 Å for PB-10.

Atomic force microscopy (AFM) studies were performed on the phenyl benzoate system to investigate the surface morphology. A Digital Instruments, Nanoscope IIIa microscope was used with a J-scanner (1-100 micron scanning size). Figure 3 shows an AFM surface plot of PB-6 on a silicon wafer with a scan size of 2 micron and a height of 10 nanometer. The image shows a smooth surface with several large clusters, whose origin is uncertain.

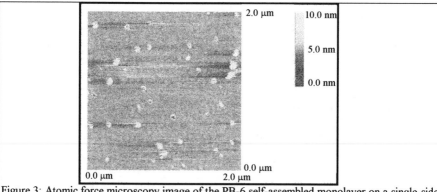

Figure 3: Atomic force microscopy image of the PB-6 self-assembled monolayer on a single-side polished silicon wafer (silicon (2)).

<u>Liquid Crystal Alignment</u>

The utility of these self-assembled monolayers incorporating octyloxy phenyl benzoate mesogenic units for liquid crystal alignment was investigated. Phenyl benzoate self-assembled monolayers formed on SiO$_2$ coated Indium Tin Oxide (ITO) coated glass were used for this study. As is shown in Table 1, the water contact angles on this surface are similar to those on the silicon wafers, and it is therefore reasonable to expect that a similar well-formed monolayer exists on these substrates. Two substrates on which the same SAM was formed were placed facing each other with a 10 micron mylar spacer to control the cell thickness. The nematic liquid crystal compound pentyl-cyanobiphenyl (5CB) was then capillary filled in the isotropic phase and the cell was brought to room temperature.

Optical polarizing microscopy was used to check the alignment of the nematic liquid crystal within the cell. The entire phenyl benzoate series resulted in homeotropic (perpendicular) alignment of the nematic liquid crystal (See Fig. 4).

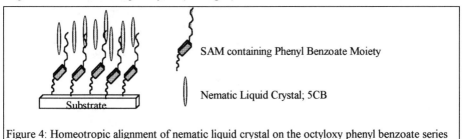

SAM containing Phenyl Benzoate Moiety

Nematic Liquid Crystal; 5CB

Substrate

Figure 4: Homeotropic alignment of nematic liquid crystal on the octyloxy phenyl benzoate series

The crystal rotation [16] technique was used to measure the pretilt angle of the liquid crystal on the surface. The pretilt angles deviated only 1 to 2 degrees away from the substrate's normal direction for each sample. This nearly perfect normal alignment of the liquid crystal molecules on the SAM surfaces was confirmed using conoscopy. The alignment of the liquid crystal on the SAM surface was thus not influenced by the tethering length of the SAM to the surface. This result is consistent with the contact angle measurements, which also revealed that the packing of the monolayers is governed by the lateral interactions between the mesogenic groups and not by the steric hindrance between the tethers. The tethers accommodate surface irregularities and, due to their strong lateral interactions, the mesogenic groups form a planarizing 'platform', so that the surface sensed by the liquid crystal overlayer is purely that of the octyloxy chain for the entire phenyl benzoate series (n = 3, 6, 10).

SUMMARY

A series of surface-active trichlorosilane compounds based on an octyloxy phenyl benzoate mesogen (Fig. 2) was successfully synthesized. The tethering lengths between the mesogenic and the silane moiety involved n = 3, 6, or 10 methylene groups for the phenyl benzoate. Self-assembled monolayers of these compounds were constructed on silicon wafers, glass, and SiO$_2$ coated ITO glass using standard methods [4, 5]. Water contact angles, ellipsometry and atomic force microscopy revealed a well organized complete monolayer for the phenyl benzoate series. The values of the water contact angles were close to those observed for polyethylene and vinyl-terminated monolayers (99-102°, Table 1) [4, 5]. This indicates that the surface is not composed of principally the terminal methyl groups but also has a strong methylene component. This is consistent with the loose packing of the octyloxy phenyl benzoate groups which have significantly larger cross-sectional areas. The contact angle measurements and the nematic liquid crystal alignment characteristics disclosed that the packing of the monolayers is governed by the lateral interactions between the mesogenic units and not by the steric hindrance between the tethers.

ACKNOWLEDGEMENTS

This work was supported by the National Science Foundation's Materials Research Science and Engineering Center on Polymer Interfaces and Macromolecular Assemblies (CPIMA) under grant number DMR94-00354.

REFERENCES

1. A. Ulman, An Introduction to Ultrathin Organic Films: from Langmuir-Blodgett to Self-Assembly (Academic Press, Inc., San Diego, 1991).
2. C. D. Bain, Ph.D. Dissertation, Harvard University, 1989; C. D. Bain and G. M. Whitesides, J. Am. Chem. Soc. 110, 3665 (1988); J. P. Folkers, P. E. Laibinis, and G. M. Whitesides, J. Adhesion Sci. Technol. 6(12), 13997 (1992), and references therein.
3. J. Sagiv, J. Am. Chem. Soc. 102, 98 (1980); R. Moaz and J. Sagiv, J. Colloid Interface Sci. 100, 465 (1984); J. Gun, R. Iscovici, J. Colloid Interface Sci. 101, 201 (1984).
4. S. R. Wasserman, Ph. D. Dissertation, Harvard University, 1988.
5. L. M. Lander, Ph. D. Dissertation, University of Akron, 1994.
6. Y. Shnidman, A. Ulman, and J. E. Eilers, Langmuir 6 1071 (1993); S. -C. Chang, I. Chao, and Y. -T. Tao J. Am. Chem. Soc. 116, 6792 (1994); S. D. Evans, E. Urankar, A. Ulman, and N. Ferris, J. Am. Chem. Soc. 113, 4121 (1991); E. Sabatani, J. Cohen-Boulakia, M. Bruening, and I. Rubinstein, Langmuir 9, 211 (1993), and references therein.
7. J. Sagiv, Isr. J. Chem. 18, 346 (1979).
8. N. Tillman, A. Ulman, J. S. Schildkraut, T. L. Penner, J. Am. Chem. Soc. 110, 6136 (1988).
9. N. Tillman, A. Ulman, and J. F. Elman, Langmuir 6, 1512 (1990).
10. T. Seki, et al., Langmuir 9, 211 (1993), and references therein.
11. L. M. Siewierski, Ph. D. Dissertation, University of Akron, 1995.
12. F. J. Kahn, G. N. Taylor, and H. Schonhorn, IEEE Proceedings 61, 823 (1973); F. J. Kahn, Appl. Phys. Lett., 22, 386 (1973).
13. C. S. Mullin, P. Guyot-Sionnest, and Y. R. Shen, Phys. Rev. A 39, 3745 (1989); K. Ogawa, et al., Langmuir 7, 1473 (1991).
14. J. Y. Yang, K. Mathauer, and C. W. Frank, Microchemistry, 441 (1994).
15. Y. Hayashi and K. Schimura, Langmuir 12, 831 (1996).
16. T. J. Scheffer and J. Nehring, J. Appl. Phys. 48(5), 1783- 92 (1977)

SUSPENDED PARTICLE DISPLAY USING NOVEL COMPLEXES

H. TAKEUCHI*, A. USUKI*, N. TATSUDA*, H. TANAKA*, A. OKADA*, K. TOJIMA**
*TOYOTA CENTRAL R&D LABS., INC., Nagakute, Aichi, 480-11, Japan,
e0910@mosk.tytlabs.co.jp
**TOYOTA MOTOR CORPORATION, Toyota, Aichi, 471, Japan

ABSTRACT

Suspended particle display (SPD) can be controlled in transparency by applying a.c. voltage. We found new synthesized particles for SPD and methods for making them fine. Three new SPD particles showing different colors have been found: δ-herapathite, 1,10-phenanthroline-iodine complex and pyrazinophenanthroline-iodine complex as showing red, blue, purple, respectively. Recrystallization and ultrasonic treatment were found to make them fine crystals. Accordingly, transparency, stability for sedimentation and thermal stability were improved.

INTRODUCTION

Suspended particle display (SPD) consists of polarizable particles and dispersion medium between two glass plates with an electrode.(figure 1) SPD can be changed its transparency by applying an a.c. voltage. When an a.c. voltage is applied, polarized particles are oriented in the direction of electric field. Consequently light can pass through the SPD cell and the SPD turns into the colorless stage. When the a.c. voltage is turned off, particles are disoriented by Brownian movement. Then light is absorbed by particles, and the SPD returns to the colored stage (Figure 1). The advantages of SPD over many light controlling materials such as liquid crystals and photochromic materials are very low cost for preparation and its constituents and mechanism of transparency change. The suspension of SPD consists only of polarizable particles contained about 1wt% in the suspension and dispersion medium such as liquid paraffin or aromatic compounds. In addition, SPD needs no other attachments such as light polarized films, electrolytes and insulating layers.

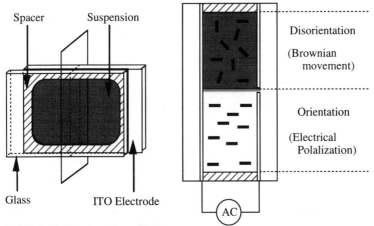

Figure 1   SPD Cell and Sectional Plane of Cell
SPD consists of polarizable particles and dispersion medium between two glass plates with an electrode. SPD can be changed its transparency by applying an a.c. voltage. When an a.c. voltage is applied, polarized particles are oriented in the direction of electric field. Consequently light can pass through the SPD cell and the SPD turns into the colorless stage. When the a.c. voltage is turned off, particles are disoriented by Brownian movement. Then light is absorbed by particles, and the SPD returns to the colored stage

Mat. Res. Soc. Symp. Proc. Vol. 424 © 1997 Materials Research Society

This paper presents new particles for SPD in focusing on iodine complex crystals like a herapathite, and discuss them in terms of their improvements in transparency, stability against sedimentation and color variation.

EXPERIMENT

Materials

Quinine, 1,10-phenanthroline, 5-methyl-1,10-phenanthroline, nitrocelullose and ditridecyl phthalate are commercially available. 5-Methoxy-1,10-phenanthroline[1], 5-amino-1,10-phenanthroline[2], pyrazino[2,3-f]phenanthroline[3] were synthesized according to the procedure published in literature.

Synthesis of α, δ and γ-Herapathite

To a mixture of ethanol (30.0g), water (30.0g), acetic acid (15.0g) and conc. sulfuric acid (0.57, 5.8mmol) was added Quinine (2.5g, 7.7mmol) and dissolved. To the solution was added a solution of iodine (0.98g, 3.9mmol) and potassium iodide (0.64g, 3.9mmol) in ethanol (5.0g) and water (20.0g) at once. After stirring for 1h and filtration gave herapathite (4.2g). Recrystallization from ethanol/water (7/3 w/w) gave α-herapathite (3.36g, 74.1%) as a greenish plate ; EA, C:40.9, H:4.4, N:4.7, O:13.2, S:4.2. C:40.8, H:4.5, N:4.8, O:13.6, S:4.1 for $C_{80}H_{106}N_8O_{20}I_6S_3$ (Quinine/HI$_3$/H$_2$SO$_4$: 4/2/3).

Recrystallization from ethanol/water (1/9 w/w) gave as a dark brown needle;EA, C:40.3, H:4.5, N:4.8, O:11.0, S:3.2. C:40.1, H:4.3, N:4.7, O:11.5, S:3.1 for $C_{140}H_{180}N_{14}O_{30}I_{12}S_4$ (Quinine/HI$_3$/H$_2$SO$_4$: 7/4/4).

Recrystallization from ethanol/water (7/3 w/w) gave δ-herapathite as a red needle; EA, C:49.0, H:5.3, N:5.4, O:11.3, S:2.7. C:48.6, H:5.2, N:5.7, O:12.0, S:2.8 for $C_{140}H_{178}N_{14}O_{26}I_{12}S_3$ (Quinine/HI$_3$/H$_2$SO$_4$: 7/4/3)

Synthesis of δ-Herapathite suspension

To ethanol (100g) was added α-herapathite (2g) and treated in a ultrasonic bath for 20 min, and resulting δ-herapathite was collected by a centrifuge (8000rpm, 30min). To ditridecyl phthalate (100g) was added δ-Herapathite (1g) and nitrocelullose (1g), and treated in a ultrasonic bath for 10h. Excess nitrocelullose in the suspension was separated by the centrifuge, dried in vacuo to afford a wine-color δ-herapathite suspension.

Synthesis of 1,10-phenanthroline-iodine complex

To a solution of 1,10-phenanthroline (0.90g, 5mmol) in water (20g) and sulfuric acid (0.124g, 1.26mmol) was added a solution of iodine (0.64g, 2.5mmol) and potassium iodide (0.42g, 2.5mmol) in ethanol (3.2g) and water (12.6g) at once. After stirring for 1h and filtration gave 1,10-phenanthroline-iodine complex (1.75g, 93.1%) as a dark green needle; mp 183.1°C; EA, C:41.0, H:2.7, N:7.4. C:40.5, H:2.8, N:7.3. for $C_{26}H_{21}N_4I_3$ (1,10-phenanthroline/HI$_3$: 2/1)

Synthesis of 5-Methyl-1,10-phenanthroline-iodine complex

To a solution of 5-methyl-1,10-phenanthroline (0.97g, 5mmol) in water (20g) and sulfuric acid (0.124g, 1.26mmol) was added a solution of iodine (0.64g, 2.5mmol) and potassium iodide (0.42g, 2.5mmol) in ethanol (3.2g) and water (12.6g) at once. After stirring for 1h and filtration gave 5-methyl-1,10-phenanthroline-iodine complex (1.78g,92.4%) as a dark green needle; mp 223.2°C ; EA, C:41.0, H:2.7, N:7.4. C:40.5, H:2.8, N:7.3 for $C_{26}H_{21}N_4I_3$ (5-methyl-1,10-phenanthroline/HI$_3$: 2/1)

Synthesis of 5-Methoxy-1,10-phenanthroline-iodine complex

To a solution of 5-methoxy-1,10-phenanthroline (1.05g, 5mmol) in water (20g) and sulfuric acid (0.124g, 1.26mmol) was added a solution of iodine (0.64g, 2.5mmol) and potassium iodide (0.42g, 2.5mmol) in ethanol (3.2g) and water (12.6g) at once. After stirring for 1h and filtration gave 5-methoxy1,10-phenanthroline-iodine complex (1.75g, 93.6%) as a dark green needle; mp 229.5°C ; EA, C:41.4, H:2.7, N:7.5, O:4.3. C:41.8, H:2.8, N:7.5, O:4.3 for $C_{91}H_{73}N_{14}O_7I_9$(5-methoxy-1,10-phenanthroline/HI$_3$: 7/3)

Synthesis of 5-Amino-1,10-phenanthroline-iodine complex

To a solution of 5-amino-1,10-phenanthroline (0.98g, 5mmol) in water (20g) and sulfuric acid (0.124g, 1.26mmol) was added a solution of iodine (0.64g, 2.5mmol) and potassium iodide (0.42g, 2.5mmol) in ethanol (3.2g) and water (12.6g) at once. After stirring for 1h and filtration gave 5-amino-1,10-phenanthroline-iodine complex (1.68g, 87.0) as a dark blue needle; mp 214.2°C ; EA, C:37.1, H:2.3, N:10.6. C:37.3, H:2.5, N:10.9 for $C_{24}H_{19}N_6I_3$ (5-amino-1,10-phenanthroline/HI$_3$: 2/1)

Synthesis of Pyrazino[2,3-f]phenanthroline-iodine complex

To a solution of pyrazino[2,3-f]phenanthroline (0.5g, 2.2mmol) in ethanol (14.0g) and water (5.0g) and sulfuric acid (0.05g, 0.5mmol) was added a solution of iodine (0.24g, 0.9mmol) and potassium iodide (0.15g, 0.9mmol) in ethanol (1.2g) and water (4.7g) at once. After stirring for 1h and filtration gave pyrazino[2,3-f]phenanthroline-iodine complex (0.72g) as a dark green needle; mp 288.3°C ; EA, C:43.3, H:2.1, N:14.5. C:43.7, H:2.2, N:14.6 for $C_{70}H_{42}N_{20}I_6$ (pyrazino[2,3-f]phenanthroline/HI$_3$: 5/2)

Synthesis of Phenanthroline derivatives-iodine complex suspension

To a solution of phenanthroline derivatives-iodine complex (0.1g) in ethanol (10g) and acetone (7g) were added nitrocelullose (0.1g) and ditridecyl phthalate (10g), and stirred for 10min. After distillation of ethanol and acetone, the resulting slurry was treated in a ultrasonic bath for 10h. Excess nitrocelullose in the suspension was separated by a centrifuge, dried in vacuo to afford a blue-color phenanthroline derivatives-iodine complex suspension.

Synthesis of Pyrazino[2,3-f]phenanthroline-iodine complex suspension

To a solution of pyrazino[2,3-f]phenanthroline-iodine complex (0.1g) in ethanol (84g) and acetone (60g) was added nitrocelullose (0.1g) and ditridecyl phthalate (10g), and stirred for 10min. After distillation of ethanol and acetone, the resulting slurry was treated in a ultrasonic bath for 10h. The suspension was separated an excess of nitrocelullose by centrifuge, dried in vacuo to afford pyrazino[2,3-f]phenanthroline-iodine complex suspension as a purple-colored suspension.

Raman Diffraction Measurement

To determine the iodine species of complexes, a laser Raman diffraction were measured for the above complexes. At 112 cm^{-1} the diffraction peak corresponding to I_3^- appeared, and the peak corresponding to I_5^- was not observed at 164cm^{-1}.

RESULTS

One of the most popular polarizers is herapathite, which consists of quinine, sulfuric acid, pentaiodic acid and water. It has used as a particle of SPD. However, the suspension of herapathite still has rooms for improvement of transparency, sedimentation and color variation.

Fine Crystallization

Herapathite as the particle of SPD must be fine crystals because fine crystals increase transparency of suspension. In particular, if the crystal size is smaller than the wavelength of visible light, scattering of light decreases and suspension becomes clear. Furthermore, it could be proof against sedimentation of particles.

In order to obtain fine crystals of herapathite, we examined two methods. First, mechanical grinding with glass beads was performed. Herapathite crystals became smaller bits to all appearances and did not show the change of quality. But the suspension of the crystals was not responded by a.c. voltage. This indicates herapathite has decomposed on mechanical grinding.

Second, we examined recrystallization of herapathite was performed. Water and ethanol were selected as recrystallizing solvents because they can dissolve herapathite crystals. As a result, we found that 4 types of herapathite could be yielded depending on the ratio of water and ethanol. They are named α- to δ-form herapathite for the sake of convenience. α-Form herapathite was greenish plane crystal obtained by recrystallization in the mixture of water/ethanol (3/7), and γ-form herapathite was dark brown needle obtained from the mixture or water/ethanol (9/1). On the other hand, δ-form herapathite of red fine needle was yielded from ethanol. β-Form herapathite was found by X-ray diffraction but it could not be isolated because it mingled with α- or γ-form herapathite. After heating these mixtures, the X-ray peaks corresponding to β-form herapathite crystal disappeared. Consequently we guess β-form herapathite has decomposed. α-, γ- and δ-herapathite could be converted to each other by recrystallization. Accordig to microscopy observation, α-form crystals were about 100 micron long and γ-form crystals were about several tens of micron long. They are larger than wavelength of visible light and therefore difficult to be kept dispersing against sedimentation. On the other hand, the average length of δ-form crystals was about 0.3 micron and the wide is about 0.02 micron, which were smaller than those of the wavelength of visible light (Figure 2).

Furthermore, ultrasonic treatment of α-herapathite was found to afford δ-herapathite and the suspension effectively. The suspension absorbed light of about 500 nm wavelength strongly (Figure 3). When 100V a.c.voltage is applied, the transparency increased and the contrast became quite striking. Next, the suspention was tested on sedimentation by keeping it in a centrifugal condition in 18000 revolutions per minute for 20 hours. After the test, no sediment was found. The condition of this centrifuge was equivalent to gravity from 2500 to 3000g.

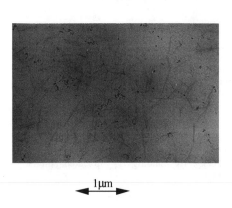

1μm

Figure 2 TEM of δ-herapathite crystals.
The average length was ca. 0.3 μm and the average width was ca. 0.02μm.

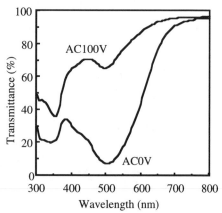

Figure 3 Transmittance chart of 1.0wt% δ-herapathite suspension filled in a cell (100μm gap). Since it absorbed light of about 500nm, the suspension presented wine colord. On applying 100V a.c. voltage, transmittance increased.

Consequently, from a simple calculation, δ-herapathite suspension is proof against sedimentation for about 6 years.

Polyfused Aromatic Compounds-Iodine Complex

Herapathite is known as a charge-transfer complex of iodine and quinine especially quinoline nuclei. If quinine is replaced by polyaromatic compounds, the color of crystal will be varied. Since longer wavelength light could be absorbed, conjugated systems of polyaromatic compounds are longer than quinine and therefore their charge-transfer are stabilized. Among many polyaromatic compounds, 1,10-phenanthroline was found to satisfy these conditions and yield blue complex with iodine.

To synthesize the suspension of 1,10-phenanthroline-iodine complex, the complex crystals must be as fine as those of herapathite. According as herapathite, recrystallization was applied to 1,10-phenanthroline-iodine complex crystals and a mixture of ethanol and acetone was found to be an effective solvent for fine crystallization.

Next, the effect of substituents of phenanthroline was investigated. Since phenanthroline functions as a donor in the complex, it is expected that electron donating substituents stabilize phenanthroline and further complex crystal. Phenanthroline derivatives were substituted at 5 positions for electron donating groups such as methyl, methoxy and amino group. Iodine complexes were then synthesized from these derivatives in the same manner. Table I were lists melting points of the derivatives. The electron donating substituent appears to be effective for stabilizing complexes.

5-methoxy-1,10-phenanthroline-iodine complex crystal suspension absorbed light of about 600 nm wavelength strongly. When 100V a.c.voltage was applied, the transparency increased and the contrast became quite striking (Figure 4). These phenanthroline-iodine complexes are novel as the particle of SPD.

Furthermore, polyfused phenanthroline derivatives were investigated because their complexes are expected to have higher thermal stability. We found pyrazino[2,3-f]phenanthroline and dipyrido[3,2-a:2',3'-c]phenazine afforded iodine complexes and purple suspension. Pyrazinophenanthroline-iodine complex crystal suspension absorbed light from 550 to 650 nm wavelength strongly. When 100 a.c.voltage is applied, the transparency increased and and the contrast became quite striking (Figure 5).

As compared with phenanthroline-iodine complex, melting points of pyrazinophenanthroline and dipyridophenandine complexes were obviously higher. This indicates that increase of fused aromatic ring is more effective for improving thermal stability.

Thus, three SPDs showing different colors, which are red, blue and purple, have been able to be prepared.

Table I Melting point of phenanthroline derivative-iodine complexes. The electron-donating substituent stabilized the complex. The complex was stabilized also by fusing aromatic ring to phenanthroline.

Phenanthroline Derivatices	mp(°C) of Iodine complex
Phenanthroline	183.1
5-Methylphenanthroline	223.2
5-Methoxylphenanthroline	229.5
5-Aminophenanthroline	214.2
Pyrazinophenanthroline	288.3
Dipyridophenazine	296.6

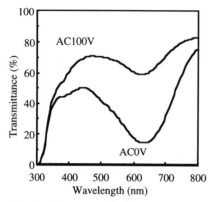

Figure 4 Transmittance chart of 1.0wt% 5-methoxy-1,10-phenanthroline-iodine complex suspension filled in a cell (100μm gap). Scince it absorbed light of about 600nm, the suspension presented wine color. On applying 100V a.c. voltage, the transparency increased.

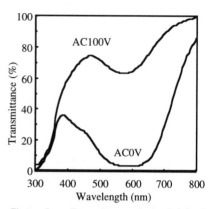

Figure 5 Transmittance chart of 1.0wt% pyrazinophenanthroline-iodine complex suspension filled in a cell (100μm gap). Scince it absorbed light from 550 to 650nm, the suspension presented purple color. On applying 100V a.c. voltage, the transparency increased.

CONCLUSIONS

SPD has advantage over many light controlling materials of low preparing cost. We have investigated improvement in transparency, stability against sedimentation and color variation of SPD. The following conclusions have been obtained.

First, preparing fine particles was found effective for improvement on transparency, stability against sedimentation. A variety of complexes were obtained during the recrystallization of herapathite, which was popular as the particle of SPD, and δ-herapathite was found by recrystallization from ethanol. The average length of δ-herapathite crystals was about 0.3 micron and the width was about 0.02 micron, which was smaller than the wavelength of visible light. δ-Herapathite afforded clear red suspension stable for sedimentation.

Second, blue and purple SPDs were obtained by using iodine complexes of polyfused aromatic compounds such as phenanthroline, dipyridophenanthroline and pyrazinophenazine. They were improved also in thermal stability.

REFERENCES

1. J. Druey and P. Schmidt, Helvetica Chimica Acta 33, 4, p1,080 (1950).
2. R. D. Gilard and R. E. E. Hill, J. Chem. Soc., Dalton p1217 (1974).
3. J. E. Dickeson and L. A. Summers, Aust. J. Chem., 23, p1023 (1970).

Cr/Cr COMPOUND BILAYERED THIN FILM BLACK
MATRIX FOR LCD COLOR FILTER

Byung Hak Lee, Iee Gon Kim, and Si-Hyun Lee
Display R & D Center, Samsung Display Devices Co., Ltd.
575 Shin-Dong, Paldal-Ku, Suwon, Kyungki-Do, Korea

ABSTRACT

The effect of process parameters on the optical property of the chromium/chromium compound multilayered thin film black matrix for liquid crystal display color filter has been investigated. A chromium compound layer was deposited reactively by DC magnetron sputtering, using gas mixture of Ar, CO_2, and N_2 on soda-lime glass substrate. Metallic chromium layer deposition was followed to enhance the light obstruction of the black matrix. X-Ray diffraction and transmission electron microscopy study showed that CrN is the major phase with unidentifiable oxides and carbides present as minor phases, in chromium compound layer, with nano-crystalline structure. Metallic chromium layer showed columnar growth structure. The ratio of CO_2 in the reactive sputtering gas greatly affected the optical property of the multilayer, despite that N_2 is the gas which contributes to the formation of the major phase. Delicate change of optical property is due to compositional variation of the compound layer, caused by reaction due to environmental changes. When processed with 10.28% of CO_2 and 4.88% of N_2 in 3.3 mTorr pressure, film with optical reflectivity of 3.31% at 655 nm of photon wavelength was obtained.

INTRODUCTION

Cr thin film, deposited on glass substrate, has been widely used for the black matrix of liquid crystal display color filters, to enhance the contrast of the display by blocking the stray light. Its high inherent reflectivity, however, hinders the effectiveness of the thin film as a black matrix, especially since the trend in flat panel display is toward a higher resolution. CrO_x/Cr multilayered thin film has been used as a higher quality black matrix for high resolution display, benefiting from lower reflectivity of the chromium oxide while maintaining the superior light barrier function of the chromium layer. Deposition of these films are normally performed by a sputtering process due to its process stability and relatively high deposition rate. Little has been published about the optical and microstructural characteristics about chromium oxide thin film. Transmission in the UV region of thermally oxidized Cr film has been reported[1]. Deposition of chromium oxide can be done either by sputtering of the oxide target or reactive sputtering of metallic chromium target. Sputtering of oxide target yields a low deposition rate, so reactive sputtering is often preferred. The reactive sputtering process, however, of elemental target with oxygen impose a health problem. During the process of reaction, a compound containing Cr^{6+} forms and it is known to be hazardous to human health. Although the inside of the chamber is in vacuum during deposition, a minute amount of hazardous material adsorbed on the surface of the chamber may present a serious problem. In this paper, we propose a new reactive sputtering process, for manufacturing a multilayered chromium/chromium compound black matrix which eliminates the health problem, and report the optical characteristics and microstructural analyses provided by the new process.

EXPERIMENTS

DC magnetron sputtering was performed for thin film layer deposition. The sputtering conditions appear in Table 1. For the formation of the chromium compound layer, a mixture of CO_2 and N_2 gas was used for reactive sputtering. No trace of Cr^{6+} was found after opening the chamber. The chromium compound layer was formed onto a 370x400 mm^2 size glass substrate, followed by a metallic chromium layer. A Tencor Alphastep profile gauge was used to measure film thickness. X-Ray Diffraction(XRD) analyses were conducted with a Phillips PW1710 diffractometer. X-Ray Photo-electron Spectro-scopy(XPS) was performed for compositional analysis. A JEOL Transmission Electron Microscope(TEM) was used for microstructural observation. An X-Rite 310TR Photo- densitometer was used for the measurement of optical density which is defined as

$$D = \log(Io/I)$$

where D is the optical density, Io is the intensity of incident light, and I is the intensity of transmitted light.   Optical density is the measure of the opaqueness of the object in interest. A photospectrometer was used for reflection and transmission spectrum analysis.

Table 1 Sputtering process parameters for chromium/chromium compound bilayer thin film black matrix

	Chromium Layer	Chromium Compound Layer
Sputtering Power Density	DC 3.70 W/cm^2	DC 3.37 W/cm^2
Base Pressure	10^{-6} Torr	10^{-6} Torr
Sputtering Pressure	8.8 mTorr	3.8 ~ 4.2 mTorr
Deposition Rate	110 Å/min.	50 Å/min.
Pre-Sputtering Time	3 min.	3 min.

RESULTS

Fig.1 shows the XRD pattern to check the composition of the chromium compound single layer. Identifiable 3 peaks are from CrN, overlapped on the hump pattern originated from the

324

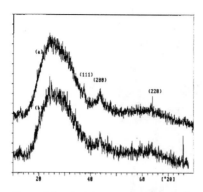

Fig.1   XRD patterns of the chromium compound single layer thin film   (a)
sputtering pressure of 4.2 mTorr, CO_2 13.1%, N_2, 4.7%,   (b) sputtering pressure of
3.9 mTorr, CO_2 10.3%, N_2,4.9%. Both patterns show 3 weak CrN peaks
superimposed on the hump shaped background signal from amorphous substrate.

amorphous glass substrate. From these XRD patterns, CrN is the major phase of the chromium
compound layer despite that a significant amount of CO_2 was introduced into the chamber for
reactive sputtering. Pattern 1(b) is the XRD pattern for a slightly different process parameter -
the amount of sputtering gas- from that of 1(a). Peaks of the CrN can be observed at the same
location, but the intensity became weak, suggesting a change in chemical composition. Fig.2 is
the TEM micrograph and Selected Area Diffraction Pattern(SADP) of the sample identical to
the one of Fig.1(a). From this TEM micrograph and SADP, the chromium compound has a very
fine polycrystalline phase with an average grain size of 20 nm, with well defined grain
boundaries. 3 evident rings from CrN can be seen in the SADP, matching results with XRD
analysis. Different from the XRD results, however, unidentifiable multiple rings can be seen,
meaning the chromium compound layer is not composed of a CrN single phase, but a mixture
of multiple phase with CrN as major phase. For further study, XPS analyses were performed.
Fig. 3 is the XPS pattern of the sample of Fig.1(a). Fig.3(a) is the pattern without any etching of
the surface, while Fig.3(b) is the pattern with 20 min. etching with an Ar ion beam. Peaks of
oxygen and carbon can be observed from both patterns (a) and (b),backing up the fact of the
existence of oxide, carbide, or oxy-carbide of chromium. Since the major phase of the
chromium compound is CrN, we call the compound layer "CrN" hereinafter. Optical
transmission spectrum analyses showed that the sample of Fig.1(a) has minimum transmission
of 6.3 %, at 425 nm wavelength region. With a small change of process parameter(sample (b) in
Fig.1), wavelength region and value for minimum transmission didn't vary much. Fig.4 is the
cross-sectional TEM micrograph and SADP of the Cr/"CrN" bilayer. Due to small fractional
volume of the "CrN" layer, SADP almost came from Cr layer. Nevertheless, SADP shows
polycrystalline rings which do not belong to Cr or CrN, showing matching results with CrN

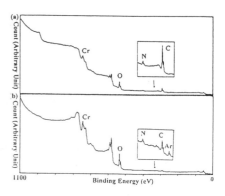

Fig.2   Plane view TEM micrograph and
SADP of chromium compound single layer
thin film.

Fig.3   XPS spectra of chromium compound
layer (a) as-deposited sample, (b) after 20 min.
etching with Ar ion beam. Insets are the
enlargement of the acorresponding area.

single layer analysis. Classical columnar growth microstructure of Cr layer can be clearly seen in Fig.4.

In order to achieve optimal property, i.e. minimum reflectivity with high optical density, sputtering conditions were varied for deposition. Since the optical property of the Cr layer other than the optical density(transmission) which is mainly determined by thickness, is not affected much by sputtering condition, emphasis was placed upon the "CrN" layer deposition. Sputtering gas composition variation, which is expected to make chemical composition variation in the film, more precisely, the change of ratio of constituent phases, was made. The variation was performed by controlling the flow rate of each gas. As mentioned earlier and contrary to normal procedure using Ar and O_2, resulting CrO_x layer, mixture of Ar, CO_2, and N_2 was used for reactive sputtering. The optical property of CrO_x is quite similar to that of CrN in terms of reflectivity and transmission. Fig.5 shows the optical reflectivity and density change of the Cr/"CrN" bilayer as a function of the content of CO_2 in the sputtering gas. Each bilayer consists of 1100Å of Cr and 550Å of "CrN" layers, as mentioned earlier. Optical density was more than 4 for all of the samples. Variation of optical density over 4 does not have much significance since it is less than 0.1 % of the optical transmission. From Fig.5, the minimum reflectivity of 3.3 % at 655 nm photon wavelength can be obtained with 10.3 % of CO_2 content in the sputtering gas. The reflectivity value includes the reflectivity of the glass substrate itself. Considering that the normal reflectivity of the glass is ~4 %, this reflectivity is a very low value. Even a true black substrate with a glossy surface has ~4 % reflectivity[2]. The fact that minimum reflectivity occurs at 655 nm means the film has a "bluish" appearance. The appearance (color) can be adjusted by controlling the thickness of the "CrN" layer. By interfering the light inside the film, one can eliminate the undesired light portion, hence

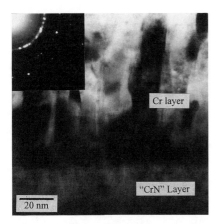

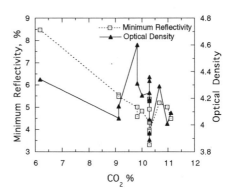

Fig.4   Cross sectional TEM micrograph and SADP of Cr/ "CrN" bilayer thin film.

Fig.5   Optical property change of Cr "CrN" bilayer due to CO_2 concentration change in the sputtering gas.

achieving true black. The change of wavelength with minimum reflectivity also appeared in Fig.5. This change is thought to be due to the refractive index change of the "CrN" film. Reflectivity and corresponding wavelength change by CO_2 content in sputtering gas is considered to be due to the compositional change of the film. The ratio change of the constituent phases determines the change in optical property of the "CrN" film. It was found that the optical property was very difficult to control with the content of N_2 in the sputtering gas. It is very interesting to note that even though CrN is the major phase of the "CrN" layer, optical property of the film can be controlled by the content of CO_2 gas which determines the minor phases. The amount of the minor phases in the film plays an important role in determining the optical property of Cr/"CrN" bilayered film.

CONCLUSIONS

Fabrication and optical characteristics of a high quality black matrix for LCD color filter by reactive DC magnetron sputtering process has been presented. Reactive sputter deposited thin film on glass substrate with CO_2 and N_2 has mixed phase nano-scale crystalline structure, with major phase of CrN and other oxide, carbide, or oxy-carbide as minor phases. Metallic chromium layer, which gives a light obstruction effect, is columnar growth nano-scale polycrystalline structure. Optical property of the Cr/Cr compound bilayer is greatly affected by the composition of sputtering gas. 3.31 % reflectivity at 655 nm photo wavelength could be obtained with sputtering gas of 10.28 % of CO_2, 4.88 % of N_2 and the rest of Ar. Constitutional ratio of phases present in the film affects the optical property, which can be controlled more effectively by the content of CO_2, rather than N_2 in the sputtering gas.

ACKNOWLEDGMENTS

Authors are grateful to Dr. G. Choe of Chonnam National University of Korea(now at Materials Research Corporation) for the help in TEM work and helpful discussion. Authors are also thankful to Mr. H. K. Park of anayltical lab of SDD for XPS analyses of the samples, and to S. K. Kang and S. S. Lee of SDD for their support during the course of this study.

REFERENCES

1. D. S. O'Grady and P. B. Wilber, SPIE Proc. **2197,** 194 (1994)
2. F. W. Billmeyer,Jr. and Max Saltzman, Principles of Color Technology (John Wesley & Sons, New York, NY 1981), p.126.

DEVELOPMENT OF ORGANIC PHOTORESIST
FOR COLOR FILTER BLACK MATRIX

C. W. Kim, S. G. Rho, D. K. Jung, E. J. Jeong, S. Y. Kim, H. G. Yang
Samsung Electronics Co., Ltd., Kiheung, Korea

ABSTRACT

Normal CrOx/Cr type color filter black matrix requires multi-layer sputtering and photo/etch processes which cause several problems such as particles generation, high production cost, and environmental issues. Various kinds of materials are being reviewed to substitute the existing Cr base materials. Especially carbon based organic type and pigment type photo-sensitive materials are actively under development. Our researches are focused on the characterization of material properties, color filter manufacturing techniques which are optimized for liquid crystal process, and the display quality of newly developed materials. Carbon based organic photoresist has excellent optical, material, and process properties which can be used for TFT-LCD color filter products. 14.2" VGA panel is fabricated by using this organic black matrix technique. It is expected that this technique will lead the color filter production market.

INTRODUCTION

In LCD (Liquid Crystal Display) production, low cost and high production yield become the essential factors as the pressure for the reduction of production cost is highly requested. Among these parts for LCD production, cost reduction of color filter is requested since it is contributed to the significant portion of the total production cost. In addition to production cost, high display quality is also highly recommended. When the reflected beam of external light at the surface of TFT-LCD panel is mixed with the transmitted display beam, the display quality which is closely related to the contrast ratio is degraded. Minimization of the surface reflection is required. Normal Cr black matrix (BM) can not meet these requests and new types of color filters are actively under developments.

Currently, CrOx/Cr double layered BM are widely used to reduce the surface reflectance of the external light. Although CrOx/Cr BM [1, 2, 3] meets the low surface reflectance requirement, it has several problems like particles generation, high production cost, and environmental issues. To evaluate various kinds of BM, process conditions and the optical properties of CrOx/Cr structure are discussed. In addition to Cr or CrOx/Cr structure, graphite [4] BM which uses lift off method, carbon based organic BM, and pigment type structures are being reviewed [5, 6]. Carbon type organic BM uses the dispersion of carbon in organic materials whereas the pigment type uses the mixture of red, green, and blue pigments. The reaction mechanism of carbon based organic photoresist is the same as pigment type R, G, B, and BM photoresists. Low cost, high quality, and simple BM process is developed by using the carbon based organic material.

EXPERIMENTAL

CrOx/Cr Type Black Matrix

Strong reflection from normal Cr type color filter comes from the high reflectance of Cr film. For the reduction of surface reflection, CrOx/Cr double layer is widely used. With this structure, the incident beam is destructively interfered by the reflected beam. This destructive interference as well as the high absorption of CrOx layer can effectively reduce the surface reflection. CrOx/Cr double layered BM deposition process is developed by using reactive sputtering method. Flow rate of reactive O2 gas in the sputtering chamber and deposition power are correlated and optimized to the transmittance and refractive index of the double layered films. CrOx/Cr double layered BM with the absolute reflectance of less than 7% is developed which is almost 1/9 of the normal Cr BM. Due to the double layer structure, BM thickness is increased by 500~600A. This does not require any extra adjustments in liquid crystal process.

Materials for Organic Black Matrix

In color mosaic processes, pigment dispersed photoresists are widely used. Two types of photo-sensitive mechanisms are known about the process of pigment dispersed photoresist. First one is radical polymerization type (RPT) which uses acryl or epoxyacryl as a binder. Components of the RPT are acryl copolymer, initiators, functional monomer, pigments and solvents. Second one is cross-linked type (CLT) which uses poly vinyl alcohol (PVA) as binder. Components of the CLT are photo-sensitized copolymer, pigments, and solvents. CLT contains PVA which is thermally unstable compared to acryl type. Also the photo sensitivity of CLT is lower than RPT. RPT is thermally stable and has very good photo sensitivity which comes from the chain reaction. So RPT materials are selected for BM process. These have the same mechanism as the pigment type R, G, B color photoresists which are currently used for our color mosaics [7].

Initiator (Ditrihalomethyl-S-triazine)	Multi-functional acrylic monomer	Binder (copolymer of benzylmethacrylate & methacrylic acid)

Table 1. Typical components of organic photoresist.

Structures of initiator , multi-functional acrylic monomer, and binder are shown in table 1. Radical polymerization mechanism can be explained as follows. When RPT material is exposed by light, ditrihalomethyl-S-triazine initiator is decomposed into radicals and polymerized by multi-functional acrylic monomer which produces various kinds of polymers. Radical reaction is terminated by the termination of light exposure. Produced polymers by radical reaction are not dissolved by developer solution. Binder which is so called resin is the copolymer of benzylmethacrylate and methacrylic acid. This binder has several functions. It makes the liquid phase monomer as film at room temperature, and it is developed by organic or non-organic type alkali surfactants. Also it stabilizes the pigment dispersion, too. Binder is closely related to thermal, chemical and photo reliabilities. In table 2, the radical polymerization process mechanism is summarized.

Step	Reaction
Initiation	R-Cl + hυ --> R : Cl -->R* + Cl*
Propagation	R' - H + Cl* --> HCl + R'*
	Monomer + R* (or Cl*, R'*) --> Polymer*
Termination	Polymer* + Polymer* (or R*, Cl*, R'*) --> Polymer

Table 2. The mechanism of radical polymerization.

In Table 3 chemical components of organic carbon and pigment type photoresists are summarized [8].

Components	Chemical Name	
	Organic Carbon Type	Pigment Type
Solvent 1	Propylene Glycol Monoethyl Ether Acetate (PEAC)	Ethyl-3-ethoxypropionate (EEP)
Solvent 2	Propylene Glycol Monomethyl Ether Acetate	Propylene Glycol Monomethyl Ether Acetate
Solvent 3	Cyclohexanone	Cyclohexanone
Resin	Methacrylate Resin	Methacrylate Resin
Monomer	Multi-Functional Acrylic Monomer	Multi-Functional Acrylic Monomer
Pigments	Carbon Black	Pigment Blue / Violet / Red / Yellow

Table 3. Chemical components of carbon and pigment type photoresists.

Material properties of organic BM should have the following properties. First, it should shield the back light beam completely. And the surface reflectance from the outside should be minimized to maintain the contrast ratio under bright environment. Normally, optical density of higher than 3.0 and surface reflectance of less than 5% is required to keep good contrast ratio. Edge profile after development needs to be vertical and the thickness uniformity of the pattern should be within 3%. There should be no residues when the pattern is developed. The surface of the color filter which is closely related to the liquid crystal process should be flat.

	Types of black matrices				
	Metal Type		Graphite	Organic Type	
	Cr	CrOx/Cr	Type	Carbon	Pigment
Materials	Cr	Cr, CrOx	Carbon Solution	Carbon + Resin	Pigments
Typical BM Thickness (μm)	0.15	0.2	0.5	0.8	1.5
Corresponding Optical Density	3.5	3.5	3.0	3.5	2.5
Surface Reflectance (%)	65%	7%	6%	4%	4%
Process Wastes	Cr waste	Cr waste	Water Soluble	Organic	Organic
Relative Cost	4/5	1	1/3	1/2	1/2

Table 4. Comparison of various kinds of black matrices.

Various kinds of BM materials are compared in table 4. Thickness and corresponding optical density, surface reflectance, process wastes, and relative costs are summarized. From the optical point of view, organic carbon type photoresist shows the best characteristics. Although graphite type material is cheapest and has best optical density property, it has several process difficulties. So our research is focused on organic type materials and compared to metal type BMs.

Process Comparison of CrOx/Cr and Organic BM

CrOx/Cr BM Process	Organic BM Process
Initial Cleaning	Initial Cleaning
CrOx / Cr Deposition	PR Coating/Exposure/Develop
Cleaning/PR Coating/Exposure/Develop	R,G,B Patterning and ITO Deposition
CrOx / Cr Etch and PR Strip	
R,G,B Patterning and ITO Deposition	

Table 5. Comparison of CrOx/Cr and organic black matrix processes.

Process steps of CrOx/Cr and organic BM are compared in table 5. As you can see in the table the organic BM process is much simpler than normal Cr or CrOx/Cr BM. Due to the process simplicity, organic BM process has the advantages of low cost and high production yield.

RESULTS

In fig. 1, coating speed dependence of film thickness and the corresponding optical density of organic carbon film are shown. Since the film thickness is inversely proportional to RPM, optical density is decreased as RPM is increased. According to this data, 0.8~0.9 μm thick organic carbon film is necessary to get the optical density of 3.5 which is the typical value of 0.15 μm thick Cr BM.

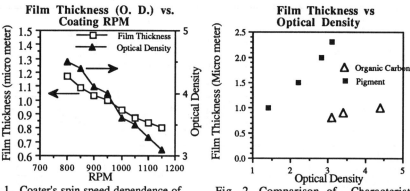

Fig. 1. Coater's spin speed dependence of film thickness and optical density of organic carbon material.

Fig. 2. Comparison of Characteristics Organic Black Matrix Types

Fig. 2 shows the relation between film thickness and optical density for organic carbon and pigment BM. To get the optical density of 3.0, the thickness of the pigment type is almost 3 three times thicker than the organic carbon type. It is quite obvious that the pigment type BM is not a good candidate for the color filter which require critical cell gap control. But in some cases like black matrix on TFT structure, the pigment type is used because it has much higher resistivity than organic carbon type.

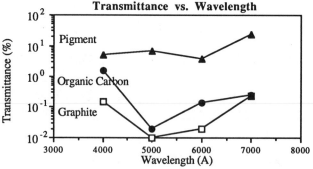

Fig.3. Transmittance vs. wavelength for various materials. (Film thickness is fixed to 1 μm.)

In fig. 3 wavelength dependence of transmittance for graphite, organic carbon, and pigment type BM are shown. Thicknesses are fixed to 1 μm for all three of the films. Graphite shows the lowest transmittance for all over the visible wavelength area which has the richest carbon content. Pigment type shows the highest transmittance. Organic carbon has similar characteristic as graphite except the short wavelength region.

In table 6, step heights for different BM types are summarized. Thickness of metal CrOx/Cr is the thinnest among them and has the step height of 0.08 μm without using any planarization layer. The step which comes from the overlap of BM and color mosaic does not affect to the liquid crystal process because the height is negligible. But for organic carbon or for pigment type BM, the situation is quite different. Organic carbon has 0.5 μm step which is too high to be used for LCD process. To reduce the step height overcoating layer is used. By using it the step height is reduced to 0.3 μm. But this value is still several times thicker than normal Cr BM. The step height is large enough to affect the liquid crystal processes like alignment layer printing step or the cell gap control step. So it is critical to minimize the step height.

BM Type	CrOx/Cr only	Organic Carbon only	Organic Carbon with Overcoat Layer
Film Thickness	0.2 μm	0.8 μm	0.8 μm
Step Height	0.08 μm	0.5 μm	0.3 μm

Table 6. Comparison of step heights. Step is the difference of heights between color mosaic and overlap area of BM and color mosaic.

There are several ways to reduce this step. First one is the adoption of overcoating layer which can effectively make the color filter surface flat. Transparent and non-conducting materials like acryl or polyimide are widely used for the overcoating layer. But still the step needs to be improved. Recently another approach is under development. It is so called the off-focus method. The basic idea of this method is as follows. Since the exposed beam profile at the edge of the pattern is as high as the profile at the center, the overlapped area is exposed two times (one for color mosaic and the other for BM) and becomes thicker than other places. If we can reduce the exposure at the edge of the pattern, the film thickness can be reduced by the reduction of exposure. To get under exposure at the edge of the pattern, off-focus patterning method is used. By applying the off-focus design rule either to the BM mask or color mosaic mask, it is possible to get almost flat color filter. Exposed beam profile as well as the overlap design rule is the key factors for this planarization process. It is expected that the planarization process is essential for the organic BM which has the disadvantage of film thickness compared to the metal type BM.

	Cr BM on C/F with Overcoating Layer	Organic Carbon BM with Overcoating Layer	Organic Carbon BM
Black	0.2882	0.3662	0.3561
White	55.90	62.58	61.64
Contrast Ratio	193.96	170.89	173.10
Viewing Angle	L/R : 48 ~ 48 U/D: 23 ~ 37	L/R: 48 ~ 48 U/D: 19 ~ 46	L/R: 46 ~ 47 U/D: 19 ~ 46
Crosstalk	6.0	5.76	3.42
Wx,Wy	0.3033, 0.2911	0.3177, 0.3024	0.3176, 0.3025
Rx,Ry	0.6218, 0.3300	0.6239, 0.3303	0.6248, 0.3286
Gx,Gy	0.3095, 0.5735	0.3167, 0.5585	0.3178, 0.5567
Bx,By	0.1468, 0.1262	0.1468, 0.1223	0.1466, 0.1297
Color Reproductivity	60.57	58.49	58.64
Reflectance	1	2/3	2/3

Table 7. Comparison of Cr B/M on C/F & Organic Carbon Type Module Characteristic

In table 7, optical properties of TFT LCD panels which use Cr, organic carbon BM with and without overcoating layer are compared. BM on color filter structure is adopted in this comparison. In this structure, BM is processed after color mosaics are patterned. In that way surface reflectance is reduced by color mosaic which has lower reflectance than Cr film. But still the surface reflectance is higher than organic carbon panels. Contrast ratio of organic BMs are a little bit lower than Cr BM case. This suggests that the cell gap control of organic BM is not perfectly optimized than Cr BM panel.

CONCLUSIONS

The key factor for organic BM material is the patterning properties. In order to enhance the patterning property, the optical density is sacrificed which is a big handicap for liquid crystal process. Material with high optical density as well as with good patterning property and wide process margin is highly requested for black matrix material. By using organic carbon type BM optical density of 3.4 is obtained from the 0.8μm thick film. Since the material itself is intrinsically black the surface reflectance is less than 4% which is better than currently used CrOx/Cr type BM. By using overcoating material or defocusing method, the step problem can be solved effectively. From these background, it is expected that the production cost of color filter is reduced by 1/2 ~ 1/3. The organic carbon type material is the good candidate for the mass production of black matrix.

ACKNOWLEDGMENTS

The authors would like to thank S .W. Lee, J. S. Lee, J. H. Souk, W. K. Chang, and S. S. Kim for their support of this work. We also thank Mr. H. Amari of Fuji-Hunt Electronics Technology Co. for his full support in conducting this project.

REFERENCES

1. K. Tsuda, Displays 14 (2), 115-124 (1995)

2. T. Kosei, T. Fukunaga, H. Yamanaga, T. Ueki, J. Res. Develop., 34-43 (1992)

3. M. K. Hur, Y. K. Lee, H. S. Song, C. O. Jung, C. W. Kim, J. G. Lee, Semiconductor Technical Journal 10 (4), 1035-1039 (1995)

4. M. Tsuboi, S. Kotera, A. Iwashita, H. Yamane, H. Chiyoda, S. Tachizono, T. Yamagishi, AMLCD 94, 84-86 (1994)

5. C. M. Hesler, E. S. Simon, L. Bacchetti, R. Brainard, SID 95 Digest, p. 446 (1995)

6. T. Koseki, H. Yamanaka, T. Ueki, M. Nayuki, M. Honjo, T. Minamiyama, A. Arai, Display Technology, IBM Japan Ltd., DMTC 95, 107-108 (1995)

7. M. Tani, T. Sugiura, International Display Research Conference, 103-111 (1994)

8. Fuji-Hunt Material Safety Data, Fuji-Hunt Electronics Technology Co., Ltd. : MSDS File No. T-050E

Part IV

Transparent Conducting Oxides

EFFECT OF BASE PRESSURE IN SPUTTER DEPOSITION ON
CHARACTERISTICS OF INDIUM TIN OXIDE THIN FILM

Byung Hak Lee, Choong-Hoon Yi, Iee Gon Kim, Si-Hyun Lee
Display R & D Center, Samsung Display Devices Co., Ltd.
575 Shin-Dong, Paldal-Ku, Suwon, Kyungki-Do, Korea

ABSTRACT

A novel approach to sputter deposit low resistivity indium tin oxide thin films in cost effective way has been made. To reduce the complication of out-gassing from organic substances that are deposited at prior process steps in the flat panel display manufacturing, DC+RF magnetron sputtering was performed with a relatively high base pressure. Introduction of impurity atoms originated from residual air in the chamber into the film produced the change in lattice as well as microstructure of the film. Sputter deposition with higher base pressure resulted in lower resistivity of the film. Increase in the mobility of the carrier resulted in resistivity decrease. Resistivity as low as 1.5×10^{-4} $\Omega \cdot cm$ has been achieved with DC 0.67 W/cm^2 + RF 0.67 W/cm^2 power density and 10^{-5} Torr base pressure. X-ray diffraction analysis showed as much as 0.2 % increase in lattice constant resulted from higher base pressure sputtering. No further improvement of conductivity was observed for the base pressure over 10^{-5} Torr, suggesting impurity introduction saturation into the film.

INTRODUCTION

Optically transparent conductor is an indispensable element in opto-electronic devices such as photovoltaic solar cells. Chopra *et al*[1] reviewed various candidates for practical use. Among many conductors, Indium Tin Oxide (ITO) thin film has been most widely used. Its high optical transmittance at visible region with low electrical resistivity, originated from highly degenerated, wide band gap(3.8 eV) semiconducting property[2], enabled such applications. Various deposition techniques, such as spray pyrolosys[3], chemical vapor deposition[4], and e-beam evaporation[5] have been utilized to deposit ITO. The most widely used technique is the sputtering deposition using oxide target, due to its process stability and low resistivity of the film. Resistivity lower than 2.0 x 10^{-4} $\Omega \cdot cm$ can be easily achieved with relatively higher deposition rate. As the resolution of the flat panel display becomes higher, lower resistivity is frequently required. High substrate temperature, often exceeding 300°C, is used to achieve the lower resistivity. In case of process involving any organic substances prior to ITO film such as color filter manufacturing, however, substrate temperature is normally

Mat. Res. Soc. Symp. Proc. Vol. 424 © 1997 Materials Research Society

limited to below 250°C. Stollenwerk *et al* proposed insertion of a thin silver layer between two ITO films to increase the conductivity and save the ITO target cost[6], but its implementation in real mass production is still in premature state. In this paper, we demonstrate a novel way to deposit low resistivity ITO films at the substrate temperature of 200°C, while saving the energy and time required to deposit the film.

EXPERIMENTAL PROCEDURE

ITO film was deposited onto 370x300 mm^2 glass substrate, either with or without prior organic substances, using a magnetron sputtering machine. Power density of DC 0.67 W/cm^2 and RF 0.67 W/cm^2 was supplied to the target simultaneously. A hot pressed oxide target contained 10 wt.% of SnO$_2$, having 95 % of the theoretical density. The sputtering chamber can accommodate 16 substrates for one batch of deposition with a rotating substrate holder to ensure film uniformity. 2.3 mTorr of Ar gas was introduced into the chamber for sputtering. The thickness of the film was determined by using a Tencor Alphastep profile gauge, with a lift-off type pattern marked on the substrate. 2000 Å of ITO films were deposited for each sample with a deposition rate of 50 Å/min. Considering that 16 samples are made at the same time, this deposition rate is quite significant. The vacuum pressure to be reached before actual sputtering begins (herein called base pressure) was varied for each deposition, ranging from 10$^-$^{7} to 10^{-5} Torr. A Balzers SPM 452 was used for Residual Gas Analysis(RGA) during the sputtering process. Deposition temperature was 200°C. Sheet resistance was measured with a 4-point probe. A Hall effect measurement system from Biorad was used with a van der Paaw method. X-Ray Diffraction (XRD) was performed using a Phillips diffractometer model PW1710. Microstructural analyses were performed with Phillips XL30 Scanning Electron Microscope(SEM).

RESULTS AND DISCUSSION

Fig.1 is the plot showing the variation of resistivity of the film due to the change of the sputtering base pressure. As the base pressure increases, the resistivity of the film decreases. The resistivity dropped from 1.95 x 10^{-4} Ω·cm with 2.5x10^{-7} Torr base pressure to 1.64 x 10^{-4} Ω·cm with 10^{-5} Torr base pressure. For the base pressure greater than 10^{-5} Torr, no further improvement in resistivity was observed. This effect was observed for substrates of bare glass as well as organic substance(color photo-resist) coated glass. From this fact, it can be found that the effect did not originate from the out-gassing of the organic substance. Shigesato *et al* [7] reported a resistivity increase as the water vapor content increases in their DC magnetron sputter deposition of ITO films, while Ishibashi *et al*[8] reported a resistivity drop with the presence of water vapor for the low temperature deposition of ITO films. For the deposition temperature exceeding 150°C, Ishibashi *et al* did not observe any improvement in conductivity

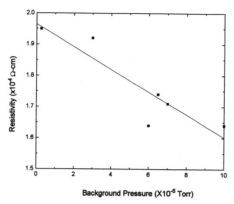

Fig.1 The resistivity change of ITO film as a function
of sputtering base pressure.

due to water vapor. In our experiment, the substrate temperature was 200°C. Fig.2 is the plot of
Hall measurement results of the same samples of Fig.1. This plot shows that the carrier
concentration remains fairly constant while the mobility increases, with increasing base
pressure. From this plot, the resistivity increase resulted from the change in mobility, rather
than higher carrier concentration. Mobility increase may be caused for various reasons, but
most notably from the improvement in crystallinity (less defects) in many ITO thin film
depositions. To check this, XRD analyses were performed. Fig.3(a) shows the XRD patterns of
the same samples. Neither apparent increase in crystallinity nor change in preferred orientation
can be detected from the patterns. Fig.3(b) is the enlargement of the 222 peak area of the same

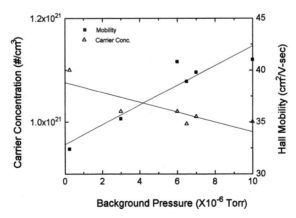

Fig.2 The change of carrier mobility and concentration of
ITO film as functions of the sputtering base pressure.

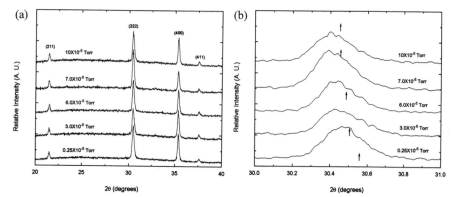

Fig.3　(a) XRD patterns of the samples of Fig.1　(b) Same XRD patterns, enlarged at the area at 222 peak, showing peak shift as the sputtering base pressure changes.

patterns. Increase in lattice constant due to higher base pressure deposition can be clearly seen. The peak position of 222 shifted as the base pressure changes, suggesting the incorporation of impurity atoms into the film. Converting from the 222 peak position, the lattice constant of the film deposited with 10^{-7} Torr base pressure was 10.520 Å, changes into 10.541 Å, 0.2 % increase, with 10^{-5} Torr base pressure. It is certain that this increase in lattice constant affected the electrical property of the film, because the change in atomic distance affects the band structure. Fig.4 is the RGA results, showing various gas partial pressure change during the sputter deposition process, for base pressure of 1.0×10^{-5} Torr. It has been known that the content of oxygen in reactive sputtering is very important due to the oxide coverage of the target surface[9]. From the plot, the amount of oxygen in the residual gas is minimal, and it is

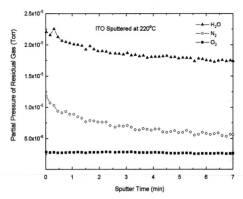

Fig.4　RGA results before and during the sputtering, with the sputtering base pressure of 10^{-5} Torr.

338

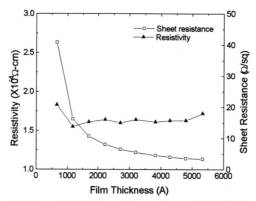

Fig.5   The change of resistivity and sheet resistance of the ITO film as functions of the film thickness, deposited with 10^{-5} sputtering base pressure.

hard to say oxygen affected the target surface. H_2O is considered the most important specie affecting the film property. Atomic structure of the ITO film in the paper of Shigesato et al was affected by H_2O, but their resistivity increased[7]. The amount of H_2O present in the chamber during the sputtering process is proportional to the base pressure before the sputtering actually begins. RGA results and the fact that the lattice constant increase with higher base pressure suggests that the impurities are incorporated into the lattice either interstitially or substitutionally. Fig.5 is the plot showing the change of the resistivity and sheet resistance as functions of the film thickness. As can be seen from the plot, the resistivity stays constant, a result which is contrary to Ishibashi et al[8]. From the RGA result, the amount of H_2O decreases as the sputtering deposition proceeds. The fact that the resistivity stays constant with increasing film thickness means that initial crystal formation of ITO plays important role, and H_2O constantly affects the crystal structure.

CONCLUSIONS

Possible H_2O introduction into ITO film by means of intentional high base pressure sputtering resulted in the lower resistivity of the film. Atomic and microstructural change resulted from the incorporation of impurity atoms in simultaneous RF and DC magnetron sputtering, produced a favorable effects on electrical property. Mobility increase resulted from band gap alteration and larger crystal grain produced lower resistivity with higher base pressure sputter deposition. This process opened the possibility to save cost and increase the productivity of the ITO film deposition process in flat pane display manufacturing using this technique, by reducing the time required to achieve the base pressure.

ACKNOWLEDGMENTS

Authors are grateful to Mr. S. W. Cho of SDD for his help in RGA and Hall measurement, and Mr. I. Y. Song of Samsung Advanced Institute of Technology for his help in SEM. Another gratitude goes to Mr. Brian Rupp of SDD, who kindly reviewed this manuscript. Authors are also thankful to Mr. S. K. Kang and S. S. Lee for their support during the course of this research.

References
1. K. L. Chopra, S. Major, and D. K. Pandya, Thin Solid Films, **102**, 1 (1983)
2. S. Naseem and T. J. Coutts, Thin Solid Films, **138**, 65 (1986)
3. R. Groth, Phys. Status Solidi, **14** ,69 (1966)
4. L. A. Ryaboya and Y. S. Savitskaya, Thin Solid Films, **2**, 141 (1968)
5. I. Hamberg and C. G. Granqvist, J. Appl. Phys. **60**, R123 (1986)
6. J. Stollenwerk, B. Ocker, and K. H. Kretschmer, Digest of Technical Papers, (Display Manufacturing Technology Conference, Jan. 1995, Society for Information Display), p.111
7. Y. Shigesato, Y. Hayashi, A. Masui, and T. Haranou, Jap. J. Appl. Phys. **30**, 814 (1991)
8. S. Ishibashi, Y. Higuchi, Y. Ota, and K. Nakamura, J. Vac. Sci. Technol. A **8**, 1399 (1990)
9. M. I. Ridge and R. P. Howson, Thin Solid Films, **96**, 121 (1982)

LOW RESISTIVITY TRANSPARENT INDIUM TIN OXIDE (ITO) FILMS SPUTTERED AT ROOM TEMPERATURE WITH H$_2$O ADDITION

Ken-ichi ONISAWA, Etsuko NISHIMURA, Masahiko ANDO, Takeshi SATOU, Masaru TAKABATAKE*, and Tetsuroh MINEMURA
Hitachi Research Laboratory, Hitachi, Ltd.
7-1-1 Omika-cho, Hitachi-shi, Ibaraki-ken, 319-12, Japan
*Electron Tube & Devices Division, Hitachi, Ltd.
3300 Hayano, Mobara-shi, Chiba-ken, 297, Japan

ABSTRACT

A new kind of amorphous indium tin oxide (ITO) film with good pattern delineation properties and mass production capability, as well as low resistivity and high transparency has been developed. The film was prepared by a cluster-type DC magnetron sputtering apparatus at room temperature with H$_2$O addition to the argon sputtering gas. The amorphous ITO film quality was improved by effective termination of oxygen vacancies with -OH species generated by enhanced decomposition from the added H$_2$O in the plasma.

INTRODUCTION

Transparent conductive oxide (TCO) films, especially indium tin oxide (ITO) thin films, are indispensable to produce flat panel display devices like liquid crystal displays, which are now enjoying an expanded market. The TCO films require high transmittance and low resistivity and these are generally considered to be primary goals when new TCO materials are developed. However, the following two factors are also very important for application to devices; pattern delineation properties and mass production capability. Since dry etching technology for ITO films is still in the R&D stage, a better wet etching property is necessary for pattern formation. In a mass production line, measures such as elimination of substrate heating during the sputtering step and increase of the wet etching rate are effective to realize higher throughput. Consequently, amorphous ITO films deposited at room temperature are suitable because their wet etching rate is two orders of magnitude higher than that of polycrystalline ITO films.

Drawbacks to introducing amorphous ITO films are their relatively high resistivity (>6x10^{-4} Ωcm), etching residue caused by micro crystallite ITO generated in the amorphous ITO matrix, and instability of properties when heat-treated (even by a photoresist baking step). In order to reduce the resistivity of the amorphous ITO films, an attempt to deposit films in the in-line type sputtering chamber with a small amount of H$_2$O addition is known to be effective[1]. It was also reported that good etching characteristics are obtained for ITO films sputtered under an atmosphere with residual H$_2$O gas[2]. However, a detailed study on the properties required for ITO films deposited by the H$_2$O addition method has not been made yet. Furthermore, by the in-line type sputtering improvement of the film quality is limited due to the increase of substrate surface temperature by exposure to a plasma.

The authors pointed out previously [3,4] that although a small amount of ITO crystallites was finely dispersed in the amorphous ITO films prepared by DC magnetron sputtering with controlled addition of H$_2$O in the Ar sputtering gas, crystal growth was suppressed; specifically crystallization temperature increased up to 200°C from approximately 150°C for the film without H$_2$O addition. This technology allows amorphous ITO films to be utilized since etching residue and instability of properties are inhibited. Based on these findings we have further developed amorphous ITO films with better film quality by employing a new type of sputtering system.

In the present paper, properties of amorphous ITO films, including their resistivity, when prepared with H$_2$O addition are discussed. A newly developed DC magnetron sputtering system was employed, in which a magnet placed behind the cathode-target is moved to obtain a uniformity of thickness for large substrate size, and the addition of H$_2$O can be precisely controlled.

Mat. Res. Soc. Symp. Proc. Vol. 424 © 1997 Materials Research Society

EXPERIMENT

Sputtering Method

A cluster type DC magnetron sputtering apparatus equipped with a precise gas introduction system and Q-mass analyzer was used to deposit amorphous ITO films. The target material used was a sintered In_2O_3 body containing 5% SnO_2; body density was higher than 95%. Glass substrates were 370x470x1.1mm. Thickness of the ITO films was approximately 130nm. Substrate were not heated during deposition. DC power, Ar pressure, and H_2O pressure were 0.5-1.3 kW, $5x10^{-3}$Torr, $2.5-10x10^{-5}$Torr, respectively. Some of the ITO films were annealed at 150 °C for 1h in N_2 atmosphere. Here, 150°C was selected because it is equal to the resist baking temperature plus a margin.

Evaluation methods

Thickness and resistivity of the films were measured by a stylus type profiler and the four-points probe method, respectively. Structures of the films were investigated by the x-ray diffraction (XRD) method, scanning electron microscopy (SEM), and thermal desorption spectroscopy (TDS). XRD measurements were done in the following manner to allow detection of a small quantity of microcrystalline ITO; the X-ray source was Cu-Kα, the tube voltage and tube current were 50kV and 200mA, respectively, and the incident X-ray beam was fixed at a 1° angle to the sample surface. In order to estimate carrier concentration and mobility in the films, transmission and reflection spectra were measured by a spectrophotometer and their infrared region was analyzed using Drude's model[5].

RESULTS AND DISCUSSION

Figure 1 shows XRD spectra after annealing at 150 °C for the ITO films deposited with different H_2O pressures. These demonstrate that the (222) diffraction peak of In_2O_3 crystal seen at around 30° on the shoulder of the broad amorphous In_2O_3 peak disappears as H_2O pressure is increased from 2.5 to $6.5x10^{-5}$Torr and then it become stronger at higher H_2O pressures. XRD peak intensity for (222) plane of In_2O_3 crystal as a function of H_2O pressure is plotted in Fig.2. From this graph it is clear that XRD peak intensity has a minimum value at H_2O pressure of $6.5x10^{-5}$Torr. The XRD peak intensity value should be smaller in order to decrease the amount of etching residue which is caused by crystalline ITO phase present in the amorphous ITO matrix. Then it is necessary to control H_2O pressure to $6.5x10^{-5}$Torr in this particular case. Examination of the film structure by SEM reveals that even in the as-deposited films ITO crystallites of approximately 20nm diameter are generated in the amorphous host, and after annealing at 150 °C, the average size of the crystallites increases for lower H_2O pressure whereas the density of the crystallites increases after annealing for higher H_2O pressure. This phenomenon can be explained by considering the kinetic model for nucleation and growth of crystalline particles in amorphous matrix[3].

The relation between resistivity and H_2O pressure is plotted in Fig.3, indicating that resistivity has a minimum at around $6x10^{-5}$Torr. This trend is in good agreement with the results for XRD peak intensity shown in Fig.2.

In Fig.4 the transmission spectra are presented with H_2O pressure as a parameter. The transmittance, which includes transmittance of the glass substrates, is found to increase with H_2O pressure in the visible range. The value is larger than 85% for the films sputtered with H_2O pressure higher than $5x10^{-5}$Torr. Furthermore, the transmittance in the infrared region increases with H_2O pressure. This is due to the decrease in reflectance of the samples with H_2O pressure. The decrease should be caused by the decrease in the free carrier concentration since the reflection is induced by plasma oscillation of free carriers in the films. Free carrier concentration is estimated by fitting these spectra using Drude's model, where the value of the effective mass is assumed to

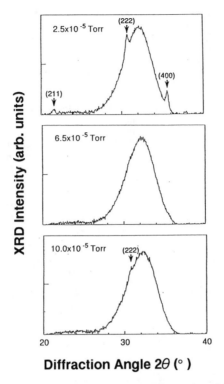

Fig.1 XRD spectra obtained for several ITO samples with different H_2O pressures. The samples were annealed at 150°C for 1h.

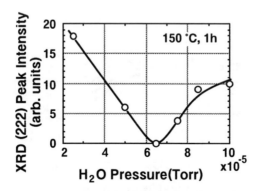

Fig.2 XRD peak intensity of (222) plane of In_2O_3 as a function of H_2O pressure.

be 0.3 times the free-electron mass and independent of samples[6]. The calculated results as a function of H_2O pressure are shown in Fig.5. It is found from the graph that the carrier concentration decreases with the increasing H_2O pressure. The estimated mobility using the data in Fig.3 and carrier concentration is also plotted in Fig.5, demonstrating that the mobility increases with H_2O pressure.

TDS measurements have confirmed that the amounts of hydrogen, oxygen, and H_2O gas desorbed from the ITO films increase with H_2O pressure[3]. In addition, it was found that the incorporated H_2O molecules in the films transformed to -OH bonds and terminated oxygen vacancies with heat treatment[4]. This means that with the increase of H_2O pressure, the number of -OH bonds increases, in other words, the number of oxygen vacancies decreased. This explains the decrease of carrier concentration since the oxygen vacancies are the origin of the free carriers in ITO films. Furthermore, the decrease in number of the oxygen vacancies should be responsible for the increase of mobility of the carriers. Termination of an oxygen vacancy by -OH bonds is also associated with the finding that the amount of ITO crystallites changes with H_2O pressure as shown in Fig.2.

The resistivity of as-deposited ITO films is plotted in Fig.6 as a function of DC power at H_2O pressure of 7.5×10^{-5} Torr, indicating that resistivity decreases with the increase of power and a value as low as $4 \times 10^{-4} \Omega$cm is obtained at 1.2kW. In the figure XRD (222) peak intensity is also plotted, showing that the intensity decreases with power like the resistivity. This demonstrates that

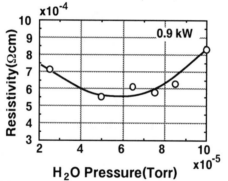

Fig.3 Resistivity of ITO films as a function of H_2O pressure.

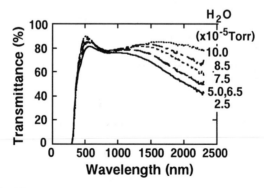

Fig.4 Transmission spectra of ITO films with H_2O pressure as a parameter.

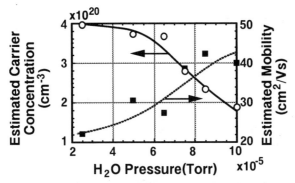

Fig.5 Estimated carrier concentration and mobility of ITO films as a function of H_2O pressure.

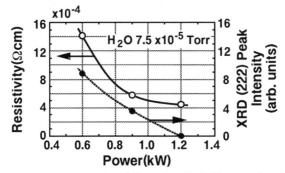

Fig.6 Resistivity and XRD peak intensity of ITO films as a function of power. The samples for XRD were annealed at 150°C for 1h to allow comparison with the data in Figs.1

increasing the power, which is essentially equal to the plasma density, enhances decomposition of H_2O and thus -OH termination. Therefore, better film quality results and this causes a better transport property. We noted that the new sputtering method in which the magnet is moved could realize increasing power without generation of ITO crystallites in the film. In conventional sputtering system like the in-line type system, increasing power brings an increase of the substrate temperature, resulting in generation of ITO crystallites in the film.

CONCLUSIONS

By using the cluster-type DC magnetron sputtering apparatus and introducing H_2O to the argon sputtering gas, amorphous indium tin oxide (ITO) films with good wet etching property as well as low resistivity and high transparency have been fabricated at room temperature. Optimizing sputtering conditions such as H_2O pressure and power realized the film structure with only a little microcrystalline ITO, which was the cause of etching residue, and the resistivity as low as approximately $4\times10^{-4}\Omega$cm. It was speculated that the amorphous ITO film quality was improved by effective termination of oxygen vacancies with -OH species generated by enhanced decomposition in plasma.

ACKNOWLEDGMENTS

The authors would like to thank Dr. François Leblanc for valuable discussion and suggestions. They are also grateful to Mr. Eiji Setoyama of Kokubu Works, Hitachi, Ltd. for encouragement.

REFERENCES

1. S. Ishibashi, Y. Higuchi, Y. Ota, and K. Nakamura, J. Vac. Sci. Technol. A8, 1399(1990).

2. H. Kitahara, T. Uchida, M. Atsumi, Y. Kaida, and K. Ichikawa, Proc. 1st Int. Symp. on Sputtering and Plasma Processes, Tokyo, 1991, p.149.

3. M. Ando, M. Takabatake, E. Nishimura, F. Leblanc, K. Onisawa, T. Minemura, Proc. 16th Int. Conf. of Amorphous Semiconductors, Kobe, 1995, p.385; also to be published in J. Non-crystalline Solids, 1996.

4. E. Nishimura, M. Ando, K. Onisawa, M. Takabatake, T. Minemura, to be published in Jap. J. Appl. Phys. May(1996).

5. H. Køstlin, R. Jost, and W. Lems, Phys. Status Solidi (a)29 (1975)87.

6. J. R. Bellingham, W. A. Philips, and C. J. Adkins, Thin Solid Films, 195(1991)23.

ATOMIC HYDROGEN EFFECTS ON THE OPTICAL AND ELECTRICAL PROPERTIES OF TRANSPARENT CONDUCTING OXIDES FOR a-Si:H TFT-LCDs

JE-HSIUNG LAN and JERZY KANICKI
Department of Electrical Engineering and Computer Science
Center for Display Technology and Manufacturing, The University of Michigan
Ann Arbor, MI 48109-2108

ABSTRACT

The effects of the atomic hydrogen treatment (H-treatment) of indium-tin oxide (ITO) and aluminum-doped zinc oxide (AZO) films have been investigated. The atomic hydrogen was generated by hot-wire chemical vapor deposition (HW-CVD) technique. Experimental results have shown that AZO films are chemically very stable under the H-treatment; almost no variation in the optical transmittance and electrical resistivity was observed. On the contrary, ITO films, either prepared by sputtering with ex-situ or in-situ thermal-annealing, have shown severe optical and electrical degradation and surface whitening after the H-treatment. SEM studies of the H-treated ITO surfaces have revealed that the surface whitening was due to the increase in surface roughness and the formation of granule-like metallic balls. Auger electron spectroscopy has indicated that the balls were mainly composed of indium atoms and the areas between balls were rich in oxygen atoms. These results were confirmed by X-ray diffraction and X-ray photoelectron spectroscopy measurements done on ITO before and after the H-treatment. Finally, we have demonstrated that a-SiO$_x$ deposited by PECVD will completely suppress the chemical reaction between ITO surfaces and atomic hydrogen generated by HW-CVD technique.

INTRODUCTION

Indium-tin oxide (ITO) having the highest optical transmittance and electrical conductivity is the only transparent conducting oxide (TCO) used in hydrogenated amorphous silicon (a-Si:H) thin-film transistor (TFT) liquid-crystal displays (LCDs). During the fabrication of a-Si:H TFT-LCD shown in Fig. 1, ITO films are exposed to hydrogen containing plasma generated during the deposition of amorphous silicon nitride (a-SiN$_x$:H) films by the plasma-enhanced chemical vapor deposition (PECVD) technique. It has been reported that ITO optical and electrical properties will be modified under the hydrogen containing plasma environment[1-5]. In our previous work, we have proposed aluminum-doped zinc oxide (AZO) films, for the first time, as an alternative transparent conducting oxide to ITO for a-Si:H TFT-LCD applications[6]. AZO films are stable under the exposure to hydrogen plasma, and they have an optical transmittance and electrical resistivity comparable to ITO films. In addition to the chemical stability, AZO films have shown better patterning characteristics than ITO films, in terms of a higher etch rate and no etching residue formation after the completion of etching[7].

To further explore the chemical stability of ITO and AZO films, we have exposed both films to the atomic hydrogen generated by the hot-wire chemical vapor deposition (HW-CVD) technique.

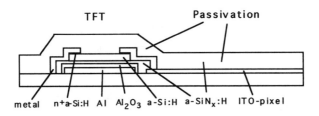

Fig. 1. A cross-sectional view of a-Si:H TFT-pixel unit cell.

EXPERIMENTAL

In this work, Corning glass code 7059 was used as substrate for all the TCO films. ITO films were prepared by sputtering methods using hot-pressed sintered targets composed of 90 wt.% of In_2O_3 and 10 wt.% of SnO_2. Two kinds of ITO films having a similar film resistivity ($\sim 2 \times 10^{-4}$ Ω-cm) but different microstructures, Fig. 4(a) and (b), were used in this study. ITO film A was deposited at room temperature and followed by a post-deposition thermal-annealing at a temperature of 230 °C (ex-situ thermally annealed film); while ITO film B was deposited at temperature of 200 °C (in-situ thermally annealed film). The thickness of ITO films was about 1200 Å. Similarly, AZO films were deposited in a DC magnetron sputtering system using a sintered target composed of 98 wt.% of ZnO and 2 wt.% of Al_2O_3. The substrate temperature during the sputtering was kept constant at 250 °C. The thickness of AZO films was about 1000 Å. The 500 Å thick amorphous silicon oxide (a-SiO_x) film was deposited by PECVD technique under the following conditions: 24 and 20 sccm for SiH_4 and N_2O flow rates, respectively; 250 °C for the substrate temperature; 100 mT for the chamber pressure; and 80 W for the R.F. power.

The degradation studies of TCO films have been carried out by exposing the TCO films to atomic hydrogen generated by the HW-CVD technique; In this case, the molecule hydrogen (H_2) was flowing through the tungsten wire kept at a temperature of 1700 °C. TCO samples were kept at a constant temperature of 300 °C on a sample holder facing down exposed to the atomic hydrogen flux. In this work, all the H-treatment conditions were kept the same, except the H_2 flow rate, which was set at 60 and 200 sccm, respectively. The chamber pressure was kept at 100 mT during the experiment. The hydrogen exposure time for all the samples was kept constant at 1 min.

Sheet resistance for all of the TCO films was determined by the four-point probe method. The crystallinity and microstructure of the films were analyzed by a Rigaku X-ray diffraction (XRD) system (Cu Kα radiation). The bonding energy and atomic concentration of ITO films were analyzed by a Perkim Elmer PHI 5400 X-ray photoelectron spectroscopy (XPS) system. The scanning Auger electron spectroscopy (AES) was also used to analyze the ITO films before and after the H-treatment. The surface morphology of the ITO films prior to and after the H-treatment were examined by JEOL scanning electron microscope (SEM). The optical transmittance spectra of the films were measured by a Varian Cary 5E UV-visible spectrum analyzer; the average transmittance in visible-light range was obtained from the average values of transmittance taken between 0.39 and 0.7 μm.

RESULTS and DISCUSSIONS

Optical Transmittance and Electrical Resistivity Variation

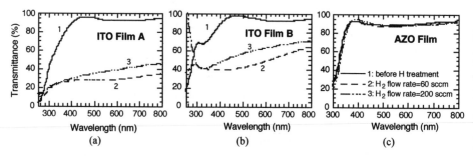

Fig. 2. UV-visible transmittance spectra for (a) ITO film A, (b) ITO film B, (c) and AZO film before and after the H-treatment. Curve 1: TCO films before the H-treatment, 2: TCO films after the H-treatment with H_2 flow rate = 60 sccm, and curve 3: with H_2 flow rate = 200 sccm.

The optical transmittance spectrum of ITO films A and B and AZO film before and after the H-treatment are shown in Figs. 2(a), (b), and (c), respectively. The average values of transmittance, within the visible-light range, are listed in Table I. As shown in Figs. 2(a) and (b), the optical transmittance of both ITO films have drastically decreased after the H-treatment, indicating that ITO films, prepared by either ex-situ or in-situ thermal-annealing method, were not chemically stable in the atomic hydrogen environment. After the H-treatment, the surface of ITO films became rough, giving the ITO a milky-like appearance. This result indicates that the surface morphology or the microstructure of ITO films has been changed by the H-treatment. In addition, it is interesting to note that (a) optical transmittance for both ITO films have slightly increased when the H_2 flow rate was increased from 60 to 200 sccm and (b) the surface appearance of ITO films became less milky for a higher H_2 flow. This phenomenon can be attributed to the surface morphology variation and it will be discussed in details in the next section. In the case of AZO films, in contrast, almost no optical transmittance variation was observed after the H-treatment, Fig. 2(c). This suggests that AZO films are chemically very stable in a highly chemically reducing environment such as atomic hydrogen flux.

Table I. TCO: sheet resistance (R_{sh}: $\Omega/\square$) / average transmittance (T_r: %)

	ITO Film A	ITO Film B	AZO Film
before H-treatment	18.5 / 92.6	15.1 / 93.9	75.0 / 91.6
after H-treatment (H_2 flow = 60 sccm)	35.9 / 28.0	27.8 / 45.6	76.1 / 91.1
after H-treatment (H_2 flow = 200 sccm)	n.m. / 36.0	n.m. / 58.6	74.3 / 91.7

Table I shows the sheet resistance variation for ITO and AZO films before and after exposure to the atomic hydrogen, respectively. After the H-treatment, the sheet resistance of ITO films increased for both H-treatment conditions. In the case of a higher H_2 flow rate (200 sccm), sheet resistance for both ITO films were so high that could not be measured by the four-point probe technique; "n.m." in the table represents non-measurable values. These results suggest that the ITO surface become highly resistive after the H-treatment. On the other hand, only a small variation in sheet resistance was observed for AZO films. Therefore, AZO films are not influenced by the atomic hydrogen from both optical and electrical points of view.

Morphological Variation of ITO films

Figs. 3(a) and (b) show the surface morphology changes of ITO films A and B after exposure to atomic hydrogen with H_2 flow rate of 60 sccm; while Figs. 3(c) and (d) show the surface morphology variation for ITO films A and B after the H-treatment with H_2 flow rate of 200 sccm. From these SEM micrographs, it is very clear that the surface of ITO films has been greatly damaged and a large amount of granules (balls) were formed on the film surfaces following the H-treatment. For lower H_2 flow rate, the size of the balls was smaller, but the density of the balls was larger. From the optical point of view, it is expected that the higher density of surface roughness will scatter more optical light, leading to a lower optical transmittance. Accordingly, the H-treated ITO film having a higher density of the balls will thus have a lower optical transmittance. As it is shown in Table I, ITO film A treated with H_2 = 60 sccm has the lowest optical transmittance (T_r = 28 %), which can be associated with its highest ball density. In addition, because the density of the balls in ITO film A is larger than in ITO film B, it is expected that ITO film A will have a lower optical transmittance than the ITO film B, Table I. The difference in the ball density between the H-treated ITO films A and B could be associated with their microstructure differences prior to the H-treatment, Fig. 4.

Increasing the H_2 flow rate from 60 to 200 sccm, the balls increased in size from 0.3 to 1.1 μm for ITO film A and from 2.5 to 3.7 μm for ITO film B. As shown in Fig. 3(d), it is interesting to point out that there are certain shapes that can be observed for the larger balls, and there is almost no small ball formation surrounding the larger balls. This observation can suggest that the larger balls were coagulated from a certain number of the smaller balls located in the vicinity of the larger balls.

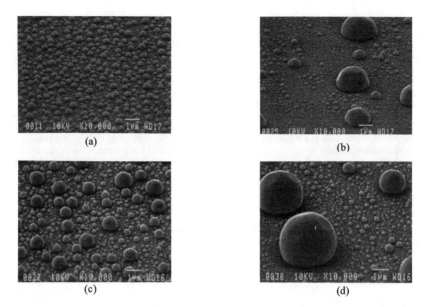

(a)
(b)
(c)
(d)

Fig. 3. SEM pictures for ITO films after hydrogen treatments with H_2 flow rate = 60 sccm: (a) ITO film A and (b) ITO film B; H_2 flow rate = 200 sccm: (c) ITO film A and (d) ITO film B.

The ball size and its density were different for the H-treated films A and B: film A had a higher density of balls of a smaller size and film B had a higher density of balls of a larger size. This difference in ball formation could be associated with the microstructure difference in films A and B. Thus, the coagulation of balls would considerably depend on the dissociation ability of the indium-oxygen and tin-oxygen bonds in the ITO structure; and the bond dissociation ability will depend on the bonding energy between the atoms. Since the strength of bonding energy depends on the atomic bonding arrangement in the ITO, it is expected that the microstructure and crystallinity of the ITO films would have an effect on the ball formations.

Microstructure and Crystallinity Variation of ITO films

To identify the surface formation of the ITO films after the H-treatment, XRD method was used to examine the microstructure variation of the H-treated ITO films. Figs. 4(a) and (b) show the XRD patterns for ITO films A and B, respectively. Before the H-treatment, ITO film A had a major peak along (222) direction; while ITO film B had a major peak along (222) direction and a minor peak along (400) direction. In the case of the H-treatment with H_2 flow rate of 60 sccm, all the major and minor peaks in (222) and (400) directions for both ITO films were suppressed and a metallic indium peak along (101) direction was detected. This XRD patterns indicate that ITO compounds were still present on the H-treated surfaces and the decrease in the major XRD peak intensity could be attributed to the decrease of the ITO film thickness following the H-treatment; i.e. the ITO films have been dry etched away by the atomic hydrogen. On the other hand, the emergence of metallic indium phase along (101) could suggest that poly-crystalline indium grain has been formed on the H-treated film surface following the reduction of indium oxide after the H-treatment.

In the case of H-treatment with a higher H_2 flow rate (200 sccm), ITO films A and B show different trends in the XRD patterns. For film A, only the indium (101) peak remained and all the other peaks have disappeared. In the case of ITO film B, in addition to the indium (101) peak, another indium peak along the (110) direction emerged. Moreover, a strong peak located at $2\theta=35.6°$ has been observed, in Fig. 4(b), which appears to be close to the position of ITO (400) peak, where $2\theta=34.9°$. Obviously, this new peak has shifted from the ITO (400) peak position,

350

indicating that sub-oxides in the form of In_2O or InO have been formed; or it could also be that ITO stoichiometry for a low resistivity film has been completely changed after the H-treatment. Therefore, the H-treated ITO films were composed of a complex compounds containing metallic indium, indium oxide, and indium suboxides. The exact nature of ITO after the H-treatment will be discussed below.

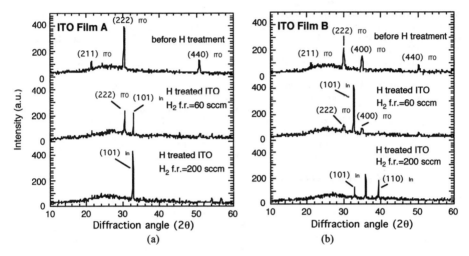

Fig. 4. XRD patterns for (a) ITO film A and (b) ITO film B before and after the H-treatment. The H-treatment was done with H_2 flow rate (f.r.) of 60 and 200 sccm, respectively.

Compositional Variation of ITO films

1. X-ray Photoelectron Spectroscopy (XPS)

The variation of atomic bonding energy (E_b) and atomic concentration (at.%) of ITO films before and after the H-treatment were investigated by XPS and they are summarized in Table II. Since the X-ray probe beam size is about 1.1 mm and film surface nonuniformity due to the ball

Table II. Atomic concentration (at.%) and bonding energy peak (E_b) for ITO films before and after the H-treatment, respectively.

XPS data for:	ITO Film A		ITO Film B	
Before the H-treatment	at. % / E_b (eV)		at. % / E_b (eV)	
	as-treated	sputtered	as-treated	sputtered
Oxygen (O 1s)	55.36 / 530.0	53.48 / 530.0	54.27 / 530.0	52.41 / 529.9
Indium (In $3d_{5/2}$)	39.78 / 444.3	41.81 / 444.3	41.29 / 444.3	43.40 / 444.3
Tin (Sn $3d_{5/2}$)	4.86 / 486.3	4.71 / 486.2	4.45 / 486.3	4.18 / 486.2
After the H-treatment (H_2 flow rate = 60 sccm)				
Oxygen (O 1s)	50.77 / 530.2	48.82 / 529.9	53.5 / 530.2	49.79 / 529.9
Indium (In $3d_{5/2}$)	45.10 / 444.5	46.48 / 444.3	42.36 / 444.5	46.48 / 444.2
Tin (Sn $3d_{5/2}$)	4.13 / 486.2	4.70 / 486.0	4.14 / 486.3	3.73 / 486.0
After the H-treatment (H_2 flow rate = 200 sccm)				
Oxygen (O 1s)	70.59 / 538.4	64.97 / 537.9	74.09 / 538.1	68.61 / 537.7
Indium (In $3d_{5/2}$)	28.46 / 451.2	32.26 / 450.3	24.71 / 451.0	28.77 / 449.9
Tin (Sn $3d_{5/2}$)	0.95 / 491.9	2.77 / 490.4	1.20 / 492.7	2.61 / 491.7

formation, the XPS results shown in this table represent an average values for E_b and at.% for the H-treated ITO films. To exclude the surface oxide formation and contamination of the films, the XPS data were taken on both the H-treated film surfaces (as-treated) and the H-treated film surface sputtered by an Ar^+ gun operated at 3 keV for 1 min (sputtered).

For all the cases shown in the Table II oxygen content was higher on the film surface before the sputtering. This observation indicates that the oxidation occurred on the ITO surface before and after the H-treatment. After the H-treatment, the atomic ratio of oxygen to indium varied. Initially, the oxygen to indium atomic ratio (O/In) was 1.28 and 1.21 for films A and B, respectively. In the case of H-treatment with a lower H_2 flow rate (60 sccm), the O/In ratio decreased to 1.05 and 1.07 for films A and B, respectively. This result indicates that the H-treated film surface became indium-rich, probably due to the reduction of indium oxide and enhancement of the concentration of indium atoms on the film surface. No significant binding energy variation for In ($3d_{5/2}$) and Sn ($3d_{5/2}$) have been observed.

It is expected that ITO films treated with a higher density of reactive H species should have a higher indium to oxygen atomic ratio. However, in the case of the H-treatment with H_2 flow rate of 200 sccm, the H-treated ITO film became more oxygen-rich. After sputtering off ~ 200 Å of the H-treated film surface, the O/In ratio increased to 2.01 and 2.38 for ITO films A and B, respectively. This observation indicates that ITO surfaces after the H-treatment with high density of atomic hydrogen have become oxygen-rich; and this stoichiometry change can be due to the ITO surface oxidation and/or it can be associated with the formation of larger balls on ITO surface.

2. Auger Electron Spectroscopy (AES)

To further identify the chemical composition of the larger ball (Fig. 5) and the area surrounding it, we have used AES assisted by SEM to analyze the ITO surface after the H-treatment. Table III shows the results of this analysis for ITO films B before and after the H-treatment with H_2 flow rate of 200 sccm; area of ball and area between the balls were analyzed separately. In order to exclude the film surface oxidation, similarly to XPS about 200 Å of films were sputtered off by Ar^+ before the AES measurements were made. The atomic ratio of indium to oxygen (In/O) for the ITO films before the H-treatment was 0.74; after the H-treatment, the In/O ratio varied from area to area. The In/O ratio became 1.27 and 0.21 for the area 1 and area 2 indicated on Fig. 5. These results suggest that the ball had a higher metal-like character and it was mostly composed of indium-rich oxide. On the other hand, for the area outside the balls, an enhancement of the oxygen concentration was observed, indicating that most of the indium atoms were segregated from these area after the H-treatment.

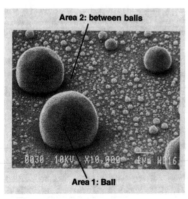

Fig. 5 ITO film B after the H-treatment; H_2 flow rate = 200 sccm.

Table III. AES compositional analysis

Region/Position	Composition (at. %)		
	O	In	Sn
Before H treatment:			
on surface:	57.8	37.2	5.0
sputtered:	55.1	40.9	4.0
After H treatment (200 sccm):			
Area 1:			
as-treated:	54.0	43.7	2.3
sputtered:	39.6	50.4	10.0
Area 2:			
as-treated:	75.1	24.9	--
sputtered:	79.2	16.6	4.2

Based on this segregation model, the ball formation mechanism can be described as follow: the chemical reaction between atomic hydrogen and ITO surface can cause the indium atoms to be reduced from the ITO film surface, resulting in ITO having a more metal-like character. When more and more indium atoms are reduced, they can coagulate together and form a small ball composed of poly-crystalline metallic indium, as indicated by the XRD patterns shown in Fig. 4. At the same time, because of the formation of small balls the surfaces of the H-treated ITO films become more and more nonuniform in terms of stoichiometry; areas of small balls become indium-rich and areas outside (between) the balls becomes oxygen-rich. In other words, the indium and oxygen atoms have become segregated from each others on ITO film surface after the H-treatment. When further chemical reaction between atomic hydrogen and ITO proceeds, the surface segregation between indium and oxygen atoms on ITO surface will become more significant. This surface segregation will consequently result in smaller balls to coagulate together and form larger balls shown in Fig. 5.

Finally, we would like to point out that the XPS results for ITO film B shown in the previous section is consistent with the AES analysis discussed in this section. Under the H-treatment with H_2 flow rate of 200 sccm, the O/In ratio for ITO film B is 2.38, indicating that most of the H-treated ITO surfaces are composed of oxygen atoms. Indeed, the H-treated ITO surfaces are mostly covered by the oxygen-rich areas (area 2) and only a small portion of the ITO surface is covered by the indium-rich area contained in the larger balls (area 1). Accordingly, the H-treated ITO surfaces are oxygen-rich.

ITO Films with a Surface Barrier Layer

We have shown that ITO films, either prepared by ex-situ or in-situ thermal-annealing techniques, were chemically very instable in the atomic hydrogen environment. To use ITO films in a-Si:H devices, a surface barrier layer on top of ITO films is required to prevent the chemical reaction between atomic hydrogen and ITO surface. In this work, a-SiO_x prepared by PECVD technique was used as a barrier layer. The thickness of the a-SiO_x was about 500 Å and was not optimized. Figs. 6(a) and (b) show the transmittance spectra for ITO films having a capping layer of a-SiO_x described above. The transmittance spectra shown in Fig. 2 were also included for comparison.

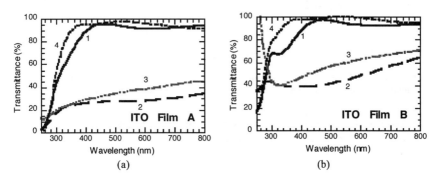

Fig. 6. UV-visible transmittance spectra for: (a) ITO film A and (b) ITO film B. Curve 1: ITO before the H-treatment, 2: the H-treated ITO with H_2 = 60 sccm, 3: the H-treated ITO with H_2 = 200 sccm, and 4: the H-treated ITO capped with a-SiO_x with H_2 = 200 sccm.

It is clear from Fig. 6 that the transmittance of the films capped with a-SiO_x did not show any degradation after the H-treatment. On the contrary, the optical transmittance in the visible light range was enhanced from 92.6 to 95.9 for ITO film A and from 93.9 to 98.8 for ITO film B. This enhancement is due to the anti-reflection effect introduced by a-SiO_x; a smaller difference in the refractive indices between silicon oxide (n ≈ 1.5) and air (n=1.0) in comparison to ITO (n ≈ 1.9) and air (n=1.0) is responsible for this effect. Therefore, a-SiO$_x$ is a very useful protection layer for ITO films that can effectively suppress the chemical reaction between ITO and atomic hydrogen. In addition, we have also observed that a-SiO$_x$ has a much higher etch rate (~ 160 Å/sec) than ITO films (etch rate is negligible) in the buffered hydrofluoric (BHF) solutions. This etch rate difference will help in removing oxide layer by wet etching without affecting the ITO films during the ITO-pixel electrode fabrication. Therefore, the deposition of a-SiO$_x$ capping layer on ITO prior to the deposition of a-SiN$_x$:H films can effectively be used in the fabrication of TFT-LCDs.

CONCLUSIONS

In summary, we have shown that ITO films prepared by either ex-situ or in-situ thermal-annealing method are not chemically stable in the atomic hydrogen environment. After the H-treatment, the ITO surface becomes milky-like and film optical transmittance is significantly reduced. In contrast, AZO films are chemically very stable under the same H-treatment conditions, showing almost no variation in optical and electrical properties. The reduction in optical transmittance as well as electrical conductivity for ITO films after the H-treatment can be attributed to the ball formation on the film surface. The formation of the balls not only causes the optical light scattering out of the film surface, but it also segregates the film surface, making the H-treated films more oxygen-rich and consequently more resistive. The balls have been shown to be indium-rich and the areas outside (between) the balls are oxygen-rich. The degradation in the optical and electrical properties of ITO films in atomic hydrogen environment could suggest that ITO may not be a suitable TCO film for TFT-LCDs. However, we have also shown that by adding a-SiO$_x$ surface protection layer on top of ITO films, the degradation of ITO films can be completely suppressed. Consequently, by using a protection layer, ITO films can still be used in fabrication of a-Si:H TFT-LCDs.

ACKNOWLEDGMENTS

The authors would like to thank Prof. E. Gulari and Ms. Shuang-Ying Yu for assistance with the hot-wire CVD system. This work was supported by the University of Michigan, Center for Display Technology and Manufacturing.

REFERENCES

1. S. Major, S. Satyendra Kumar, M. Bhatnagar, and K.L., Chopra, Appl. Phys. Lett. **49**, 394 (1986).
2. H. C. Weller, R.H. Mauch, and G.H. Bauer, Proc. of 22th IEEE PVSC, 1290 (1991).
3. E. Kimura, G. Kawachi, N. Konishi, Y. Matsukawa, and A. Sasano, Jpn. J. Appl. Phys. **32**, 5072 (1993).
4. R. Banerjee, S. Ray, N. Basu, A.K. Batabyal, and A.K. Barua, J. Appl. Phys. **62**, 912 (1987).
5. T. Minami, H. Sato, H. Nanto, and S. Takata, Thin Solid Films. **176**, 277 (1989).
6. J.-H. Lan and J. Kanicki, Proc. of AMLCD '95, 54 (1995).
7. J.-H. Lan and J. Kanicki, J. Electronic Materials (submitted).

PROPERTIES AND STABILITY OF BISMUTH DOPED TIN OXIDE THIN FILMS DEPOSITED ON VARIOUS TYPES OF GLASS SUBSTRATES

V. FOGLIETTI[1], A. D' ALESSANDRO[2], A. GALBATO[2], A.ALESSANDRI[2], R. BECCHERELLI[2], F. CAMPOLI[2], S. PETROCCO[1], M. WNEK[3], P.MALTESE[2].
(1) Istituto di Elettronica dello Stato Solido del CNR,Via Cineto Romano 42, 00156 Roma
(2) Universita' degli studi di Roma " La Sapienza", Dip. Ingegneria Elettronica, Via Eudossiana 18, 00185 Roma
(3) Institute of Physics, Jagellonian University, ul. Reymonta 4, 30-059 Kracow, Poland

ABSTRACT

This work reports on deposition of transparent semi-insulating tin oxide thin films deposited by spray-pyrolysis over large area glass substrates. The precursors used are based on diluted chloride solutions. Bi-doped high resistivity tin oxide thin films are required in the fabrication process of an analog grey scale ferroelectric liquid crystal matrix display. Various parameters have been optimized in order to achieve good reproducibility and uniformity of the electrical and optical properties. Deposition temperature and the properties of the glass substrates have been found as the most critical variables which affect the deposition rate and the quality of the deposited films. The different substrates used, borosilicates and soda-lime glasses, lead to different electrical properties. A further investigation on the stability of the film properties versus different kind of aging procedures have been conducted. Preliminary results of the operation of the complete display making use of our films are shown.

INTRODUCTION

The work described in this paper is finalized to use a transparent high resistivity layer to develop an analog gray shades passive matrix display using surface stabilized ferroelectric liquid crystal (SSFLC). Such a display can provide small cost solutions for TV and various multimedia applications. The construction is made by coupling a high resistivity plate to a conventional Indium-Tin-Oxide plate. The transparent high resistivity layer is electrically connected by metal stripes and the voltage drop between them provides continuous grey levels through continuous changes in the switched area[1]. This work is supported by the ESPRIT project "PROFELICITA" in the development of a range of analog grey scale ferroelectric liquid crystal displays. Among various possibilities examined, the addition of Bismuth to the intrinsically n-type doped tin-oxide results to an increase of the electrical surface resistivity up to tens of $M\Omega/\square$ required for our application, without degrading significantly the optical transmittance of the device. Furtherly high resistivity SnO_2:Bi have been chosen for the possibility of being deposited over large areas with good uniformity and the relatively small cost of the deposition equipment. The films are deposited by spray-pyrolysis[2,3] using a novel deposition system that improves uniformity over large areas[4]. In this paper we report the electrical and morphological properties of Bismuth-doped Tin Oxide thin films, deposited on different kinds of glass substrates. A careful investigation has been made concerning the stability of these films versus the various steps of the ferroelectric liquid crystal display fabrication procedure. Finally preliminary results from our complete device are shown in the last paragraph.

FILM DEPOSITION AND PROPERTIES

Figure 1 shows the spray-pyrolysis deposition chamber. It consists of a flat rotating hot plate that can house substrates up to 18 cm in diameter and a diffusor, consisting of a pyrex glass ring, surrounds this hot plate. The aerosol is injected radially in the deposition chamber through the many small holes that are regularly arranged in the inner surface of the glass ring, close by the hot plate. During the deposition the chamber is covered by a pyrex glass plate that enables the aerosol to saturate the volume over the hot plate, thus resulting in a good uniformity of the film thickness over the substrate area. Further details on the deposition system configuration can be found in a previous paper[4]. A dilute solution of stannic chloride and bismuth chloride is properly atomized using nitrogen as a carrier gas. The diluted solution in isopropyl-alcohol is prepared in large quantities such as 500 ml. It is further diluted in DI water immediately before the start of the run.

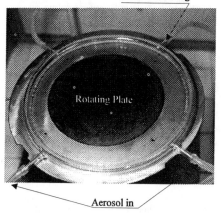

Fig. 1) Photograph of the deposition chamber. The pyrex diffusor has many holes located in the inner surface of the glass ring.

The films were deposited on Corning 7059 glass substrates and on soda-lime glass substrates with and without a SiO_2 buffer layer. Films without Bi were also deposited at two different deposition temperatures, in order to compare doped and undoped film results. Figure 2 shows the values of the surface resistivity for various films deposited at 310 °C and 340 °C. The thickness of the films has been estimated to be in the range 1000-1500 Å. Values of surface resistivity for Bi doped films deposited at 340 °C are out of range of the measuring instrument, >1000 MΩ/□. A downward parabolic-like temperature dependence of the deposition temperature has been usually found with our SnO_2:Bi films, with a minimum around 310 °C[4]. This minimum resistance value can be varied according to the water concentration of the solution and adapted to the required surface resistivity. In the case of undoped SnO_2 the resistivity shows a gradual decrease by raising the deposition temperature. This result is fully consistent with previously published results[5]. The difference in the resistivity obtained between the Corning substrates and the soda-lime glass may be related to diffusion of impurities which also affect nucleation and growth of the films, that also results in the observed change in the grain size of the polycrystalline films. Figure 3 are SEM micrographs showing the morphology of polycrystalline films of a) SnO_2 and b) SnO_2:Bi films deposited on Corning substrates. It is evident on both cases a microcristalline structure, with a grain size an order of magnitude smaller in the SnO_2:Bi films. A slight growth of grain size on increasing the film deposition temperature has been systematically observed in our experiments; the Lifshitz-Slezov model of the growth mechanism of polycrystalline films provides good agreement with this experimental evidence [6]. The increase in surface resistivity in the higher temperature deposition range in the SnO_2:Bi films is related to a pronounced increase in Bismuth concentration, as evident from a compositional analysis of Bismuth and Tin previously reported[4] with a Bismuth-Tin weight ratio ~0.3-0.4. Various measurements already published[4,5] and the strong mismatch between Bi and Sn ion size supports the hypotesis that grain boundaries act as segregation regions for Bismuth where it easily oxidizes, being the

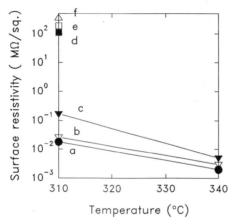

Fig.2). Surface resistivity for various films deposited at 310 °C and 340 °C. a) SnO_2films deposited on Corning 7059. b) SnO_2 films deposied on soda-lime glass with a buffer layer of SiO_2. c) SnO_2 deposited on soda lime glass. d) SnO_2:Bi on buffered soda-lime glass e) SnO_2:Bi on Corning glass f)SnO_2:Bi on soda-lime substrate.

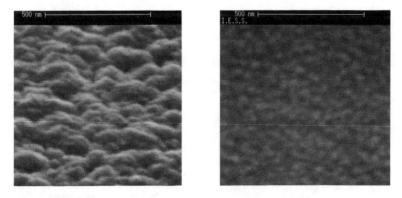

Fig. 3. SEM micrographs of a) SnO_2 films and b) f) SnO_2:Bi films deposited on Corning substrates at 310 °C .

Gibbs free energy of Bi_2O_3 formation ΔG = -493 KJmol compared to ΔG = -520 KJmol for SnO_2[7]. In this situation the tunneling across grain boundaries may be the dominant effect on the carriers transport, as already reported by Kojima et al [8]. SnO_2:Bi films with thicknesses of 1000-1500 Å show a mean value of the optical transmittance equal to 0.8, measured in the wavelength range from 0.4 and 1.0 μm.

After deposition films are exposed to various thermal, photolithographic and deposition steps which contribute to build an SSFLC display device. Further steps, like harsh thermal treatments were added in the fabrication process. This further steps accelerate aging processes in order to study the film stability. One of the main concerns in using such a high resistivity films is related to resistance variations which can degrade the performance of the final device. The influence of oxygen surface species on semiconductor oxides has been clearly identified in studying mechanism of gas sensing[9]. Physisorbtion of oxygen and various species which get in touch with the film during the various process steps lead systematically to a variation of the resistance between metal tracks. Despite most of the steps performed during the device fabrication are wet processes, which could dramatically increase the conductivity of our films, the total variations observed in the films deposited on Corning and buffered soda-lime substrates are relatively small and can be accounted for using different starting values of surface resistivity films. Metal track made of Silver over Cromium, serving as column electrodes were patterned on the high resistance film. Figure 4 shows measurements of track to track, "t-t" resistance, in two different cells on Corning and buffered soda-lime substrates versus two main steps required for the fabrication of the device and a further aging thermal treatment. Unreproducible results were obtained in the non-buffered soda-lime substrates and consequently were not considered for device fabrication. The initial value of the t-t resistance measured immediately after the stripes photolitography is a close to the calculated value, which is directly related to the original surface resistivity measurement. A nylon film, employed to align the FLC, is spin-coated on top of the metal tracks. The nylon film deposition requires a prebake at 90 °C for 20 minutes and a recrystallization annealing which is performed at a temperature of 160 °C for 4 hours. The two steps produce a factor of two increase in the resistance. The resistance is then further increased by a 17 hours aging thermal treatment at 120 °C. The steps were performed in atmospheric pressure and in vacuum oven without appreciable differences.

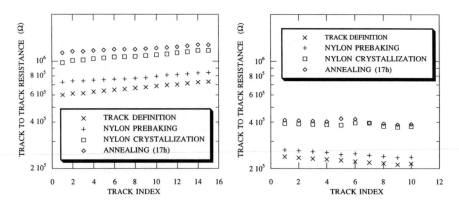

Fig. 4. Track to track resistance measured on (left) Corning glass and (right) on SiO_2 buffered soda-lime substrate.

SSFLC DISPLAY WITH ANALOG GREY SCALES

As previously described the SnO_2:Bi films are used as high resistance transparent electrodes in order to obtain gradation in surface stabilized ferroelectric liquid crystal panel displays[1]. The display operates by controlling the liquid crystal area switched between adjacent metal tracks, hence this grey scale technique can be named switching area control (SAC). The main role is played by the high resistivity of the SnO_2:Bi coating layer which induces a gradient of the electric field from one powered metal track to the other. Once a certain row electrode has been selected the switching area on that row, besides the metal tracks of the columns, is controlled by modulating the amplitude or the position of the data voltages applied to the tracks, during the writing pulse of the selection waveform applied to that row. In this way the electric field gradient is responsible for domain creation and switching process growing gradually from the metal tracks. Fig. 5 shows 4 pictures of an area of the display of about 10 mm^2 between two crossed linear polarizers. The dark lines represent the column tracks, Fig 5a and 5d show black and white states, while Fig. 5b and 5c are intermediate grey shades, obtained by means of four different data voltages applied to the columns. According to this approach a single pixel can be defined as the transparent space on a row electrode besides a single column metal track either up to half way to neighboring tracks or up to them. Such a former configuration does not imply neither loss in resolution nor an augmented complexity in the connections with respect to a standard display with same number of row and column electrodes. In the latter configuration all odd tracks are connected to a data independent waveform and this implies a reduction of resolution by 50%, even though a better control of the switching areas is guaranteed. An interesting characteristic of SAC technique combined with bistability of FLC's is that the gradual switching area is made up of fully switched domains hence the grey shades are stable and non volatile even after unplugging the display. In other words each pixel can memorize any grey shade, so that an image can be stored in the display at any time and even after disconnecting the power supply.

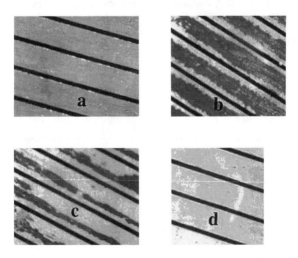

Fig. 5. Four different grey shades in an analog SSFLC display using SnO_2:Bi as high resistance transparent film: (a) black, (b) and (c) intermediate greys, (d) white [1].

CONCLUSIONS

Bismuth doped high resistivity tin oxide has been deposited by spray-pyrolysis on various type of glass substrates. The film deposited on Corning glass and on soda-lime glass buffered with SiO_2 are suitable for fabrication of analog passive matrix display using ferroelectric liquid crystals. The stability of the electrical properties have been investigated. Different grey shades in an analog FLC display have been shown.

ACKNOWLEDGMENT

We are in debt to Giovanni Petrocco for his technical assistance in providing SEM pictures.

REFERENCES

[1] P. Maltese, F. Campoli, A. D' Alessandro, V. Foglietti, A. Galbato, A. Galloppa, G. Rafaelli and M. Wnek, FLC 95 Proceedings of the International Ferroelectric Liquid Crystal Conference, Cambridge 1995, Ferroelectrics, in press.
[2] J.C. Manifacier, L. Szepessy, J.F. Bresse and M. Perotin, Mater. Res. Bull. **14**, 163 (1979).
[3] P. Grosse, F.J. Schmitte, G. Frank and H. Kostlin, Thin Solid Films **90**, 309 (1988).
[4] V. Foglietti, A. Galbato, A. Bearzotti, A. Galloppa and P. Maltese, in Polycrystalline thin films, Mater. Res. Soc. Proc., MRS (1995), in press.
[5] A. De Acutis, M.C. Rossi, M. Barbetti, in Flat Panel Display Materials , Mater. Res. Soc. Proc., MRS Vol. 345, 235 (1994).
[6] I.M. Lifshitz and V.V. Slezov, J. Phys. Chem. Solids, **19**, 35 (1961).
[7] The data are taken from TAPP, thermochemical and physical properties, ES microware.
[8] M. Kojima, H. Kato, A. Imai and A. Yoshida, J. Appl. Phys., **64**, 1902 (1988).
[9] P.T. Moseley, D.E. Williams, in Techniques and mechanism in gas sensing, edited by P.T. Moseley, J.O.W. Norris and D.E. Williams, Adam Hilger(1991)ISBN 0-7503-0074-4.

Part V

Field Emission Display Materials and Technology

THE EFFECT OF TECHNOLOGY AND MATERIALS CHOICE ON FIELD EMISSION DISPLAY PERFORMANCE

MARKO M. G. SLUSARCZUK
Silicon Video Corporation, 6580 Via del Oro, San Jose, CA 95119, mslusarczuk@svc.com

ABSTRACT

The Field Emission Display (FED) is often portrayed as the emerging alternative to the Active Matrix Liquid Crystal Display (AMLCD). FED displays are based on cold cathodes that emit electrons which are accelerated across a potential in a vacuum to strike a phosphor and emit light. There are numerous material and technology choices at each point of the display system design. Each alternative is characterized by advantages and consequences that affect subsequent choices and system performance. This paper discusses the alternative approaches to cathode material and structure, electron beam focusing, support structure, phosphor and system integration. These choices are related to ultimate display performance and cost.

INTRODUCTION

The Field Emission Display (FED) is a complex arrangement of components and materials that must function together as a system. The key elements of this system are: the planar electron source or cathode; the phosphor coated faceplate that generates the light; and the physical structures that maintain vacuum integrity and provide the mechanical strength to withstand atmospheric pressures. Each choice of material that comprises each element of this system carries with it an associated performance, cost, and manufacturability trade off. In many cases, once a particular decision is made, it automatically eliminates other options and sets the constraints on subsequent decisions. Together these choices determine whether the resulting product is competitive in the marketplace or simply remains a laboratory curiosity. This paper analyzes the higher order decisions and their impact on FED performance.

As a basis for the analysis, it is important to establish at the outset the criteria against which subsequent decisions are evaluated. In the case of an FED, these criteria are set by the market, and today the market is determined by the Active Matrix Liquid Crystal Display (AMLCD). The final FED product will be measured by performance criteria and product opportunities that are defined by the AMLCD. "Close" and "almost good enough" are virtual guarantees of market failure.

The evaluation of each trade decision must be carried out on the basis of its impact on three dimensions: cost, performance and manufacturability. In the competitive market place, cost is perhaps the most critical dimension. Cost, however, is not set by today's prices of comparable AMLCDs, but rather, by the anticipated price of comparable AMLCDs when the FED enters the market.

In today's market, performance is likewise set by the AMLCD, and consists of such parameters as power consumption, weight, form factor, brightness, contrast ratio, viewing angle, color gamut, and lifetime. Most of these individual parameters can be enhanced at the expense of others, i. e. brightness can be improved at the expense of power consumption. The market place, however, has established a demand trend for portable applications. This in turn, means that power consumption, weight, and brightness (sunlight readability) become the dominant parameters.

The third dimension, manufacturability, is one that is most often overlooked at the research level. It includes process robustness, particle and defect sensitivity, and the number of critical masking operations. The processes must be robust, that is, the choice of material and process must be such that it can tolerate fluctuations likely to be encountered during manufacturing without significantly affecting final system performance. Second, the less sensitive the process and materials are to particulate contamination and defects, the higher the final yield. Third, critical masking operations reduce yield and throughput, and increase overall product cost.

ANALYSIS

Given the above basis for analysis, let us now proceed to look at the effect of material choices on FED system design and performance. The analysis will proceed along the subsystems of cathode, faceplate and physical structures.

Cathode

The cathode is an addressable planar source of electrons. A number of different approaches and materials have been developed that accomplish this task, such as gated tips (Spindt emitter), edge emitters, volcanoes, and diamond films. The most commonly used of these technologies is the gated emitter, and for that reason will be the technology discussed in this paper.

The gated emitter consists of a sharp tip, usually metal, that is fabricated so as to be centered on an aperture in a metal electrode that forms the gate. When, in a vacuum, a sufficient voltage is applied between the tip and the gate, the electric field will cause electrons to emit from the tip surface towards the gate. Some electrons strike the gate, but others pass through the aperture into the vacuum. The voltage that is required to extract the electrons depends on a number of parameters, including the distance of the tip to the edge of the gate, the sharpness of the tip, the properties of the tip material, and the surface properties of the tip.

Generally, the electric field necessary for electron emission to occur is on the order of 10^7 V/cm. This implies that the smaller the gate aperture, the lower the voltage that is required to stimulate electron emission. Lower voltage is desirable for a number of reasons:

- **Lower Power** - parasitic capacitance between the address electrodes results in a switching power loss that is equal to $1/2\ C\ V^2$. The lower the voltage that must be switched, the lower the switching power loss.

- **Lower Cost Drivers** - at a gate aperture of one micron, about 80 Volts is needed between the gate and emitter tip. At a gate aperture of 200 nanometers, less than 20 Volts is required. At the lower voltage, conventional CMOS electronics can be used. At the higher voltage, more expensive drivers are necessary.

- **Lower Beam Divergence** - the voltage on the gate imparts kinetic energy onto the electrons that has components orthogonal to, and in the plane of the gate electrode. The voltage component in the plane of the gate causes the electron stream to diverge. By reducing the overall gate voltage, this beam spreading effect can be minimized. A narrower spatial distribution of electrons is easier to focus onto the appropriate phosphor dot on the faceplate, thereby helping to maintain color purity.

There are additional benefits of a smaller gate aperture size:

- **Higher Redundancy** - with a smaller gate aperture, more emitter structures can be fit in the sub-pixel space. If a particle or other defect causes a certain number of emitters not to function, there will be a sufficient number of emitters left so that the pixel remains functional. This translates to higher defect tolerance and higher yield.
- **Higher Uniformity** - with more emitters per sub-pixel, fluctuations in emission properties of individual tips average out, improving pixel-to-pixel uniformity.
- **Lower Current per Emitter Tip** - with a larger number of emitter tips per sub-pixel, each emitter needs to contribute less to the pixel beam current. This in turn lowers the possibility of tip damage, improving display life and reliability.

Given that small gate apertures are desired what are the materials and processing challenges of achieving such structures?

- **High Quality Dielectric** - the gate and addressing electrode must be separated by a dielectric. As the aperture decreases in size, the dielectric layer must be fabricated thinner and thinner. This in turn calls for a higher quality, pinhole free dielectric layer.
- **Low Dielectric Constant Material** - decreasing the space between the electrodes increases the capacitance and thereby the switching power loss due to parasitic capacitance. A lower dielectric constant material can help compensate for this effect.
- **High Aspect Ratio Emitter Tips** - the emitter tips are formed by evaporating material through the gate hole aperture. Typically, tips fabricated in this manner have an aspect ratio of 1:1. As the gate hole dimension shrinks, the dielectric layer must be made thinner so that the tip remains in close proximity to the gate hole edge. The limitations on dielectric thickness described above can be overcome if the emitter tip can be formed with an aspect ratio such that it is taller than it is wide at the base.
- **Novel Lithography Techniques** - the traditional method of exposing large substrates, the large area stepper, can achieve resolutions only as low as one micron. To achieve sub-micron features, a novel tool, such as holographic lithography, is required. Such new tools and processes require innovations in resists, developers and other processes. Any innovations in this area, however, must consider tool cost, throughput, and tool maintenance requirements

Faceplate

The phosphor on the faceplate converts the energy of the impinging electrons into light. The choice of phosphor material sets many of the design criteria for the rest of the display. Specifically, phosphors can be classified as those that require electrons that have been accelerated through a high voltage potential and those that can function with electrons that have been accelerated through a low voltage potential. Although there is no specific demarcation that divides these two classes of phosphors, low voltage is usually at about 500 Volts, while the high voltage regime is generally considered to be above 4 000 Volts.

The choice of the type of phosphor material has major implications for system design and performance. A discussion of the ramification of the two choices follows.

High Voltage Phosphors

Color high voltage phosphors have been developed for the CRT and have been used commercially for over fifty years. They are well characterized, power efficient, with proven long life, optimized colorimetry, and are available from multiple vendors in quantities that can sustain high volume display production.

High voltage phosphors can be "aluminized." Electrons that have been accelerated through a several kilo Volt potential can pass through a thin layer of aluminum and still retain sufficient energy to excite a phosphor. There are several reasons why one would choose to place a thin layer of aluminum between the electron source and a phosphor:

- When an electron strikes a phosphor particle light is generated in all directions. Normally the light that is generated towards the rear of a display and away from the viewer is lost and represents wasted energy. The aluminum layer acts as a mirror and reflects light that is generated towards the rear of the display back to the viewer;
- The aluminum acts as a protective layer that minimizes the possibility of materials leaving the phosphor and contaminating the emitter tip;
- Some form of conductor is necessary to uniformly distribute across the faceplate the electric potential that attracts the electrons. The aluminum layer serves as this electrode, and also provides a return path for the electrons. Without such a layer, a transparent electrode, such as indium tin oxide (ITO) must be deposited on the glass faceplate. Such an electrode is not perfectly transparent and will attenuate some of the light generated by the phosphor.

Although high voltage phosphors have numerous advantages, their use in a field emission display presents a number of significant engineering challenges. Because the cathode and faceplate are at a several kilo Volt potential difference, the space between the two must be sufficient to prevent arcing and breakdown. To withstand a 5 kV potential, the gap must be about 1.2 mm. This large gap carries two consequences. First, the electrons diverge sufficiently within that distance to require some form of active focusing. This is an added design complexity and cost. Second, the design of the internal support structure becomes difficult.

The market dictates that the display be lightweight. This in turn calls for thin glass or other materials for the faceplate and cathode substrate. But given the vacuum inside the display, some form of periodic internal support is necessary to keep the faceplate and cathode apart. The requirements on this structure are:

- **Optically invisible** - the viewer of the display cannot see the support structure. The structure must be thin enough to fit in the approximately 100 microns (at 80 lines per inch) between pixels. Thus, the support material itself must be on the order of 50 microns.
- **Mechanically strong** - the support material must withstand at a minimum atmospheric pressure, and more desirably the stresses and shocks that a display would be subjected to during normal use. Given that the structure requires an aspect ratio of about 25:1 and can only be 50 microns thick, the material must be very strong.
- **Somewhat electrically conductive** - the surface of the material must have some level of electrical conductivity. If it were not conductive, electrons that strike the surface would cause it to charge and thereby would affect the path of the electron beam. The degree of conductivity, however, must be sufficiently low to prevent significant power drain through the supports.

- **Low secondary electron coefficient** - if an electron strikes the support structure, it should generate as few secondary electrons as possible. Such spurious electrons can affect the color purity by exciting phosphor patches other than the desired one.
- **Vacuum compatible** - the material must be capable of operating in vacuums of 10^{-8} and better without outgassing or otherwise contaminating the vacuum.
- **Compatible with phosphors** - certain materials, such as copper, "poison" phosphors and cause their emissivity to change.
- **Proper coefficient of thermal expansion** - the FED requires a thermal bake cycle to desorb gases and ensure a good vacuum. The CTE of the supports must match that of the faceplate and cathode to prevent damage during this thermal cycling.

The engineering challenge of identifying a material that simultaneously meets all of the requirements is significant. Another factor that enters into the design equation is that the material must be sufficiently low cost and easy to fabricate into the desired shape so that the resulting display still meets the price objective.

Low Voltage Phosphors

Low voltage phosphors offer several key engineering advantages over high voltage phosphors. Since the gap between the faceplate and cathode can be much smaller, the electron beam does not diverge to the point that active focusing is required - proximity focusing is adequate. Furthermore, because the gap is much smaller and the voltage is much lower, simpler materials, such as small glass spheres, can be used as spacers.

Notwithstanding the significant engineering advantages that low voltage phosphors offer, they also carry significant drawbacks. First of all, low voltage phosphors are still in the research and development stage. They are not yet available in production quantities from qualified vendors. Nor does their spectral output meet the colorimetry requirements of displays. Most significantly, their power efficiency is much lower than that of high voltage phosphors.

The lower luminous efficiency has significant implications for both the power consumption of the display and the lifetime of the phosphor itself. Under electron bombardment phosphors suffer from cumulative damage and their luminous output decreases. Table 1 presents a comparison of high and low phosphors driven to generate an equal amount of light.

Electron Energy at Phosphor	500 V	4,000 V
Phosphor efficacy (l/W)	5	10
Power to faceplate (W)	3.32	1.66
Average current density (mA/cm^2)	17	1.1
Time to 1/2 brightness (hours)	3,300	51,000

Table 1: Power and lifetime implications of phosphor choice for a 11.3-inch diagonal display yielding 20 fL through a 50% filter.

As is apparent from Table 1, low voltage phosphors consume more power for the same light out as high voltage phosphors. Furthermore, because of the increased beam current to the phosphor, the Coulomb damage process is accelerated resulting in a much lower expected time to half brightness.

Table 2 summarizes the implications of the phosphor material choice.

	High Voltage	**Low Voltage**
Advantages	- Can be aluminized (capture more light) - High efficiency (lower power) - Long life - Good colorimetry - Qualified products available in quantity	- Do not require focusing - Simple spacer/support structure
Penalties	- Tall thin, strong internal support spacers - Spacers must be electrically invisible - Spacers must be optically invisible - Focusable emitters	- Short phosphor life - Low efficiency (higher power) - Unknown colorimetry - Materials still in research stage

Table 2: Comparison matrix for high and low voltage phosphors.

System Integration

The FED system is a vacuum system containing numerous materials and components. As such, the various components must be assembled, aligned, and evacuated for the system to function as a display. The evacuation and seal processes present several significant materials challenges.

First of all, the individual display components must be assembled and sealed with some form of adhesive. Since during the vacuum processing steps the display will be subjected to temperatures up to 400 C, this adhesive material must:

- Have a coefficient of thermal expansion that matches the other components;
- Be able to withstand high temperatures without degrading;
- Provide a leak tight seal;
- Not outgas materials that impair the vacuum;
- Have high strength;
- Be an insulator (cathode and faceplate are at a high potential difference);
- Not react with metals or other system components.

As discussed previously, the market requires the display to have a thin profile. This in turn, implies that the device must be assembled and evacuated without a tubulation port through which the vacuum is drawn and which is then sealed to perfect the vacuum. Assembly, evacuation, and sealing operations must be developed that allow a thin profile.

The vacuum must be drawn and maintained within a device that has capillary flow dimensions, and has a very high surface to volume ratio. Such a system requires a getter to adsorb any gaseous species that evolve during device operation. Absent a tubulation port into which the getter material can be inserted, and because of the market's requirement for a compact form factor, there is very little space within the envelope for the getter material. Whatever material is used must have high capacity and high pumping speed. The getter material and method for activating it presents a significant materials challenge.

CONCLUSIONS

The FED is a complex system of interrelated components. Each materials or technology decision carries with it an impact on overall system performance and cost. The challenge for the scientist and engineer is to perform the trade off decisions while maintaining a system perspective across technical disciplines that may be outside his/her area of expertise. Local optimization of subsystem performance cannot be carried out at the expense of overall system performance and cost.

EMISSION CHARACTERISTICS OF THE MO-COATED SILICON TIPS

Heung-Woo Park, Byeong-Kwon Ju, Jae-Hoon Jung, Yun-Hi Lee, Jung-Ho Park*,
In-Jae Chung**, M.R.Hascard** and Myung-Hwan Oh
Dept. of Electronics and Information Technology, KIST, 130-6, P.O.Box 131, Cheongryang,
Seoul, Korea, bkju@willow.kist.re.kr
* Dept. of Electronics Eng., Korea Univ., Seongbuk-Ku, Seoul 136-701, Korea
** Microelectronics Centre, Univ. of South Australia, S.A., 5098, Australia

ABSTRACT

Uniform and reproducible silicon tip arrays were fabricated using the reactive ion etching followed by the re-oxidation sharpening. Molybdenum was then coated on the some of the silicon tip array. Current-voltage characteristics and current fluctuations were measured in the high vacuum environment. Field emission currents were proved by the Fowler-Nordheim plot studies.

INTRODUCTION

The performance improvement of cold emitters using Fowler-Nordheim tunneling phenomena has been studied. Materials, such as molybdenum, and single crystalline silicon are often used as materials for field emitters. Field emitter array is applicable to flat panel displays, sensors and analytic equipments like SEM, AFM and STM.

Silicon wafers have an advantage of compatibility for standard I.C.process, but they have some shortcomes of instability of emission characteristics and destruction of tip array. To overcome these problems, the coating of some materials like oxide, refractory metal, silicides, diamond and diamond like carbons on the silicon emitter is studied.[1],[2],[3],[4]

Effects of the coating of proper materials on the silicon tip field emitter array are as follows.

• Emission stability of the tip can be increased by coating of physically and chemically stable materials.

• Thermal stability of the tip can be increased by improvement of the thermal conduction characteristics owing to the getting rid of thermal energy effectively from the field emission array.

• The lower turn-on voltage and the higher emission currents can be expected by coating the materials of low work function on the silicon tip array.

• Life times of the silicon tip array can be increased by coating the materials of large hardness on the silicon tip array preventing from the destruction of tip from ion bombardments.

In this study, uniform and reproducible silicon tip array was fabricated and molybdenum was coated on some of the silicon tip array. Electrical characteristics of them were measured.

EXPERIMENT

Heavily doped n-type (100)-silicon wafer was used in this experiments. After RCA cleaning, silicon wafers were thermally oxidized to get 1μm-thick oxide layer. Through the photolithography process, pattern of 220x220 array of 1.5μm diameter circles is transferred on the oxide layer. They were etched using reactive ion etching method and re-oxidized at 950°C. Oxide layer was etched using buffered oxide etchant to obtain the silicon tips. Molybdenum thin film(100Å thick) was evaporated with electron-beam evaporator on the very tip of some of the silicon tips.

Tip geometries of bare silicon tips and Mo-coated silicon tips were inspected by SEM (scanning electron microscope). Electrical characteristics of bare silicon tips and Mo-coated silicon tips were measured in the ultra high vacuum chamber.

371

RESULTS

Fig. 1 shows a fabricated silicon tip array and molybdenum-coated silicon tip array. By using the reactive ion etching and re-oxidation sharpening, very good uniformity of the tip array could be obtained. The geometric change by the molybdenum coating on the silicon tip array was insignificant.

Current-Voltage Characteristics

Fig. 2 shows current-voltage characteristics of the silicon tip array and the molybdenum-coated silicon tip array. Measurement was done in the ultra high vacuum chamber with a turbo pump and an ion pump. The vacuum level in the chamber was maintained under 10^{-7} Torr, and the distance between cathode and anode plate was set to $1\mu m$. For stable emission characteristics, a sample was annealed for 1 hour at 200°C in the vacuum chamber. Stabilization of the emission currents was observed after the slow increase of emission current for 30 minutes. Turn-on voltage of the silicon tip array was about 35V and maximum emission current was about 1mA, while turn-on voltage of the molybdenum-coated silicon tip array was 15V and maximum emission current was about 6mA. Emission current was significantly increased by the molybdenum coating.

Current Fluctuations

Current fluctuations of the silicon tip array and molybdenum coated silicon tip array are shown in Fig. 3. Bias voltage was $1/\sqrt{2}$ times the voltage for the maximum emission current. While emission current of the coated tip was 4mA and stable, current of the uncoated tip was 500μA. Furthermore, the uncoated tip was destroyed after 13 minutes and rapid lowering of the emission current was observed. It is thought that these results originated from the properties of the molybdenum. Melting point and hardness of the molybdenum are higher than those of silicon, and work function of the molybdenum is lower than that of silicon. By molybdenum coating, tip became harder and more stable in thermally, physically and chemically. Properties of the often used refractory metals and metal silicides for coating were shown in table 1.

Table 1.   Properties of refractory metals and refractory metal silicides

	Melting point [°C]	Resistivity [Ω -cm]	Thermal expansion Coefficient [ppm/°C]	Thermal conductivity [W/cm · °C]
Si	1412	$\sim 3.15 \times 10^5$	2.33	1.5
W	3382	5.3	4.5	-
Mo	2617	5.5	5	1.38
Ti	1690	43 ~ 47	8.5	-
Ta	2996	13 ~ 16	6.5	-

Fowler-Nordheim Plots

Fowler -Nordheim plots of the uncoated tip and the coated tip are shown in Fig. 4. The linearity of the plot proves that the current is obtained by the field emission. The effective emission area(α) and the field conversion factor(β) can be calculated from the Fowler-Nordheim equation using the slope and intersection with vertical axis in Fig. 4. Variables used in the calculation, effective emission area and field conversion factor are shown in the table 2. The table shows that the effective emission area of the coated silicon tip is hundred times lager than that of uncoated tip. This result somewhat explains why the emission current of the coated silicon tip is lager than that of the uncoated tip.

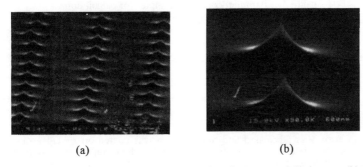

Fig. 1. SEM photographs of Si tip array(a) and Mo-coated Si tip array(b)

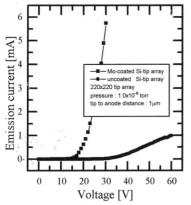

Fig. 2. I-V characteristics of Si tip array
and Mo-coated Si tip array

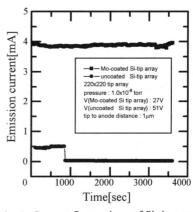

Fig. 3. Current fluctuations of Si tip array
and Mo-coated Si tip array

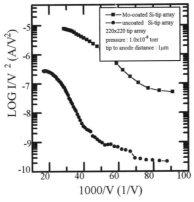

Fig. 4. Fowler-Nordheim plots of Si tip array
and Mo-coated Si tip array

Table 2. Parameters of each samples for calculation

	Voltage[V]	Current[mA]	Φ[eV]	α[m^2]	β[m^{-1}]
uncoated Si-tip array	40	0.227	4.312	1.061×10^{-15}	3.568×10^6
	50	0.644	4.312	1.326×10^{-15}	3.568×10^6
	60	0.985	4.312	2.101×10^{-15}	3.568×10^6
Mo-coated Si-tip array	20	0.62	4.24	2.633×10^{-15}	3.254×10^6
	25	2.33	4.24	3.127×10^{-15}	3.254×10^6
	30	5.75	4.24	3.907×10^{-15}	3.254×10^6

CONCLUSIONS

Silicon tip array was fabricated and molybdenum was coated on it. Tip array was very uniform and reproducible owing to the reactive ion etching and re-oxidation sharpening process. Turn-on voltage of the uncoated tip was 35V and maximum current was 1mA while turn-on voltage of the coated tip was 15V and maximum current was 6mA. Current fluctuations of the uncoated tip and coated tip were compared. While uncoated silicon tip was destroyed after 13 minutes of operation and showed rapid lowering of emission current, coated tip showed stable emission characteristics. By coating molybdenum film on the very silicon tip, tip became harder and more stable.

ACKNOWLEDGMENTS

This work was suppoted by the Korea Institute of Science and Technology. The authors are very grateful to Ms. T.Y. Kim for helpful discussions and analyses of the samples.

REFERENCES

1. D.W. Branston and D. Stephani, IEEE E.D. **38**, 2329 (1991)
2. J.C.Jiang and R.C. White, SID '93 Digest, 596 (1993)
3. R.A. King et al., J. Vac. Sci. Technol. B **13**(2), 603 Mar/Apr (1995)
4. J. Liu et al., J. Vac. Sci. Technol. B **13**(2), 422 Mar/Apr (1995)

ELECTRON EMISSION FROM DIAMOND AND CARBON NITRIDE GROWN BY HOT FILAMENT CVD OR HELICAL RESONATOR PECVD

Eung Joon Chi, Jae Yeob Shim, Soon Joon Rho, and Hong Koo Baik
Department of Metallurgical Engineering, Yonsei Univ., Seoul 120-749, Korea

ABSTRACT

To improve silicon field emitters, diamond and carbon nitride coatings were applied by hot filament CVD and helical resonator CVD, respectively. Helical resonator CVD was first proposed as a new method to grow carbon nitride on silicon tips without damage. Diamond and carbon nitride coatings lowered turn-on voltage and increased emission current. Microstructural and electronic investigations of carbon nitride film were performed. The negative electron affinity of carbon nitride is suggested for enhancing emission current.

INTRODUCTION

To enhance the emission current and chemical stability of silicon field emitters, various materials have been coated on silicon. Among these materials, diamond has attracted much attention due to its unique properties such as negative electron affinity[1], high thermal conductivity, extreme hardness and good chemical stability. It is also regarded as a possible candidate for a cold cathode in field emission displays. Many researchers reported that diamond could emit electrons at very low electric field[2, 3]. But the emission mechanism is not clearly understood. In addition, control of the uniformity and microstructure is needed for field emission device applications.

In this study, carbon nitride film by helical resonator PECVD was coated on the silicon tips and tested as a new field emitter. Carbon nitride is a material that, if it can be made to have stoichiometry of β-C_3N_4, would have the same structure as β-Si_3N_4 and shows properties comparable to those of diamond[4]. Also, the thermal stability of carbon nitride is expected to be even better than that of diamond. By helical resonator PECVD, amorphous carbon nitride films were synthesized. Though having an amorphous structure, it shows high hardness, low friction coefficient and chemical inertness. Helical resonator PECVD is a deposition process in which an intense plasma can be sustained below 10mTorr[5]. This process offers low energy particles at low pressure and allows an uniform coating of carbon nitride without damaging the sharpness of the silicon tips. The use of carbon nitride as a field emitter as reported here is a new development in this area. Here we will show the I-V characteristics and microstructural analysis of carbon nitride field emitters, together with those of diamond emitters.

EXPERIMENT

Silicon tips were prepared from vapor-liquid-solid grown whiskers. Diamond was deposited on silicon tips by hot filament CVD. Details of the fabrication process are described elsewhere[6].

Carbon nitride was also deposited on silicon tips by helical resonator PECVD of which a schematic diagram is shown in Fig. 1. Main deposition parameters are summarized in Table I.

The field emission tests were performed with a diode structure in an ultra high vacuum chamber. The measurements were carried out at a pressure less than 5×10^{-9} Torr. The tip heights of the various emitters used for current-voltage measurements were almost the same(about 100μm high). The distance between emitter and anode was kept within 200μm. For evaluation of material characteristics of carbon nitride film, some samples were prepared on planar substrates of silicon or glass. SEM and Raman spectroscopy were used to examine the tip morphology and the structural quality of the films, respectively. With a UV-visible spectrophotometer, the optical bandgap of carbon nitride film was calculated.

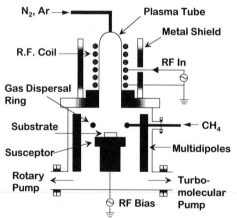

Fig. 1 Schematic diagram of helical resonator
PECVD

Table I. Deposition conditions for carbon
nitride by helical resonator PECVD

RF frequency	13.56 MHz
RF power	100 W
Base pressure	5×10^{-6} Torr
Working pressure	9×10^{-4} Torr
CH_4 flow rate	3 sccm
N_2 flow rate	10 sccm

RESULTS AND DISCUSSION

SEM micrographs of silicon and diamond coated silicon field emitter arrays are shown in Fig. 2 (a) and (b), respectively. The silicon tips were very uniformly arrayed and diamond particles were continuously deposited on silicon tips. The field emission of the above samples showed typical characteristics as reported earlier by others[7, 8]. That is, lower turn-on voltage and higher emission current of diamond coated emitter were exhibited compared with silicon emitters (see Fig. 3).

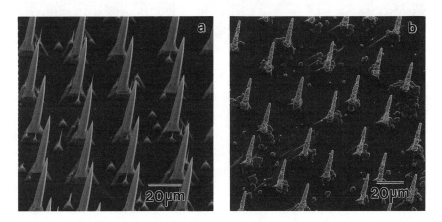

Fig. 2 SEM micrographs of (a) silicon and (b) diamond coated silicon field emitters

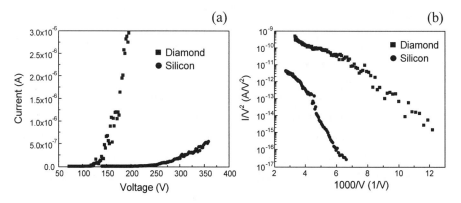

Fig. 3 (a) I-V curves and (b) F-N plots for silicon and diamond coated silicon field emitters

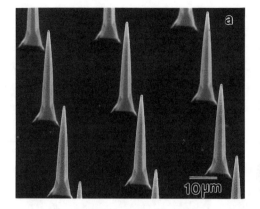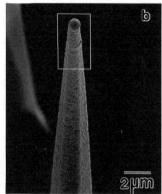

Fig. 4 SEM micrographs of silicon field emitter after carbon nitride coating
(a) array and (b) magnified view of single tip

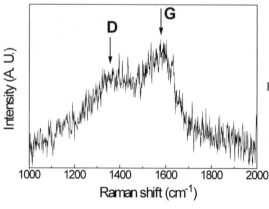

Fig. 5 Raman spectrum of carbon
nitride film grown by helical
resonator PECVD

Fig. 4 shows SEM micrographs of silicon tips coated by carbon nitride films. Though the apex of tip became a little blunt after coating, the total geometry did not change from the original shape. The films were coated very uniformly showing smooth surface morphology. By Raman spectroscopy, the structural quality of carbon nitride also was investigated (Fig. 5). The peaks at about 1360 and $1590 cm^{-1}$ are ones generally observed in amorphous carbon nitride films. Though not analyzing the accurate concentration of nitrogen in the film, it can be known that the film had amorphous structure and contained in it chemical bonding between carbon and nitrogen atoms.

An I-V curve and a F-N plot for a carbon nitride coated emitter are shown in Fig. 6. The turn-on voltage of this emitter was comparable to that of a diamond coated emitter, about 150V, and the emission current for carbon nitride was even higher than for diamond. Detailed investigation of this behavior is still progressing.

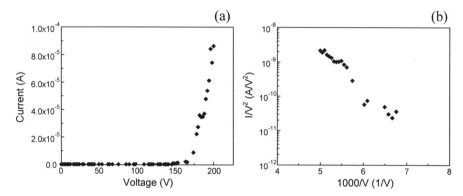

Fig. 6 (a) I-V curve and (b) F-N plot for carbon nitride coated emitter

One possible mechanism of enhanced emission current would be a negative electron affinity(NEA) of carbon nitride film. So far, no study has reported whether carbon nitride exhibits NEA or not. But many physical properties of carbon nitride, such as hardness, thermal conductivity and band gap, are very similar to those of diamond which has NEA. In addition, many nitrides like AlN and GaN also show NEA characteristics. Therefore, it might be expected that carbon nitride also has NEA. To examine the electronic structure of the carbon nitride film, UV-visible absorption spectroscopy was carried out. From this result(not shown in this paper), the optical bandgap of carbon nitride in this study was calculated at 1.75eV. This value is much larger than 1.12eV of bandgap of silicon. Since the carbon nitride film was uniformly coated on the silicon tip and emitted many electrons, it should have a certain feature that can enhance field emission. But what the feature is like is not clear and further investigation is needed.

CONCLUSIONS

Diamond and carbon nitride were used as coating materials for enhancing silicon field emitter. Both diamond and carbon nitride lowered turn-on voltage and increased emission current. In the case of carbon nitride, more emission current was obtained than with diamond. For this phenomenon, negative electron affinity of carbon nitride was suggested. It has been found that carbon nitride grown by helical resonator PECVD has an amorphous structure and an optical bandgap of 1.75eV.

ACKNOWLEDGMENT

The authors wish to acknowledge E.I. Givargizov and V.V. Zhirnov for providing the silicon emitters and the diamond coated silicon emitters.

REFERENCES

1. F. J. Himpsel, I. A. knapp, J. A. Van Vechten and S. E. Eastman, Phys. Rev., B(20), p. 624 (1979)

2. C. Wang, A. Garcia, D. C. Ingram, M. Lake and M. E. Kordesch, Electronics Lett., 27, p. 1459(1991)

3. M. W. Geis, J. C. Twichell, C. O. Bozler, D. D. Rathman, N. N. Efremow, K. E. Krohn, M. A. Hollis, R. Uttaro and T. M. Lyszcarz, Abs. 6th Int. Conf. Vacuum Microelectronics, p. 160 (1993)

4. A. Y. Liu and M. L. Cohen, Phys. Rev., B(41), p. 10727 (1990)

5. J. M. Cook, D. E. Ibbotson and D. L. Flamm, J. Vac. Sci. Technol., B(8), p. 1 (1990)

6. E. I. Givargizov, J. Vac. Sci. Technol., B(11), p. 449 (1993)

7. E. I. Givargizov, J. Vac. Sci. Technol. , B(13), p. 414 (1995)

8. D. Hong and M. Aslam, J. Vac. Sci. Technol., B(13), p. 427 (1995)

PROPERTIES OF DIAMOND-LIKE CARBON FOR THIN FILM MICROCATHODES FOR FIELD EMISSION DISPLAYS

J Robertson and W I Milne,
Engineering Dept, Cambridge University, Cambridge CB2 1PZ, UK.

ABSTRACT

Diamond-like carbon is a strong candidate for field emission microcathodes for field emission displays because of its low electron affinity and chemical inertness. The field emission properties of various types of diamond-like carbon such as a-C:H and ta-C are reviewed in the framework of a bonding model of their affinity.

INTRODUCTION

The flat panel display market is presently dominated by a-Si based active matrix liquid crystal displays (AMLCDs). However, some deficiencies of AMLCDs such as poor viewing angle and high power consumption make it worth investigating alternative technologies. A very promising alternative is the field emission display (FED). In an FED, electrons are emitted from a matrix of microcathodes and accelerated across a narrow vacuum gap to excite a screen of phosphor pixels. The advantages of FEDs are that they could retain the high picture quality of cathode ray tubes, are thin, as the electron beam is not scanned, and have low power consumption because they use field emission rather than thermionic emission and are an emissive not subtractive display, so each pixel only passes current when it is on.

Existing FEDs use 'Spindt' tips of Mo or Si as shown schematically in Fig. 1(a) [1]. Tip radii of order 20 nm are used to give the high electric fields ≈ 10 V/nm needed for field emission from materials like Mo and Si, which have high electron affinities of 4 - 5 eV. Recently, Pixtech have demonstrated working FEDs based on this principle. However, tips can suffer from reliability and lifetime limitations due to the erosion and poisoning.

An alternative is to use thin films of materials which naturally emit electrons at a low field - low electron affinity materials. Hydrogenated diamond surfaces have a negative electron affinity (NEA) [2-5] and in principle can emit electrons with a minimal barrier. However, the high 800°C growth temperature of CVD diamond is incompatible with glass substrates. Amorphous, diamond-like carbon (DLC) is a low electron affinity material which can be deposited cheaply at room temperature onto glass or plastic substrates. The absence of tips and the chemical inertness of DLC minimises the erosion and poisoning of the cathodes.

Figure 1(b) illustrates a schematic of the simplest DLC -based FED. The DLC is deposited as a series of thin film microcathodes connected as the rows of the matrix. The phosphor pixels are anodes and are connected as columns. This design is relatively simple and economic and is being developed by Si Diamond [6,7]. Pixel addressing in this design requires the switching of the full anode voltage.

An alternative is to use a triode configuration, Figure 1(c). Here the emission field is generated by gate electrodes placed close to the cathodes, which can therefore be addressed by low voltages compatible with Si integrated circuits, while the phosphor voltage is unswitched.

Mat. Res. Soc. Symp. Proc. Vol. 424 © 1997 Materials Research Society

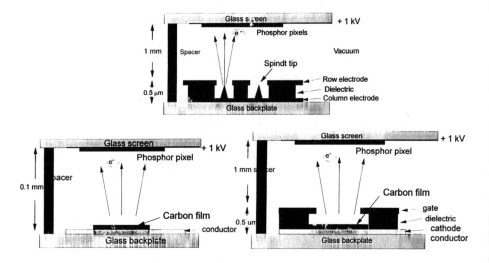

Fig 1. (a) Field emission displays using Mo Spindt tips, (b) DLC in diode design, (c) DLC in triode design.

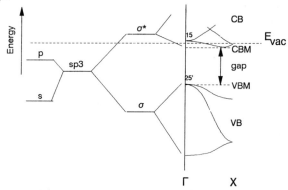

Fig. 2. Development of bands in covalent systems showing how the conduction band minimum (CBM) can lie above the vacuum level, giving a negative electron affinity.

ELECTRON AFFINITY

Electron emission from natural and chemical vapour deposited (CVD) diamond surface have received considerable attention because of its negative electron affinity [2-5,8-12]. However, as diamond has a wide band gap with no shallow donors, it has been found that the efficiency of field emission scales with the inverse of CVD diamond quality [10], suggesting that the low resistance grain boundary paths were needed for current flow.

The origin of the low electron affinity in diamond and other wide band gap covalent solids like AlN and cubic boron nitride (c-BN) have been shown to derive from their

382

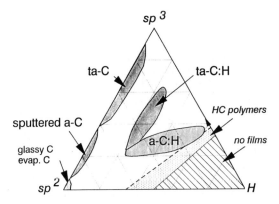

Fig. 3. Ternary phase diagram of C:H system showing types of diamond-like carbon

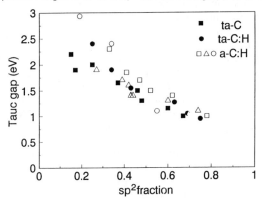

Fig. 4. Optical gap versus sp^2 fraction for various types of diamond-like carbon

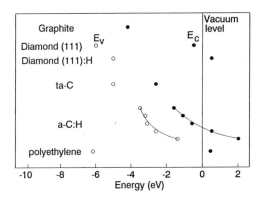

Fig. 5. Experimental band energies with respect to the vacuum level for various carbon systems.

electronic bonding [13], Figure 2. In diamond, bonds are formed by the interaction of sp^3 hybrids, giving rise to bonding (σ) and antibonding (σ^*) states, which develop into the valence and conduction band, respectively. The valence band maximum (VBM) level must lie below the vacuum level, but there is no a-priori reason that the conduction band minimum (CBM) must. Thus, the origin of the NEA in diamond and c-BN is primarily their wide band gap.

DIAMOND-LIKE CARBON

Diamond-like carbon (DLC) is a form of amorphous carbon (a-C) or hydrogenated amorphous carbon (a-C-H) containing a significant fraction of sp^3 bonding [14-16]. The amount of sp^3 bonding and hydrogen in any DLC depends on the deposition process used and can be displayed on a ternary phase diagram, Fig. 3. The DLCs are in the upper part of the figure. The formation of sp^3 sites in DLC require that deposition occurs from a source of medium energy ions. The most popular process to make a-C:H is plasma enhanced chemical vapour deposition (PECVD). This a-C:H is very useful but has too much hydrogen for some applications. A modest amount of sp^3 bonding can be generated by using ion assisted sputtering, while reactive sputtering can generate a-C:H.

The highly sp^3 bonded forms of a-C, which we denote tetrahedral amorphous carbon or 'ta-C', can be formed by deposition from highly ionised ion beams of appropriate energy. Three techniques are typically used, the mass selected ion beam (MSIB)[17], the laser ablation of graphite [18] and the filtered cathodic vacuum arc (FCVA)[19]. The laser ablation method directs a high powered UV laser at a graphite target. The FCVA technique is arguably most appropriate for industrial scale up. It consists of an arc stuck on a graphite target, which produces a highly ionised plasma beam, which is passed through a magnetic filter to remove particulates and neutral species, and then condensed on the substrate. The ion energy of the incident beam is varied by applying a negative bias voltage to the substrate. A variant of this method uses a compacted graphite powder target, and can be operated without filter as it emits many fewer particulates [20]. This method is able to coat large areas without rastering. The optimum ion energy for the FCVA is found to be about 100 eV, for which the sp^3 content can reach 80-85% [19]. A type of a-C:H with rather low hydrogen content and up to 75% sp^3 bonding has recently been prepared from a plasma beam source (PBS)[21]. This material is denoted ta-C:H, by analogy to ta-C.

The electronic structure of a-C depends on its sp^3 content. sp^3 sites form only σ bonds to adjacent sites, while sp^2 sites form both σ and the weaker π bonds [14]. The π states lie closest to the Fermi level and so they determine the band gap. An experimental compilation found that the optical band gap of a-C:H, ta-C:H and ta-C depends principally on the sp^2 content [22], as shown in Fig. 4. The hydrogen content is less important.

The arguments of the previous section suggest that the electron affinity (EA) of covalent systems (without surface layers) depend primarily on their bonding [13]. As band energies depend on the bonds, not the existence of periodicity, they apply equally well to amorphous carbons. Figure 5 shows a compilation of the band energies of carbon systems with respect to the vacuum level. Dehydrogenated diamond surfaces have a small positive EA. The hydrogenated surface has an NEA of order -1 eV [5]. It has been argued that this difference arises from an internal charge transfer within the C-H bond which gives rise to a small surface dipole term [13]. Fig.

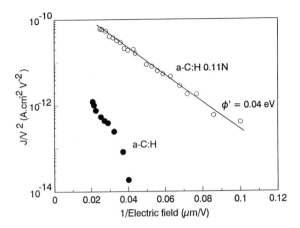

Fig. 6. Fowler-Nordheim plot of field emission current density vs. electric field for a-C:H and a-C:H:N after [28].

5 also shows the EA of ta-C derived from I-V measurements of a ta-C/c-Si heterojunction [23]. This gives a sizeable positive EA, but this may be too large, in the light of our field emission measurements. The VBM energies of a-C:H have been derived from photoemission partial yield measurements [24], and the CBM values found by adding the optical band gaps. These show that a-C:H can develop an increasingly negative EA as the optical gap increases. It shows that wide gap a-C:H has a sizeable NEA. These wide gap a-C:H have high sp^3 contents and high H contents. They also have very high resistivities, well above 10^{12} Ω.cm, which could limit overall emission currents. The overall data for C-H systems suggest that the EA for a-C:H decreases with both increasing sp^3 content and increasing H content, that is away from the sp^2 corner of Fig. 3.

On the other hand, a-C has an important advantage over diamond in that it is possible to dope it n-type. There are no shallow, soluble n-type dopants for diamond itself; nitrogen forms a deep level ≈ 1.4 eV below the CBM. This level is now shallow in a-C:H because of its narrower band gap. Thus the a-C:H can be doped both n- and p--type [14]. The n-type doping of ta-C has recently been demonstrated [25,26].

ELECTRON EMISSION FROM DLC

There have been a number of measurements of field emission from a-C and a-C:H. Feng et al [27] described field emission from ta-C deposited by laser ablation. This work was principally concerned with the areal homogeneity of the emission, which will be an important issue for FEDs. A direct study of field emission from a-C:H was performed by Silva and Amaratunga [28]. They found a turn on field of about 26 V/μm for a-C:H of band gap 2.0 eV. The emission was greatly increased for a-C:H modified by nitrogen. 11 atomic % of N was found to reduce the turn-on field to 10 V/μm. Similar experiments have been carried out on ta-C deposited at an ion energy of 100 eV. A turn-on field of order 12 V/μm was found for these preliminary measurements. This was reduced to ≈ 5 V/μm for N-doped samples.

The field emission current densities should obey the Fowler-Nordheim relationship

$$J = a.E^2. \exp(- b\phi^{3/2}\beta/E) \qquad (1)$$

where ϕ is the barrier height, a,b are constants and β is the so-called field-enhancement factor for sharp geometries. Fig 6 shows the emission currents plotted for a-C:H and a-C:H:N. The slopes of the curves are found to correspond to $\phi = 0.05$ eV and 0.04 eV for $\beta = 1$. These low effective barrier heights account for the low turn-on voltages of a-C:H systems. They show that a-C:H and ta-C will have great promise for FED microcathodes.

However, the interpretation of the emission data still needs development. Firstly, the change in conductivity activation energy of a-C:H with N-doping suggests that doping raises the Fermi level by about 0.4 eV. Thus, the change in effective barrier height should be well over 0.01 eV. This requires β to be well over 1. The need for $\beta > 1$ is somewhat unexpected for DLC films, as these are usually very smooth, of the order 1 nm. Thus, the details of the emission process still require further study.

REFERENCES

1. I Brodie, Proc IEEE **82** 1006 (1994)
2. F J Himpsel, J A Knapp, J A vanVechten, D E Eastman, Phys Rev B **20** 624 (1979)
3. B B Pate, Surface Sci **165** 83 (1986)
4. J van der Weide, Z Zhang, P K Baumann, M G Wensell, R J Nemanich, Phys Rev B **50** 5803 (1994)
5. C Bandis, B B Pate, Phys Rev B **52** 12056 (1995)
6. N Kumar, H K Schmidt, Society for Information Display Digest '94 43 (1994)
7. N Kumar, H K Schmidt, C Xie, Solid State Technol **38** p71 (May 1995)
8. W S Xu, Y Tzeng, R V Latham, J Phys D **26** 1776 (1993)
9. W S Xu, R V Latham, Y Tzeng, Electronics Lett **29** 1596 (1993)
10. W Zhu, G P Kochanski, S Jin, L Seibles, J Appl Phys **78** 2707 (1995)
11. W Zhu, et al, Appl Phys Lett **67** 1157 (1995)
12. M W Geis, J C Twichell, J Maculey, K Okano, Appl Phys Letts **68** 1328 (1995)
13. J Robertson, Diamond Related Mats **5** xxx (1996)
14. J Robertson, Adv Phys **35** 317 (1986)
15. J Robertson, Surface Coatings Technol **50** 185 (1992)
16. J Robertson, Pure Appl Chem **66** 1789 (1994)
17. Y Lifshitz, S R Kasi, J W Rabalais, Phys Rev Lett **68** 620 (1989)
18. D L Pappas,et al, J Appl Phys **71** 5675 (1992)
19. P J Fallon, V S Veerasamy, C A Davis, J Robertson, G Amaratunga, W I Milne, Phys Rev B **48** 4777 (1993)
20. M Chhowalla, M Weiler, C A Davis, B Kleinsorge, G Amaratunga, Appl Phys Lett **67** 894 (1995)
21. M Weiler, S Sattel, H Ehrhardt, J Robertson, Phys Rev B **53** 1594 (1996)
22. J Robertson, Phys Rev B (to be published 1996)
23. V S Veerasamy, et al, Solid State Electronics **37** 319 (1994)
24. J Schafer, J Ristein, L Ley, J Non-Cryst Solids **164** 1123 (1993)
25. V S Veerasamy, et al, Phys Rev B **48** 17954 (1993)
26. J Robertson, C A Davis, Diamond Related Materials **4** 441 (1995)
27. F Zeng, I G Brown, J W Ager, J Mater Res **10** 1585 (1995)
28. G Amaratunga, S R P Silva, Appl Phys Letts, to be published

Gold Overcoatings on Spindt-Type Field Emitter Arrays

S. L. Skala*[a], D. A. Ohlberg**, A. A. Talin** and T. E. Felter**
*Coloray Display Corporation Fremont, CA 94539
**Sandia National Laboratories, Livermore, CA 94551

ABSTRACT

The electron emission properties of a Spindt-type field emitter array have been measured before and after deposition of approximately 100 Å of gold. The workfunction of the emitter decreased by 5% after gold deposition resulting in an 11% reduction in turn-on voltage. Emission stability as measured by RMS current noise improved by 40%. Improvements in emission do not withstand exposure to air. However, baking at moderate temperatures (200°C) restores the emission improvements obtained with the gold overcoating. Fowler-Nordheim plots show that the enhanced emission after baking is due to a increase of the Fowler-Nordheim intercept and not a decrease in slope. Additionally, the gold over coatings resist poisoning as a 50,000 L dose of oxygen only slightly affects emission.

INTRODUCTION

Many people have used various coatings and treatments to improve the emission characteristics of field emitters. Coatings of electropositive elements can result in very low extraction voltages, but are typically poisoned by as little as a few Langmuirs of oxygen or other residual gasses.[1,2] Cleaning techniques such as field forming or hydrogen plasma treatments[3] significantly improve emission, but these must be carried out *in situ* of an evacuated and sealed enclosure. This requirement adds considerable cost to manufacturing by either requiring long field-forming times or equipment necessary to perform *in situ* plasma treatments. Work a few years ago by Ishikawa *et al*[4] showed that a gold overcoating significantly reduces the extraction voltage and improves the emission stability of a single tip field emitter. In this work, we show that gold over coatings not only improve emission characteristics of Spindt cathode arrays but that the emission improvements can be restored after air exposure by moderate baking in vacuum.

EXPERIMENTAL

A 10,000 tip Spindt cathode obtained from SRI International was used for the experiment. Gold was evaporated from a gold-coated tungsten wire located approximately 2 cm from the cathode. The thickness of the gold film deposited on the cathode was estimated to be 100 Å from the gold coating on a glass witness slide adjacent to the cathode.

Measurements were taken with aid of a computer running a IEEE-488 interface. A cylindrical anode located 1 cm from the cathode and biased to 500 V collected the cathode current. Chamber pressure during all current measurements was 1×10^{-9} Torr. Numerous I-V's were taken at each data point and all Fowler-Nordheim slope and intercept data was calculated using an average of at least 10 individual I-V plots. Noise measurements were taken by recording anode current at a sampling frequency of 5.7 Hz. The plots of current noise shown in Figure 2 were obtained by normalizing the anode current vs. time data by the average anode current. RMS noise values were determined by averaging the RMS variation of 10 equal segments of the normalized anode current plots. This averaging eliminated the effect of current variations with a time period greater than ~35 seconds from the reported RMS noise values.

Mat. Res. Soc. Symp. Proc. Vol. 424 © 1997 Materials Research Society

The cathode was tested for durability immediately prior to the initial emission measurements by drawing approximately 5 μA emission current for 2 hours.

RESULTS

Figure 1 shows the I-V relation and Fowler-Nordheim plots of the Spindt cathode before gold deposition, after gold deposition, after venting the gold-coated cathode to air and after moderate baking of the gold-coated cathode. I-V, Fowler-Nordheim and noise data are summarized in Table 1. Gold coating of the Spindt cathode moderately improved emission performance. The turn-on voltage, as determined by the gate bias necessary to obtain 1 uA anode current, decreased by 8%. A decrease in the slope of Fowler-Nordheim plots following gold deposition implies a slight decrease in workfunction of ~4.5% assuming no change in the field enhancement factor.

Gold overcoating significantly reduced emission current noise. Figure 2 shows plots of the normalized emission current over time and indicates the magnitude of emission noise. It is readily apparent from Figure 2 that emission stability is enhanced after deposition of gold. Comparison of the RMS noise measurements shows nearly a 40% reduction in current noise following gold deposition.

TABLE 1. Summary of I-V, Fowler-Nordheim and noise measurements for the Spindt cathode at various points before and after gold deposition. V@ I = 1μA is used as an estimation of turn-on voltage and I @ V = 64 V is used to compare emission currents. RMS noise measurements were taken at an avarege current of approximatly 7.5 μA.

Description of Data	F-N Slope	F-N Incpt.	V@ I=1μA	I@ V=64V	Norm. RMS Noise
Before Gold Deposition	-675	5.60x10-5	56.5 V	6.1 μA	0.0429
After Gold Deposition	-629	7.34x10-5	52 V	18.2 μA	0.0248
After Vent to Air	-712	1.28x10-4	56 V	7.7 μA	0.0455
After Bake to180 ^{O}C	-760	8.25x10-4	51.5 V	25.7 μA	0.0172
After 50,000 L O$_2$ Dose	-690	2.18x10-4	52.5 V	18.1 μA	0.0193

After gold deposition and emission measurements, the cathode was exposed to air to determine if air will poison a gold-coated cathode. The vacuum chamber and cathode were vented to air for ~5 minutes, pumped down and baked to 85 °C over night. Emission and noise characteristics returned to nearly those of the uncoated cathode after exposure to air, thus confirming that air exposure does poison a gold-coated cathode. However, baking the cathode at 180 °C for approximately 10 minutes under vacuum (1 x 10^{-9} Torr) restored and even slightly enhanced the improvements in emission characteristics obtained by gold overcoating. Analysis of the Fowler-Nordheim plots indicated that baking results in a

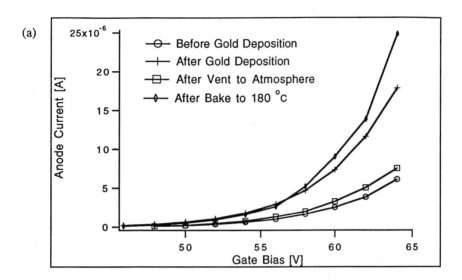

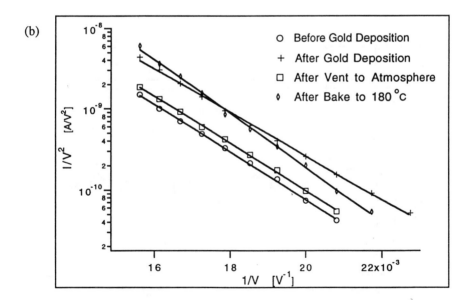

FIG. 1. (a) Anode current vs. gate bias (b) Fowler Nordheim plots for the Spindt array at various points before and after gold deposition.

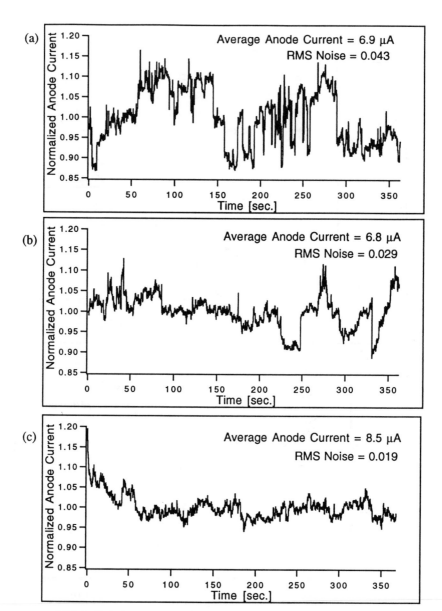

FIG. 2. Emission current vs. time for the Spindt cathode (a) before gold deposition, (b) immediatly after gold deposition and (c) after vent to air and 180°C bake.

workfunction higher than that of an untreated cathode. Consequently, emission improvements after baking to 180 °C are due to a increase in the F-N intercept, indicating an increase in emission area.

After baking, the cathode was exposed to a moderate dose (50,000 L) of oxygen to estimate the ability of a gold-coated cathode to withstand contaminates that a sealed cathode will be exposed to during the expected lifetime of a display. As shown in Table I, emission performance does decrease slightly, but still remains significantly better than that of an uncoated cathode.

DISCUSSION

Air exposed field-emission cathodes are known to exhibit poor stability due to the presence of an oxide layer and other surface contaminants. Emission stability improves after running a cathode for long periods of time, which presumably desorbs most surface contaminants, or by cleaning techniques such as hydrogen plasma treatments. Hydrogen plasma treatments have also resulted in a slight decrease in workfunction.[3] Deposition of gold in vacuum is another way of obtaining a clean emitting surface. Furthermore, the inert nature of gold inhibits oxidization of the tip surface when exposed to air. Hence, contaminants are only physisorbed to gold after air exposure and can be desorbed by baking at moderate temperatures.

The behavior of gold-coated cathodes after baking deserves closer attention. Analysis of the F-N data indicates that the emitting area increases significantly after baking. An increase in emission area of ~20x from the untreated cathode was estimated from the product of the F-N intercept and F-N slope squared[5]. This differs from the initial gold deposition where emission improvements were almost exclusively due to a decrease in workfunction. It is probable that baking not only desorbs physisorbed contaminants but induces a morphological change of the emitting surface. The large increase in emitting area is technologically very beneficial since emission noise is reduced and cathode reliability improved. However, recent work by Schwoebel et al[6] indicates that overcoatings of group IVa transition metal elements increases the Fowler-Nordheim intercept but does not increase emission area. Further experiments will be useful in elucidating the physical effects of transition metal overcoatings.

CONCLUSION

We have shown that emission characteristics of Spind-type molybdenum field emitter arrays are improved by deposition of a gold overlayer. Moreover, baking to ~200 °C reverses poisoning effects of air exposure and appears to induce a morphological change to the emitting surface.

ACKNOWLEDGMENTS

This work was funded in part by DOE CRADA 1148. We are grateful for fruitful discussions and the support from Heinz Busta and the rest of the Coloray staff.

(a) Current address VLSI Technology, Inc. San Jose, CA 95131

1 J. M. Macaulay, I. Brodie, C. A. Spindt and C. E. Holland, Appl. Phys. Lett. **61**, 997 1992.

2 L. W. Swanson and N. A. Martin, J. Appl. Phys. **46**, 2029 1975.

3 P. R. Schwoebel and C. A. Spindt, J. Vac. Sci. Technol. **B12**, 2414 1994.

4 J. Ishikawa et al, J. Vac. Sci. Technol. **B11**, 403 1993.

5 C. A. Spindt, I. Brodie, L. Humphrey and E. R. Westerberg, J. Appl. Phys. **47**, 5248 (1976).

6 P. R. Schwoebel, C. A. Spindt and I. Brodie, J. Vac. Sci. Technol. **B13**, 338, (1995).

EMISSION CHARACTERISTICS OF ARRAYS OF DIAMOND-COATED SILICON TIPS

DOUGLAS A.A. OHLBERG, ALEC A. TALIN, AND THOMAS E. FELTER
Sandia National Laboratories, Livermore, California 94550
E. I. GIVARGIZOV
Institute of Crystallography, Leninsky pr. 59, 117333 Moscow, Russia

ABSTRACT

The electron emission behavior of diamond-coated silicon whisker arrays was examined with a point-probe scanner and compared with the behavior of uncoated silicon whiskers as well as with diamond crystallites on a flat Si surface. Emission in all cases was inhomogeneous, and the emission site density was less than the whisker density. Both the diamond-coated and uncoated whiskers displayed a distribution of turn-on fields which peaked at around 10 V/μ, but the uncoated whiskers were observed to burn out if emission outputs exceeded 0.3 μA.

I. INTRODUCTION

Considerable effort has gone into the design and construction of flat panel displays based on field emission devices. In the earliest approach[1], microlithography is used to produce regular arrays of very sharp, micron sized tips. The sharpness of the tips concentrates the field lines sufficiently to permit electron tunneling at relatively low voltage. More recently, thin planar films of diamond or other carbon allotropes deposited onto flat substrates have been found to produce random arrays of rugged emitters.[2], even though the films are rather smooth.

In the present work, we pursue the combination of geometrical field enhancement (sharpened tips) with carbon films. Specifically, arrays of silicon whiskers have been coated with diamond films[3]. The arrays are grown using the vapor liquid solid (VLS) technique developed by Wagner and Ellis,[4] and the diamond is deposited using conventional hot filament CVD of methane seeded in hydrogen. Because the coated tips are expected to be rather inert, they should also be resistant to poisoning.

Previous studies on the emission properties of diamond-coated Si whiskers,[3] were done on single tips. In the present work, we report emission behavior of large ensembles of these tips using a point-probe scanner.

II. EXPERIMENTAL

The samples were fabricated at the Institute of Crystallography. SEM micrographs showed that although the diamond deposition occurred preferentially on the tips in the form of crystallites and agglomerated globules, patches of these features were also observed along the whisker shanks and interstices, as well as at the faceted sample peripheries. The samples were 1 cm^2 in area and contained whiskers in the central region surrounded by a faceted periphery. The whiskers were 130 microns in height.

The point-probe scanner consists of a tungsten tip with a radius of curvature ranging from 1 - 10 mm. Motion of the probe along the x, y, and z axes was controlled by computer-driven stepper motors with a lateral step size of 0.3 mm. The probe was mounted inside a UHV chamber above a stationary platform which docked samples introduced into the chamber via a loadlock. The base pressure of the scanner chamber was 5×10^{-10} Torr. A video microscope allowed visual monitoring.

The gap between the tops of the whiskers and the anode (probe tip) was initially set by lightly touching the sample periphery and then withdrawing the tip by the estimated whisker height and the desired gap. The gap distance could then be more accurately determined by extrapolating from a curve of turn-on voltage vs (relative) distance. Note that the slope of such a plot gives the field required for a given current, since the turn on voltage is the voltage required to produce a specified current.

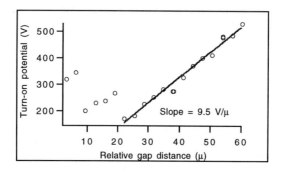

Fig. 1 The potential required for 10 nA of current as a function of relative gap distance. The discontinuity at lower gaps is thought to be due to a break in the whisker caused by probe-whisker contact

III. RESULTS

With the exception of a few missing whiskers at random spots, the arrays imaged by the SEM were quite periodic, see Fig. 2. Nevertheless, the emission maps show no underlying periodicity. Furthermore, the emission site density is much less than the whisker density. SEM images taken at higher resolution were unable to explain the inhomogeneities: no obvious morphological features related to areas of good or poor emission site density. Clearly, emission sites are relatively uncommon, and not associated with electron transport directly along a whisker axis.

The observed inhomogeneity and random distribution of the emission maps is due to the relatively low density of emission sites and the random orientation of these sites about the whisker shanks. Electrons emitted from the shank first experience a radial field driving them away from the axis of the whisker at an azimuthal angle set by the position of the emitter site. After traveling a radial distance on the order of a whisker diameter, the field lines begin to curve, directing the electrons towards the anode.

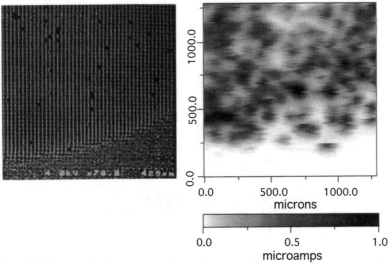

Fig. 2 SEM image of whisker array and emission map of sample over same region. Both image and emission map are on the same scale. The probe-emitter gap was 60 microns, and the gap bias, 1500V.

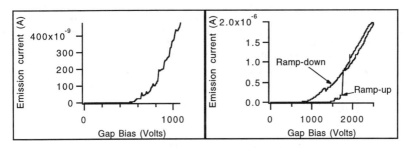

Fig. 3 Examples of initial IV behavior of emission sites. The first example exhibits a smooth turn-on while the second displays an abrupt increase in current after a smooth turn-on.

Measurements of current vs. voltage, taken over random sites above whiskered regions which had undergone no prior exposure to an electric field, revealed a variety of behaviors. In some cases, emission was observed to proceed abruptly after some "event" at the emission site, while in others, smooth curves were seen with no "event". Typical examples are shown in Fig. 3. Because of this range of behaviors, preconditioning an area to high electric field optimizes emission site density as illustrated in Fig. 4. At no point during the conditioning of the emission regions was sparking or arcing, a phenomenon commonly associated with abrupt emission onsets, observed in the video microscope.

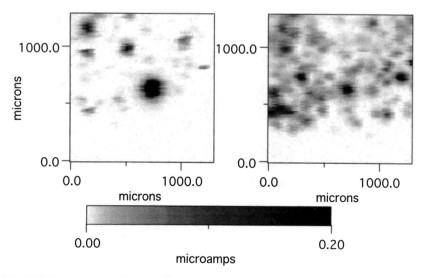

Fig 4. Emission maps made at gap distance of 60 micron and gap potential of 900 V before and after exposure to 1500 V at 60 micron gap.

The emission behavior was compared for diamond coated whisker arrays, uncoated whisker arrays, and diamond crystallites found on the faceted peripheries of the samples. For each case, a region was selected containing twenty to thirty emitters, and for each emitter, a turn-on field for 10 nA of current was determined. The resulting distributions are shown in Fig. 5. Although both the coated and uncoated whisker arrays peak at turn-on fields of approximately 10 V/μ, the diamond coated arrays display a narrower distribution. In either case, the range of turn-on voltages was significantly lower than that displayed by the diamond crystallites at the sample periphery. Clearly, geometrical field enhancement is aiding emission from the whiskers.

Although only minor differences were seen between their respective turn-on fields, diamond-coated Si whiskers proved to be considerably more durable than bare Si whiskers. When driven to emission currents exceeding 0.3 mA, most Si whiskers were observed to burn out immediately, whereas stable emission behavior was seen with the diamond-coated whiskers at current outputs of a few microamps. Subsequent examination of the burned-out Si whisker arrays under an SEM revealed no morphological changes.

IV. CONCLUSION

Diamond coating of silicon whiskers enables an order of magnitude higher current to be obtained before device failure. While the coating does not reduce the onset voltage, it does appear to reduce the distribution of onset voltages. Emission appears to be from the shank of the whisker, with an approximate voltage reduction of four times when compared to coated flat silicon in the periphery of the sample.

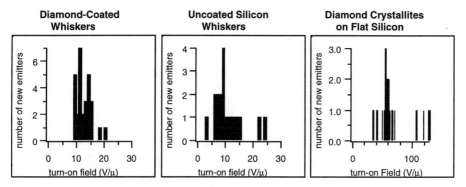

Fig. 5. Distribution of turn-on voltages required to produce 10 nA of emission current for three different types of emitters.

V. REFERENCES

1. I. Brodie and C. Spindt in P. W. Hawkes, ed., <u>Advances in Electronics and Electron Physics: Microelectronics and Microscopy</u>, Academic Press, Inc., San Diego, 1992, pp. 6,7.
2. N. Kumar, H. K. Schmidt, M. H. Clark, A. Ross, B. Lin, L. Fredin, B. Baker, D. Patterson, W. Brookover, C. Xie, C. Hilbert, R. L. Fink, C. N. Potter, A. Krishnan, and D. Eichman, <u>SID 94 Digest</u>, 43 (1994).
3. E. I. Givargizov, V. V. Zhirnov, A. N. Stepanova, E. V. Rakova, A.N.Kiselev, P. S. Plekhanov, Appl. Surf. Sci. **87/88**, 24 (1995).
4. Wagner and Ellis, Appl. Phys. Lett. **4,** 89 (1964).

STUDY ON THE DIAMOND FIELD EMITTER
FABRICATED BY TRANSFER MOLD TECHNIQUE

Byeong-Kwon Ju, Seong-Jin Kim, Jae-Hoon Jung, Yun-Hi Lee, Beom Soo Park,
Young-Joon Baik, Sung-Kyoo Lim*, and Myung Hwan Oh

Div. Electronics and Information Technology, KIST, Cheongryang P.O. Box 131, Seoul, Korea
*Dep. Electronics Engineering, Dankook University, #29 Anseo-dong, Cheonan, Korea

ABSTRACT

Strip-shaped diamond-tip field emitter array was fabricated by using the transfer mold technique. The sharp turn-on characteristic was observed from the current-voltage measurement of the fabricated diamond-tip field emitter array. The turn-on characteristic of the diamond-tip field emitter array was compared with that of a flat diamond film. High emission current density was obtained from the diamond-tip field emitter array. The threshold voltage of the diamond-tip field emitter array was lower than that of a flat diamond film.

INTRODUCTION

The desirable properties of a field emitter array are low operating voltage, high emission current density, and mechanical hardness. Silicon field emitter arrays have been used as electron sources due to simple fabrication processes. But silicon field emitter arrays have some problems due to chemical and physical weakness. In order to resolve the problems, new material systems have been required. Diamond film has the desirable properties and is considered as an attractive candidate for field emitters[1],[2].

The diamond emitter has been generally fabricated in the shape of a flat film. Although electrons can be emitted from the flat diamond film, a tip-shaped diamond has an advantage for better field emission.

In this paper, the strip-shaped diamond-tip field emitter array fabricated by transfer mold technique was investigated[3],[4].

EXPERIMENT

The fabrication procedure of the diamond-tip field emitter array is shown in the Fig.1.

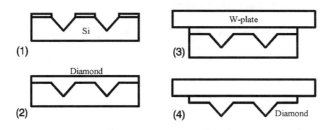

Fig.1 The fabrication procedure

A 2000Å-thick SiO_2 film was grown on a silicon wafer by dry oxidation. 5 μm-wide and 5 mm-long SiO_2 islands were formed by photolithography. The distance between SiO_2 islands was 3 μm. V-grooves were formed on silicon substrates by Ethylenediamine-Pyrocatechol-Water(EPW) solution. A polycrystalline diamond film was deposited by plasma-enhanced chemical vapor deposition(PECVD) on the silicon mold. 2% CH_4 diluted in H_2 was used as a reactant gas. The substrate temperature was kept at 950°C. The deposited diamond was attached to a tungsten plate with silver conducting epoxy. Then, the silicon mold was removed by etching in $HF:HNO_3$ = 1:1 solution and only the diamond-tip field emitter array remained on tungsten plate.

The electrical characterization of the fabricated diamond-tip field emitter array was performed in a vacuum test station under the pressure of $2x10^{-6}$ Torr. The tip-to-anode distance was 100 μm in this experiment. A 100 μm-thick polyimide film was placed between the anode plate and the cathode surface.

RESULTS

The shape of mold

Fig.2 (a) shows the cross section of mold before diamond deposition. The mold has the shape of a V-groove. The walls of the groove are {111} planes which were formed by orientation dependent etching(ODE).

Diamond film was deposited on the fabricated mold as shown in Fig.2 (b). It was found that the mold were completely filled with the diamond.

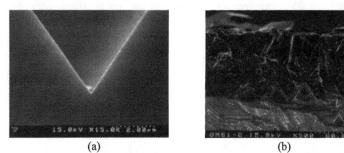

(a)   (b)

Fig.2 (a)A silicon mold and (b)the diamond film deposited on the mold

Geometry of the fabricated diamond-tip emitters

Fig.3 (a) shows the fabricated diamond-tip emitter array after removing the silicon mold. As shown in this figure, a uniform array of the diamond-tip could be obtained by transfer mold technique. Fig.3 (b) shows one of the diamond-tip emitters. The tip radius of the diamond-tip was about 300Å.

Diamond quality of the emitter array

The diamond quality of the tip array was assured by Raman spectroscopy. Strong peak was observed at 1337 cm^{-1} as shown in Fig.4.

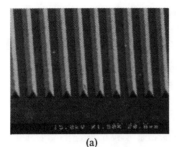 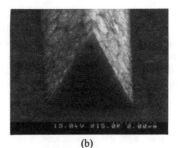

(a) (b)

Fig.3 The fabricated diamond-tip field emitter
(a)diamond-tip field emitter array (b)diamond-tip

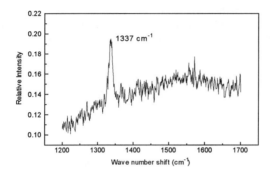

Fig.4 Raman spectra of the diamond-tip emitters

Electrical characterization

Fig.5 (a) shows the current-voltage characteristic of the diamond-tip emitter array fabricated by transfer mold technique.

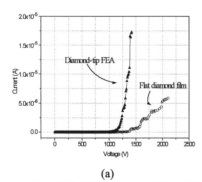 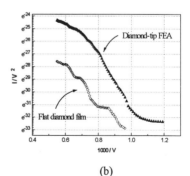

(a) (b)

Fig. 5 Electrical characteristics of the diamond-tip FEA and the flat diamond film.
(a) I-V curves (b) Fowler-Nordheim plots.

The I-V characteristic of a flat diamond film is also shown for comparison in this figure. The sharp turn-on characteristic was observed from the current-voltage measurement of the diamond-tip field emitter array. As shown in the figure, the diamond-tip emitter array has lower threshold voltage than that of the flat diamond film.

Fig.5 (b) shows Fowler-Nordheim plot of the diamond-tip emitter array. As shown in the plot, diamond-tip emitter array has linear Fowler-Nordheim characteristic.

CONCLUSIONS

Strip-shaped diamond-tip field emitter array was fabricated by transfer mold technique. High current density under 2×10^{-6} Torr environment was obtained from the diamond-tip field emitter array. It is shown that the diamond-tip field emitter array has very sharp turn-on characteristic and linear Fowler-Nordheim characteristic. It was found from this experiment that the threshold voltage of the diamond-tip emitter array fabricated by transfer mold technique was much lower than that of the flat diamond film. The doping conditions and the thickness effects of the diamond film are under investigation. It may be possible to use the strip-shaped diamond-tip field emitter array for flat panel displays.

REFERENCES

[1] M. W. Geis, N. N. Efremow, J. D. Woodhouse, M. D. McAleese, M. Marchywka, D. G. Socker, and J. F. Hochedez, IEEE Electron Device Lett., 12, 456, (1991).
[2] W. P. Kang, J. L. Davidson, Q. Li, J. F. Xu, D. L. Kinser, and D. V. Kerns, The 8th Intl. Conf. Solid-State Sensors and Actuators, and Eurosensors IX, 182, (1995).
[3] M. Nakamoto, K. Ichimura, T. Ono, and Y. Nakamura, IVMC'95 Tech. Dig., 186, (1995).
[4] M. Sokolich, E. A. Adler, R. T. Longo, D. M. Goebel, and R. T. Benton IEDM Tech. Dig., 159, (1990).

INVESTIGATION OF ELECTRON EMISSION FROM Si AND HOT FILAMENT CVD DIAMOND

J. Y. Shim, E. J. Chi, S. J. Rho, H. K. Baik

Department of Metallurgical Engineering, Yonsei University, Seoul 120-749, Korea

ABSTRACT

The field emission characteristics of the Si emitters and the diamond coated Si emitters are investigated. The Fowler-Nordheim plots of the two types of Si emitters show linear slopes. It means that the I-V characteristics follow the Fowler-Nordheim relation. Field emission for the two types of diamond coated Si emitters exhibits significant enhancement both in turn-on voltage and total emission current. The Raman spectrum shows that the high intensity graphite peak is observed with diamond peak and thereby large amounts of graphite may be included in the diamond grain boundary. It seems to be thought that the graphite participates in the low field emission. However, further investigations are needed to understand whether the graphite may enhance the emission characteristics of diamond or not.

INTRODUCTION

In spite of high turn-on voltage and low current density, Si emitter is a promising candidate for electron sources in microelectronic devices due to its uniformity, reliability, and compatibility with integrated circuit. Therefore, many researches currently have been made on Si field emitter.

On the other hand, diamond has been also received much attention as a cold cathode material for field emission devices due to its unique properties: negative electron affinity(NEA), good thermal conductivity, chemical and mechanical stabilities[1]. Especially, the fact that the field emission takes place at very low electric fields(3-40V/μm) than Si or metal emitter is attracting a great deal of attention. Many researches are thus focused on the low field emission behavior of flat diamond film and diamond coated tip[2-3]. So far, several papers have been reported to explain low field emission from diamond: negative electron affinity, grain boundary inclusions, high defect density, and microspikes are considered as the origin of low field emission[4-10]. However, it is not clearly understood how electron emission from diamond occurs at such low fields. As the case of Si field emitter, an understanding of diamond field emission characteristics is necessary to develope diamond field emitter device with excellent structural and electrical properties.

In this paper, we report a fundamental study with the diamond and Si field emitters. Field emission was characterized with the emitters of several different heights. Emission characteristics of the diamond and Si field emitters were also discussed.

EXPERIMENT

The uncoated and diamond coated Si emitters were used as a cold cathodes. Si emitters were prepared by vapor-liquid-solid growth method and sharpened by chemical etching and dry oxidation with subsequent HF solution. The Si emitters were p-type with resistivity of about 10Ω-cm. Polycrystalline diamonds were coated by hot-filament chemical vapor deposition (HFCVD). The heights of Si emitters were about 45 and 155μm, respectively. The heights of diamond emitters were 60 and 120μm, respectively. Diameters of Si and diamond tips were below 50μm. Tip-to-tip distance were about 5-30μm. Fig. 1 shows the SEM morphology of the as-prepared samples of uncoated and diamond coated Si emitters.

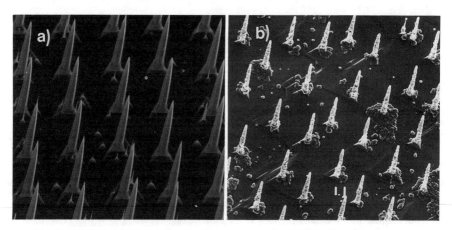

Fig. 1. SEM micrographs of (a) the uncoated Si emitter arrays and (b) the diamond coated Si emitter arrays.

Field emission from Si and diamond emitter was measured by using ultra high vacuum(UHV) I-V test system. UHV I-V test system is consisted of ion pumped main chamber and turbo molecular pumped load-lock chamber. After obtaining base pressure better than 3×10^{-9} torr, field emission measurements were carried out. Distance from a cathode to a tungsten anode was kept within 200µm. Up to 1kV voltage was applied to a tungsten anode to collect electrons emitted from the cathode. The field was varied with increasing the applied voltage at a fixed cathode-anode separation. The emitters were not intentionally pre-annealed before measurement.

Raman spectroscopy was used to evaluate the structural quality of the diamond emitter. SEM was used to observe the surface morphology of the emitters before and after field emission measurement.

RESULTS AND DISCUSSION

A. Si emitters

Fig. 2 shows the relative change in emission currents of the Si emitters versus applied voltage. The data indicated that the field emission occurs in the 45µm Si emitters at 220V and the emission current increases steeply to 0.52µA with increasing the applied voltage. In 155µm Si emitters, the field emission current is detected at about 240V and has a maximum value of 0.41µA. Fig. 3. shows the Fowler - Nordheim characteristics corresponding to the two types of

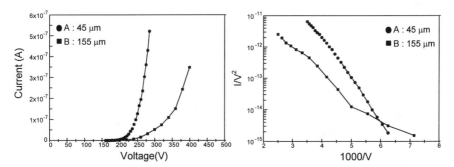

Fig. 2. Field emission characteristics of the
          Si emitters

Fig. 3. F-N plots of the Si emitters

Si emitters shown in Fig. 2. Type A sample has the typical Fowler-Nordheim behavior with a straight slope, which indicates that the I-V characteristics follow the Fowler-Nordheim relation. Type B sample also has a less steep slope than type A.

B. Diamond coated Si emitters

Fig. 4 shows the relative change in emission currents of the diamond coated Si emitters versus applied voltage. The field emission occurs in the 60μm(not shown here) and 120μm diamond emitters at about 100V. The current emitted from 60μm diamond emitters increases up to 0.99μA with applied voltage, while the emission current from 120μm diamond emitters increases to 50 μA. Fig. 5 shows the Fowler-Nordheim characteristics corresponding to the diamond coated emitter shown in Fig. 4. The 120μm diamond coated emitters exhibits characteristic Fowler-Nordheim behavior.

Fig. 6 shows the Raman spectrum of diamond coated Si emitters. The Raman spectrum of diamond coated Si emitters is divided into two vibration modes. The broad Raman peak observed at 1333cm^{-1} is originated from the sp^3 bond of the diamond coated on the Si emitters and may be due to laser scattering at the diamond surface. Typical graphite peak is detected at about 1560cm^{-1} and the graphite peak intensity is comparable to the diamond peak intensity. It seems that large amounts of graphite particles are included in the diamond grain boundary.

Latham et. al. [7-8] suggested that the'antenna effect' is a mechanism for low field cold

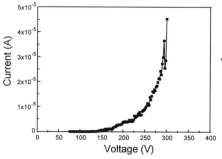

Fig. 4. Field emission characteristic of
the diamond coated Si emitters

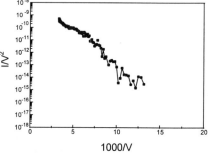

Fig. 5. F-N plot of the diamond coated Si
emitters

emission of diamond. In this theory, the insulating material and the conductive particles participate in the electron emission. First, the field enhancement occurs at the surface due to charge accumulation on the conductive particles embedded in the insulating material and then a eventual dielectric breakdown that allows electron emission. Zhu et. al.[10] reported that the Raman peak width of diamond is strongly correlated with the emission field. High quality diamond with low defect densities and a narrow peak width (FWHM<5cm⁻¹) did not emit electrons due to the highly insulating nature of the high quality diamond. However, the defective diamond with a peak width of 7-11cm⁻¹ exhibited characteristic F-N behavior. Our diamond emitters might have a considerable amount of graphite causing dielectric breakdown of diamond. But, further experiments are necessary to explain whether the graphite inclusions enhance the emission characteristics of diamond or not.

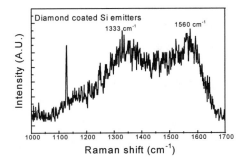

Fig. 6. Rama spectrum of the diamond coated Si emitters

SUMMARY

The Fowler-Nordheim plots of two types of Si emitters show linear slopes, indicating that the I-V characteristics follow the Fowler-Nordheim relation. The diamond coated Si emitters exhibits significant enhancement of field emission characteristics both in turn-on voltage and total emission current than the Si emitters. The Raman spectrum shows that the diamond peak is observed with the high intensity graphite peak. It seems that the graphite is considerably included in the diamond grain boundary and participates in the low field emission. But, further investigations are needed to confirm whether the graphite inclusions enhance the emission characteristics of diamond or not.

ACKNOWLEDGMENTS

The authors wish to acknowledge V. V. Zhirnov and E. I. Givargizov for providing the Si emitters and the diamond coated Si emitters.

REFERENCES

1. F. J. Himpsel, I. A. Knapp, J. A. Van Vechten and D. E. Eastman, Phys. Rev., **B20**, 624 (1979).

2. T. Asano, T. Maruta, T. Ishikura, and S. Yamashita, 8th Int. Vac. Microelectronics Conf. Portland, Oregon 283(1995).

3. W. B. Choi, J. Liu, M. T. McClure, A. F. Myers, J. J. Cuomo, and J. J. Hren, 8th Int. Vac. Microelectronics Conf. Portland, Oregon 315(1995).

4. Z.-H. Huang, P. H. Cultler, N. M. Miskovsky, and T. E. Sullivan, Appl. Phys. Lett. **65**(20), 2562 (1994)

5. C. Wang, A. Garcia, D. C. Ingram, M. Lake, and M. E. Kordesch, Electron. Lett. **27**(16), 1460 (1991).

6. N. S. Xu, R. V. Lantham, and Y. Tzeng, Electron. Lett. **29**(18), 1596 (1993)/4.

7. S. Bajic and R. V. Latham, 2th Int. Vac. Microelectronics Conf. IOP Publishing Ltd. 315(1995).

8. S. Bajic and R. V. Latham, J. Phys. D. **21**, 200(1988).

9. V. V. Zhirnov, E. I. Givargizov, and P. S. Plekhanov, J. Vac. Sci. Technol. **B13**(2), 418(1995)

10. W. Zhu, G. P. Kochanski, S. Jin, and L. Seibles, J. Appl. Phys. **78**(4), 2707 (1995).

PHOSPHOR SYNTHESIS ROUTES AND THEIR EFFECT ON THE PERFORMANCE OF GARNET PHOSPHORS AT LOW-VOLTAGES

L. E. Shea*, J. McKittrick*, and M. L. F. Phillips**
*Department of Applied Mechanics & Engineering Sciences and Materials Science Program, University of California, San Diego, La Jolla, CA 92093-0411
**Sandia National Laboratories, Albuquerque, NM 87185

ABSTRACT

Garnet phosphors have potential for use in field emission displays (FEDs). Green-emitting $Gd_3Ga_5O_{12}$:Tb (GGG:Tb) and $Y_3Al_5O_{12}$:Tb (YAG:Tb) are possible alternatives to ZnO:Zn, because of their excellent resistance to burn, low-voltage efficiency, (3.5 lm/W from GGG:Tb at 800 V), and saturation resistance at high power densities. Hydrothermal and combustion synthesis techniques were employed to improve the low-voltage efficiency of YAG:Tb, and $Y_3Ga_5O_{12}$:Tb (YGG:Tb). Synthetic technique did not affect low-voltage (100-1000 V) efficiency, but affected the particle size, morphology, and burn resistance. The small particle size phosphors obtained via hydrothermal (<1 μm) and combustion reactions (<1 μm) would benefit projection TV, high-definition TV (HDTV), and heads-up displays (HUDs), where smaller pixel sizes are required for high resolution.

INTRODUCTION

Field emission displays (FEDs) require phosphors that are highly efficienct at low voltages (<1 kV). To compete with liquid crystal displays (LCDs) in terms of power consumption, the red, green, and blue phosphor compositions in an FED must have efficiencies of 4, 6, and 1 lumens/watt (lm/W) respectively [1]. Phosphors under consideration for use in FEDs include Y_2O_2S:Eu (red), ZnS:Cu,Al (green), and ZnS:Ag,Cl (blue). Oxide phosphors are preferred in these devices since sulfides are known to contaminate the emitter tips by the release of corrosive, sulfur-related gases (e.g., SO_2, SO) under prolonged electron bombardment [2]. Unfortunately, the cathodoluminescence efficiencies of oxides at low voltages are usually lower than those of sulfides. Because the cathode-to-screen distance in FEDs is much shorter than in a conventional cathode-ray tube (CRT), the phosphor is exposed to higher current densities. Therefore, in addition to low-voltage efficiency, other major requirements for FED phosphors include resistance to saturation and Coulombic aging at high current densities. Garnet host lattices such as $Y_3Al_5O_{12}$ (YAG) are known to resist degradation caused by high power density, and exhibit good chromaticity. Tb^{3+} - doped YAG and related garnet phosphors are typically synthesized by solid-state reaction of yttrium and aluminum oxides at temperatures in excess of 1550°C [3]. Conventional high temperature synthesis yields coarse particles that require grinding to achieve the desired particle size. It is possible to degrade the luminescent properties by the mechanical action of grinding, therefore the process must be precisely controlled [4,5]. The improvement of synthesis technique has been investigated as a possible means of increasing the efficiency of oxide phosphors for application in FEDs. It was the goal of this study to prepare various garnet phosphors via alternative synthesis routes such as hydrothermal and combustion synthesis and compare their low voltage efficiency data to those of solid-state reacted garnets.

Hydrothermal synthesis is a method to produce fine grained phosphors directly without grinding. Oxide phosphors prepared hydrothermally include Zn_2SiO_4:Mn [6], YVO_4:Eu [7], and $Y_3Al_5O_{12}$:Tb [8]. The hydrothermal synthesis of submicron YAG:Tb powders with cathodoluminescent properties similar to those of YAG:Tb prepared via high temperature routes, has been reported [1]. Combustion synthesis is a technique commonly used to produce multicomponent oxide ceramics. The process involves the highly exothermic redox reaction of metal nitrate-carbonaceous fuel precursors. We are the first to apply this technique to the synthesis of luminescent oxides [9]. In this work, we report the hydrothermal synthesis and combustion synthesis of the following garnet phosphors: $Y_3Ga_5O_{12}$:Tb (YGG:Tb) and YAG:Tb.

Mat. Res. Soc. Symp. Proc. Vol. 424 © 1997 Materials Research Society

EXPERIMENT

Synthesis of Phosphors

The starting materials used for the hydrothermal synthesis of YAG:Tb and YGG:Tb were $Y(NO_3)_3 \cdot 4H_2O$ (Aldrich), $Tb(NO_3)_3 \cdot 5H_2O$ (Johnson Matthey), $AlCl_3 \cdot 6H_2O$ (Baker) and/or $Ga(NO_3)_3 \cdot 9H_2O$ (Aldrich). The appropriate nitrates were dissolved in water and 27% aqueous NH_3 was added until a pH of 10 was reached. The resulting gels were filtered, dried at 150°C, and fired at 500°C for 6h in N_2 gas. The amorphous precursors were then ground, individually welded in Au tubes with an equivalent weight of water, and autoclaved (Leco Tem-Pres) at 600°C and 3.2 MPa. Products were recovered by filtration, using Gelman 0.22 μm membrane filters, and heat treated at 1200°C.

For combustion synthesis, the starting materials $Y(NO_3)_3 \cdot 6H_2O$ (Alfa REacton 99.99%), $Al(NO_3)_3$ (Alfa Aesar 99.999%), $Ga(NO_3)_3 \cdot 6H_2O$ (Aldrich) and $Tb(NO_3)_3 \cdot 9H_2O$ (Alfa Puratronic 99.999%) were mixed with the appropriate amount of carbohydrazide, CH_6N_4O fuel (Aldrich), dissolved in deionized water, placed in a 300 mL Pyrex dish, and introduced into a muffle furnace preheated to 500°C. After combustion, powders were crushed in a mortar and pestle and heat treated for 1-2h at 1000-1600°C.

Phosphors were also prepared via solid-state reaction of stoichiometric amounts of Y_2O_3 (Aldrich), $Tb(NO_3)_3 \cdot 5H_2O$ (Johnson Matthey), Al_2O_3 (Ceralox), or Ga_2O_3 (Fluka) at 1200°C for 18h, followed by a grinding step and a final heat treatment at 1450°C for 6h.

Phosphor Characterization

Particle size was determined by scanning electron microscopy (SEM) using a Cambridge 360 microscope. Samples were sputtered with Au before analysis. Cathodoluminescence data were collected from samples packed approximately 1mm deep into stainless steel cups. The samples were placed in a vacuum chamber which was evacuated to a pressure of less than 5×10^{-6} Pa. A hot filament, low energy electron gun (Kimball Physics) was the source of the beam, which was steered and focused using external Helmholtz coils. During analysis the sample cup was maintained at a potential of +100 V with respect to ground. The electron beam was focused onto a spot 5.6 mm in diameter. Luminous intensity of the light emitted from the sample was measured with a Minolta CS-100 photometer.

RESULTS

As a result of hydrothermal processing, the amorphous precursor gels crystallized to form pure YAG:Tb and YGG:Tb powders. The YAG:Tb powder consisted of well formed crystallites with a garnet habit as shown in the SEM micrograph of Figure 1. Particle diameters were between 0.2 and 2 μm, with a median diameter of 0.55 μm. The garnet morphology of YGG:Tb was less defined. This is most likely a result of over-nucleation of the precursor gel, due to firing at an excessively high temperature. The particle size is similar to that of YAG:Tb, with a median particle diameter of 1 μm. Combustion synthesis yielded a continuous, porous network of powders. The particles which make up this network are typically interconnected, forming agglomerates that are easily broken up with light grinding. The presence of pores and voids in the as-synthesized powders is a result of the escaping gaseous products during combustion. Figure 2 shows as-synthesized YGG:Tb particles that are not part of the porous network. These particles have diameters less than 1 μm. After heat treatment to 1400°C for 2h, the individual particles began to sinter and form necks, as shown in Figure 3.

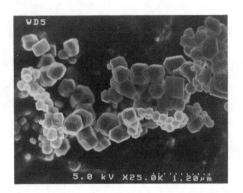

Figure 1. SEM micrograph of hydrothermally synthesized YAG:Tb powder.

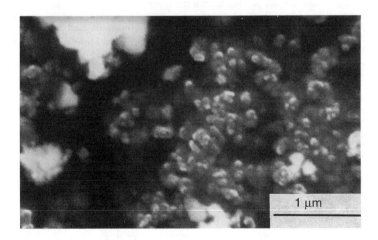

Figure 2. SEM micrograph of combustion synthesized YGG:Tb particles in the as-reacted state.

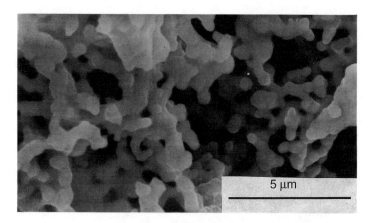

Figure 3. SEM micrograph of combustion synthesized YGG:Tb heat treated at 1400°C 2h.

Low-voltage (200-1000 V) cathodoluminescent efficiency data were collected for solid-state, hydrothermal, and combustion synthesized YAG:Tb and YGG:Tb, at a constant power of 15 μW. Figure 4 shows the efficiency in lumens per watt (lm/W) as a function of electron accelerating voltage for YAG:Tb made by each of the three techniques. The efficiencies for all three phosphors were essentially the same at voltages below 600 V. At these voltages, the penetration depth of the incident electron beam is low, exciting the surface layer of the phosphor particles. In the higher voltage regime, (>600 V), the penetration depth of the electron beam is greater and therefore penetrates deeper into the phosphor particles. The efficiencies of solid state and hydrothermal YAG:Tb at these voltages were approximately 0.5 lm/W greater than combustion synthesized YAG:Tb. It is likely that the combustion synthesized phosphor particles are disordered below the surface layer, hence there is a slight loss of efficiency. This disorder may be a result of the non-equilibrium nature of rapid exothermic reactions.

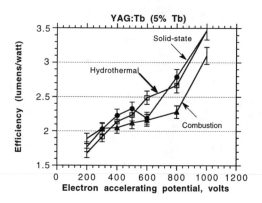

Figure 4. Effect of synthesis route on the CL efficiency of YAG:Tb phosphors.

The saturation resistance of hydrothermal garnets is superior to that of solid-state garnets. Saturation data at high power densities (up to 0.15 W/cm^2) were collected for YGG:Tb at a constant voltage of 600 V. These data are pictured in Figure 5. Though less efficent than the solid-state reacted powder, hydrothermal YGG:Tb retains a larger fraction of its original effciciency after exposure to maximum beam power. Solid-state YGG:Tb sustained more permanent damage, as measured by the loss of efficiency at low power densities. It is possible that YGG:Tb crystallites distribute heat and surface charge more uniformly, reducing the intensity of hot zones.

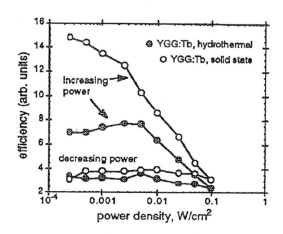

Figure 5. Saturation/burn data for YGG:Tb at 600 V.

CONCLUSIONS

YAG:Tb and YGG:Tb were prepared by hydrothermal, combustion, and solid-state synthesis. The low-voltage efficiency was not appreciably affected by synthetic route at 600 V and below. The differences in efficiency of the YAG:Tb prepared by the three techniques was more pronounced in the 600-1000 V regime. Each process yields phosphors with different physical properties. Smaller, well formed crystallites that are more resistant to burn at high power densities are obtained via hydrothermal synthesis. Combustion synthesized powders are porous networks in the as-synthesized condition. Fine grain size phosphors attainable via hydrothermal synthesis would benefit deposition of screens for FEDs. For optimal screen performance, combustion synthesized powders must be separated prior to deposition.

ACKNOWLEDGMENTS

This work was supported by the California MICRO program and Sandia National Laboratories, Albuquerque, NM. The contributions of Robert Walko, Robert Mays, and Gary Schuster are appreciated.

REFERENCES

1. M. L. F. Phillips and B. G. Potter, Jr., *Proceedings of the ACerS Symposium on the Synthesis and Application of Lanthanide-Doped Materials,* **67**, (1996).

2. S. Itoh, T. Kimizuka, and T. Tonegawa, *J. Electrochem. Soc.*, **136** [6], 1819-1823 (1989).

3. V. B. Glushkova, V. A. Krzhizhanovskaya, O. N. Egorova, Yu. P. Udalov, and L. P. Kachalova, *Inorganic Materials*, **19** [1], 80-84 (1983).

4. N. A. Diakides, *Proceedings of the Society of Photo-Optical Instrumentation Engineers 17th Annual Technical Meeting*, **42**, 83-91 (1973).

5. E. Sluzky, M. Lemoine, and K. Hesse, *J. Electrochem. Soc.*, **141** [11], 3172-3176 (1994).

6. T. Takahashi, Japanese Patent 63,196,683 (1988).

7. R. C. Ropp and B. Carroll, *J. Inorg. Nucl. Chem.*, **39** 1303-1307 (1977).

8. T. Takemori and L. D. David, *Am. Ceram. Soc. Bull.*, **65** [9], 1282-1286 (1986).

9. L. E. Shea, J. McKittrick, O. A. Lopez, and E. Sluzky, accepted to *J. Am. Ceram. Soc.*, 1996.

CHARACTERIZATION OF Y$_2$SiO$_5$:Ce, YAG:Tb AND YAG:Eu RGB PHOSPHOR TRIPLET FOR FIELD EMISSION DISPLAY APPLICATION

A.G.CHAKHOVSKOI*, M.E.MALINOWSKI**, A.A.TALIN**, T.E.FELTER**,
J.T.TRUJILLO*, C.E.HUNT* and K.D.STEWART**
* Electrical and Computer Engineering Dept., University of California, Davis CA 95616
** Sandia National Laboratories, Livermore CA 94551

ABSTRACT

The spectral response and outgassing characteristics of the three new, low-voltage phosphors combustion synthesized and electrophoretically deposited for application in field-emission flat-panel displays, are presented. The phosphors, forming a candidate Red-Blue-Green (RGB) triplet are YAG:Eu, YAG:Tb and Y$_2$SiO$_5$:Ce. These cathodoluminescent materials are tested with electron-beam excitation at currents up to 50 μA within the 200-2000V (eg. "low-voltage") and 3000-8000V (eg. "medium voltage") ranges. The spectral coordinates, as compared with industrially-manufactured P22 phosphors in low voltage operation, are reasonable; however, there is considerable difference in the green coordinates, and the red and green materials show significant satellite intensities. Phosphor outgassing, as a function of time, is measured by a residual gas analyzer at fixed 50 μA beam current in the low-voltage range. We find that after two hours of excitation, levels of outgassed CO, CO$_2$ and H$_2$ stabilize to low values.

INTRODUCTION

The choice of phosphor materials is one of the key factors which determines the future success of field emission display (FED) technology. The proper phosphor must perform satisfactorily with the currents and voltages accessible in a flat panel display. Most commercially available cathodoluminescent phosphors have been developed for application in cathode ray tubes which operate in the 10 to 30 kilovolt range.

Although recently there has been relatively little work done on synthesizing cathodoluminescent materials for low and medium voltage ranges, new materials are now being developed and tested [1-3]. In this work, we examine three candidate phosphors (an RGB "triplet") for application in FEDs: YAG:Eu (red, peak intensity at 712 nm), YAG:Tb (green, peak at 550 nm) and Y$_2$SiO$_5$:Ce (blue, peak at 436 nm).[†] We characterize luminance and chromaticity and the outgassing products and rates of the phosphors as excited in vacuum by an incident electron beam. We characterize the RGB triplet over two voltage ranges: one defined as "low voltage", spreads from 200 to 2,000 V; the other, designated as "medium voltage", extends from 3 kV to 8 kV. The operating voltage in a field emission display can depend on factors such as (1) the construction of the field emitter array (single- or multiple-gated emitters), (2) focusing technique (proximity-focusing, focusing grid, co-planar focusing) and addressing type (addressable pixels in the emitter array or on the screen), (3) the anode-cathode spacing (~ 0.1 - 5.0 mm), and (4) the excitation energy of the phosphor. Limitations and constraints concerning the applicability of low- and high-voltage phosphors to flat panel displays were discussed earlier in [4,5].

PHOSPHOR SCREEN FABRICATION

The phosphor powders were fabricated using a combustion synthesis technique which involved an exothermic reaction of a metal nitride and an organic fuel. The as-synthesized foamy porous powders were heat treated for 1 hour at 1600° C, and electrophoretically deposited onto conductive ITO-coated glass substrates at a constant voltage of 400 V. After deposition, the screens were desiccated at 425° C for 1 hour. [6]

EXPERIMENTAL TECHNIQUE

Two separate ultrahigh vacuum (UHV) systems were used for these experiments. Medium-voltage experiments (3 to 20 kV) were performed at a base pressure of 10^{-9} Torr using a system with a quick-access load-lock mechanism and a thermionic electron gun. We used an unfocused

† - Phosphors were received from AMES Dept., UC San Diego

electron beam with current up to 50 μA. A spot diameter of about 3 mm was used for brightness and chromaticity measurements.

Low-voltage and outgassing experiments were performed in a second UHV system equipped with a low-energy electron gun with voltages up to 1,500 V and a Balzers residual gas analyzer (RGA). This entire system, including the phosphor screen samples, was baked at 150° C for three days prior to taking desorption/optical measurements. After bakeout, the background system pressure (all electron guns on) was ~ 5×10^{-10} Torr. The phosphors samples were loaded in a special carousel [7] allowing the examination of up to 8 phosphors under identical vacuum conditions. Two slots in the carousel were assigned for control samples (an ITO-coated glass and an outgassed Ni foil anode) for background measurements. The Ni anode was also used for preliminary outgassing of the electron gun during a 2-hour warming-up period.

During the gas-desorption measurements, the anode current was maintained at ~ 50 μA for all the samples. The screens were biased to + 100 V with respect to ground, and the low-voltage electron gun was kept at - 650 volts with respect to the ground (total beam energy 750 eV). During the low-voltage brightness measurements, the combined effective energy of the beam was varied from 200 to 2,000 V.

Prior to exposing each sample to the electron beam, a background RGA spectrum, from 1 to 101 AMU was recorded with the electron gun turned on but without applying the bias voltage to the screen sample. After the background was taken, potential was adjusted such that approximately 50 μA of current was collected on the sample. Additional background outgassing occurred when the beam was transported from the gun to the sample. Electron beam desorption of residual gas (typically hydrogen, carbon monoxide and carbon dioxide) is a common effect in UHV systems.

This outgassing, which was an instrument effect, initially hampered attempts to take meaningful data. In order to circumvent this undesired effect, a different desorption technique was developed. In this technique, the beam was initially undeflected and was kept on one spot of each phosphor screen for a period of time required such that the outgassing rates were level (normally, within an hour or two). The beam was then electrostatically deflected by approximately 3-4 mm from its original, undeflected position. Assuming that background desorption did not increase during this minor deflection, the new position of the beam would produce a different outgassing rate, an increase of a desorption due to the phosphor alone. This method was effective at detecting prompt outgassing from samples at rates above the detectability limits of the system. We note that small desorption rates, well below the detectability limits, can also influence acceptability of a certain phosphor for FED applications.

The estimated minimum detectability, defined as a signal approximately 2x the maximum peak-to-peak noise, for each of the three main gas species are estimated to be as follows (1 Torr-l/s at 20° C corresponds to ~ 3.3×10^{19} molecules/s):

H_2: ~4×10^{-9} Torr-l/s (corresponds to ~1×10^{-7} Torr-l/s-cm^2 for a 0.004 cm^2 sample);

CO, CO_2: ~5×10^{-10} Torr-l/s (corresponds to ~1.3×10^{-8} Torr-l/s-cm^2 for a 0.004 cm^2 sample);

EXPERIMENTAL RESULTS

a) Luminance spectra
Luminance spectra for the three phosphors were measured at an acceleration voltage of 7.5 kV and a beam current of ~ 50 μA. Peaks of intensity correspond to the following wavelengths: 712 nm for red YAG:Eu, 550 nm for green YAG:Tb, and 436 nm for blue Y_2SiO_5:Ce. Fig. 1 shows the luminance spectra for the examined RGB triplet. (The spectrum for the YAG:Eu phosphor was measured in a different set-up and was rescaled for a better visibility).

b) Brightness and Cromaticity
Brightness and chromaticity characteristics of the phosphors were measured using a digital chroma meter with a data processor, enabling direct readings of the brightness in candelas per square meter, as well as the color space of the phosphors. The changes in brightness were

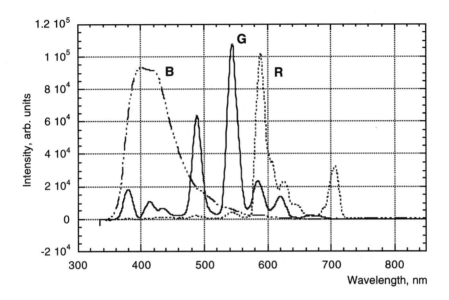

Fig.1: Luminance spectra: R-red (YAG:Eu), G-green (YAG:Tb), B-Blue phosphor (Y_2SiO_5:Ce)

measured as a function of the acceleration voltage, keeping the total power of electron gun constant during the measurements. Fig. 2a shows the changes in brightness for the 3 phosphors in the low voltage range, the same characteristics in the medium voltage range are shown in Fig.2b. No brightness saturation was observed in these ranges, the characteristics are linear in the medium voltage range; some luminance was observed at 200 Volts.

Fig. 3. shows the color space of the tested triplet compared with the phosphor standard reference of the Phosphor Technology Center of Excellence (PTCOE) [8]. The colors are reasonable; however, there is considerable difference in the green coordinates.

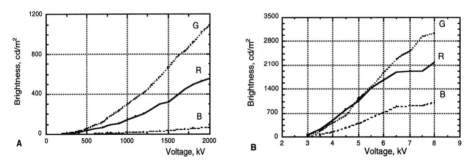

Fig.2. Brightness as a function of beam energy for low (a) and high (b) acceleration voltages. R - red phosphor (YAG:Eu), G - green phosphor (YAG:Tb), B- Blue phosphor (Y_2SiO_5:Ce)

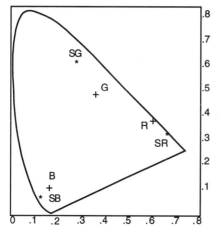

Fig. 3: Color coordinates of the examined phosphors on the 1931 Chromaticity Diagram (shown by + symbols): R - red (YAG:Eu), G - green (YAG:Tb), B - blue (Y_2SiO_5:Ce).

The coordinates for PTCOE standard low-voltage RGB triplet (shown by * symbols) manufactured by Osram Sylvania [8] are given as a reference (SR - Y_2O_2S:Eu, SG - ZnS:Cu,Al; and SB - ZnS:Ag). The colors of the examined phosphors, specifically the green, are clearly inside the PTCOE phosphor ranges, indicating the lower spectral range for the RGB triplet.

c) Outgassing

During the electron-beam excitation, a considerable amount of outgassing from the phosphor screens was detected. All three phosphors showed similar spectra of outgassing and intensity of gas desorption. Fig. 4a shows the mass-spectrum of the outgassing products from one of the phosphors - YAG:Eu. A fresh phosphor surface was exposed to electron beam having a spot area of 9.47 mm^2. Fig 4b shows the outgassing spectrum for the same spot after 2 hours of continuous operation. We used the "seasoned" spectrum obtained from outgassed spot as an instrument background reference and compared it to outgassing from a new, "fresh" spot exposed to the deflected electron beam. The background spectra can be subtracted from the "fresh" spectra giving information on the phosphor material alone. A spectrum resulting from such subtraction is given in fig. 5.

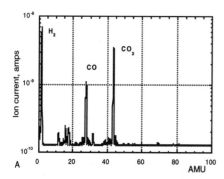 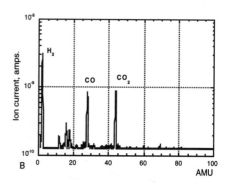

Fig 4: Sample of outgassing spectrum from the YAG:Eu phosphor: a - fresh spot; b - same spot after 2 hours of desorption. Major detected peaks correspond to the following ions: H_2, CO, CO_2

The measured RGA spectra show the degree of outgassing as an intensity of ion current detected by the quadrupole mass-spectrometer corresponding to certain masses of the residual gas ions. In order to show how much of a gas product is actually desorbed from a certain phosphor, the RGA data was re-calculated to give the outgassing values in Torr-liters per second. The RGA response to a certain gas flow was calibrated by backfilling the vacuum chamber with a standard

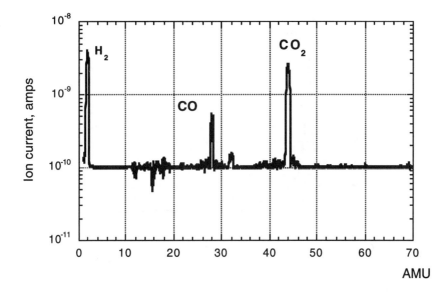

Fig 5: A sample of "real" outgassing from YAG:Eu phosphor: instrument background spectrum is subtracted from the measured desorption spectrum.

gas (H_2, CO, or CO_2) through a known leak at a series of different pressures. This procedure enabled a conversion ratio between the ion detector current and actual gas flow to be derived, and also determined the system's sensitivity. The actual phosphor area exposed to electron beam was estimated for each phosphor and for every beam configuration using video frames which showed the exposed area of phosphor samples; known dimensions of the mounting hardware were used to set the scale on the video frame.

Fig. 6 shows the amount of gas released by a square millimeter of the phosphor surface versus time for one of the phosphor screens. The actual time necessary for the outgassing to reach the initial (background) reading varied from sample to sample due to the thickness of the phosphor layer and to the nature of the phosphor material. Quantitative outgassing data for phosphor screens deposited using different technologies will be subject of future work.

CONCLUSION

The spectral response (and peak intensities) of YAG:Eu (red- 712nm), YAG:Tb (green - 55nm) and Y_2SiO_5:Ce (blue-436nm), electrophoretically-deposited on ITO for application to field-emission flat-panel displays, have been measured. The colors are reasonable, however there are satellite emissions in red and green, and a broad-band emission in the blue. The brightness is linear in the low and medium voltage ranges, with visibility initiated near 200V in each case. The electrostimulated outgassing products, containing none of the phosphor elements, stabilizes at low values after two hours. Followup research includes (1) improving the monochromaticity and correcting the color space for natural color gamut, (2) analysis of variation in outgassing as a function of phosphor layer thickness and encapsulation, (3) assessment of the interaction of these phosphors (as well as the PTCOE P22 references) with field-emission cathodes, and (4) examination of the long-term excitation (eg. 3000+ hr.) reliability of these materials on screens.

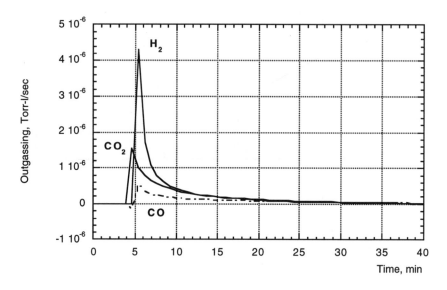

Fig 6. Outgassing versus time for YAG:Eu phosphor per square millimeter of surface exposed to electron beam. The measured beam spot size is 9.47 mm^2.

REFERENCES

1. T. E. Felter, A. A. Talin *et al.* Characterization of ZnGa$_2$O$_4$ and YAG:Cr Phosphors for Field Emission Displays, SID 95 Digest, p.466-469

2. S.M.Jacobsen, Phosphors for full-color Low-Voltage FEDs, Technical digest of the First International Conference on Display phosphors, San Diego, USA (1995), p. 213-215

3. D.W.Smith, A.Vecht, C.S.Gibbons, D.Morton, C.Walding, Thin-Film Phosphors for FED Application, SID Digest, 27-2, p. 619-622, (1995)

4. A.G.Chakhovskoi, W.D.Kesling, J.T.Trujillo, C.E.Hunt, Phosphor selection constraints in application of gated field-emission microcathodes to flat panel displays, Journal of Vacuum Sci. & Technology B, **12**, 2, (March-April), p.785-9, (1994)

5. W.D.Kesling Field-Emission Device Modeling and Application to Flat-Panel Displays, Ph.D. Dissertation, The University of California, Dept. of Electrical and Computer Engineering, Davis, California, 1995.

6. L.Shea, J.McKittrick, E.Sluzky, Advantages of Self-Propagating Combustion Reactions for Synthesis of Oxide Phosphors, Technical digest of the First International Conference on Display phosphors, San Diego, USA (1995), p. 241-243

7. M. E.Malinowski, K. D.Stewart *et al.* Gas Desorption from FEA - Phosphor Screen Pairs, IVMC-95, Technical Digest, Portland, OR, Aug. 1995, p. 202-206.

8. S.M.Jacobsen, C.Stoffers, S.Yang and C.J.Summers, Low voltage Phosphor Data Sheets, Phosphor Technology Center of Excellence, GTRI, publication of Jan. 1996.

CHARACTERIZATION OF LOW-VOLTAGE PHOSPHOR SCREENS FOR FED APPLICATIONS

B.S. Jeon, S.W. Kang, J.S. Yoo and J.D. Lee*
Department of Chemical Engineering, Chung-Ang University, Huksuk-Dong 221, Dongjak-ku, Seoul 156-756, Korea
*School of Electrical Engineering, Seoul Nat'l University, Shinlim-Dong San 51, Kwanak-ku, Seoul 151-742, Korea

ABSTRACT

A phosphor screen which operates at low excitation energy must show good emission efficiency and sustain stabilities against high vacuum and high current density operation. In this study, low voltage phosphors suitable for FED applications were experimentally addressed and screened onto ITO-glass by the electrophoretic deposition which offers several advantages, including easy control of film thickness and high packing density. Process variables such as deposition rate, salt concentration, etc., in electrophoretic deposition method were optimized to achieve high quality of the screens. Correlations of process variables with particle size, surface morphology and luminescent properties of phosphors will be presented. Particularly, much attention was put on the optimization of phosphor screens for the maximal CL intensities and adhesion of particles over substrate when excited by field emitter arrays.

INTRODUCTION

Recently, field emission display (FED) have begun to show promise for commercialization. One of the serious impediments which limits commercialization is the low efficiency of the phosphor screen at low voltage excitation. FED phosphor synthesis, screening and cathodoluminescent properties have been intensively studied for years[1].

The main problems which arise peculiarly from the FED application and resultant requirements for FED phosphors are summarized in Table 1.

Table1. Problems of FED phosphors and their requirements for FED application

Problems	Requirements	Remarks
Low range of cold electrons	Perfection of phosphor surface Large specific area No Al-protective layer	Controllable
High current density excitation - Coulombic aging - Device environment	Conductivity (electrical, thermal)	Screening
	Stability	Material

Large efforts has been put on the identification of phosphor materials to satisfy the requirements for FEDs application. As of now, reliable data had been published by Kishino and Itoh [2].

Table 2. Present status of FED phosphors

Color	Material	Efficiency(lm/w)	Achieved(lm/w)	%
Red	Y_2O_3:Eu	3.3	0.57	17
Green	$ZnGa_2O_4$:Mn	4.2	1.50	36
Blue	Y_2SiO_5:Ce	1.4	0.30	21

1/480 duty, 400V, 75mA/cm^2

Here, target efficiency is for 300 cd/m^2. We can easily notice that the efficiency should be increased by 4-5 times for full color display. Now, the improvement of optical performance of low-voltage phosphors have been carried out by a new synthetic route[3], controlling distribution of activator concentration in a phosphor[4], and even optimizing the geometric structure of phosphor particle[5].

In this work, Screening technology for the FED applications have been studied based on the requirements shown in Table 1.

EXPERIMENT

Electrophoretic method for phosphor screening has been intensively developed by the Talbot et al.,[6]. The experimental apparatus for the electrophoretic deposition was made by putting vertically stainless steel anode plate and sodalime glass cathode plate that is coated by Indium-Tin Oxide (ITO) in the bath container. They are placed parallelly at a distance of approximately 3 cm. The electrodes were connected to a high voltage power supply. Deposition solution was prepared with 1~2 g/ℓ ZnO:Zn phosphors and $Mg(NO_3)_2$ in IPA solvent. Here, $Mg(NO_3)_2$ · $6H_2O$ concentrations range from approximately 10^{-5} M to 10^{-3} M. It was stirred for 6 hrs before deposition to assure complete dissolution of the $Mg(NO_3)_2$ and dispersion of phosphor particles. The process variables examined were electric field strength, deposition times, and $Mg(NO_3)_2$ concentration. The deposition experiments were performed by setting 400 V as anode voltage while monitoring the current. The deposition times ranged from 2~10 min. The samples after deposition were dried at 50 °C and then baked at 450 °C for 10 min, which converted the hydroxides to oxides. After the phosphor particles were deposited onto 3x3 cm^2 substrate, the deposited weight of samples was determined by weighing. Cathodoluminescence measurements were made with an evacuated system that employed thermionic electron gun, two solenoids to focus the beam a 3.2 mm spot, and a micrometer to monitor the current at the sample. The phosphor screens were mounted on a multi-stage sample holder that could be manipulated inside the UHV chamber. Luminance was measured with a Minolta CS-100 meter equipped with close-up lens number 122 (focus length 323mm). Three to ten measurements were taken and luminance was averaged for anode voltage.

RESULTS AND DISCUSSION

Quality of the phosphor film in electrophoretic deposition may be dependent on the process variables and salts to be used. First, we observed that the cathodoluminescence strongly relied on the deposition rate when ZnO:Zn phosphors were deposited on the ITO glass. Figure 1 shows the dependence of deposition rate on cathodoluminescence for the given layer thickness within a margin of experimental error.

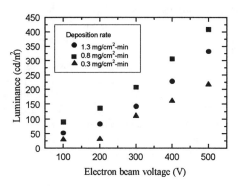

Fig 1. Deposition rate vs. luminance.

Here, deposition rate was controlled by concentration of phosphors in a bath and deposition time. Now weight of deposited phosphors were experimentally determined. Figure 2 indicates the relationship between the packing density and the deposition weight. Film thickness was also included in Figure 2.

Figure 3 represent the cathodoluminance as a function of deposited weight. From this figures, brightness was obtained by optimization of packing density.

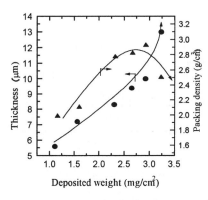

Fig 2. Thickness and packing density vs. deposited weight

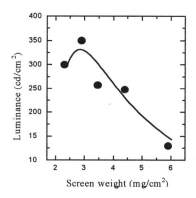

Fig 3. luminance vs. screen weight

As commented in introduction, the range of penetration depth of electron is very shallow at low-voltage excitation. So, the effective volume for cathodoluminescence may be confined on the surface of a phosphor. This might be one of the reasons that the phosphor screens are thin and have large tranmittance without the loss of photons due to the thick film. Our recent work also showed that small particle size of the phosphor might be much favorable to the low-voltage cathodoluminescence[7].

One of the key issues in low-voltage phosphor screen is the charge build-up. Some of investigators have tried to develop the phosphors which are intrinsically conductive. But good results have not been reported yet. Most popular method is to add conductive materials to standard phosphors. Even special grids and leakage paths are formed on the anode plate for reduction of charge build-up.

When we deposited ZnO:Zn phosphors on ITO glass sheet resistance due to the binder, MgO was increased by 40%, which was, of course, dependent on the amount of MgO.

From this point of view, conductive binder, such as SnO could be a good candidate for the binder of low-voltage phosphor screen. Tin Oxide(SnO) could also be a host material for red-emitting phosphor. When we deposited red-emitting Y_2O_3:Eu phosphors, $SnCl_4 \cdot 5H_2O$ as a salt was tried. But, charging for phosphors was not good enough for electrophoretic deposition. The effect of the various salt on the electrophoretic deposition should be studied more.

CONCLUSIONS

For the FED applications, electrophoretic deposition was found to be very valuable technique because of its controllability for packing density and uniformity. Besides, since the chance to be contaminated by organic materials is very low, the outgassing might be minimized.

The formation of SnO as binder for red-emitting phosphor(Y_2O_3:Eu) screen was tried for reduction of charge build-up, which was not successful. We believe that the proper selection of a salt for conductive binder should read to the improvement of low-voltage screen efficiency.

REFERENCES

[1] See extended abstracts of the first international conference on the science and Technology of display phosphors (Nov. 14-16, San Diego, California) 1995.
[2] T. Kishino and S. Itoh, IVMC'95 Proceeding (Portland, Oregon, July 30-Aug. 3) 1995.
[3] L.E. Shea, J. Mckittrick, M.L.F. Phillips, This proceeding, 1996.
[4] S. Itoh, Y. Yonezawa, H. Toki, and Y. Kagawa, Proceedings of the 15th International Display research conference (Asia Display '95),(Hamamatsu, Japan. Oct. 16-18) 1995.
[5] A. K. Albessard, M. Tamatani, M. Okumura, N. Matssuda, H. Hattori, S. Motoki, Proceedings of the 15th international display research conference (Asia Display '95),(Hamamaatsu, Japan. Oct. 16-18) 1995.
[6] D.C. Chang, R.P. Rao and J.B. Talbot, 1st international conf. Sci. and Tech. of display phosphors, 291,1995.
[7] J.S. Yoo and J.D. Lee, submitted in J. Electrochem. Soc. for publication.

DEGRADATION MECHANISMS AND VACUUM REQUIREMENTS FOR FED PHOSPHORS

Paul H. Holloway,[#] Joe Sebastian,[#] Troy Trottier,[#] Sean Jones,[#] Hendrik Swart[#,*], and Ronald O. Petersen[@]
#-Dept. of Materials Science & Engineering, University of Florida, Gainesville, FL 32611-6400
@-Motorola, Inc., Phoenix Corporate Research Laboratory, 2100 East Elliot Road, Tempe, AZ 85284
*-Visiting Scientist from Dept. of Physics, University of the Orange Free State, Bloemfontein, South Africa

ABSTRACT

Degradation of phosphors in field emission displays (FED's) is described and related to electron beam stimulated surface reactions between ZnS and residual gas in the vacuum system. The requirements for producing and maintaining vacuums in FED packages is reviewed and limitations associated with the size of the FED's are discussed. It is concluded that vacuum production and maintenance is critical to the performance of FED's, and this is not a simple task.

INTRODUCTION

Flat panel displays are attracting considerable attention because the US has decided to develop a manufacturing capability to meet the needs of military customers. Defense requirements range from high definition tabletop map displays to helmet mounted and "heads up" displays. Consumer electronics displays should benefit from this emphasis for applications such as wrist watches, electronic equipment status displays, and lap top portable computers.

Liquid crystal displays (LCD's) are by far the most popular flat panel displays.[1] They are very successfully used in many military systems, portable computers, and consumer electronics. However LCD's have some drawbacks, such as a poor viewing angle and temperature sensitivity (operation normally limited to between 0°C and 50°C). Equally important is the need to constantly change the battery in a lap top computer; the LCD consumes 4 watts of power to provide only 60 candela/m^2 of brightness. The normal TV picture tube produces an average brightness of more than 150 cd/m^2. Thus there are opportunities for competitive display technologies.

Most alternative display technologies are light emissive, rather than the light "valve" characteristic of LCD's. One competitive flat panel display technology is the plasma displays (PD's).[2] The traditional amber color from a neon gas discharge has been improved using xenon gas discharges with multiple phosphors to yield full color, meter diagonal displays. Similarly, thin film alternating current electroluminescent displays are very attractive, exhibiting full color and VGA performance without viewing angle or temperature sensitivity problems.[3] However for both technologies, power consumption is still too high for portable military or consumer electronics.

For many years, vacuum florescent displays (VFD's) have been a mainstay for digital arrays. The common blue-green emission of the low voltage ZnO phosphor gives high brightness (100 cd/m^2). Long lifetimes and good stability are claimed to result from operation in a vacuum of $< 1 \times 10^{-7}$ Torr. While VFD's dominate the monochrome digital display market, lack of full color and lack of high resolution has limited their market penetration.[1]

Recently there has been a great deal of excitement about field emission displays (FED's).[4,5] Like the VFD, the FED is a vacuum display and features all of the advantages of cathode ray tubes (CRT's). In contrast to the CRT, the FED is flat. Compared to the active matrix LCD's, FED's generate three times the brightness with greater viewability and temperature range at the same power

425

level. Some aspects of FED's are compared to TFT-LCD's in Table 1.[5] Currently, two companies are demonstrating prototype FED's-Pixel and Micron Displays. Pixel reports a brightness of 100 cd/m^2 at 1.7 watts for their 6" color displays. The promise of better brightness at lower power consumption is driving the development of FED's.

An FED is shown schematically in Fig. 1, and consists of an array of metal (e.g. molybdenum, tungsten, platinum, etc.) or semiconductor (silicon, diamond) tips with radii of curvatures of <100 nm with an intertip spacing of a few micrometers, a dielectric layer between the tips and the gate electrode, and a screen with a phosphor (white) or phosphors (red, green and blue). In operation, electrons tunnel from the array of tips and are accelerated towards (and focused on) the phosphors placed only a few hundred micrometers away from the tip. While FED's have strong attributes, there are also problems in the areas of: 1) field emitter and field emitter array (FEA's); 2) phosphor screen; 3) vacuum requirements; and 4) FED packaging. The focus of the current research has been on the second and third aspects of the FED's. However, these areas are not independent since the vacuum aspects of the FED's can affect the FEA's through changes in the emitter work function versus gas type and pressure, and through discharge at higher presssures ($\geq 10^{-7}$ Torr) because of the close spacing in the FED's. Obviously good packaging is critical to maintenance of the vacuum.

Table 1. Comparison of targeted versus demonstrated properties of FED and TFT-LCD, respective

Property	FED (Target Values)	TFT-LCD
Thickness	6-10 mm	23 mm
Weight	<0.2 kg	0.33 kg
Contrast Ratio	>100:1	60:1 to 100:1
Viewing Angle	>80°C	±45° H, ±30° V
Maximum Brightness	>200 cd/m^2	60 cd/m^2
Power Consumption @ 60 cd/m^2	<1 W	4 W
Operating Temperature	-50°C to +80°C	0°C to 50°C

DEGRADATION OF PHOSPHOR SCREENS

As indicated above, FED's offer the possibility of high brightness at low power. However, the brightness of a cathodoluminescent phosphor increases linearly with increasing voltage over the range of ≈2-20 kV.[5] One objective of FED technology development is to reduce the accelerating voltage for the device to as low as possible for several reasons. First, as discussed above, the dimensions between the cathode (emitter tip) and ground plane (anode plus other areas in the device) is on the order of a few hundred micrometers, and this can lead to corona discharge and dielectric breakdown, even at modest voltages. Second, focusing of electrons onto phosphors become more difficult at high voltage. Finally, high voltage drivers are more expensive, and portable displays tend toward lower voltage drivers. However, lower voltages come at the price of lower luminous efficiency.

In addition, the luminance from the phosphors is known to age.[6,7] The degree of aging is commonly found to be related to the "coulombic loading" of the phosphor, i.e. current multiplied by time and expressed as coulombs of charge per unit area. Many phosphors have been found to exhibit lifetimes to half brightness of about 200 coulombs/cm^2. The mechanism(s) for this degradation are only poorly understood. However, it is apparent that the conditions of the surface not only dictates the initial luminous efficiency, but also affects the rate at which the luminous efficiency will degrade.

Penetration of low energy electrons ($\approx$1keV) into the luminescent solid is on the order of a few tens of angstroms. The surface of luminescent materials are known to be defective, or have coatings which act as non-luminescent dead layers. Improving the luminescence of this surface is one of the keys to efficient low voltage phosphors. A second key is a crystal lattice that will not degrade under electron bombardment. As electrons strike the luminescent materials, they are absorbed in the lattice creating secondary electrons and holes by ionization. The secondary electrons/holes transfer energy through the lattice to the luminescent center. The expression for radiant efficiency is

$$\eta = (1\text{-}r)[\ Sq(hv/E)]$$

where r is the amount of luminescent radiation which is absorbed by the sample, hv is the average energy of the emitted radiation, E is the average energy required to create an electron-hole pair, S is the transfer efficiency of electron-hole pair energy to the luminescent center, and q is the quantum efficiency of the luminescent center. When the surface of the luminescent lattice is defective and/or less ionic, the interaction between electrons is reduced and the value of S is reduced. In addition, not all incoming energy results in ionization and the balance is lost to phonon excitation in the crystal lattice. As the crystal becomes more defective, greater energy is lost to non-radiative transitions and lattice vibrational heat, resulting in reduced luminous efficiencies.

Recently it has been shown that electron bombardment of a standard P22 phosphor (ZnS doped with various transition metals) leads to electron-stimulated surface reactions.[6,7] By following these reactions using *in situ* Auger electron spectroscopy and cathodoluminescent intensity from the same primary beam, changes in the surface chemistry and luminous efficiency can be observed. An Auger electron spectrum from the phosphor before and after electron bombardment is shown in Fig. 2. Notice that electron bombardment by a 2 keV primary electron beam (2 mA/cm^2) for 1320 min. resulted in a decrease of the carbon peak intensity and an increase in the oxygen and zinc Auger peak intensities. The carbon signal is from surface contamination by "adventitious" organic and hydrocarbon molecules from the air and handling. The changes in the cathodoluminescent (CL) intensity and Auger peak heights are plotted versus electron dose (C/cm^2) for a typical phosphor in Fig. 3. The electron beam is postulated to stimulate the dissociation of water (or other residual gas molecules such as H$_2$ or CO$_x$) into atomic species which subsequently react with the ZnS to form ZnO, as illustrated in Fig. 4. In either case, the result is a decrease in the cathodoluminescent intensity, as shown also in Figs. 3 and 5.

The correlation of degradation of CL brightness with vacuum is shown in Fig. 5, where the intensity is plotted versus coulombs/cm^2 for residual vacuum pressures ranging from 6 x 10^{-9} to 7 x 10^{-8} Torr. These are total residual gas pressures measured in the stainless steel bell jar ultra-high vacuum Auger system. The residual gases were dominated by water, with minor peaks from CO and CO$_2$. The rate of surface degradation is dependent upon the type of gas, since we have observed a slower rate of degradation in O$_2$ as compared to residual gas dominated by water at equal pressures.

Thus, it is clear that electron-stimulated surface reactions lead to degradation of the phosphor, and is one mechanism causing the loss of 50% CL intensity at a loading of 200 C/cm^2. The fact that such degradation reactions are a function of pressure emphasizes the need for good vacuums in FED's for high brightness and long lifetimes. It is clear that total pressures of <1 x 10^{-7} Torr would be desired, and partial pressures of reactive gases such as water should be much lower.

VACUUM ISSUES IN FED's

Obviously, the residual gases in a FED vacuum device are critical to its performance and lifetime. The common residual gases in a sealed vacuum are hydrogen, oxygen, water, methane,

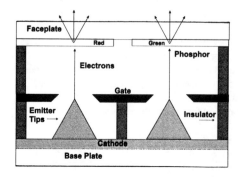

Fig. 1 Schematic of a Field Emission Display (FED).

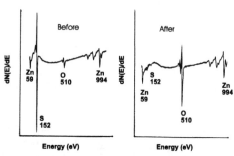

Fig. 2 Auger electron spectra from before and after 2 keV electron bombardment at 2 mA/cm² for 1320 min. in residual vacuum of 1 x 10⁻⁸ Torr.

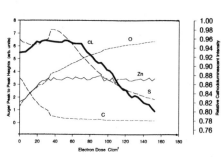

Fig. 3 Cathodoluminescent intensity and Auger peak heights versus electron dose for 2 keV electrons in residual gas.

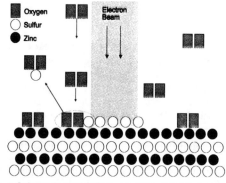

Fig. 4 Schematic model of electron stimulated surface reaction between residual gas molecules and the surface of ZnS resulting in removal of S and growth of a surface ZnO layer.

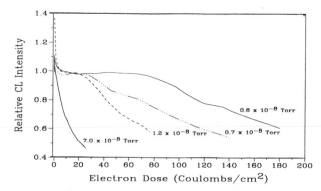

Fig. 5 Degradation of CL intensity versus dose for residual vacuum pressures ranging from 9 x 10⁻⁹ to 7 x 10⁻⁸ Torr.

carbon monoxide, and carbon dioxide.[8] While there are a number of sources of these gases in the vacuum, the most common ones in sealed FED devices are outgassing, vaporization, leaks, permeation, and evolution of gases from bulk solutions. [8]

The gas load from these sources must be removed and the vacuum maintained by a vacuum pump. As for CRT's, this pump will almost certainly use a getter mechanism. There are two types of getter pumps-evaporable and non-evaporable. The principle of operation for evaporable getter pumps is deposition of a thin film of barium by flash evaporation. The thin film then reacts with active gases such as oxygen, nitrogen, water and hydrogen forming solid solutions by absorption of the gases or forming solid oxides, nitrides and/or hydrides. This approach to vacuum pumping has the advantage of very high pumping speeds, however the pumping capacity may saturate after a certain quantity of gas throughput. In this event, the getter pump must be refresh by deposition of another metal film. This is not always convenient/possible with the sealed vacuum anticipated for FED's.

An alternative pumping scheme is non-evaporable getters. In this approach, alloys of zirconium-aluminum-vanadium-iron are heated to temperature between 200°C and 600°C to dissolve any surface oxide that may have formed, then maintained at temperatures between 50°C and 400 C to dissolve any subsequent gas that strikes the surface. Because gas is continuously dissolved into the bulk of the non-evaporated getter if the pump is kept hot, there is not a saturation problem. To reduce power requirements however, these pump can also be operated in a periodic re-activation mode, similar to evaporable getter pumps.

The gas load in the vacuum from vaporization and permeation is controlled by proper selection and processing of materials for construction and packaging. For example, the packaging materials should all be inorganic since organic materials are highly permeable to atmospheric gases such as water, oxygen, nitrogen and helium. Neither should a material be selected which has a high vapor pressure. For example, zinc, mercury, cadmium, sulfur and phosphorous all have very high vapor pressure and therefore should not be allowed on the manufacturing floor. Care should be taken in joining materials since low melting point silver solders may contain one or more of these elements. Finally, the materials should be stable against dissociation. For example, lead solder-glasses are commonly used to seal glass faceplate and glass emitter bases in displays. At sealing temperature, well above the solder glass transition temperature, some lead oxide compounds may dissociate producing oxygen partial pressures in excess of 10^{-2} Torr. This oxygen must be removed from the vacuum, since it may degrade the emitter and phosphor, as discussed above. The resulting metallic lead may evaporatively coat the surfaces of the display at the sealing temperatures, again degrading performance.

Beyond these sources of gas, leaks can result from poor welding practices, scratches across seals, poor bonding at sealed interfaces (e.g. ceramic or glass/metal seals), or use of porous materials with long, narrow conductance paths. For example, improper sealing of a piece of glass to oxidized steel may lead to weak metal/oxide bonds which cause a slow leak. This problem may be solved by preoxidizing the nickel-chromium-iron alloys in wet hydrogen to form a chromium-rich oxide which exhibits excellent bonds to the metal. Checking for leaks using a helium detection in either the sniffing or "bombing" mode is one way to locate such leaks.

Evolution of gas from bulk solution is a common problem with hydrogen in metals, particularly stainless steel and aluminum. Over long periods of time at room temperature, prodigious quantities of gas may be evolved. This gas source may be controlled by careful handling of the metal prior to assembly. Proper cleaning procedures, particularly with acids and electropolishing, must be carefully followed to avoid charging the metal with gas. In addition, the parts may be vacuum fired prior to assembly in order to reduce this gas load.

Permeation is the other mechanism by which diffusion can cause a gas load. By avoiding elastomer seals in packaging, the remaining permeation problem is from helium through glass. This problem is controlled largely through selection of materials. For example, pure silica has the highest

He permeation constant, followed by borosilicate glass and soda-lime glass. Based on accumulation from atmosphere at room temperature, the He partial pressure will increase from 10^{-16} to 10^{-6} Torr in 3 days for silica, one month for borosilicate, 100 years for soda-lime, and 1000 years for hard 1720 glass. Obviously, proper selection of the type of glass is critical. Of course, He is not a reactive gas and therefore doesn't present the same problem with respect to degradation of the phosphor or emitter. However, the vacuum pumping in FED's will be getter pumps of the evaporable or non-evaporable types, which have no pumping speed for inert gases. Thus permeation will allow a pressure buildup to the point where corona discharge will cause problems as discussed above.

The dominant source of gas in the FED's will be outgassing from the glass and possibly cleaning solvents from the surfaces. The FED's will present a special problem relative to minimizing the outgassing since the ratio of surface to volume of the display will be very large compared to CRT's. As a result, the total gas throughput necessary for the getter pumps to handle (normally expressed as Torr-liters/sec) will be large. The outgassing rates for pyrex glass after being exposed to air for one month then pumped in vacuum for 10 hours was 1.6×10^{-10} Torr-liters/cm^2-sec.[8] Assuming a 6 inch diagonal display with a moderate amount of interior surface area, a gas load of $»10^{-7}$ Torr-l/sec would be anticipated. Thus an effective pumping speed of only 1 l/sec would maintain a vacuum of 10^{-7} Torr. However, the effective pumping speed, S_e, is equal to:[8]

$$S_e = S_p C/(S_p + C)$$

where S_p is getter pump speed and C is the conductance of the channel between the gas source and the getter pump. As a worst case analysis, assume that the spacing between the emitter tip and the phosphor was only 100 μm wide and was the only conductance path to the pump. For the 6" diagonal display, the conductance would be $\approx 10^{-2}$ l/sec. Since the effective pumping speed is always lower than the smallest value of either S_p or C, the vacuum could not be maintained with this configuration. In other words, the pumping for FED's must be distributed with channels allowing minimum conductances greater than 1 l/sec. This is quite different from CRT's with their large open tube structure which creates very large conductance values, and the distributed area of the getter pumps creates large pumping speeds to maintain good vacuums.

For a total gas load, after proper degassing and with minimal leak and permeation rates, of 3×10^{-9} Torr-l/sec from a 20 cm X 20 cm panel with a volume of 1 liter, the pressure rate of rise without a pump would be $\approx 10^{-5}$ Torr/hour or $\approx 7 \times 10^{-3}$ Torr/month. To maintain a pressure of 10^{-7} Torr, the quantity of gas pumped per month would be $\approx 7 \times 10^{-3}$ Torr-l. For a flash evaporated barium film, the sorption capacity is ≈ 0.02 Torr-l/g or ≈ 0.07 Torr-l/cm^3. Thus, about 0.1 cm^3 of Ba will be consumed per month of operation. For deposition over an area of 100 cm^2, this amounts to a film 10 μm thick. Over a 10 year lifetime, a film 1 mm thick would be required. Obviously placement of Ba over this large an area to this thickness would require some careful engineering.

Similar arguments may be used to evaluate the quantity of non-evaporable getter required. The throughput capacity of a Zr-V-Fe getter is ≈ 0.05 Torr-l/g, therefore the same gas load rates and lifetime will require ≈ 4 cm^3 of material, and it must be distributed to achieve the necessary pumping speeds. In addition, periodic re-activation will be required over the 10 year lifetime.

SUMMARY

FED's offer an exciting potential for a major advance in display technology. However, many problems remain to be solved before this technology is viable for reliable products with long lifetime in portable military and consumer electronic devices. Degradation of the phosphors, particularly at low voltages is an obvious problem. This is intimately connected with the vacuum level generated and maintained in the FED package. Thus lower sensitivity phosphors are needed, and very good vacuum levels are required to minimize degradation of either the phosphor or the emitter tips.

ACKNOWLEDGMENT

The work at the University of Florida was supported by an ARPA grant to the Phosphor Technology Center of Excellence.

REFERENCES

1. L. B. Tannas, Jr., Ed., *Flat Panel Displays and CRT's* (Van Nostrand Reinhold, NY, 1985).
2. L.E. Weber and J.D. Birk, MRS Bulletin **21**, 65 (1996).
3. P.D. Rack, A. Naman, P.H. Holloway, S.-S. Sun and R.T. Tuenge, MRS Bulletin **21**, 49 (1996).
4. J.E. Jaskie, MRS Bulletin **21**, 59 (1996).
5. P.H. Holloway, J. Sebastian, T. Trottier, S. Jones, H. Swart and R.O. Petersen, Solid State Tech. p.47 (August, 1995).
6. J. Sebastian, S. Jones, T. Trottier, H. Swart and P.H. Holloway, J. of the Soc. Information Display **3/4**, 147 (1995).
7. H.C. Swart, J.S. Sebastian, T.A. Trottier, S.L. Jones and P.H. Holloway, J. Vac. Sci. Technol. in press (1996).
8. J.F. O'Hanlon, *A User's Guide to Vacuum Technology* 2nd Ed. (John Wiley & Sons, NY, 1989).

NATURE OF THE GREEN LUMINESCENT CENTER IN ZINC OXIDE

K. VANHEUSDEN*, W. L. WARREN*, C. H. SEAGER*, D. R. TALLANT*
J. CARUSO**, M. J. HAMPDEN-SMITH**, T. T. KODAS**
*Sandia National Laboratories, Albuquerque, New Mexico 87185-1345
**Nanochem Research Inc., 11102 San Rafael NE, Albuquerque, New Mexico 87122

ABSTRACT

We apply a number of complementary characterization techniques including electron paramagnetic resonance, optical absorption, and photoluminescence spectroscopies to characterize a wide range of different ZnO phosphor powders. We generally observe a good correlation between the 510-nm green emission intensity and the density of paramagnetic isolated oxygen vacancies. In addition, both quantities are found to peak at a free-carrier concentration n_e, of about 1.4×10^{18} cm^{-3}. We also find that the green emission intensity can be strongly influenced by free-carrier depletion at the particle surface, especially for small particles and/or low doping. Our data suggest that the green PL in ZnO phosphors is due to the recombination of electrons in singly occupied oxygen vacancies with photoexcited holes in the valence band.

INTRODUCTION

In order to guide the design of new phosphor materials, or improve the efficiency of established phosphors, the physical mechanisms behind their luminescence must be understood. ZnO material for example, a high-efficiency low-voltage phosphor, has recently regained much interest because of its potential use in new low-voltage luminescence applications, such as the field emissive display technology. It is now quite commonly accepted that the green luminescence in ZnO arises from a radiative recombination involving an intrinsic defect center [1-5]. Surprisingly however, the exact nature of the recombination center and mechanism responsible for the green luminescence in ZnO are still a matter of controversy.

In the present work, we use a number of complementary characterization techniques such as electron paramagnetic resonance (EPR), optical absorption, and photoluminescence (PL) spectroscopies to obtain a better understanding of the physics underlying the luminescent process in ZnO phosphor powders. Samples studied include different commercially available ZnO powders and high purity ZnO powders fabricated using spray pyrolysis [6]. Our data show a strong correlation between the green luminescence, the paramagnetic oxygen-vacancy concentration, and the free-carrier concentration. Many features of our data are explained by taking into account the effects of the Fermi level position in the ZnO gap and free-carrier depletion at the particle surface. We conclude that the defect center responsible for green luminescence is the isolated singly-ionized oxygen vacancy center.

EXPERIMENTAL TECHNIQUES

Sample preparation

Two different commercial phosphor powders: ZnO (powder A) and ZnO:Zn (powder B) were obtained from Fisher and Sylvania, respectively. A third type of ZnO powder (powder C) was synthesized using a bench scale spray pyrolysis reactor, utilized by *Nanochem Research, Inc.* All aerosols were generated by an ultrasonic transducer from an aqueous zinc nitrate (99.999 %) solution (10 wt. % Zn). The ultrasonically generated aerosol was passed through the reactor with either air, nitrogen or forming gas [N$_2$:H$_2$; 93:7 (by volume, 99.999% pure)] at various furnace temperatures (700-900 °C).

Where indicated, additional post-synthesis reduction and oxidation treatments were performed in a flow of forming gas [N$_2$:H$_2$; 95:5 (by volume, 99.999% pure)] or pure O$_2$

433

(99.999% pure), respectively, through a single-wall quartz tube, with a flow rate of $\approx 1 \; l$ /min. The tube was inserted into a resistance-heated tube furnace, maintained at a constant (± 1 °C) temperature. The powders were placed on a quartz boat and inserted into the preheated furnace. Following the anneals, the boat was pulled out of the tube furnace and quenched in air to room temperature. The reduction treatments in forming gas (FG) ambient were observed to cause significant release of Zn vapor. The amount of Zn released increased strongly with temperature. The release of Zn was not observed when annealing the powders in pure N_2, a less reducing ambient. Powder particle sizes were determined from scanning-electron-microscopy imaging, ranging from 50-400 nm for powder A, 500-1500 nm for powder B, and 400-1000 nm for powder C.

Analysis techniques

To obtain accurate optical absorption data for powders normal transmission techniques are not appropriate because of strong scattering of the transmitted beam. To minimize these scattering effects a calorimetric technique was employed known as photothermal deflection spectroscopy (PDS). This technique monitors the optical energy absorbed in the sample by detecting the temperature rise in an adjacent liquid medium (Fluorinert) with a laser deflection method [7,8]. The absorption coefficient $\alpha(E)$ obtained by this technique has been calibrated by using powder obtained from single crystal ZnO. The free-carrier concentration (n_e) can be calculated from the free-carrier absorption [9]. While the absolute accuracy of deducing the free-carrier concentration is estimated at $\pm 50\%$, the error in deducing changes in n_e from sample to sample is considerably smaller.

Two different excitation energies were used to obtain photoluminescence (PL) emission spectra: continuous-wave, near UV excitation at 370 nm from a Xe lamp as filtered through a double monochromator; and pulsed far-UV laser excitation at 232 nm from a Nd:YLF laser/ optical parametric oscillator source. Excitation intensities were in the mW/cm^2 range. Fluorescence spectra were recorded using a charge-coupled detector linked to a 0.6-m spectrograph.

EPR measurements were performed using an X-band ($\approx$ 9.430 GHz) Bruker ESP-300E spectrometer between 4.3 and 294 K. Cylindrical quartz tube containers were used to insert the powders into the microwave cavity. Photo-EPR measurements were performed as well by using thin powder films between two quartz platelets and an optical access microwave cavity.

EXPERIMENTAL RESULTS

In all samples two predominant paramagnetic defects were observed. The EPR signals are spectrally close to each other: one broad resonance with g = 1.9596 and one narrower line with g = 1.9564. Both signals are observable from 4.3 K to 294 K and are attributed to the singly ionized oxygen vacancy ($V_O^{\bullet}$) defect. Since both signals are spectroscopically so close to each other, the broader g = 1.9596 signal has been attributed to some distortion of the original narrow signal [1], probably caused by a nearby interstitial oxygen (Frenkel pair) [10]. We will refer to the latter defect as the $V_O^{\bullet}$-complex, as opposed to the narrow signal which is simply an isolated $V_O^{\bullet}$. Only the isolated-$V_O^{\bullet}$ signal was observed to be largely affected by the oxidation and reduction anneals. Also, in certain samples the isolated-$V_O^{\bullet}$ signal grows strongly upon illumination while the $V_O^{\bullet}$-complex signal shows no photosensitivity. The rest of this work will focus on the isolated-$V_O^{\bullet}$, which wil be simple refered to as $V_O^{\bullet}$. As explained above, n_e was derived from the optical absorption spectrum. The 510-nm green fluorescence intensities were obtained by integrating the PL emission spectrum (using 370 nm excitation). In addition to the green PL, UV fluorescence peaking near 390 nm, could also be observed using the 232-nm excitation source. No direct correlation could be established between the intensity of the green and the UV PL bands. This suggests that they originate from different sites in the ZnO phosphor.

Figure 1 shows the dark concentration (no illumination) of paramagnetic $V_O^{\bullet}$, the free-carrier electron concentration (n_e), and the intensity of the green emission peak for the type-A ZnO powder as a function of FG anneal temperature. These data show a correlation between the intensity of the green emission, the dark density of $V_O^\bullet$, and n_e. Notice the large difference in absolute scale between the $V_O^\bullet$ density and n_e. This strongly suggests that the $V_O^\bullet$ center is not the primary donor dopant in ZnO. The $V_O^\bullet$ EPR signal was observed to be photosensitive for photon energies as low as 2.3 eV, as has been observed previously [11,12]. After extended broad-band illumination, the $V_O^\bullet$ density was observed to saturate at $\approx 2.5 \times 10^{15}$ cm^{-3} in all the type-A samples studied.

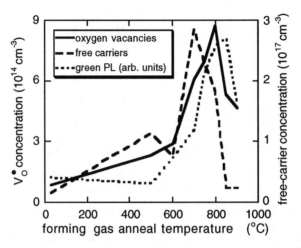

FIG. 1. Intensity of the green emission peak, free-carrier (electron), and singly-ionized oxygen vacancies ($V_O^\bullet$) concentration as a function of reduction anneal temperature for ZnO powder A (Fisher). Anneals were performed in a flow of forming gas [N_2:H_2; 95:5 (by volume)] for 20 min at a flow rate of ≈ 1 l/min. All data were recorded at 294 K.

The initial (untreated) type-B ZnO:Zn powder starts with much more intense green PL and higher n_e than the type-A ZnO powder (approximately 50 times greater), so the effects of oxidation treatments was investigated on this phosphor. Figure 2 shows the $V_O^\bullet$ (dark) density and the n_e values, together with the intensity of the green emission peak for the ZnO:Zn powder (type B) as a function of isochronal oxidation temperature. Again, a reasonable correlation is observed between the density of $V_O^\bullet$, green PL, and n_e, i.e., the three parameters drop roughly one order of magnitude between 700 and 1050 °C. As for the type-A ZnO, 232-nm excitation results in a UV band in addition to the green band, again both showed no correlation. However, the UV band in the type-B samples is typically one order of magnitude smaller than in the type-A powders. Note that this is opposite to the trend observed for the green band intensities; the green PL is roughly 50 times greater in the type-B phosphors. It is also important to note that the $V_O^\bullet$ signal did not display any significant photosensitivity in the type-B powders (as-prepared or after oxidation). This suggests that essentially all of the oxygen vacancies are in their singly-ionized paramagnetic $V_O^\bullet$ state.

Finally, type-C spray-pyrolysis ZnO powders were analysed. By varying the carrier-gas and temperature during spray pyrolysis and post-processing reduction annealing, a wide range of n_e values were obtained: n_e was found to be enhanced by spray pyrolysis performed using forming gas as the carrier gas and/or by post synthesis reduction anneals in forming gas. Figure 3 shows the concentration variations of the paramagnetic $V_O^\bullet$ defect and the 510-nm green-emission efficiency for 370-nm excitation as a function of the free-carrier concentration in the various ZnO samples. Both the $V_O^\bullet$ defect density and the green emission intensity peak at $n_e \approx 1.4 \times 10^{18}$ cm^{-3}, decreasing at lower and higher free-carrier concentrations. An analogous behavior was observed for the green luminescence at 230-nm excitation. Finally, the isolated-$V_O{}^{\bullet}$ signal in the type-C powders was enhanced upon illumination., but this effect strongly decreased with increasing n_e.

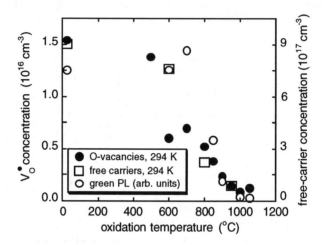

FIG. 2. Intensity of the green emission peak, free-carrier (electron), and singly-ionized oxygen vacancies ($V_O{}^{\bullet}$) concentration vs. oxidation anneal temperature for type-B ZnO (Sylvania). Anneals were performed in a flow of O_2 for 1 h at a flow rate of $\approx$ 1 l /min. All data were recorded at 294 K.

DISCUSSION

One general conclusion from Figs 1, 2, and 3 is the close correlation between the paramagnetic isolated-$V_O{}^{\bullet}$ centers and the green PL intensity for all the 3 different samples studied. It is thus suggested that the obvious mechanism responsible for the green emission is recombination of $V_O{}^{\bullet}$ electrons with photoexcited holes in the valence band

Oxygen vacancy centers and free-carrier concentrations

Oxygen vacancies in ZnO can occur in three different charge states: the $V_O{}^{0}$ state which has captured two electrons and is neutral relative to the lattice, the singly ionized $V_O{}^{\bullet}$ state, and the $V_O{}^{\bullet\bullet}$ state which did not trap any electrons and is doubly positively charged with respect to the lattice. Only the $V_O{}^{\bullet}$ state is paramagnetic, and consequently observable by EPR. Because the $V_O{}^{0}$ state is assumed to be a very shallow donor [11], it is expected that the oxygen vacancies will predominantly be in their paramagnetic $V_O{}^{\bullet}$ state under flat band conditions. As can be seen from Figs. 1, 2, and 3, there exists a good correlation between n_e and the isolated-$V_O{}^{\bullet}$ concentration (without illumination) for n_e values below 1.4×10^{18} cm^{-3}. Although the oxygen vacancy is an intrinsic donor [13] in ZnO, the coexistence of another donor, probably excess Zn [12] at interstitial sites [14], with a much higher density must be inferred to explain the large differences between the $V_O{}^{\bullet}$ and the free-carrier densities in all of the powders studied here.

From the dependence of the green PL efficiency on the free-carrier (conduction-electron) concentration as shown in Fig. 3, it seems that there exists an optimum n_e value for green PL efficiency which is $\approx 1.4 \times 10^{18}$ cm^{-3}. The decline in green PL efficiency at higher n_e values is tentatively attributed to a Fermi-level effect. The effective density of states in the ZnO conduction band can be calculated to be about 5.7×10^{18} cm^{-3}. Since the conduction-electron concentrations in the type-C powders reach or even exceed this value, the Fermi-level will be located very close to the conduction band. As a result, the isolated singly-ionized $V_O^\bullet$ center may well change its ionization state to the neutral, diamagnetic (not observable by EPR) V_O^0 state in the powders with the highest free-carrier concentration, thus decreasing the density of the species responsible for the green PL. The latter assumption is in agreement with the steep drop in the isolated-$V_O^\bullet$ signal intensity with increasing n_e as shown in Fig 3. Finally, the decline in green PL efficiency for n_e values above 1.4×10^{18} cm^{-3} suggests that the donor site itself (excess or interstitial Zn related) is not the active green luminescence center, as has been suggested in the past [13].

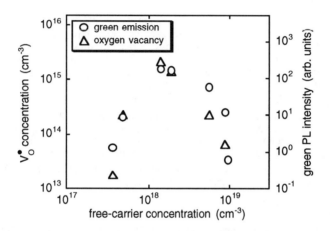

FIG. 3. Concentration of $V_O^\bullet$ defects and 510-nm green-emission efficiency as a function of the free-carrier concentration in type-C ZnO samples. The free-carrier concentration was obtained experimentally from the optical absorption spectrum.

Free-carrier depletion at particle surfaces

Band bending at ZnO powder particle surfaces should be considered [10,15] to explain correlations between n_e and the ionization state of the oxygen vacancy and the resulting variations in PL intensity. Band bending will create an electron depletion region of width W at the particle surfaces:

$$W = \sqrt{\frac{2\,\varepsilon_{ZnO}\,V_{bi}}{e\,N_D}}$$

where V_{bi} is the potential at the boundary, e is the electronic charge, N_D is the donor density, and ε_{ZnO} is the static dielectric constant of ZnO. Assuming that the depletion layer charge density (N_D) is roughly equal to the measured free-carrier concentration, and V_{bi} is of the order of 1 V, the depletion layer width can be calculated to be roughly 100 nm for type-A and 30 nm for type-B powders (unannealed). Due to the difference in particle size, the depletion region will cover a significant fraction of the particles in the type-A ZnO powder (200-nm average particle size), whereas, for the larger-grained type-B ZnO powders (1-µm average particle size) with a higher free-carrier density, it only represents a few percent of the total volume. In the fraction of this region where the Fermi level passes below the $V_O^\bullet/V_O^{\bullet\bullet}$ energy level, all oxygen vacancies will be in the diamagnetic $V_O^{\bullet\bullet}$ state. Under sufficiently strong illumination, band bending can be reduced as minority carriers are attracted to the surface and become trapped; this converts the $V_o^{\bullet\bullet}$ centers to the paramagnetic $V_o^{\bullet}$ state. The fact that the $V_o^{\bullet}$ signal is not photosensitive in the type-B powders, but highly photosensitive in the much smaller-grained type-A powders, supports the notion that surface band bending controls the photosensitivity of the $V_o^{\bullet}$ EPR signal. The particle size for the type-C powders is intermediate between type-A and type-B, and their depleted fraction depends on their free-carrier concentration. The width of the depletion region, and hence, the fraction of oxygen vacancies that is diamagnetic ($V_o^{\bullet\bullet}$) or PL-inactive, is inversely proportional to the square root of the free-carrier concentration in the particles. This explains why the green PL is found to increase with increasing n_e, for n_e values below 1.4×10^{18} cm^{-3}.

SUMMARY AND CONCLUSIONS

Our data provide new information about the PL mechanisms responsible for green emission in ZnO. We observe a good correlation between the green emission intensity and the singly-ionized isolated oxygen vacancy density in a wide range of different ZnO powders. Based on these observations, we suggest that the green PL in ZnO phosphors is due to the recombination of electrons in singly occupied oxygen vacancies ($V_o^{\bullet}$) with photoexcited holes in the valence band. Both the green PL efficiency and the isolated vacancy density are observed to peak as a function of free-carrier concentration at a concentration of about 1.4×10^{18} cm^{-3}. It is argued that at free-carrier densities below 1.4×10^{18} cm^{-3}, free-carrier depletion at the particle surface plays a major role in the correlation between the free-carrier concentration, the density of singly-ionized isolated oxygen vacancies, and the green emission intensity. The decline of the green PL and paramagnetic isolated oxygen vacancy density at higher free-carrier densities is tentatively attributed to a change in the charge state of the luminescence center: $V_o^{\bullet} \rightarrow V_o^{0}$, as the ZnO becomes degenerate due to the high free-carrier density.

ACKNOWLEDGMENT

The part of this work performed at Sandia National Laboratories was supported by the US Department of Energy under contract DE-AC04-94AL85000 and by DARPA. Nanochem Research Inc. acknowledges the Office of Naval Research (STTR contract N00014-95-C-0278).

REFERENCES

[1] P. H. Kasai, Phys. Rev. 130, 989 (1963).
[2] N. Riehl and H. Ortman, Z. Elektrochem. 60, 149 (1956).
[3] F. A. Kröger and H. J. Vink, J. Chem. Phys. 22, 250 (1954).
[4] L. Ozawa, Cathodoluminescence Theory and Applications, (VCH Publishers, New York, NY, 1990), p. 255.
[5] K. Vanheusden, C. H. Seager, W. L. Warren, D. R. Tallant, and J. A. Voigt, Appl. Phys. Lett. 68, 403 (1996).
[6] Y. Senzaki, J. Caruso, M. J. Hampden-Smith, T. T. Kodas, and L-M Wang, J. Am. Ceram. Soc. 78, 2973 (1995); C. Roger, T. Corbitt, C. Xu, D. Zeng, Q. Powell, C. D. Chandler, M. Nyman, M. J. Hampden-Smith, and T. T. Kodas, Nanostructured Mater. 4, 529 (1994); A. Gurav, T. Kodas, T. Pluym, Y. Xiong, Aerosol Sci. and Technol. 19, 411, (1993).
[7] W. B. Jackson, N. M. Amer, A. C. Boccara, and D. Fournier, Appl. Opt. 20, 1333 (1981).
[8] C. H. Seager and C. E. Land, Appl. Phys. Lett. 45, 395 (1984).
[9] D. G. Thomas, J. Phys. Chem. Solids 9, 31 (1958).
[10] K. Hoffmann and D. Hahn, Phys. Status Solidi A 24, 637 (1974).
[11] A. Pöppl and G. Völkel, Phys. Status Solidi A 125, 571 (1991).
[12] G. D. Mahan, J. Appl. Phys. 54, 3825 (1983).
[13] M. Liu, A. H. Kitai, and P. Mascher, J. Luminescence 54, 35 (1992).
[14] G. Heiland et al., Solid State Phys. 8, 193 (1959).
[15] K. Vanheusden, W. L. Warren, J. A. Voigt, C. H. Seager, and D. R. Tallant, Appl. Phys. Lett. 67, 1280 (1995).

Part VI

Other Emissive Display Materials

VARIATION OF LUMINESCENT EFFICIENCY WITH SIZE OF DOPED
NANOCRYSTALLINE Y_2O_3:Tb PHOSPHOR

E. T. GOLDBURT*, B. KULKARNI*, R. N. BHARGAVA*, J. TAYLOR** and M. LIBERA**
* Nanocrystals Technology, PO Box 820, Briarcliff Manor, NY 10510, etgoldburt@aol.com
** Materials Science Department, Stevens Institute of Technology, Hoboken NJ

ABSTRACT

In doped nanocrystalline Y_2O_3:Tb phosphors the luminescent efficiency was found to increase with the decrease in the particle size from 100 Å to 40 Å. This correlation was obtained from microstructural studies performed using transmission electron microscopy and luminescent measurements.

INTRODUCTION

In the performance of low voltage field emission displays (FED) the size of the phosphor particles plays much more critical role than in conventional cathode ray tube (CRT) displays. In fact for low voltage operation (<5 kV) the phosphors with average size less than 0.1 μm are needed. Attempts made to decrease the particle size below 1,000 Å resulted in decrease of the luminescent efficiency of the phosphors due to increase in the non-radiative contribution associated with the surface states of the phosphor. The current phosphors used in CRT and flat-panel displays were developed several decades ago and designed for high voltage CRT applications. Phosphors with better performance at low excitation voltages should find immediate application in various displays. In particular, the practical realization of FED should occur if low voltage phosphors became available.

Recent discovery of doped nanocrystalline (DNC) luminescent materials in the size range of 30 - 50 Å provides a promising alternative to the applications such as low voltage FED phosphors. The additional benefit of using the nanocrystalline phosphor is that the non-radiative contribution decreases relatively with the decrease of the particle size. The original result of doped quantum-dot materials or doped nanocrystals implemented through room temperature organometallic synthesis yielded ZnS:Mn nanocrystals possessing high luminescent efficiency and ultrafast decay time as was reported by Bhargava et al [1]. The efficiency in nanocrystals was measured at 18% as compared to 16% in the bulk whereas the decay times were 4 - 20 nsec as compared to the msec range in the bulk material. In manganese doped zinc sulfide nanocrystals

Mat. Res. Soc. Symp. Proc. Vol. 424 © 1997 Materials Research Society

Mn^{2+} replaces Zn^{2+} and its local symmetry and structure has been studied in detail by EPR[2] and EXAFS[3]. These properties manifest atomic states of an impurity in a quantum confined nanocrystal.

In the past decade a large amount of work has been done in the preparation of the nanometer size particles, and their optical properties have been extensively studied [4-7]. However, the lack of incorporation of dopants into these nanocrystals have prevented these materials from being seriously considered for practical devices. In particular, activators are needed in quantum dots to have well defined chromaticity in these phosphors.

In addition to Mn doped ZnS work we reported previously on preparation and optical spectroscopy of Mn, Eu, and Tb doped nanocrystals of zinc sulfide and Eu and Tb doped nanocrystals of yttria [8-9]. The original synthesis was modified to improve the stability of the nanocrystals. Novel sol-gel processing techniques were developed to synthesize Y_2O_3:Eu and Y_2O_3:Tb nanocrystals. A new synthesis was developed to incorporate Tb^{3+} and Eu^{3+} in zinc sulfide nanocrystals. The characteristic green emission of Tb^{3+} and red emission of Eu^{3+} ($d^5 - f^7$ transition) respectively, has been observed. In addition, the comparison was made between the intensity of Tb^{3+} emission in standard terbium doped LaOBr phosphor and terbium doped nanocrystalline yttria phosphor. To date the efficiency of the nanocrystalline phosphor is about 50 percent of that of the standard phosphor.

In this work we report the variation of luminescent efficiency with size of doped nanocrystalline Y_2O_3:Tb phosphor for the first time. Both the optical and microstructural properties of the samples were studied. The microstructural properties of the doped nanocrystals were studied using Transmission Electron Microscopy (TEM). The sizes determined from electron microscopy were correlated with the optical properties of the doped nanocrystalline phosphors. We observe that the decrease in particle size of the doped nanocrystalline phosphor results in the increase of the photoluminescent efficiency. This data should help in engineering of the low voltage FED devices.

EXPERIMENT

Preparation of Tb Doped Yttria Nanocrystalline Phosphor

A new approach to incorporate rare earth impurities in metal oxide hosts was developed. A room temperature organometallic synthesis was used for the precursors. The principal difference with conventional preparation techniques is in the temperature of the reaction. The temperatures in excess of $1,000\,^\circ C$ are typically used for conventional phosphor synthesis which limits the concentration of the activator ions to less than 5% to avoid luminescence self-quenching. For room temperature synthesis of the doped nanocrystalline (DNC) phosphors this limitation is lifted because the activator ions are separated via the nanocrystal boundaries.

Incorporation of terbium in the yttria nanocrystals was achieved to assess the luminescence efficiency of these DNC phosphors. Sol-gel processing techniques were utilized to achieve the best efficiency. The near room temperature synthetic techniques were used yielding high level of Tb incorporation in nanocrystalline phosphors.

Optical Measurements and TEM Characterization

Direct measurements of the nanoparticle size were made via high resolution transmission electron microscopy. Specimens were prepared by dispersing small dilutions of nanocrystals in water, sonicating the solution, and then placing a drop of solution on a holes-carbon support film on a copper TEM grid. After evaporation of the aqueous solvent, specimens were examined in a Philips CM30 superTwin TEM. Images were recorded on Kodak SO-163 film at typical magnifications of 700,000X. All nanocrystalline tend to agglomerate into clusters on the order of a few microns in size. Many clusters would have nanocrystals distributed in approximately monolayer form. These were used to analyze the particle size distributions. A given TEM grid contained many hundreds of such clusters and image data were collected from several different clusters at different locations within the grid.

Nanoparticle sizes were determined from enlarged prints by tracing particle edges defined by the termination of sets of lattice fringes and measuring the diameter. In cases where the particle images were elliptical, a characteristic size was determined by averaging the magnitudes of the major and minor axes for each particle. The optical measurements including photoluminescence (PL), photoluminescence excitation (PLE) and relative efficiency were carried out as described in detail elsewhere [9].

RESULTS AND DISCUSSION

Optical Properties of Y_2O_3 DNC Phosphor

The optical properties of the three samples A, B and C were measured and reported here. Figure 1 shows the photoluminescent emission (PL) and photoluminescent excitation (PLE) data for terbium doped yttria. PL spectra differ only in their intensity (see Fig. 1a) for the samples A, B and C. The highest and lowest luminous output are 3.2 and 0.8×10^6 count per second in samples A and C, respectively. Microstructural data for the three samples are described in the

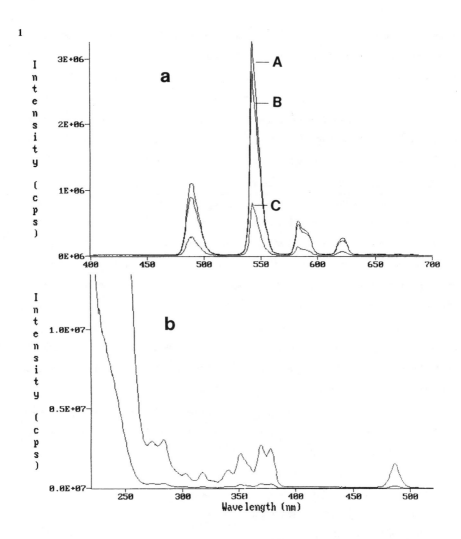

Figure 1. Photoluminescent emission spectra of A, B and C samples (a) and excitation spectrum of A sample (b). The PL spectra is magnified 10X to show the Tb^{3+} absorption manifold.

next section. The PL emission spectra in Figure 1a show the features of the well known $^{5}D_4$ - $^{7}F_n$ transitions of the corresponding Tb^{3+} levels. The individual $^{5}D_4$ - $^{7}F_n$ transition peaks are broader than the corresponding transitions in the bulk samples. Figure 1b shows the PLE spectrum of Y_2O_3:Tb (sample A) with detector set at the green line (~544 nm). The absorption at higher energy is due to charge transfer states of the Y_2O_3 host. The absorption between 300 - 500 nm is due to Tb^{3+} ion in all the PLE spectra. The PLE and PL spectra of DNC samples are similar to the standard bulk LaOBr:Tb spectra. The PL peak at 544 nm is the dominant peak responsible for the green color in the terbium doped yttria nanocrystalline phosphors.

<u>TEM and Nanoparticle Size Measurements</u>

Figure 2 shows a typical TEM micrograph of terbium doped yttria (sample B) with the

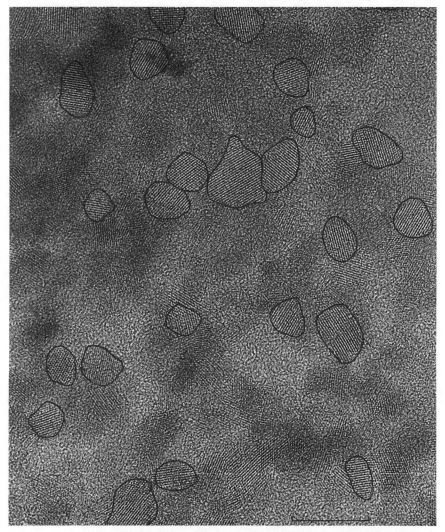

Figure 2. TEM micrograph of sample B (average size 50 Å) used for determining particle size distribution. The bar size is 10 nm.

size of the nanocrystals varying from 40 to 80 Å. The borders of the nanocrystals are traced to identify the particles which were used for determining the particle size and its distribution. The magnified TEM images of the samples A, B and C are shown in Figure 3. The nanocrystalline particles in sample A, B and C are respectively in the range of 40 to 50 Å, 50 to 60 Å and 70 to 100 Å and the micrographs of these samples are shown in Figure 3a, 3b and 3b. These sizes illustrate the characteristic sizes of the particles prepared and studied for understanding luminescent properties of the DNC materials.

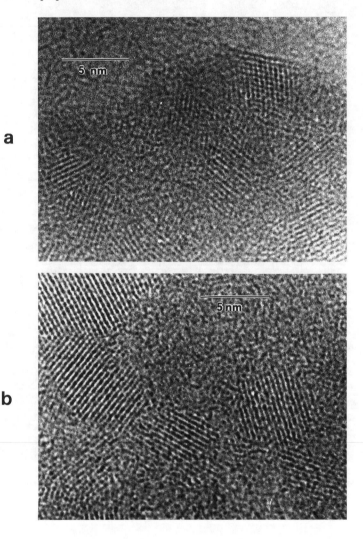

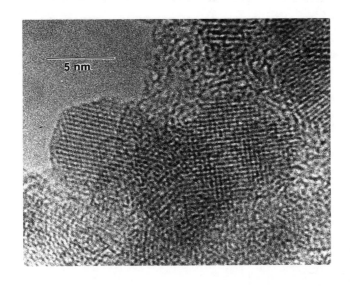

c

Figure 3. Typical TEM micrographs of the samples A (a), B (b) and C (c).

Figure 4 shows the size distributions, as obtained from TEM, corresponding to the samples A, B and C. The bar height represents the number of the particles. It is clear that the particles are the smallest in the sample A and the largest in the sample C. The analyzed data from TEM is summarized in Table 1 below.

Table 1. Microstructural and optical properties of DNC Y_2O_3 phosphors

Sample	A	B	C
Number of particles	46	51	36
Average size, nm	4.4	5.0	9.1
Standard deviation, nm	1.1	1.0	2.2
PL intensity, $\times10^6$ cps	3.2	2.8	0.8

Size Distribution of Nanocrystals

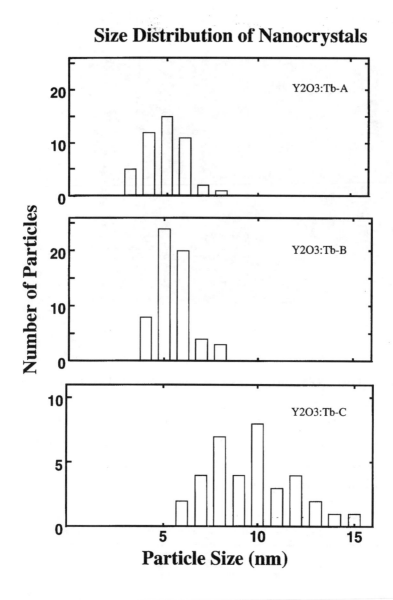

Figure 4. Particle size distribution of Y_2O_3:Tb nanocrystals in samples A, B and C.

Luminescent Efficiency vs. Particle Size

The original concept of doped quantum-dot materials or doped nanocrystals came out of band gap engineering ideas. The band gap of ZnS, for instance, increases as the size of ZnS particle decreases. The increase of this size dependent band gap energy ΔE_g is proportional to $1/R^2$, where R is the radius of the particle [5]. This allows one to vary the band gap to match the electronic fixed levels of the activators. This type of band gap engineering helps to develop phosphors which may have enhanced efficiency due to resonant energy transfer.

In Y_2O_3:Tb DNC phosphor the host is an insulator. The mechanism of the energy transfer from the host to the activator is different from the semiconductor DNC phosphors. To engineer these DNC phosphors it is extremely important to understand the dependence of luminescent efficiency on the particle size in the DNC insulator phosphors. Figure 5 shows the efficiency versus particle size of DNC Y_2O_3:Tb phosphor for the samples. The data for Figure 5 was taken from the Table 1. The data in Figure 5 suggests that the efficiency varies inversely as the square of the particle size. This is expected from quantum confinement model, however, more work is needed to test this hypothesis.

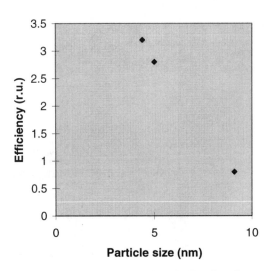

Figure 5. Luminescent efficiency vs. DNC Y_2O_3:Tb phosphor particle size.

This work is the first attempt in quantifying the DNC particle size and size distribution and their correlation with the luminescent efficiency. The subsequent progress in this work should allow a better engineering of the DNC phosphors.

CONCLUSIONS

Doped nanocrystalline yttria doped terbium phosphors were prepared using room temperature sol-gel processing. It was shown that the luminescent efficiency of the DNC insulator phosphors is inversely proportional to the phosphor particle size. This was established using TEM and luminescent measurements for particle in the size range of 40 to 100 Å.

To date, we have prepared DNC Y_2O_3:Tb^{3+} phosphors with efficiencies which are about 50% of the standard LaOBr:Tb^{3+} phosphor [9]. The lowering of processing temperature in the bulk materials from 1,000°C to near room temperature as well as phosphor size reduction from 1 μm in the bulk to 50 Å in DNC materials, offer distinct advantages for the use in FED devices. Further reduction and control of the particle size and particle size distribution should increase the light output significantly.

ACKNOWLEDGMENT

This work was partially supported by ARPA contract # N61331-95-0013 .

REFERENCES

1. R. N. Bhargava, D. Gallagher, X. Hong and A. Nurmikko, Phys. Rev. Lett. **72**, 416 (1994).

2. T. A. Kennedy, E. R. Glaser, P. B. Klein, and R. N. Bhargava, Phys. Rev. B**52,** R14 356 (1995).

3. Y. L. Soo, Z. H. Ming, S. W. Huang, Y. H. Kao, R. N. Bhargava and D. Gallagher, Phys. Rev. B **50**, 7602 (1994).

4. R. Rosetti, R. Hull, J. M. Gibson and L. E. Brus, J. Chem. Phys. **82**, 552 (1985). L. E. Brus, Appl. Phys. A**53,** 465 (1991), and references therein.

5. Y. Wang and N. Herron, J. Phys. Chem. **95**, 525 (1991), and references therein.

6. S. H. Tolbert and A. P. Alivisatos, Science **265**, 373 (1994) and A. P. Alivisatos, MRS Bulletin, Volume XX, p.23, August 1995.

7. B. O. Dabbousi, M. G. Bawendi, O. Onitsuka, and M. F. Rubner, Appl. Phys. Lett. **66**, 1316 (1995).

8. E. Goldburt and R.N. Bhargava, J. Elec. Soc., in publication.

9. E. Goldburt and R.N. Bhargava, J. SID, in publication.

10. Van Uitert, p. 485 in Luminescence of Inorganic Solids, ed. P. Goldberg, Academic Press, New York (1966).

IMPROVED GROWTH OF SrGa$_2$S$_4$ THIN FILM
ELECTROLUMINESCENCE PHOSPHORS

D. BRAUNGER[*], T. A. OBERACKER[*], U. TROPPENZ[**], and H.-W. SCHOCK[*]
[*]Universität Stuttgart, Institut für Physikalische Elektronik, Pfaffenwaldring 47, 70569 Stuttgart, braunger@ipers1.e-technik.uni-stuttgart.de, GERMANY
[**]Heinrich-Hertz-Institut, AG Elektrolumineszenz, Hausvogteiplatz 5-7, 10117 Berlin, GERMANY

ABSTRACT

Cerium activated ternary strontium thiogallate phosphor thin films have been deposited by multi-source evaporation of elemental strontium, gallium and sulfur as source materials. Cerium chloride ($CeCl_3$) has been used as activator starting material. In order to avoid SrS as well as Ga phase segregations elemental In as well as compounds, such as LiF, NaF, $LiCl$, and $SrCl_2$, have been investigated as possible growth enhancing agents.

Coevaporation of indium resulted in an improved crystallization with the drawback of an increased reabsorption of the cerium emission. The other additives lead to improved crystal quality combined with true-blue cathodoluminescence (CL) performance (C.I.E. color coordinates of up to 0.14/0.12) which proves phosphor layers without any SrS phase segregations.

Thin film electroluminescent (TFEL) devices with $SrGa_2S_4:CeCl_3$ phosphor layers prepared with LiF as additive exhibit highest ever reported L_{40} luminances of 78 cd/m² and luminous efficiencies of 0.017 lm/W at a transferred charge of 1 µC/cm² with 1 kHz / 50 µs bipolar square wave excitation.

INTRODUCTION

Commercialization of thin film electroluminescence flat panel displays is still limited by a lack of an efficient blue emitting phosphor material. Cerium activated strontium thiogallate thin films exhibit true blue electroluminescence, and are therefore highly promising candidates for such an application.

The first TFEL devices incorporating such active layers and prepared by sputtering and post-deposition annealing were reported by Barrow et al [1] and Bénalloul et al [2] in 1993. The first films fabricated by physical vapor deposition (PVD) and molecular beam epitaxy (MBE) followed by Oberacker et al [3] and Y. Inoue et al [4], respectively, in 1994.

Films with composition sufficiently close to stoichiometry prepared by PVD- and MBE-methods so far could only be obtained by a high excess of Ga_2S_3-compound as Ga- and S-containing source material [3][4][5] during evaporation. The use of elemental strontium, gallium and sulfur as starting materials in a thermal evaporation process does not lead to single phase films. However, the additional coevaporation of a flux-agent leads to improved cystallization of $SrGa_2S_4$ thin films without the necessity of an exact rate control and of such a high Ga-surplus [6][7]. An enhancement of crystal quality of strontium thiogallate can also be obtained by partially replacing strontium by calcium [2] or gallium by indium.

Mat. Res. Soc. Symp. Proc. Vol. 424 © 1997 Materials Research Society

EXPERIMENT

The $SrGa_2S_4:Ce,Cl$ thin films investigated in this work were deposited by reactive thermal evaporation of the three elements of the host material, cerium chloride ($CeCl_3$) as dopant, and one of the additives indium (In), lithium fluoride (LiF), sodium fluoride (NaF), or strontium chloride ($SrCl_2$) onto heated $ZnO:Al/Si_3N_4$- or indium tin oxide/aluminum titanum oxide-coated glass substrates. The substrate temperature ($\approx 600°C$) as well as the evaporator temperatures were kept constant during the deposition.

The transparent conductive oxides $ZnO:Al$ and ITO, the bottom insulator Si_3N_4 as well as the stacked top insulators Si_3N_4 and Ta_2O_5 were fabricated by rf-sputtering processes, the bottom insulator aluminum titanum oxide (ATO) by atomic layer epitaxy (ALE) [8]. The TFEL devices were completed by e-gun evaporating an array of 2 mm² Al-dots on the top insulator.

The films were characterized by X-ray diffraction (XRD) and energy dispersive X-ray (EDX) analysis with regard to elemental composition. Transient photoluminescence (PL) measurements show the quality of the incorporation of the luminescent center cerium. Cathodoluminescence (CL) spectroscopy is very sensitive with respect to SrS phase segregations, due to the efficient green cerium emission in the SrS.

RESULTS and DISCUSSION

Different approaches for improvement of cerium doped strontium thiogallate phosphor layers are possible. Four ways which might be successful are shown in Table 1. Variation of the source materials, partially replacing the matrix elements, exchanging the activator compound, and using a flux agent, respectively, are described.

Table 1:   Approaches for improvement of phosphor layers for blue emitting $SrGa_2S_4:Ce$ TFEL devices.[*)]Experiments reported in this paper.

matrix element source materials	replacement of matrix elements	activators	additives
$Sr^{*)}$, $SrCl_2^{*)}$	Ca [2]	$CeCl_3^{*)}$, CeF_3 [9]	$LiF^{*)}$, $LiCl^{*)}$
$Ga^{*)}$, GaS, Ga_2S_3 [3]	$In^{*)}$		$NaF^{*)}$
$S^{*)}$			$SrCl_2^{*)}$

Evaporation of Ga or GaS as gallium-containing source materials result in mixed phase films with significant elemental Ga-droplet-segregations [3]. Partially replacing elements of the host matrix by other elements with similar electron configuration could be a way to improve crystallization conditions. Ca as well as In are candidates for such a replacement. However, change of the host could influence the emission of the activator, e.g. Ca is known to shift EL-spectra towards longer wavelength [2].

Coevaporation of In in addition to the elements Sr, Ga and S leads to a reduction of Ga-droplet segregations. Furthermore, the process window for single phase films is widened. However, substantial improvement of crystal quality could only be achieved at In/Ga-ratios higher than unity. An adverse effect is an increased reabsorption of the cerium emission due to the narrower bandgap of the host ($E_g(SrGa_{1.07}In_{0.93}S_4) \approx 2.5$ eV) as can be shown in CL-spectra (Figure 1). The photoluminescence decay times of around 22 ns reveal a good incorporation of the luminescent center cerium. The resulting TFEL devices exhibit weak luminances and instability under high electric fields.

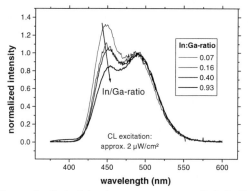

Figure 1: Cathodoluminescence spectra of *Sr(In,Ga)S₄* films with varying indium content indicating reabsorption of the host.

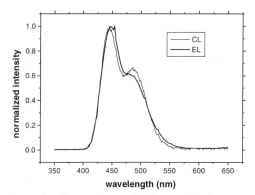

Figure 2: Comparison of CL- and EL-spectra of *SrGa₂S₄:CeCl₃:LiF* thin films.

Regarding literature data on *SrS*-production [10] flux-assisted growth of thiogallates seems to be of interest even in PVD-processing. Extensive systematic studies have been made using *LiF* as flux agent [7]. CL- and EL-measurements (Figure 2) as well as XRD-analysis (Figure 3b) prove that *SrGa₂S₄* films can be fabricated without significant *SrS* phase segregations. A variation of the *LiF*-flux in the vapour resulted in an optimum *LiF/Sr*-ratio of about 20 % at a *Ga*-surplus of more than 100 % compared to stoichiometric growth. A reduction of the *LiF*-rate leads to a decreased necessity of the *Ga*-surplus but is accompanied by an enhanced tendency to *Ga*-rich phase segregations (Figure 3a). In this case the deposition of stoichiometric thiogallate films is only possible with perfect rate control of the *Sr/Ga*-ratio in the vapor. *SrS*-phase segregations can be detected (Figure 3c) already at small *Sr*-excess during a short period of deposition time. High *LiF*-rates lead to LiF- segregations as detected by XRD.

The luminance versus applied voltage (L-V) characteristic of the TFEL device including an active layer by *LiF*-assisted growth is shown in Figure 4. L_{40} luminances of 78 cd/m² and luminous efficiencies of 0.017 lm/W at a transferred charge of 1 μC/cm² with 1 kHz / 50 μs bipolar square wave excitation were achieved.

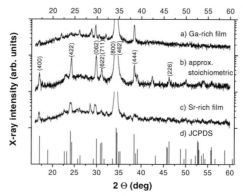

Figure 3: XRD patterns of $SrGa_2S_4:Ce,Cl$ TFEL devices, incorporating active layers grown with *LiF* as flux agent.

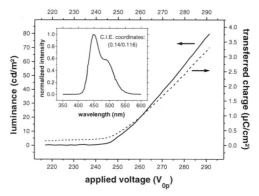

Figure 4: Luminance versus applied voltage curve of the best $SrGa_2S_4:Ce,Cl$ TFEL device with active layer grown with *LiF* as flux agent.

Similar experiments with coevaporation of the other compounds *NaF*, *LiCl* and $SrCl_2$ instead of *LiF* resulted in improved polycrystalline thin films compared to non flux supported deposition. The flux/*Sr*-ratios in the vapor so far have been kept to 20 %, the optimum of the *LiF*-assisted growth. Especially $SrGa_2S_4:CeCl_3$ films prepared with *NaF* as flux agent exhibit more efficient CL-spectra than *LiF*-assisted grown films. No *SrS:Ce* emission could be detected. EL-devices exhibit similar spectral performance but reduced efficiency and stability. L_{40} luminances of 12.3 cd/m² and luminous efficiencies of 0.0033 lm/W at 1 µC/cm² have been achieved. This weak EL performance may be due to *NaF*-segregations, which influence the electrical behavior. In comparison to *LiF*-codoping the CL intensity exceeds by far the EL response. Therefore, the high CL efficiency indicates the possibility of further improvements in particular by further optimizing the *NaF/Sr*-ratio in the vapor.

In order to investigate the influence of the "flux" agents on the preparation conditions during strontium thiogallate film growth, gallium precursor films have been deposited on glass/TCO/insulator substrates. The deposition of *Ga* without any addition on a substrate heated to 600 °C led to gallium droplets as

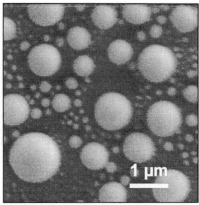

Figure 5: SEM image of Ga precursor w/o evaporating any flux agent.

can be seen in a secondary electron microscopy (SEM) image (Figure 5). With additionally coevaporating *LiF* or *NaF*, respectively, no condensation on the substrate was found. It can be concluded, that a reevaporation of gallium and flux agent from the heated substrate takes place. The same experiments carried out with the substrate at room temperature a condensation of flux agent and gallium is found. This experiments explain why *Ga* droplet segregation is avoided by additionally coevaporating a flux agent like *LiF* or *NaF*.

The influence of the $CeCl_3$ doping rate was examined with respect to the crystallization of the host material. *Ce* was found to be on regular sites without forming clusters. The incorporation of *Ce*

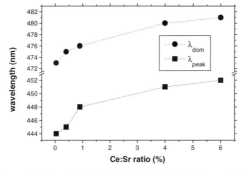

Figure 6: Spectral shift of $SrGa_2S_4:Ce,Cl$ based TFEL devices due to activator concentration.

activators was varied in the range of *Ce/Sr*=0.03 to 6 at%. No significant change of brightness was found in the whole range of cerium concentration. As has been supposed [11] a spectral shift occurs. The whole emission spectrum (i.e. dominant, center and peak wavelength) shifts to longer wavelength as can be seen in Figure 6. This result for low *Ce* concentration is in good accordance with measurements on powder samples [12].

CONCLUSIONS

This work has shown that reactive thermal evaporation of the elements *Sr*, *Ga*, *S*, and additional coevaporation of *In* or a flux agent leads to high quality polycrystalline strontium thiogallate thin films. *LiF* and *NaF* used as flux agents seem to be the most favourable compounds for further improvements of the quality of the active layer. The process window is widened, i.e. no exact rate control except *Ga*-rich (and *S*-rich) composition of the vapour is necessary. Furthermore, no high Ga_2S_3 excess as reported for PVD and MBE processing is used.

The improved crystal quality of the films is accompanied by an improved luminance. With this PVD-process best ever achieved L_{40} luminances of up to 78 cd/cm² have been achieved. These investigations show that thermal evaporation is one of the most suitable methods for preparing ternary thiogallate compounds.

ACKNOWLEDGMENTS

This work has been supported by the European Union under contract BRE.CT93.0452. The authors would like to thank U. Rühle and M. Storz for performing the EDX measurements. PL and EL measurements have been carried out by B. Hüttl and G. Gers, Heinrich-Hertz-Institut/Berlin, Germany.

REFERENCES

1. W.A. Barrow, R.C. Coovert, E. Dickey, C.N. King, C. Laakso, S.S. Sun, R.T. Tuenge, R. Wentross and J. Kane, SID '93 Digest, 761 (1993).
2. P. Bénalloul, C. Barthou, J. Benoit, I. Eichenauer and A. Zeinert, Appl. Phys. Lett. **63**, 1954 (1993).
3. T.A. Oberacker, K.O. Velthaus, R.H. Mauch, H.W. Schock and R.T. Tuenge, EL '94 Digest, Beijing, 1994, p. 160.
4. Y. Inoue, K. Tanaka, S. Okamoto and K. Kobayashi, Proc. Int. Display Research Conf., Monterey, 1994, p. 169.
5. K. Tanaka, Y. Inoue, S. Okamoto and K. Kobayashi, J. Cryst. Growth **150**, 1211 (1995).
6. T.A. Oberacker, D. Braunger and H.W. Schock, presented at the ICTMC10, Stuttgart, 1995 (to be published).
7. D. Braunger, T.A. Oberacker and H.W. Schock, accepted for publication in J. Cryst. Growth, (1996).
8. Substrate delivered by Planar International.
9. T.A. Oberacker and H.W.Schock, II-VI-conference, Edinburgh, 1995 (to be published in J. Cryst. Growth).
10. in: Gmelin Handbook of Inorganic Chemistry, Strontium, System Number **29**, 8th Edition (Verlag Chemie, Weinheim, 1960) p. 245.
11. Y. Inoue, K. Tanaka, S. Okamoto, Y. Tsuchiya and K. Kobayashi, Asia Display '95, 1995 (unpublished).
12. C. Fouassier (private communication): The shift to longer wavelengths at high concentration is less pronounced. This would indicate lower concentration in trivalent cerium (3 % instead of 6 %) in these thin films compared to powder samples.

INVESTIGATION OF THE NUCLEATION AND GROWTH OF THIN-FILM PHOSPHORS

T.S. Moss, R.W. Springer, N.M. Peachey, and R.C. Dye
Los Alamos National Laboratory, M.S. E549, Los Alamos, NM 87545

ABSTRACT

The deposition of $CaGa_2S_4$:Ce has been accomplished using a commercial liquid delivery system on two substrate surfaces: ZnS and SrS. However, the film deposited on ZnS was not of satisfactory quality because of the formation of an amorphous layer and a high amount of residual porosity within the deposition. The use of a SrS layer on top of the ZnS improved the nucleation by reducing the interfacial energy between the substrate and deposition. It greatly reduced the porosity in the coating and reduced the formation of the amorphous layer. The crystallinity of the $CaGa_2S_4$ 400 peak was also increased by a factor of ten when a layer of SrS was used. Further, the FWHM of the 400 peaks from the two depositions was not significantly different, indicating that the crystallite size and strain was approximately the same. The B_{40} was increased by a factor of two, from 1.84 cd/m^2 for ZnS to 3.67 cd/m^2 for SrS. This increase is an improvement in the performance of the films and is attributable to the increase in the crystallinity.

INTRODUCTION

The development of an efficient blue phosphor for use in thin film electroluminescent displays (TFEL) is an important milestone in their deployment. It has been recently reported that Ce^{+3} doped alkaline earth thiogallate materials show emission with chromaticities within the blue region[1]. Cerium-doped calcium thiogallate ($CaGa_2S_4$:Ce) is such a material whose production has been and continues to be actively pursued. In the recent past, a full color VGA TFEL panel has been fabricated and tested[1]. The deposition of $CaGa_2S_4$:Ce has been reported from several methods, including sputtering[1], molecular beam epitaxy[2], and multisource deposition[3]. However, problems inherent to these techniques have prevented their widespread adaptation to a manufacturing scale. The primary limitation has been in the required extra annealing step to produce the crystallographic properties necessary for optimum luminance. The extra processing step also limits glass substrates to the high temperature compositions, adding to the overall cost of the display.

The process of metal-organic chemical vapor deposition (MOCVD) is well suited to overcome the limitations of the other techniques because it is capable of rapidly producing high quality, crystalline material at temperatures below 600°C. The relatively simple deposition equipment is easily integrated into a manufacturing process, and several of the layers necessary for the TFEL device may also be produced using MOCVD, further simplifying the production scheme.

The MOCVD of $CaGa_2S_4$:Ce has been previously reported by this group[4-7]. These papers have dealt with the deposition using two reagent systems: traditional thermal bubblers and a commercial liquid delivery system. However, analyses of the electroluminescent (EL) properties of samples produced from these two techniques have not produced coatings with optimal EL emissions. Further studies of the samples have shown that the nucleation and growth characteristics have not been producing the expected dense films. As a result, the processes influencing the nucleation of $CaGa_2S_4$:Ce have been investigated with the intent of better understanding how to positively influence the EL properties.

EXPERIMENTAL

The deposition of the $CaGa_2S_4$:Ce phase was done using mixtures of $Ca(tmhd)_2$, $Ga(tmhd)_3$, $Ce(tmhd)_4$, and H_2S. The reagent delivery was performed using a liquid delivery system from Advanced Technology Materials, Inc., model LDS-300B. A description of the reagent delivery process has been described elsewhere[7]. Depositions were done at a temperature of 582 °C and a pressure of 5 torr. The substrates were glass slides with deposited conductive,

459

dielectric, and nucleation layers and were furnished by Planar Systems, Inc. In this study, two nucleation layers were used: ZnS and SrS. In the case of SrS surfaces, the SrS was deposited on top of the ZnS. This arrangement could be completed into a device by the subsequent deposition of a top dielectric layer and metallic contact and permitted EL measurements to be made on the as-deposited material.

Samples were characterized for chemical composition, and microstructure by x-ray fluorescence (XRF), x-ray diffraction (XRD), and transmission electron microscopy (TEM). Where possible, EL spectra were taken to determine the color and brightness of the emission. XRF was done on a Tracor unit, and XRD measurements were taken on a rotating anode Rigaku diffractometer using a position sensitive detector and an incident angle of 20°. SEM was done on a Hitachi S-4200, and a JEOL 2010 was used for TEM micrographs of sample cross-sections.

RESULTS AND DISCUSSION

The initial work on the deposition of the $CaGa_2S_4$:Ce was accomplished using the ZnS deposition surface. The ZnS surface was considered to be desirable as a nucleation surface for $CaGa_2S_4$:Ce, and prior experience showed that its presence increased the EL brightness. However, an examination of the quality of material that was produced when the ZnS surface was used shows that the deposition was far from ideal. An example of the problems is illustrated in a cross-sectional TEM micrograph of deposited $CaGa_2S_4$:Ce, shown in Figure 1. This micrograph shows that the grains are long and thin with a width of 50 to 80 nm. The grains appear to emanate from an interface with no evidence of renucleation. A closer inspection of the deposition grown near the interface showed that there was a narrow nucleation zone with little competitive growth between grains near the interface. Also in this micrograph, it is possible to see that there is a large amount of included porosity. The location of this porosity appears to be within the deposition and on top of an amorphous looking layer, as if the crystalline material had renucleated on top of an amorphous $CaGa_2S_4$:Ce layer.

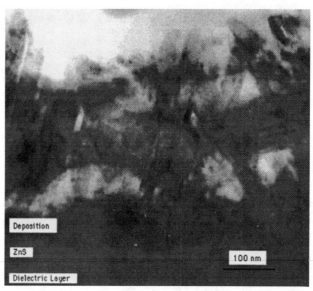

Figure 1. The deposition onto ZnS surfaces has included porosity within the deposition.

To clarify some of the issues surrounding the quality of the initial layer, a second sample was prepared for TEM from a sample with a deposition time of 2.5 minutes, as opposed to 20 minutes. This micrograph is shown in Figure 2. From this picture, it is possible to observe the initial nuclei that formed on the surface of the ZnS. They appear as dark, rectangular areas, growing away from the ZnS and have a width of approximately 10 nm and a height of 40 nm. They appear to be crystalline by virtue of their uniform dark color which lighten when tilting the sample. The density of the nuclei on the surface appears to be quite low (on the order of 1 every 100 nm as observed on the lower magnification pictures, not shown here). Moiré fringes also are present in an area near the center of the micrograph, as well as two other areas which are not as clearly visible. This would indicate that there are crystalline grains within the deposition. However, the rest of the film has an amorphous texture to it. The diffraction pattern of the entire coating did not show any defined spots or rings; further indicating that the bulk of the initial deposition was amorphous.

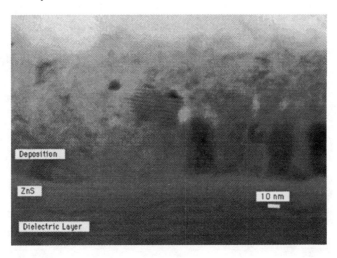

Figure 2. Cross-section TEM of initial deposition shows an amorphous layer with small crystallites.

It is possible to speculate that the deposition is occurring when an initial layer composed of a mixture of small crystalline grains and amorphous material. At some point after this layer is deposited, a second layer is nucleated, containing the large and highly crystalline grains which grow for the duration of the experiment. The nucleation of this second layer results in large amounts of porosity because the growth rate is so much faster than the nucleation rate. This type of a nucleation and growth mechanism is not an ideal situation for the deposition of high quality materials because of the high amounts of amorphous material and included porosity. Among the easier ways to alter the nucleation mechanism is to change the deposition surface such that the interfacial energy between the nucleus and the surface is reduced. If the interfacial energy is lowered by a different substrate surface, the nucleation rate is expected to increase because the critical radius to form a nucleus will decrease.

The choice for a new substrate surface was SrS. It was anticipated that the SrS would provide a better nucleation surface by the reduction of the interfacial energy. Further, if a boundary layer was formed between the substrate and the coating, SrS would improve the EL properties. In the case of the ZnS, the boundary layer would be either $Zn_xCa_{1-x}Ga_2S_4$:Ce or $ZnGa_2S_4$:Ce; neither of these compounds would improve the luminance of the film and could shift the EL color towards the green. The use of SrS would form either $Sr_xCa_{1-x}Ga_2S_4$:Ce or

SrGa$_2$S$_4$:Ce as a boundary layer. These compounds have the added benefit of having a blue or a blue-green emission and are also fairly bright.

The change in the nucleation behavior was quite dramatic when the SrS substrates were used. A TEM cross-section is shown in Figure 3. This figure shows a sample which was deposited using the same deposition conditions as the sample seen in Figure 1. The width of the grains of this sample ranged from 75 to 125 nm, an increase above that seen from the ZnS substrate. Also, it is worth noting that there appears to be a wider distribution of the grain sizes with the SrS substrate, indicating that there is a high amount of competitive grain growth. The increase in the degree of competitive growth would be expected from a sample with a higher nucleation rate because the increased number of nuclei must vie for space to grow into with greater frequency. Of particular interest is the reduced amount of porosity which is seen in the film. This observation is attributed to the reduced interfacial energy between the nucleus and the substrate which led to an increased nucleation rate. Further, there does not appear to be the large amount of amorphous deposition which was observed in the deposition onto ZnS

Figure 3. The use of SrS as the deposition surface greatly reduced the amount of porosity in the coating.

The use of SrS substrates also affected the crystallinity of the deposition. The 400 peak in the CaGa$_2$S$_4$ pattern (JCPDS Card 25-134) was used to gauge the crystallinity since it was present in most every pattern when CaGa$_2$S$_4$ was deposited. Because the patterns were taken with different sampling times, the count rate rather than the peak height is of interest. The count rate for the 400 peak for the sample deposited on ZnS and shown in Figure 1 was 0.59 counts/second. This rate is compared with the SrS substrate shown in Figure 3 which had a count rate of 8.29 counts/second. Based on this difference, it is possible to determine that the use of SrS as the substrate surface either increased the crystallinity of the deposition or produced a material with much higher preferred orientation. However, taking into account the difference in the apparent porosity and reduced amount of amorphous material, it is probable that depositing onto SrS produced a sample with more crystalline grains.

The full-width, half maximum (FWHM) of the 400 peak is also of interest as to how the change in substrate surface affects the deposition process. Visual measurements of the peaks were made to determine there was a difference in the crystallite size or strain. While these measurements were made without taking in account the broadening due the diffractometer, they did serve as a qualitative technique to evaluate relative effects. The results show that there was no determinable difference in the FWHM of the 400 peak between the ZnS (FWHM=0.13 °2θ) and the SrS

(FWHM=0.12 °2θ). The difference between the two values was not resolvable within the error of the diffractometer, indicating that the quality of the material deposited on the two surfaces was not determinably different.

The EL of these two samples was expected to show an increase in the brightness which would correspond to the increased crystallinity. The brightness of the deposition on ZnS at 40 volts above the turn-on voltage (B_{40}) was 1.84 cd/m^2; B_{40} for the deposition on SrS was 3.67 cd/m^2. The result is roughly a factor of two improvement when using the SrS surface. This increase in the brightness was welcomed as an improvement to the deposition and was attributed to the increase in the crystallinity, as evidenced by the changes in the 400 CaGa$_2$S$_4$ peak. Further increases in the brightness are expected to come from optimizing the Ce doping and minimizing the amount of Ga$_2$S$_3$ in the deposition. However, because the SrS substrates were layered on the borosilicate glass, the XRF measurements of the samples have not been able to accurately determine the quantity of Ca and Ce in the films due to peak overlap with elements contained in the glass. As a result, the degree of Ce doping could not be accurately resolved nor could the amount of Ga$_2$S$_3$ be estimated. Both the Ce doping and the Ga$_2$S$_3$ could affect the value of B_{40}. Changing to a different composition glass or a different analytical technique should alleviate these problems.

CONCLUSIONS

The deposition of CaGa$_2$S$_4$:Ce has been accomplished using a commercial liquid delivery system. However, the deposition onto ZnS surfaces was not ideal because of problems in the nucleation and growth of the film. The deposit was defined by two apparent layers: an amorphous initial layer and a renucleated crystalline layer. A closer observation of the initial layer showed some small areas of crystalline grains, including several nuclei, but the bulk of the coating was amorphous in appearance. A high amount of this porosity was observed at the interface between the two layers. This indicated that the second layer grew at a much higher rate than it nucleated. A change to SrS resulted in a continuous coating with greatly reduced porosity and increased crystallinity. The crystallinity of the CaGa$_2$S$_4$ on SrS was also increased by a factor of over ten based on the 400 peak. The FWHM of the 400 peaks between the two substrates was not significantly different, indicating that the amount of crystallite size and strain was about the same. The B_{40} was increased by a factor of two, from 1.84 cd/m^2 for ZnS to 3.67 cd/m^2 for SrS. This increase in the brightness is an improvement in the performance of the films and is attributable to the increase in the crystallinity.

ACKNOWLEDGMENTS

We wish to acknowledge the work of Chris Schaus and Mark DelaRosa at Planar Systems, Inc., for providing the substrates and EL measurements and the work of Ms. Kerry Siebein for her work in providing several of the TEM micrographs.

REFERENCES

1. W.A. Barrow, R.C. Coovert, E. Dickey, C.N. King, S. Sun, R.T. Tuenge, R. Wentos, and J. Kane, S.I.D. 93 Digest, **24**, 761 (1993).
2. K. Tanaka, Y. Inoue, S. Okamoto, and K. Kobayashi, J. Cryst. Growth, **63**, 1954 (1995).
3. T.A. Oberacker, K.O. Velthaus, R.H. Mauch, H.W. Stock, and R.T. Tunge, 1994 Workshop on Electroluminescent Displays, Digest of Technical Papers, (1994).
4. N.M. Peachey, J.A. Samuels, B.F. Espinoza, C.D. Adams, D.C. Smith, R.C. Dye, R.T. Tuenge, C.F. Schaus, and C.N. King in Novel Techniques in Synthesis and Processing of Advanced Materials, edited by J. Singh and S.M. Copley (The Minerals, Metals & Materials Society, 1995), p. 365.
5. T.S. Moss, J.A. Samuels, D.C. Smith, R.C. Dye, M.J. DelaRosa, and C.F. Schaus in Diversity into the Next Century, edited by R.J. Martinez, H. Arris, J.A. Emerson, and G. Pike (Society for the Advancement of Materials and Processes Engineering, Covina, CA, 1995),.**27**, p. 507,

6. T.S. Moss, D.C. Smith, J.A. Samuels, R.C. Dye, M.J. DelaRosa, and C.F. Schaus, submitted to S.I.D. Digest (1995).
7. T.S. Moss, R.C. Dye, D.C. Smith, J.A. Samuels, M.J. DelaRosa, and C.F. Schaus, accepted for <u>MOCVD of Electronic Ceramics II</u>, Materials Research Society (1995).

PHOTOLUMINESCENCE PROPERTIES OF δ-DOPED ZnS:Mn GROWN BY METAL-ORGANIC MOLECULAR BEAM EPITAXY

W. Park, T. K. Tran, W. Tong, M. M. Kyi, S. Schön, B. K. Wagner, and C. J. Summers
Phosphor Technology Center of Excellence, Manufacturing Research Center,
Georgia Institute of Technology, Atlanta, GA 30332-0560
e-mail: chris.summers@gtri.gatech.edu

ABSTRACT

A detailed study is reported of the optical properties of homogeneously and δ-doped ZnS:Mn thin films grown by metal-organic molecular beam epitaxy. Fine structure in the Mn luminescence was observed at low Mn concentrations. The energy of the zero-phonon $^4T_1 \rightarrow ^6A_1$ transition of the Mn ion was determined to be 2.215eV, and the phonon energies associated with the transition to be 10meV and 37meV for the TA and TO phonons, respectively. It was found that the PL intensity depended on the Mn concentration and the δ-doped ZnS:Mn showed much brighter luminescence than the homogeneously doped samples with comparable Mn concentrations. For homogeneously doped ZnS:Mn with low Mn concentrations, the temperature dependence of the linewidth and the peak position was well described by the configuration coordinate model. For δ-doped ZnS:Mn, the antiferromagnetic interaction between Mn ions was invoked to explain the low temperature behavior. The luminescence lifetime measurements suggested that the δ-doping technique resulted in better incorporation of Mn ions in the ZnS host and less defect formation.

INTRODUCTION

ZnS:Mn is the most efficient and most widely used phosphor for electroluminescent (EL) devices and has potential applications to other displays with high brightness and resolution requirements. To achieve good device performance, it is essential both to grow high quality films in order to reduce the defect center concentration and to understand the luminescence mechanisms that control optical properties. Recently, metal-organic molecular beam epitaxy (MOMBE), chemical beam epitaxy (CBE), and metalorganic chemical vapor deposition (MOCVD) techniques were used to grow high quality ZnS films at low temperature.[1,2,3] Also, recent studies have shown that the luminescence efficiency of ZnS:Mn EL devices depends strongly on Mn concentration, reaching a maximum at a Mn concentration of 1% and then decreasing rapidly at higher concentrations.[4] This concentration quenching was attributed to (1) Mn-Mn interactions and (2) interactions with native defect centers. The δ-doping technique where Mn ions are incorporated in separate planes can suppress such interactions and hence achieve brighter luminescence. In this paper, we report a detailed study on the photoluminescence (PL) properties of homogeneously and δ-doped ZnS:Mn/GaAs thin films grown by MOMBE.

EXPERIMENT

The experimental details on the PL measurements can be found elsewhere.[5] ZnS growth was carried out using diethyl zinc (DeZn) and H_2S. The δ-doping was accomplished by simultaneously applying Mn and H_2S fluxes to the ZnS growth surface for a period of ~60s without the DeZn flux. The doped layers were placed with a periodicity of 70 to 140Å in the undoped ZnS matrix. Secondary ion mass spectroscopy (SIMS) measurements were used to determine the Mn concentrations. Figure 1 shows the SIMS depth profile of a 4-layer δ-doped sample indicating planar incorporation of Mn ions. It also shows diffusion of Mn ions out of the doping plane.

RESULTS

Photoluminescence Results of Homogeneously and δ-doped ZnS:Mn

The Mn doped ZnS films showed the characteristic orange luminescence originating from the $^4T_1 \rightarrow ^6A_1$ ionic transition of Mn^{2+}. Figure 2 shows the 10K PL spectrum of a 0.85μm thick ZnS:Mn/GaAs film with a Mn concentration of 0.2%. A highly asymmetric emission band was observed at 2.12eV with a linewidth of 143meV. Fine structure in the Mn emission band was also observed with the zero-phonon line located at 2.215eV and its acoustic and optical phonon replicas. The phonon emission lines were separated by 10meV for transverse-acoustic (TA), and 37meV for transverse-optical (TO) phonons, respectively, and were in good agreement with previously reported values.[6] The band edge emissions of ZnS were also observed from this lightly doped sample, indicating the high optical quality of the epilayers. At 300K, the Mn emission band exhibited a more symmetric gaussian lineshape with the peak at 2.08eV and a linewidth of 202meV. For the δ-doped ZnS:Mn films, a symmetric emission band was observed at 2.102eV with a linewidth of 196meV at 300K. The emission band shifted to 2.075eV with a reduced linewidth of 152meV at 10K. The peak position at 10K was shifted to lower energies by 45meV compared to the homogeneously doped samples. The luminescence intensity of the δ-doped ZnS:Mn was about 6-7 times greater than the homogeneously doped ZnS:Mn with a comparable number of Mn ions.

Temperature Dependent Photoluminescence Properties

As shown in Figure 3, for the 0.2% homogeneously doped sample, the peak position remained constant in the low temperature region and then shifted towards lower energies as the temperature was further increased. This behavior was similar to that observed from KBr[7] and ZnS,[8] and described by invoking the configuration coordinate (CC) model. According to this

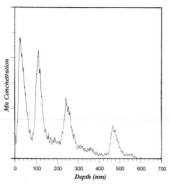

Figure 1. SIMS depth profile of a 4-layer δ-doped ZnS:Mn

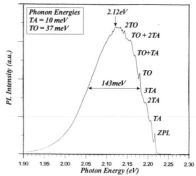

Figure 2. 10K photoluminescence spectrum of a 0.86μm thick ZnS:Mn/GaAs film with a Mn concentration of 0.2%

466

model, when the excited state has greater phonon energy than the ground state, the emission band shifts to lower energy with increasing temperature and vice versa. The red shift with increasing temperature observed for the 0.2% homogeneously doped sample indicates that the phonon energy is greater when the Mn ion is in the excited state than when it is in the ground state. However, for the δ-doped sample, Figure 3 clearly shows that the CC model fails. A similar temperature dependence has been observed in $Cd_{1-x}Mn_xTe$ and explained by the negative thermal expansion of the lattice at low temperatures, a well known property for zincblende semiconductors.[9,10] However, according to this theory, the temperature dependence of the peak position should be independent of Mn concentration, which clearly contradicts our experimental observation. Therefore, the observed temperature dependence was attributed to the interaction between Mn ions. In ZnS, Mn ions are known to order antiferromagnetically at low temperatures. When the antiferromagnetic interaction is present, the excitation of a Mn ion requires an additional spin-flip in the neighboring Mn ion. This effect raises the energy of the excited state. As the temperature is increased, the antiferromagnetic interaction becomes weaker due to thermal fluctuations. Thus, the peak position shifts to lower energy with increasing temperature, as observed at T<50K. Similar arguments have been applied to $Zn_{0.5}Mn_{0.5}Se$ alloys and the model was found to correctly predict that the temperature, T_{min}, at which the peak position reaches minimum. Note that T_{min} does not depend on the Mn concentration whereas the magnitude of the red shift in the low temperature region does.[11] In this model, T_{min}, which defines the temperature where the exchange coupling is completely destroyed, is given by,

$$kT_{min} = \Delta E_{pair} \tag{1}$$

Here, ΔE_{pair} is the exchange coupling energy for Mn pairs which is in turn given by,

$$\Delta E_{pair} = 2JS_1 \cdot S_2 = 7|J| \tag{2}$$

where J is the exchange integral and the second equality was obtained by using $S_T = 0$, and $S_1 = S_2 = 5/2$ for the ground state, and $S_T = 1$, $S_1 = 5/2$, and $S_2 = 3/2$ for the excited state. Also, J was assumed to be the same for both the ground and excited states. With J = -1.1meV, determined by susceptibility measurements,[12] T_{min} was calculated to be ~90K, which is in fair agreement with the observed T_{min}=50K, considering that the J values for the excited states can be totally different from that for the ground state.[13] Taking rigorously into account the different J values could give a better agreement, but the J value for the $^4T_1$ state is not presently known. No immediate

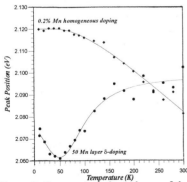

Figure 3. Temperature dependence of the Mn peak position for homogeneously and δ-doped ZnS:Mn/GaAs films

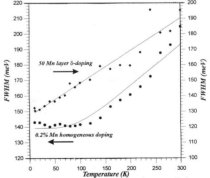

Figure 4. Temperature dependence of the Mn emission band linewidth for homogeneously and δ-doped ZnS:Mn/GaAs films

explanation could be given for the blue shift observed in the high temperature region. For $Zn_{0.5}Mn_{0.5}Se$[11] and $Cd_{1-x}Mn_xTe$[14], it was attributed to the thermal expansion of the lattice. However, our estimate using the data on the lattice constant, compressibility, and the pressure derivative of the $^4T_1$ level predicted a shift of less than 10meV for the lattice expansion from 10K to 300K. This value was much smaller than the observed shift of 40meV. Therefore, an additional mechanism, unknown at present, is responsible for the blue shift. It should be noted that the temperature dependence of the δ-doped samples was similar to that of $Zn_{0.5}Mn_{0.5}Se$ and $Cd_{1-x}Mn_xTe$ alloys which have very high Mn concentrations.

Figure 4 shows the thermal broadening of the Mn emission band for homogeneously and δ-doped ZnS:Mn/GaAs epilayers. According to the CC model, the temperature change causes changes in the distributions of phonons and hence broadening of the emission band. By averaging over the thermal distribution of the initial vibrational states, the actual lineshape at finite temperatures has been calculated, and the full width at half-maximum (FWHM) was found to behave as follows.[15]

$$\Gamma(T) = \Gamma(0)\sqrt{\coth\left(\frac{\langle\hbar\omega\rangle}{2kT}\right)} \qquad (3)$$

Here, $\Gamma(0)$ is the intrinsic linewidth at 0K and $\langle\hbar\omega\rangle$ is the average energy of all phonons participating in the transition. As shown in Figure 4, a good fit was obtained for the sample with 0.2% Mn with parameter values of $\Gamma(0) = 139$meV and $\langle\hbar\omega\rangle = 15$meV, respectively. The small value obtained for the average phonon energy indicates that a large number of acoustic phonons were involved in the transition. For the δ-doped sample, the CC model could not be applied.

Decay of the Mn Luminescence

In order to further investigate the enhanced PL efficiency in the δ-doped samples, luminescence lifetime measurements were carried out. Figure 5 shows the luminescence decay of homogeneously and δ-doped ZnS thin films. The peak height was normalized to unity for all samples. The total PL efficiency is given by,[16]

$$\eta_{PL} = \eta_{exc}\eta_{rad}\eta_{o-c} \qquad (4)$$

where η_{exc}, η_{rad}, η_{o-c} are the excitation, radiative and the out-coupling efficiencies, respectively. It is a good assumption that η_{o-c} is the same for all samples. η_{exc} is defined by,

$$\eta_{exc} = \frac{\alpha N_{Mn}}{\alpha N_{Mn} + \beta} \qquad (5)$$

Here, N_{Mn} is the Mn concentration, α the capture rate of electron-hole pairs by Mn ions, and β the volume rate of energy transfer to killer centers, respectively. η_{rad} is given by the ratio between the integrals of the decay curves of the actual decay and the purely radiative decay. A purely radiative decay should be a single exponential and hence the area under the curve is simply the radiative lifetime, τ_R. Then, the radiative efficiency can be written as,[17]

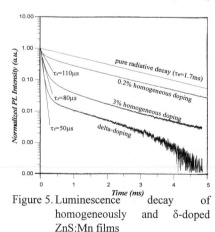

Figure 5. Luminescence decay of homogeneously and δ-doped ZnS:Mn films

$$\eta_{rad} = \frac{1}{L(0)\tau_R} \int_0^\infty L(t)dt \qquad (6)$$

As shown in Figure 5, the 0.2% Mn homogeneously doped sample showed only a small fast-decaying component which was attributed to a non-radiative process. The actual integration of the decay curve for the 0.2% sample gave a radiative efficiency of 0.58. The 3% Mn homogeneously doped sample contained a larger fast-decaying component with a smaller lifetime, exhibiting a radiative efficiency of 0.11. However, this sample showed brighter luminescence due to the larger Mn concentration. The δ-doped sample showed an even larger fast-decaying component, resulting in an even smaller η_{rad}=0.05. It seems a fair assumption that β is the same for all homogeneously doped samples. Then, from the measured η_{rad} and the observed PL intensities of the 0.2% and 3% homogeneously doped samples, the ratio α/β was calculated to be 5.2 (%)$^{-1}$. Assuming the same α/β ratio for the δ-doped sample, the expected PL intensity for the δ-doped sample was estimated to be 53% of the 3% homogeneously doped sample. Considering that the observed PL intensity of the δ-doped sample was the same as that of the 3% homogeneously doped sample, it was concluded that the α/β ratio was greatly enhanced in the δ-doped sample. The enhanced α/β ratio directly indicates less defect or killer concentrations in the δ-doped sample and hence demonstrates that the high crystalline quality of the undoped ZnS films is maintained even in the Mn doped samples by using the δ-doping technique. It turned out, however, that the enhancement of the α/β ratio, and hence η_{exc}, alone could not account for all of the observed enhancement in PL intensity, and therefore other possible origins were explored. An underlying assumption in the above discussion was that the fast-decaying components were due to non-radiative processes, as expressed for η_{rad} in Eq.(6). However, it is possible that there exists some contribution from a radiative process, such as the Mn pair emission. The Mn pair emission shows a fast decay due to the relaxation of the spin selection rule. The lifetime of the pair emission depends on the type of the pair, ranging from 90μs to 600μs.[18] The fast lifetime component in the δ-doped sample showed, however, an even faster decay of 50μs. The origin of the smaller lifetime is not clear at present, but was attributed to the structural asymmetry caused by planar Mn incorporation. It is difficult to discriminate single and pair emissions because the energies of the two transitions are very close together.[19] Thus, η_{rad} for the δ-doped sample was underestimated by simply using Eq.(6). This effect will later be quantified by selectively exciting the pair lines in the excitation spectrum. Another possibility is that the δ-doped ZnS:Mn films have larger "effective" Mn concentrations. In other words, δ-doping could offer better incorporation of Mn ions, resulting in a greater portion of the Mn ions actively participating in the luminescence process and less Mn ions forming defect centers. This was suggested by the temperature dependence of the δ-doped samples, as stated earlier, and will be validated when the effect of the pair emission is quantified.

CONCLUSIONS

Photoluminescence properties of ZnS:Mn epilayers grown by MOMBE were reported. The well-known $^4T_1 \rightarrow ^6A_1$ transition of the Mn^{2+} ion was observed, and the zero-phonon transition energy and the associated phonon energies were determined. The δ-doped ZnS:Mn showed much brighter luminescence than homogeneously doped ZnS:Mn with comparable Mn concentrations, indicating that the luminescence efficiency was greatly enhanced. Temperature dependent PL measurements revealed a drastic difference between homogeneously and δ-doped ZnS:Mn. For the homogeneously doped ZnS:Mn with a Mn concentration of 0.2%, the CC model could properly explain the behavior. However, for the δ-doped samples, the peak position and the linewidth showed different behavior. A qualitative explanation was proposed to describe the low temperature peak shift by considering the antiferromagnetic interaction between the Mn ions. The luminescence lifetime measurements showed that the δ-doped ZnS:Mn contained a larger portion of fast-decaying component. From the observed PL intensity and the luminescence efficiency obtained from the lifetime measurements, it was found that δ-doped ZnS:Mn had less defects and a larger "effective" Mn concentration and also a greater contribution from Mn pairs.

ACKNOWLEDGMENT

This work was supported by the Phosphor Technology Center of Excellence under ARPA contract No. MDA972-93-1-0030.

REFERENCES

1. W. Tong, X. Shen, B. K. Wagner, T. K. Tran, W. Ogle, W. Park, T. Yang, and C. J. Summers, *Proceeding of the SPIE*, Vol.**2408**, 182 (1995)
2. C. J. Summers, W. Tong, T. K. Tran, W. Ogle, W. Park, and B. K. Wagner, *Proceedings of the Seventh International Conference on II-VI compounds and Devices*, to be published in J. Cryst. Growth (1996)
3. A. Abounadi, M. DiBlasio, D. Bouchara, J. Calas, M. Averous, O. Briot, N. Briot, T.Cloitre, R.L. Aulombard, and B. Gill, Phys. Rev. **B50**, 11677 (1994)
4. M. Katiyar and A. H. Kitai, J. Lumin. **46**, 227 (1990)
5. T. K. Tran, W. Park, J. W. Tomm, B. K. Wagner, C. J. Summers, P. N. Yocom, and S. K. McClelland, J. Appl. Phys. **78**, 5691 (1995)
6. D. Langer and S. Ibuki, Phys. Rev. **138**, A809 (1965)
7. W. Gebhardt and H. Kuhnert, Phys. Lett. **11**, 15 (1964)
8. S. Shionoya, T. Koda, K. Era, and H. Fujiwara, J. Phys. Soc. Japan **19**, 1157 (1964)
9. S. Biernacki, M. Kutrowski, G. Karczewski, T. Wojtowicz, and J. Kossut, Semicond. Sci. Technol. **11**, 48 (1996)
10. S. Biernacki and M. Scheffler, Phys. Rev. Lett. **63**, 290 (1989)
11. J. F. MacKay, W. M. Becker, J. Spalek, and U. Debska, Phys. Rev. **B42**, 1743 (1990)
12. W. H. Brumage, C. R. Yarger, and C. C. Lin, Phys. Rev. **133**, A765 (1964)
13. D. S. McClure, J. Chem. Phys. 39, 2850 (1963)
14. M. M. Moriwaki, W. M. Becker, W. Gebhardt, and R. R. Galazka, Phys. Rev. **B26**, 3165 (1982)
15. T. Keil, Phys. Rev. **140**, A601 (1965)
16. B. Huttl, U. Troppenz, K. O. Velthaus, C. R. Ronda, and R. H. Mauch, J. Appl. Phys. **78**, 7282 (1995)
17. P. Benalloul, J. Benoit, R. Mach, G. O. Muller, and G. U. Reinsperger, J. Cryst. Growth **101**, 989 (1990)
18. U. W. Pohl and H. E. Gumlich, Phys. Rev. **B40**, 1194 (1989)
19. H. E. Gumlich, J. Lumin. **23**, 73 (1981)

TiO$_2$:Ce/CeO$_2$ HIGH PERFORMANCE INSULATORS FOR THIN FILM ELECTROLUMINESCENT DEVICES

A. R. Bally*, K. Prasad*, R. Sanjinés*, P. E. Schmid*, F. Lévy*, J. Benoit**, C. Barthou**, P. Benalloul**
*Institute of Applied Physics, EPFL, Lausanne, Switzerland, schmid@ipasg.epfl.ch
**Laboratoire d'Acoustique et Optique de la Matière Condensée, Université P. et M. Curie, Paris, France, cbp@aomc.jussieu.fr

ABSTRACT

The electrical properties of titanium dioxide thin films have been stabilised by cerium doping. These films have a high permittivity between 35 to 45 and withstand 650°C. Multilayer TiO$_2$:Ce/CeO$_2$ insulators have been fabricated. The breakdown voltage is increased by a factor 10 with a modest decrease in the permittivity (30 - 35 instead of 35 - 45).

Electroluminescent devices (ELDs) with a classical ZnS:Mn phosphor have been prepared using TiO$_2$:Ce as the first insulator and a TiO$_2$:Ce/CeO$_2$ multilayer as the second insulator. Compared with a standard ELD based on Y$_2$O$_3$ insulators, devices with the new insulators show a significant decrease of the threshold voltage along with a notable increase of the brightness. An important increase is also achieved in the total device efficiency which is maintained over a large range of brightness and transferred charge. Consequences of rapid thermal annealing and conventional thermal treatments on device performance have also been investigated.

INTRODUCTION

Thin film electroluminescent devices (TFELD) provide an attractive alternative for solid-state flat panel displays[1]. Dielectric materials play a decisive role in the performance of these Metal-Insulator-Semiconductor-Insulator-Metal (MISIM) structures[1,2,3,4]. Under actual operating conditions, very high electrical fields of the order of 100-200 MV/m are applied to the phosphor layer. The role of the dielectric is to prevent the device breakdown under these high fields and to keep charges suitably trapped in interface states so that they do not leak out into the phosphor at lower fields. Most important of all, a dielectric with a high permittivity lowers the operation threshold voltage as it increases the electric field in the emitting layer for a same applied voltage.

A number of studies have reported attempts at finding suitable dielectrics for TFELD applications having the highest figure of merit, which is defined as the product of dielectric constant and breakdown electric field $\varepsilon_o \, \varepsilon \, E_{bd}$[5]. This figure of merit, which indicates the maximum trapped charge density for an insulator material, is of the order of 4-6 μC/cm^2 in classical TFELDs[6]. Because of its high permittivity, TiO$_2$ is a potential candidate provided its electrical properties can be controlled. Undoped TiO$_2$ is always slightly substoichiometric and oxygen vacancies modify its dielectric properties. A figure of merit of only 1 μC/cm^2 has been reported corresponding to a high ε (60) but a very low E_{bd} (20 MV/m)[6].

The electrical properties of titanium dioxide have been stabilised by cerium doping. Ce doping leads to an increase of the resistivity, a reduced variation of the permittivity with frequency and a higher breakdown voltage. After a description of cerium doping effects on TiO$_2$ thin films, a comparison between a standard TFELD (with Y$_2$O$_3$ insulators and ZnS:Mn

471

phosphor) and a device with $TiO_2:Ce/CeO_2$ insulators will show the improvement due to the new insulators.

EXPERIMENTAL DETAILS

Two types of structures were prepared in the present study. For electrical measurements, thin films of undoped TiO_2, Ce-doped TiO_2 and CeO_2 were deposited on ITO-coated glass by reactive RF sputtering. 2 mm^2 gold dots were evaporated as top contacts. For TFELD investigations, MISIM structures were fabricated on ITO-coated glass. The standard Y_2O_3 insulators were deposited by e-gun evaporation and the Mn-doped ZnS phosphor layers were prepared using co-evaporation of ZnS by e-gun and MnS by thermal evaporation. Aluminium or gold dots with an area of 2 mm^2 were evaporated as top metal contacts.

Devices were subjected to a long thermal annealing (LTA) for 2 hours in vacuum at 650°C after the deposition of ZnS:Mn. Such post-deposition heat treatment of ZnS:Mn is widely used to improve the electroluminescent brightness by increasing both excitation and radiative efficiencies[6]. The device structure is given in Table I.

Table I: Electroluminescent Device configurations

Device	Transparent electrode	Bottom insulator	Phosphor	Top Insulator	Reflecting electrode
A	ITO	Y_2O_3 330 nm	ZnS:Mn 520 nm	Y2O3 360 nm	Al
B	ITO	TiO2:Ce 270 nm	ZnS:Mn 520 nm	TiO2:Ce 200 nm CeO2 200 nm	Au

An impedance analyser (HP 4192A) was used to perform ac electrical measurements on the MIM structures. The frequency was swept in the range from 100 Hz to 1 MHz. An amplitude of 100 mV was selected for the oscillation level of the signal. For luminescence measurements, a sinusoidal 1 kHz voltage source was used as the excitation signal. The brightness was measured using a Minolta Photometer. The threshold voltage is defined as the voltage required (or extrapolated) for a brightness of 1 cd/m^2. The transferred charge as a function of applied voltage was determined by measuring the stored charge in the sense capacitor of a Sawyer-Tower circuit[7].

It should be noted that in this work the phosphor was not optimised and therefore the total brightness level is less than what is generally reported[6]. Valid comparison can nevertheless be obtained since the phosphor is the same in both devices.

RESULTS

The dielectric properties of the doped and undoped films were investigated in the frequency range from 10^2 to 10^6 Hz. The permittivity and resistivity were deduced from the ac impedance measurements while taking into account the series resistance due to the ITO electrode. In the case of undoped films, the results reveal large differences in the dielectric constant ε and the ac resistivity ρ depending on the preparation and thermal treatments. At room temperature, the films exhibit values of $\varepsilon = 50 - 400$ and $\rho = 10^7 - 10^9$ Ωm at 10^2 Hz.

The effects of the cerium doping on the dielectric properties of the TiO_2 are shown in Fig. 1. Cerium doping shifts the resistivity towards higher values while the high permittivity values and the frequency dispersion is much weaker than in undoped films. The dielectric constant undergoes a sharp drop as the Ce concentration is increased from 0.4 to 0.7 at.%. Films with the largest Ce doping (2.1 at.%) exhibit a frequency independent value of the permittivity.

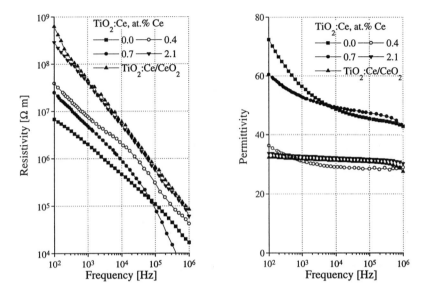

Figure 1: Dielectric properties of Ce-doped TiO$_2$ thin films.

Fig. 2 shows the current density J as a function of the electric field E for TiO$_2$ thin films having different concentrations of Ce and different thicknesses. In undoped films the current at low fields is more than two orders of magnitude higher than in the doped films. This confirms the higher resistivity of the doped material. The breakdown threshold is identified as the abrupt bending off (knee) of the J-E curve at a field of about 3 MV/m. Beyond the knee the avalanche-like current increase results in a destructive breakdown of the insulator.

In films which contains 0.4 at.% Ce, the current at low fields is lower than in the undoped films indicating an increase of its resistivity. However, the breakdown knee is not modified. In the remaining curves one observes a shift of the breakdown knee towards higher electric fields: it is situated at 7 MV/m for the films containing 1.2 and above 10 MV/m for the most highly doped sample (2.1 at.%). Thus the breakdown strength of Ce-doped TiO$_2$ increases with increasing Ce concentration.

CeO$_2$ is reported to have a higher breakdown voltage then TiO$_2$. Its value varies between 200 and 2000 MV/m in thin films[8]. The formation of multilayer insulator with TiO$_2$:Ce and CeO$_2$ shows a breakdown voltage of 100 MV/m, which is 30 times higher than the breakdown voltage of undoped TiO$_2$ thin films. The resistivity and permittivity curves represented in Fig. 1 show that the dielectric properties are quite close to than those of TiO$_2$:Ce values because CeO$_2$ thin films also have a high permittivity of about 26[8].

It is worth noting that preliminary oxide breakdown measurements indicate that Au contacts are self healing and more stable at higher voltage than Al contacts. Unlike Al contacts lead to propagative breakdown.

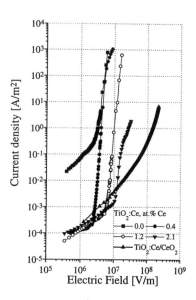

Figure 2: Electrical current density versus electrical field for different levels of doping of TiO₂:Ce.

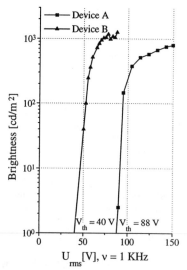

Figure 3: Brightness versus applied voltage for two different devices.

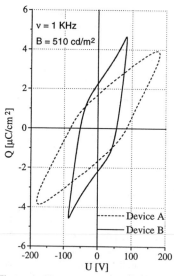

Figure 4: Charge versus applied voltage for two different devices.

The brightness dependence on the applied voltage for two different devices with the same phosphor is presented in Fig. 3. The high permittivity of the TiO_2:Ce/CeO_2 insulator compared to the Y_2O_3 insulator results in a substantial decrease of the threshold voltage for device B. In this case, threshold voltage goes down from 88 V_{rms} to 40 V_{rms}. The highest brightness saturation of the device B and the steep increase of the brightness show that the interface between TiO_2:Ce and ZnS:Mn does not leak carriers, as seen in the well-defined curve of Fig. 4. The incorporation of a CeO_2 layer in the insulators protects the device against breakdown, so that it can be driven to brightness saturation.

The Lissajou curves for the both devices at a same brightness level are shown in Fig. 4. The steeper slopes of the device B curve show the higher permittivity of our new insulators. A permittivity value of 11.8 is found for the Y_2O_3 insulator and 42.7 for the TiO_2:Ce/CeO_2 insulator. The power dissipated in the device is proportional to the area under the Lissajou curve. At the same brightness level, device B shows a better efficiency than device A.

CONCLUSIONS

The results presented in this paper show that TiO_2:Ce/CeO_2 insulators are suitable for thin film electroluminescent devices based on a ZnS:Mn phosphor. The improvements, when comparing with the standard Y_2O_3 insulator, are a decrease of the threshold voltage and an increase of both brightness and efficiency.

ACKNOWLEDGMENTS

The authors would like to thank M. Gouy, S. Gamper, H. Jotterand and A. Nappey for their valuable assistance and technical support. This work was supported by the Priority Program of the Board of the Swiss Federal Institutes of Technology: Optical Science, Applications and Technology and by the Swiss National Science Foundation.

REFERENCES

1. T. Inoguchi, and S. Mito, Electroluminescence, Topics in Applied Physics 17, eds. J. I. Pankove (Springer, New York, 1982), p. 197.
2. W. E. Howard, IEEE Trans. Electron Devices ED-24, 903 (1977).
3. S. K. Tiku and G. C. Smith, IEEE Trans. Electron Devices ED-31, 105 (1984).
4. C. N. King, Society for Information Display (1985), Seminar Lecture Notes 4.1.
5. Y. Fujita, J. Kuwata, M. Nishikawa, T. Tohda, T. Matsuoka, A. Abe and T. Nitta, Japan Display, 76 (1983).
6. Y. Ono, Electroluminescent Display, Series on information displays (World scientific Publication Co., London, 1995).
7. H. Xian, P. Benalloul, C. Barthou and B. Benoit: Jpn. J. Appl. Phys. 33 (1994) 5801.
8. H. Bergmann, Gmelin Handbuch der Anorganische Chemie, Seltenerdelement Teil C1 (Springer-Verlag, New York, 1974), p. 244.

SYNTHESIS OF CdSe/ZnS QUANTUM DOT COMPOSITES FOR ELECTROLUMINESCENT DEVICES

J.RODRIGUEZ-VIEJO[*], B.O.DABBOUSI[#], M.G.BAWENDI[#], K.F.JENSEN[*].

Massachusetts Institute of Technology, Departments of Chemistry[#] and Chemical Engineering[*], Cambridge, Massachusetts 02139.

Quantum dot composite films, consisting of II-VI nanocrystals imbedded in a ZnS matrix, are candidate phosphor materials for electroluminescent flat panel displays. The optical properties of such composites can be tailored across the visible spectral region by selecting the composition and size of the nanocrystals. We present combined solution chemistry and electrospray organometallic chemical vapor deposition (ES-OMCVD) methods for realizing such composites. Size selected, CdSe quantum dots with an overlayer of ZnS are synthesized in solution. This surface derivatization produces a large enhancement of the photoluminescence efficiency. The quantum dot composites are subsequently formed by introducing the quantum dot solution by electrospray into an OMCVD ZnS thin film process. Photoluminescence and cathodoluminescence properties of the quantum dot composites are reported.

INTRODUCTION

Semiconductor crystallites which are small compared to the exciton Bohr radius exhibit unique optical properties due to confinement of the electronic excitations [1]. The optical properties, such as absorption and emission, can be tuned across the visible by changing the size of the nanocrystals. Highly monodisperse CdSe quantum dots passivated with trioctylphosphines have quantum yields between 10-15% [2]. Recently, Hines *et al.* [3] have shown an improvement of the quantum yield up to 50% for 27-30 Å CdSe particles capped with a higher bandgap inorganic material, such as ZnS.

The incorporation of high luminescence quantum dots into an inorganic matrix of higher bandgap, provides a wide spectral window for exploring the quantum size effect and the ability of electrical integration of the composites into optoelectronic devices. Danek *et al.* [4] showed that such composites could be formed by electrospray organometallic chemical vapor deposition (ES-OMCVD). Films of CdSe/ZnSe were grown by this method, but their luminescence characteristics were too low to be useful in optoelectronic devices [5].

In this work we describe a new approach to the synthesis of size-selected overcoated CdSe/ZnS quantum dots ranging in size from 20-60 Å diameter. We also show the subsequent integration of these dots into a ZnS inorganic matrix and the excellent optical properties of these new quantum dot composites. Optical and preliminary structural characterization on both the quantum dots and thin film composites is presented.

EXPERIMENTAL

CdSe nanocrystals of selected size were prepared according to a previously described procedure [2]. The nanocrystals obtained were isolated and purified by flocculation with methanol and centrifugation. The polydispersity of the nanocrystals was narrowed using hexane/methanol mixtures [2]. The size-selected nanocrystals are redispersed in hexane and injected into a solution of TOPO at 70°C. In a glove box equimolar amounts of DEZn and (TMS)$_2$S were prepared and mixed with 5 ml TOP. The [CdSe]/[ZnS] ratio was adjusted to obtain between 3-4 monolayers ZnS coverage. The mixture was added dropwise to the stirred solution containing the nanocrystals at a rate of 1ml/min. The addition time was typically around 5 min and the temperature ranges from 140 to 220°C, depending on the size of the nanocrystals. After the addition, the temperature was quickly dropped to 100°C and the solution was stirred for over an hour. The overcoated nanocrystals were precipitated from the mixture with methanol, collected and redispersed in hexane. Surface exchange with pyridine was performed by flocculation with hexane and redispersion in pyridine.

The CdSe/ZnS nanocrystals surface exchanged with pyridine were filtered with a 0.2 μm PTFE filter, and mixed with acetonitrile in 2:1 ratio to enable electrospraying of the dots [5]. The thin film composites have been synthesized according to the technique developed by Danek et al [4]. In this method the dots were injected into a capillary electrode at a rate of 15 μl/min. H$_2$S and DEZn were used as gas precursors to form a ZnS thin film. Flow rates of 25 and 2.5 μ mol/min were used respectively. The composites were grown on glass and Si substrates at temperatures between 80 and 250°C. The pressure was maintained at 600 Torr.

Luminescence experiments were carried out at room temperature and the absolute quantum yield for the CdSe/ZnS quantum dots was obtained by comparing the integrated emission from Rhodamine 560, 590 or 640 in methanol. The nanocrystals studied by XPS were surface exchanged with pyridine and spin cast onto Si substrates at speeds of 3000 rpm. CL was performed on CdSe/ZnS QD composites deposited on Si.

RESULTS AND DISCUSSION

We have used the technique of Murray et al. [2] to produce CdSe nanocrystals ranging in size from 20 to 60 Å diameter. The work of Hines et al. [3] provided us some guidance in the choice of the organometallic precursors. DEZn is chosen as the Zn source and (TMS)$_2$S as the chalcogen source. Our approach enables the selection of any size for the nanocrystals and therefore improves the luminescence properties in the whole visible spectrum from blue, through green, to red. Figure 1 shows the UV/Vis absorption spectra of bare and ZnS overcoated CdSe dots of different sizes. The results depend strongly on the addition temperature. For the smaller sizes temperatures larger than 150°C resulted in broadening and red shift of the UV/Vis absorption spectrum reflecting the increase of both size and polydispersity. The larger sizes could be grown at temperatures near 220°C without significant changes in the UV/Vis absorption spectrum.

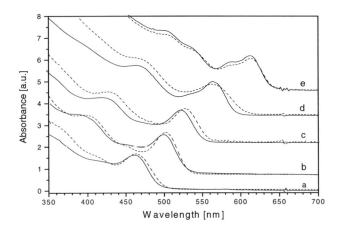

Figure 1. UV/Vis absorption spectra of bare (solid line) and ZnS overcoated (---) CdSe nanocrystals. The overcoating was made at different temperatures on the different sizes: a) r=10.4 Å, T=140°C, b) r=14.5 Å, T=160°C, c) r=16.5 Å, T=180°C, d) r=21.5 Å, T=200°C, e) r=28 Å, T=220°C.

The ZnS overcoating dramatically enhances the luminescence of the nanocrystals. Figure 2 shows typical room temperature band-edge luminescence for bare and ZnS overcoated crystallites. The increase of the quantum yield from values around 10-15% to 30-45% for samples capped with TOPO and dispersed in hexane clearly indicates the efficiency of the ZnS overlayer in electronically passivating the crystallites. The full-width-at-half-maximum of the emission peak (25-40 nm) and the absence of deep trap emission in the overcoated samples also reflects that the ZnS cap has greatly reduced the surface traps. However, surface exchange of the TOPO/TOP molecules by pyridine produces a decrease of the quantum yield down to values around 10%.

XPS measurements on 16.5 Å radius CdSe nanocrystals show the presence of C and O coming from absorbed gaseous molecules, Si from the substrate, and Cd,Se, Zn and S from the quantum dots. P could also be detected on the bare CdSe dots showing the incomplete exchange with pyridine of the nanocrystal surface in agreement with previous results [6]. An estimation of the Cd and Se content taking into account the appropiate escape depths of the electrons gives a concentration of 54 ±5% for Cd and 46 ±5% for Se. The Auger parameter of Cd in the bare and overcoated samples is 467.0 eV and corresponds exactly to the expected value for bulk CdSe. The Zn Auger parameter value is 759.2 eV which is slighlty higher than the 758.1 eV for bulk ZnS. Some of the samples have been analyzed after exposure for 24h to air. In the case of bare CdSe, a small peak develops on the right hand side of the Se 3d peak (figure 3). Its position, 59 eV, corresponds to the peak position for oxides on bulk CdSe[6]. The Se concentration drops to 30% pressumably due to formation and desorption of SeO_2. In comparison, for the overcoated samples no changes in the XPS spectra were recorded on the Se lines. Furthermore, the Cd-Se ratio of the overcoated samples is still 55-45%, indicating the absence of Se desorption and therefore suggesting the good coverage of the ZnS coating.

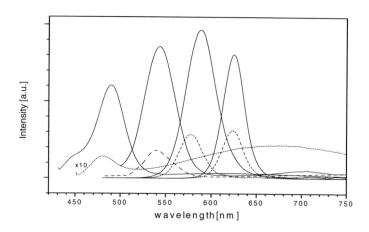

Figure 2. Comparison of the room temperature PL of bare CdSe (---) and CdSe/ZnS nanocrystals of different sizes. a) r=10.5 Å, QY=32%, b) r=16.5 Å, QY=44%, c) r= 21.5 Å, QY=44%, d) r=28.0 Å, QY=28%.

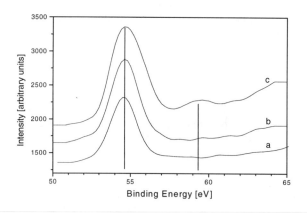

Figure 3. Se-3d core lines of 16.5 Å radius CdSe nanocrystals. a) CdSe as prepared. b) ZnS overcoated CdSe oxidized for 24 h., c) CdSe oxidized for 24 h.

The CdSe/ZnS QD composites exhibit absorption and photoluminescence spectra characteristic of CdSe/ZnS nanocrystallites. Figure 4 show the luminescence of QD composites of bare CdSe and CdSe/ZnS nanocrystals. The bare CdSe shows a weak band-edge emission and a dominant deep level emission. The incorporation of the ZnS overlayer enhances the photoluminescence properties of the films. The spectrum is now dominated by a strong band edge emission. The increase of temperature from 100 to 250°C did not produce any significant deterioration of the photoluminescence properties of the composites. The high quantum yield of the nanocrystals makes the thin film appear to glow with the color of the quantum dot emission when viewed under UV light. The absolute quantum yield of the composites is close to 10%, the value obtained for the CdSe/ZnS in pyridine

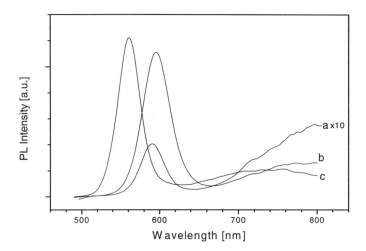

Figure 4. Room temperature photoluminescence spectra of quantum dot composites. a) QD composites with bare CdSe dots, r=21.5 Å, b) QD composites with CdSe/ZnS dots, r=21.5 Å, c) QD composites with CdSe/ZnS dots, r=16.5 Å.

The viability to integrate the quantum dot composites into optoelectronic devices has been tested by cathodoluminescence experiments. Figure 5 shows the CL and PL of a CdSe/ZnS quantum dot composite grown on a Si substrate. The CL spectrum was taken at 5kV and 1 nA. Higher voltages resulted in a decrease of the CL signal over time pressumably due to degradation of the quantum dots. There is a red-shift and broadening of the band edge peak in the CL spectrum. This behaviour has alreday been observed for electroluminescence of quantum dot/polymer structures [7]. The similarity between both curves show that the CdSe/ZnS particles cathodoluminesce and photoluminesce from the same state.

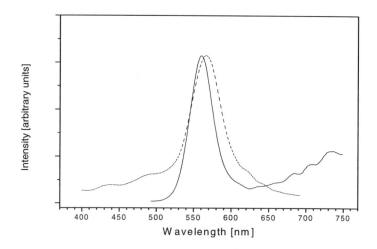

Figure 5. Cathodoluminescence (---) and Photoluminescence (solid line) of a CdSe/ZnS quantum dot composite deposited on Si. The initial CdSe dots were 16.5 Å radius.

Conclusions

The passivation of CdSe nanocrystals with ZnS provides an efficient way to increase the quantum yield over a wide range of the visible spectrum. ES-OMCVD allows efficient integration of the quantum dots into ZnS thin films. The high photoluminescence achieved on the quantum dot composites and the cathodoluminescence results obtained so far indicate the feasibility of using the quantum dots as phosphors in electroluminescent materials.

Acknowledgments

JRV and BOD thanks the *Direccio General de Recerca* from Catalonia and Saudi Aramco respectively, for fellowships. The authors would also like to thank Dr. Jurgen Michel for the CL measurements.

References

1.L.Brus, Appl. Phys., **A53**, 465 (1991).

2. C.B.Murray, D.J.Norris, M.G.Bawendi, J.Am.Chem.Soc., **115**, 8706 (1993).

3. M.A.Hines, P.Guyot-Sionnest, J.Phys.Chem., **100**, 468 (1996).

4. M.Danek, K.F.Jensen, C.B.Murray, M.G.Bawendi, Appl. Phys. Lett., **65**, 2795 (1994).

5. M.Danek, K.F.Jensen, C.B.Murray, M.G.Bawendi, Chem. of Mater., **8**, 173 (1996).

6. J.E.Bowen Katari, V.L.Colvin, A.P.Alivisatos, J.Phys.Chem., **98**, 4109 (1994).

7. B.O.Dabbousi, M.G.Bawendi, O. Onitsuka, M.F.Rubner, Appl. Phys. Lett., **66** 1316 (1995).

SILICON NANOCLUSTERS IN SILICA: A LUMINESCENCE STUDY OF VISIBLE LIGHT EMISSION FROM A SILICON-BASED MATERIAL

P. F. TRWOGA, A. J. KENYON and C. W. PITT

Department of Electronic and Electrical Engineering,
University College London, Torrington Place, London WC1E 7JE, UK

ABSTRACT

Considerable interest has been generated recently in the application of Porous Silicon to the production of light emitting devices, and a number of demonstration structures have been produced. However, the poor stability of the material pose limitations on its usefulness. We report studies of the luminescence properties of a related, but stable material: silicon-rich silica. This material consists of silicon clusters embedded in a host matrix of silica, and as such is considerably more robust than porous silicon.

We have produced a number of samples of silicon-rich silica thin films by Plasma Enhanced Chemical Vapour Deposition on to doped silicon substrates. We report photoluminescence studies that investigate the nature of the radiative process, and show evidence of two separate luminescence mechanisms: defect luminescence and exciton confinement in nanoscale silicon clusters. Preliminary electroluminescence from silicon-rich silica fabricated by Plasma Enhanced Chemical Vapour Deposition will also be presented.

INTRODUCTION

Recent development of novel forms of silicon that emit visible light have renewed interest in the production of silicon-based luminescent devices and displays and there is currently significant activity in the field of light emission from porous, amorphous (a-Si) and nanocrystalline silicon. Following Canham's initial report of porous silicon [1] there has been an explosion of interest into its photoluminescence (PL) and electroluminescence (EL) properties, and typical figures for PL quantum efficiency are currently around 5% [2]. For EL the figures are up to 1% for wet contacts [3] and 0.16% for solid contacts[4]. This is therefore a very promising material for silicon-based electroluminescent devices. However, reduction in light output after prolonged exposure to the atmosphere and light-induced degradation continue to pose problems [5,6].

A related material which has recently begun to generate interest is silicon-rich silica (SiO_x, x<2), a material with similar optical properties to porous silicon but significantly less susceptible to damage. It consists of nanometre-sized silicon aggregates suspended in a silica matrix. Depending on the deposition technique and post-process annealing, these inclusions can be either amorphous or crystalline: initially amorphous clusters tend to crystallise on annealing. Although luminescence was reported in this material some time ago [7,8], it is only recently that its optical properties have begun to be thoroughly studied. Recent results have demonstrated broad-band light emission from the blue to near infra-red from material containing silicon aggregates of varying sizes [9,10]. A number of possibilities present themselves regarding the nature of the luminescence mechanism from this material. Radiative recombination of confined excitons within silicon clusters ('quantum dots') [11,12], defect luminescence [13], interfacial effects at cluster surfaces [14] and luminescence from novel siloxene molecule [15] have all been proposed. However, photoluminescence studies carried out to date have failed to conclusively identify a unique luminescence mechanism and there are a number of contradictory reports in the literature. As with porous silicon, however, there is a growing consensus that quantum confinement plays a major role in light emission. In this paper we suggest that the PL band produced by silicon-rich silica consists of two components: a quantum confinement band and a contribution from defects associated with oxygen vacancies.

Of crucial importance to the deployment of silicon-rich silica in flat panel displays are the electroluminescence properties of the material. To date, there have been a limited number of reports of weak visible electroluminescence from SiO$_x$ based devices [16,17,18]. In this paper we present results for a simple EL test structure and compare the electroluminescence obtained with photoluminescence for the same material. We also present current-voltage characteristics for the device and discuss conduction mechanisms.

EXPERIMENTAL

For photoluminescence studies, a number of silicon-rich silica films were deposited on doped p and p+ type <100> FZ silicon wafers using PECVD. The sample wafers were first thoroughly cleaned under clean-room conditions and cleaned again *in situ* with an argon plasma for 5 minutes. Precursor gases were a silane/argon mix (5% silane (SiH$_4$) in argon) and nitrous oxide (N$_2$O) as silicon and oxygen sources respectively. The nitrous oxide flow rate was fixed at 10 sccm and the silane/argon mixture flow rate changed from run to run between 240 sccm and 120 sccm. A 13.56 MHz RF generator was used to dissociate the precursor gases, the power being varied from run to run between 5W and 50W. Substrate temperature was set in the range 30°C to 350°C. After deposition, each film was divided into a number of pieces which were then annealed under argon for 90 minutes at a series of temperatures between 200 °C and 1000 °C.

Photoluminescence spectra were measured for each film using a Coherent argon-ion laser operating at 351 nm, 457 nm or 476 nm as the excitation source.

Samples grown for electroluminescence studies were restricted to a thickness of approximately 0.1μm. Thin gold pads around 50nm thick and 5mm diameter were evaporated onto the surface of the as-grown films, along with a 1μm thick layer of aluminium deposited on the underside of the wafers to ensure good electrical contact. Contact was made to the gold pads using a needle probe and a dc voltage applied to the device. Current-voltage (I-V) curves were measured and the device placed in the spectroscopic set-up for measurement of the electroluminescence spectrum.

Luminescence was detected using a photomultiplier tube and a 0.5m single-grating monochromator.

RESULTS AND DISCUSSION

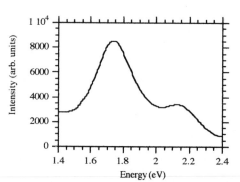

Figure 1: Photoluminescence spectrum of SiOx sample showing bimodal structure.

A typical photoluminescence spectrum of an as-grown SiO$_x$ film excited using the 476 nm line of the argon ion laser is presented in figure 1 (gas flow ratio SiH$_4$: N$_2$O: 12:1, substrate temperature 200 °C). A bimodal structure is immediately apparent, with transitions centred at 1.74 eV and 2.1 eV.

PECVD processing is complicated by the interdependence of the growth parameters. The main object of the series of films grown was to achieve high PL intensity. Hence the value of RF power, pressure, gas flow ratio and substrate temperature were optimised by examining only the intensity of measured PL in the films. Using this technique of limiting the output of the film growth programme to one important measurable, the most important parameters could be identified. The following paragraph outlines the influence of each parameter on the as-grown PL.

Gas flow ratio was the most important variable, with significant PL only seen with ratios above 10:1 and below 26:1. The range between 18:1 to 22:1 seemed to offer the strongest PL, particularly in the lower energy band. Little visible PL was detectable as the stoichiometry approached that of SiO_2. As the composition approached a-Si, an infrared band centred around $1.3 \mu m$ was detectable but no visible PL. Substrate temperature had a strong influence on the ratio of the PL strength of the two bands, with lower substrate temperature producing bright PL in the 2eV region and high substrate temperatures producing bright PL in the 1.7eV band. At 200°C both bands were generally similar in intensity with further annealing required to enhance the lower energy band. However, at 300°C the lower energy band luminescence was greatly enhanced and larger than the 2eV band by a factor of 4. Increasing the substrate temperature to 350°C (maximum operating temperature) did not make any discernible difference to the PL from that at 300°C. We selected films grown in this temperature range for electroluminescence studies. Films that were grown at higher RF powers (>20W) tended to have macroscopic inclusions of amorphous silicon and silica that appeared to have been created in the gas phase. These inclusions tended to increase the defect luminescence. It is not clear whether the silicon nanoclusters are formed in the gas phase or, as is more usual in PECVD of silica and SiO_x, at the substrate. This is an area under further study since it may offer a method of controlling the size of the clusters in the gas phase. Changing the operating pressure by choking back the rotary pump on the equipment yielded inconclusive results. This was largely because even a doubling of the operating pressure was difficult to study since it required much larger RF powers to maintain a plasma in the chamber.

Post Processing

On annealing, the higher energy band initially increases in intensity at temperatures up to 400 °C, and thereafter falls until it is no longer detectable following annealing above 600 °C. The most striking feature of the behaviour of the low-energy band on annealing is its progressive shift to lower energies with increasing annealing temperature (see Figure 2). From an initial as-grown energy of 1.7-1.8 eV, this can shift to 1.4 eV on annealing at 1000 °C. It has been well established by a number of groups that annealing silicon-rich silica films at temperatures in excess of 600 °C results in accretion of excess silicon atoms and aggregates to form progressively larger clusters [14]. If annealed at a high enough temperature for a sufficient time, the larger silicon clusters will tend to 'swallow-up' the smaller ones, resulting ultimately in macroscopic inclusions of 'bulk' silicon within the SiO_2 matrix. This is referred to as 'Ostwald ripening' [18]. An increase in cluster size produces a reduction in confinement energy and a consequent narrowing of the optical bandgap. Observation of an annealing-related red-shift in luminescence spectra is therefore generally taken to be good evidence in support of the quantum confinement hypothesis. The quantum confinement model also predicts a strong increase in luminescence intensity with decreasing cluster size. This may help to explain the observed reduction in photoluminescence intensity accompanying the red-shift of the low energy band on annealing at T_a>600°C [19].

For the case of the high energy band, the absence of an annealing-related red-shift makes it unlikely that this band is the result of recombination of confined excitons. However, luminescence around 1.9 and 2 eV has been noted before in both amorphous silica and quartz. This is generally associated with defect states produced by implantation or irradiation. Shimizu-Iwayama's work suggested the presence of defects in implanted samples and assigned an emission band around 2.2 eV to defect luminescence from states at the silicon cluster/SiO_2 interface [15].

There has been considerable debate in the literature over the origin of the 1.9 eV and 2 eV luminescence bands in silica [21,22], but the current model suggests that they are due to emission from non-bridging oxygen hole centres (NBOHCs) [21]. Such defects are usually associated with the radiolysis of hydroxyl groups or the cleavage of strained silicon-oxygen bonds by irradiation:

$$\equiv Si\text{-}O\text{-}H \rightarrow \equiv Si\text{-}O\uparrow + H. \qquad (1)$$
$$\equiv Si\text{-}O\text{-}Si\equiv \rightarrow \equiv Si\text{-}O\uparrow + Si\equiv \qquad (2)$$

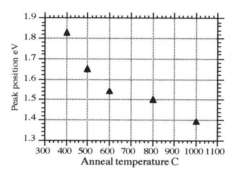

Figure 2: Peak energy of low energy band as a function of anneal temperature

However, there is evidence of the presence of NBOHCs in as-grown CVD oxide films [21]. On the basis of our data, we can not conclusively identify the defect responsible for this emission but the NBOHC is the most likely candidate. The observation that in all cases the 2 eV photoluminescence band is completely quenched following anneals in excess of 600°C indicates that the quenching mechanism may be tied in with the filling of oxygen vacancies in the silica matrix by diffusion of interstitial oxygen and water [19].

It would initially be expected that removal of hydrogen from hydrogen-rich SiO_x films that accompanies annealing would produce a large number of dangling bonds which would in turn act as non-radiative recombination centres and quench luminescence. However, there is a clear increase in the photoluminescence intensity from the 2 eV band following annealing at $T_a < 600°C$. FTIR data certainly indicate almost complete removal of hydrogen by this point [19]. This suggests that the evolution of hydrogen from the films at $T_a < 600°C$ serves to increase the population of optically active defects at a rate which is greater than that of their annihilation by thermally excited interstitial oxygen atoms. This is perhaps unsurprising when the differences in mobility between oxygen and hydrogen are considered. However, the nature of the defect responsible for the 2 eV band is not clear. We assume that it is produced by the cleavage of Si-H or O-H bonds and may be generically similar to the NBOHC, although our studies indicate that it does not appear to be ESR active [19].

Electroluminescence

In this study DC electroluminescence was only observed from thin samples of SiO_x, although samples with layers up to 1 μm thick were tested. The test device fabricated is essentially an MOS type structure (shown in the inset in figure 3) using an evaporated Au semi-transparent top electrode and evaporated Al to form an ohmic contact on the underside. The EL observed could only be seen in a darkened room, but was quite easily observed in that environment. The EL device ran with a stable output for over 6 hours. For the EL spectrum measurement a voltage of 9V and a current of 330mA were used. However, output could was visible at a current of 77mA. A current-voltage curve for the device is shown in figure 4. The electron transport under forward bias conditions is due to the tunnelling of energetic electrons from the Au layer and holes from the p-type substrate through the SiO_2 layers and into the silicon clusters. The electron-hole pairs generated in the clusters can undergo two types of radiative recombination; the exciton can recombine in the cluster with an average energy corresponding to the cluster size, or alternatively recombination can occur at a luminescent centre at the Si/SiO$_2$ interface. Thus if the confined exciton model is favoured we would expect to see a red shift in the EL spectrum accompanying an increase in annealing temperature and consequent growth of silicon clusters. In addition we may also expect to see a reduction the EL intensity. We are currently studying this. However, Qin et al. in fact observe both of these phenomena [17]. They have demonstrated EL of thin SiO$_x$ films grown by Magnetron sputtering. The I-V characteristics are very similar to those presented here, as is their EL spectrum. In their study they ascribe the observed luminescence to the presence of several types of luminescent centre in the SiO$_x$ matrix and favour the idea of a quantum confinement/ luminescent centre model in which electron-hole pairs created in the silicon clusters recombine at luminescent centres in the SiO$_2$. We are currently attempting to assess this hypothesis by examining the EL spectrum of thin SiO$_x$ films as a function of the number of luminescent defect centres present. However, to date from our own studies it is those films that exhibit mainly lower energy band PL luminescence (i.e. size effects with fewer defects) that have been most efficient in the DC EL studies.

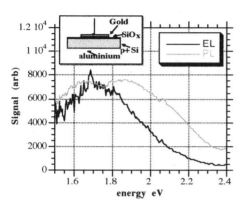

Figure 3: Plot of PL and EL (normalised) for a thin SiO$_x$ film. Inset shows test device.

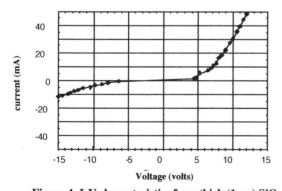

Figure 4: I-V characteristics for a thick (1μm) SiO$_x$ film

The EL spectrum produced in this study shows similarity to the lower energy band of the PL spectrum, although this spectrum has not been corrected for the absorption of the thin gold top electrode. This may suggest that the EL is due mainly to the silicon clusters and is not associated with the defects present in the film but a further study of the EL

under various bias conditions and anneal temperatures is required.

CONCLUSIONS

We have shown that the photoluminescence from silicon-rich silica films is the result of two separate luminescence mechanisms: defect luminescence, possibly from non-bridging oxygen centres or related oxygen vacancies, and radiative recombination of confined excitons. Our results are in broad agreement with those of Shimizu-Iwayama, Komoda and Fischer *et al.*

We have demonstrated electroluminescence from a simple SiO_x based device and have measured its spectrum and current-voltage characteristics. These are very much preliminary results and we are currently working on improving the efficiency of visible emission from such devices by optimising structure and introducing rare-earth dopants.

We have demonstrated that it is possible to access the excited states of rare-earth ions within silicon-rich silica via the absorption of the silicon clusters [24,25]. Recent results published by Polman's group on erbium-doped amorphous silicon have confirmed this [26]. This may open up the possibility of obtaining efficient, narrow-band visible emission from rare-earth doped silicon-rich silica films by careful choice of rare-earth dopant. We are in the process of investigating this further and will publish our results soon.

ACKNOWLEDGEMENTS
We wish to thank the Engineering and Physical Sciences Research Council of the United Kingdom for supporting this work.

REFERENCES
[1] L.T. Canham, Appl. Phys. Lett. **5 7**, p1046 (1990).
[2] J.C. Vial, S. Billat, A. Bsiesy, G. Fishman, F. Gaspard, R. Herino, M. Ligeon, F. Madeore, I. Mihalcescu, F. Muller and R. Romestain, Physica B. **1 8 5**, p593 (1993).
[3] A. Loni, A. J. Simmons, T. I. Cox, P. D. J. Calcott, L.T. Canham, Electron. Lett. **3 1**, p1288 (1995).
[4] A. Bsiesy, J.C. Vial, F. Gaspard, R. Herino, M. Ligeon, F. Muller, R. Romestain. Semiconductor-Silicon 1994 Ed. H.R. Huff, W. Bergholz, K. Sumino, p475 (1994)
[5] M. Lannoo, C. Delerue, G. Allan and E. Martin (Mater. Res. Soc. Proc. **3 5 8**, Boston, MA 1994) p13-p24
[6] B. Delley and E.F. Steigmeier, Appl. Phys. Lett. **6 7**, p2370 (1995)
[7] P.M. Derlet, T.C. Choy and A.M. Stoneham, J. Phys. Condens. Matter. **7**, p2507 (1995)
[8] J. Penczek, A. Knoesen, H.W.H. Lee and R.L. Smith (Mater. Res. Soc. Proc. **3 5 8**, Boston, MA 1994) p641
[9] A. Loni, A.J. Simons, T.I. Cox, P.D.J. Calcott and L.T. Canham, Elect. Lett. **3 1**, p1288 (1995)
[10] W. Lang, P. Steiner, F. Kozlowski and H. Sandmeier, J. Luminesc. **5 7**, p169 (1993)
[11] T. Ito and A. Hiraki, J. Luminesc. **5 7**, p331 (1993)
[12] I.M. Chang, S.C. Pan and Y.F. Chen, Phys. Rev. B. **4 8**, p8747 (1993)
[13] D.J. DiMaria, J.R. Kirtley, E.J. Pakulis, D.W. Dong, T.S. Kuan, F.L. Pesavento, T.N. Theis, J.A. Cutro and S.D. Brorson, J. Appl. Phys. **5 6**, p401 (1984)
[14] T. Komoda, J.P. Kelly, A. Nejim, K.P. Homewood, P.L.F. Hemment and B.J. Sealy, Mat. Res. Symp. Proc. **3 5 8**, p163 (1995)
[15] T. Shimizu-Iwayama, K. Fujita, S. Nakao, K. Saitoh, T. Fujita and N. Itoh, J. Appl. Phys. **7 5**, p7779 (1994)
[16] G. S. Tompa, D.C. Morton, B. S. Sywe, Y. Lu, E. W. Forsythe, J. A. Ott, D. Smith, J. Khurgin, and B.A Khan, (Mater. Res. Soc. Proc. **3 5 8**, Boston, MA 1994) p701-p706
[17] G. G. Qin, A. P. Li, B. R. Zhang, and B. C. Li, J. Appl. Phys. **7 8** (3), p2006, (1995)
[18] M. I. Veksler, I. V. Grekhov, S.A. Solv'ev, A. G. Tkachenko, and A. F. Shulekin, Tech. Phys. Lett. **2 1** (7), p530, (1995)
[19] A.J. Kenyon, P.F. Trwoga, C. Pitt and G. Rehm, J. Appl. Phys. *in press.*
[20] T. Fischer, T. Muschik, R. Schwarz, D. Kovalev and F. Koch (Mater. Res. Soc. Proc. **3 5 8**, Boston, MA 1994), p851-p856
[21] M.A. Stevens Kalceff and M.R. Phillips, Phys. Rev. B, **5 2**, p3122, (1995)
[22] S. Munekuni, T. Yamanaka, Y. Shimogaichi, R. Tohmon, Y. Ohki, K. Nagaswawa and Y. Hama, J. Appl. Phys. **6 8**, p1212 (1990)
[23] H. Koyama, J. Appl. Phys. **5 1**, p2228 (1980)
[24] A.J. Kenyon, P.F. Trwoga, M. Federighi and C.W. Pitt, J. Phys. Condens. Matter. **6**, p319, (1994)
[25] A.J. Kenyon, P.F. Trwoga, M. Federighi and C.W. Pitt (Mater. Res. Soc. Proc. **3 5 8**, Boston, MA 1994) p117-p122
[26] J.H. Shin, R. Serna, G.N. van den Hoven, A. Polman, W.G.J.H.M. van Sark and A.M. Vrendenberg, Appl. Phys. Lett. **6 8**, p997, (1996)

Electron Injection Effects in Electroluminescent Devices Using Polymer Blend Thin Films

C.C. Wu[1], J.C. Sturm[1], R.A. Register[1], L. Suponeva[2] and M.E. Thompson[2]
[1]Advanced Technology Center for Photonic and Optoelectronic Materials (ATC/POEM) Princeton University, NJ 08544
[2]Department of Chemistry, University of Southern California, Los Angeles, CA 90089

ABSTRACT

In this paper, we discuss the effects of electron injection on electroluminescence (EL) efficiency in single-layer EL devices using blends of poly(N-vinylcarbazole) (PVK) and poly(3-n-butyl-p-pyridyl vinylene) (Bu-PPyV). Pure Bu-PPyV thin films suffer from formation of excimers due to the strong interchain interaction. Diluting Bu-PPyV with the high-energy-gap and hole-transporting polymer PVK suppresses excimer formation and substantially raises both photoluminescence (PL) and EL efficiencies. The emission color is converted from red to green, indicating a transition from excimer emission to monomer emission of Bu-PPyV. The electron injection and EL efficiency can be futher improved by adding small-molecule oxadiazoles into the blends as electron-transport materials. In optimized devices, an external EL quantum efficiency (backside emission only) as high as 0.8% photon/electron and practical brightnesses of 100 cd/m^2 and 1000 cd/m^2 can be achieved at ~9V and ~13.5V, respectively.

INTRODUCTION

The performance of polymer LEDs is limited by several basic concerns. First, an efficienct luminescent polymer does not necessarily have both good hole and electron injection and transport abilities. Also, due to self-quenching or interchain interaction, the luminescence efficiencies of these polymers in solid thin films are usually lower than those in dilute solutions [1]. To overcome the carrier transport problems, various approaches have been adopted, including the use of low work function metal electrodes [2] and the incorporation of carrier transport layers to form a multilayer structure [3]. Making devices with several spin-coated polymer layers is not always possible because the solvent carrying the second layer might dissolve the layer already deposited. An alternative strategy is to blend together all carrier transport and emissive materials in a solvent and then deposit them in a single step [4,5,6]. It has also been reported that diluting a luminescent polymer with another inert high energy gap polymer can reduce the interaction between chromophores and increase the photoluminescence efficiency of the luminescent polymer [1]. In this work, we studied the effects on the optical properties and device performance of blending the luminescent polymer Bu-PPyV with the high-energy-gap, hole-transporting polymer PVK and electron-transporting oxadiazoles.

EXPERIMENTAL

The chemical structures of materials used in this study are shown in Fig. 1. Bu-PPyV was synthesized by cross-coupling the corresponding 2,5-dibromopyridine and E-1,2-bistributylstannylethylene in the presence of a palladium catalyst, as described elsewhere [7,8,9,10]. As a result of the butyl side group, Bu-PPyV is soluble in its conjugated form in conventional organic solvents. Two oxadiazoles used are 2-(4-biphenyl)-5-(4-tert-butylphenyl)-1,3,4-oxadiazole (PBD) and 2,5-bis(4-naphthyl)-1,3,4-oxadiazole (BND). Solutions containing different amounts of Bu-PPyV, PVK and oxadiazoles in chloroform were prepared to give PVK:Bu-PPyV or PVK:oxadiazole: Bu-PPyV blends of different ratios. The blend thin films were then spin-coated onto quartz slides for optical measurements and onto ITO coated glass substrates for device fabrication. Devices were finished by thermally evaporating top metal contacts in a vacuum of 10^{-6} torr through shadow masks to form an array of 2mm by 2mm devices. Metal alloys, such as Mg:Ag, Mg:In and Li:Al alloys were deposited by co-evaporation from two separate sources. All processing and device characterization was carried out under dry nitrogen atmosphere in a glove box. Absorption and photoluminescence spectra of thin films were measured on an AVIV 14DS spectrophotometer and a Perkin-Elmer LS50 luminescence spectrometer, respectively. The EL spectra were recorded by an EG&G monochromator connected to an Electrim CCD camera.

RESULTS AND DISCUSSION

Fig. 2 shows the absorption spectra of Bu-PPyV, PVK and PBD. It is clear that both PVK and PBD have shorter absorption onset wavelengths than Bu-PPyV and therefore higher energy gaps. The absorption spectra of the blends, which are not shown in Fig. 2, are simply weighted superpositions of the spectra of the individual components. The PL spectra of PVK:Bu-PPyV blends with different Bu-PPyV contents are shown in Fig. 3, using an excitation wavelength of 420 nm. This wavelength is used because it is close to the absorption peak of Bu-PPyV and below the absorption onset energy of PVK and PBD. In this way, only Bu-PPyV chromophores are excited. In blends with low Bu-PPyV content, a strong green PL is observed. With the Bu-PPyV content increased above several percent, the PL spectra gradually become broadened and shift to red. By normalizing the integrated PL intensity by the absorption of the same blend film at 420 nm, the relative PL efficiencies of Bu-PPyV chromophores diluted in PVK or PVK:PBD are obtained and displayed in Fig. 4. The efficient green PL in dilute Bu-PPyV blends is the emission of isolated Bu-PPyV chromophores, i.e. Bu-PPyV monomer emission, and the weak red PL in concentrated Bu-PPyV blends and pure Bu-PPyV thin films is from excimer emission due to the strong interchain interaction of Bu-PPyV. The PL efficiency starts dropping rapidly above about 2 wt% Bu-PPyV. It is clear that diluting Bu-PPyV with the high energy gap materials eliminates the interaction between Bu-PPyV chromophores and enhances the PL efficiency of Bu-PPyV chromophores. Similar effects are also observed in THF solutions of Bu-PPyV [10]. Transmission electron microscopy of I_2-stained blend films floated off the substrate reveals no phase separation between the Bu-PPyV and PVK in the as-spun films, consistent with the PL data.

Fig. 5 shows the EL spectra of PVK:Bu-PPyV blends. Pure PVK emits purple light. With extremely low additions of Bu-PPyV, even 0.1 wt%, there is a transition in the EL spectrum from emission characteristics of PVK to emission characteristics of the isolated Bu-PPyV units. Further increase of the Bu-PPyV content above several percent leads to a red-shift of the EL, as observed in PL. The dependence of the EL external quantum efficiency on the content of Bu-PPyV in PVK:Bu-PPyV blends is shown in Fig. 6. The EL efficiency first increases with the Bu-PPyV content and peaks around 1 wt% with an efficiency of ~0.3% photon/electron. The efficiency then drops with further addition of Bu-PPyV. From the results of Fig. 4 and Fig. 6, it is clear that at low concentration, the Bu-PPyV units function as efficient emission centers, and therefore the EL efficiency first rises with the density of Bu-PPyV in the blends and peaks at about 10 times the efficiency of pure PVK. At higher concentrations, due to the interaction of Bu-PPyV chromophores, their luminescence efficiency drops rapidly as seen in Fig. 4, leading to a drop in EL efficiency as well.

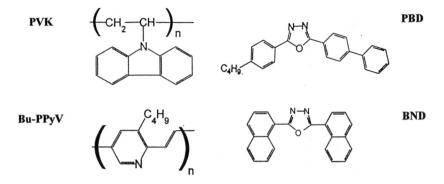

Fig. 1 Chemical structures of PVK, Bu-PPyV, PBD and BND.

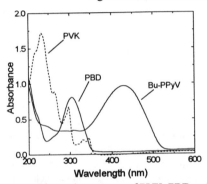

Fig. 2 Absorption spectra of PVK, PBD and Bu-PPyV.

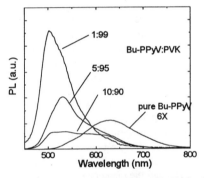

Fig. 3 PL spectra of PVK:Bu-PPyV blend thin films, excitation wavelength 420 nm.

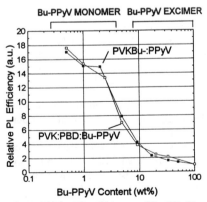

Fig. 4 relative PL efficiency of Bu-PPyV chromophores in blend thin films.

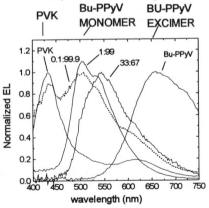

Fig. 5 EL spectra of PVK:Bu-PPyV blend devices

The effect on EL efficiency of adding electron-transporting molecules PBD to the PVK:Bu-PPyV blend is also shown in Fig. 6. A similar dependence of EL efficiency on Bu-PPyV content is also observed, but PBD can enhance the efficiency severalfold. As shown in Fig. 7, the EL efficiency first increases with added PBD. The highest external EL efficiency 0.8% is achieved with a PVK to PBD ratio of about 100:50 (by weight). Adding more PBD than this actually reduces the EL efficiency. The addition of PBD also has a tremendous effect on the I-V characteristics of the devices, as shown in Fig. 8. All the devices have similar film thicknesses. The rise of forward current with voltage first gets sharper with increasing PBD addition, leading to a reduction in the operating voltage of the devices. However, increasing the PBD:PVK ratio beyond 50:100 does not enhance the current further. Instead, the I-V curve is bent down again, as in the case of the EL efficiency. The current vs. voltage and brightness vs. voltage curves of the optimal blend device with PVK:PBD:Bu-PPyV ratios of 100:50:0.8 are shown in Fig. 9 and 10. These values correspond to an external EL quantum efficiency of 0.8% photon/electron. A practical brightness of 100 cd/m^2 is achieved at ~12V and 1000 cd/m^2 at ~16V.

We have also made blend devices using another oxadiazole, BND. It has been reported that BND has a much higher electron mobility than PBD when dispersed in a polymer binder [11]. The comparison of the EL efficiencies of BND and PBD devices is made in Fig. 7. Though the optimal EL efficiency is about the same for both PBD and BND devices, the optimization of BND devices actually occurs at a lower oxadiazole content than for PBD. Also, the optimal BND devices require lower operating voltages than PBD devices, as shown in Figs. 9 and 10. With the optimal BND devices, 100 cd/m^2 can be achieved at ~9V and 1000 cd/m^2 at ~13.5V.

Most of the device data presented so far are from devices with Mg:Ag alloy as the metal contact. We have also tried other metals and alloys. The dependence of EL efficiency on metal used is shown in Table 1. We found Mg:In alloy, Li:Al alloy and Ca gave similar EL efficiency as Mg:Ag. For Al, there is a big difference in efficiency between devices with or without oxadiazole additives. Without oxadiazole, the EL efficiency of Al devices is at least one order of magnitude lower than that of Mg:Ag devices. With oxadiazoles, the EL efficiency of Al devices is just two to three times smaller than Mg:Ag devices. It seems that with oxadiazoles, the electron injection is much easier and the metal contact is not so critical. For PVK, the electron injection is much more difficult and the metal contact is critical.

Table I. Effects of metals on EL efficiency

	Mg:Ag 10:1	Mg:In 10:1	Al	Li:Al 2.5:100	Li:Al 5:100	Ca
PVK:PBD:PPyV 100 : 100 : 5	0.4%	0.4%	0.2%	0.45%	0.35%	0.4%
PVK:PPyV 100 : 2	0.1%		0.01%			

In all optimal two-component or three-component blend devices, the emission is from the small amount of isolated Bu-PPyV chromophores and is much more efficient than that of pure PVK or PVK/oxadiazole devices. To understand the mechanisms for the emission from these efficient emission centers, we have estimated the energy levels of all materials by electrochemical spectroscopy and electronic absorption spectra [12,13]. The estimated LUMOs for Bu-PPyV, PVK, PBD and BND are -2.9 eV, -2.25 eV, -2.35 ev and -2.16 eV, respectively. The estimated HOMOs for Bu-PPyV, PVK, PBD and BND are -5.4 eV, -5.75 eV, -6.0 eV and - 5.6 eV, respectively. Comparing the energy levels suggests at least two possible mechanisms for emission from Bu-PPyV chromophores, i.e. the carrier trapping mechanism and the energy transfer

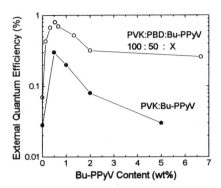

Fig. 6 dependence of EL efficiency on Bu-PPyV content (ITO/blend/Mg:Ag)

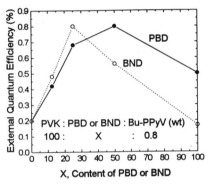

Fig. 7 dependence of EL efficiency on content of PBD or BND (ITO/blend/Mg:Ag)

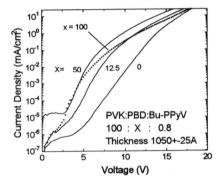

Fig. 8 effects of PBD content on forward I-V characteristics of ITO/blend/Mg:Ag devices.

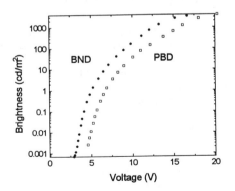

Fig. 9 brightness vs. voltage curves for optimal ITO/blend/Mg:Ag devices

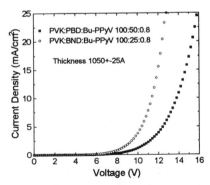

Fig. 10 forward I-V curves for optimal ITO/blend/Mg:Ag devices

mechanism [14]. In the first case, they trap both holes and electrons and form excitons on themselves. In the latter case, they trap excitons formed on carrier transporters.

If one believes that the optimization of the EL efficiency with oxadiazole content is due to better balance of hole and electron currents, the reduction of EL efficiency beyond the optimal oxadiazole content can be explained by the dominance of electron current. That the optimal content of BND is lower than that of PBD can be explained by its better electron transport/injection ability. Also, the sharper rise of the I-V curves in the optimal devices can be related to some hole-electron recombination mechanism. When either hole or electron current is dominant, the balance of both currents and therefore the hole-electron recombination is worse, leading to a higher operating voltage and lower EL efficiency. To support these hypotheses, more experiments and device modeling, taking into account other factors such as space charge effects, are necessary.

CONCLUSIONS

We have presented the results from EL devices made from Bu-PPyV blended with the hole-transporting polymer PVK and the electron-transporting oxadiazole molecules. Diluting Bu-PPyV with the high-energy-gap PVK in thin films reduces the interaction of Bu-PPyV chromophores and enhances both PL and EL efficiency. Adding oxadiazoles into the PVK:Bu-PPyV blend further improves the EL efficiency and reduces the operating voltage. The oxadiazole BND is found to give better devices than PBD, requiring lower operating voltages and smaller oxadiazole additions.

ACKNOWLEDGEMENTS

The authors would like to thank POEM for the support of this work.

REFERENCES

1. S.A. Jenekhe and J.A. Osaheni, Science **265**, 765 (1994)
2. D. Braun, A. J. Heeger, Appl. Phys. Lett. **58**, 1982 (1991)
3. A.R. Brown, D.D.C. Bradley, P.L. Burn, J.H. Burroughes, R.H. Friend, N. Greenham, A.B. Holmes, and A. Kraft, Appl. Phys. Lett. **61**, 2793 (1992)
4. J. Kido, M. Kohda, K. Okuyama and K. Nagai, Appl. Phys. Lett. **61** (7), 761 (1992)
5. C. Zhang, S. Hoger, K. Pakbaz, F. Wudl, and A.J. Heeger, J. Electron. Mater. **23**, 453 (1994)
6. G. E. Johnson, K.M. McGrane and M. Stolka, Pure & Appl. Chem. **67**, 175 (1995)
7. M.J. Marsella, T.M. Swager, Polymer Preprints **33** (1), 1196 (1992)
8. J. Tian, M.E. Thompson, C.C. Wu, J.C. Sturm, R.A. Register, M.J. Marsella, T.M. Swager, Polymer Preprints **35** (2), 761 (1994)
9. J. Tian, C.C. Wu, M.E. Thompson, J.C. Sturm, R.A. Register, Adv. Mater. **7**, 395 (1995)
10. J. Tian, C.C. Wu, M.E. Thompson, J.C. Sturm, R.A. Register, Chem. Mater. **7**, 2190 (1995)
11. H. Tokuhisa, M. Era, T. Tsutsui, and S. Saito, Appl. Phys. Lett. **66**, 3433 (1995)
12. M.R. Anderson, M. Berggen, O. Inganäs, G. Gustafsson, J.C. Gustafsson-Carlberg, D. Selse, T. Hjertberg and O. Wennerstrom, Macromolecules **28**, 7525 (1995)
13. J. Pommerehne, H. Vestweber, W. Guss, R.F. Mahrt, H. Bässler, M. Porsch and J. Daub, Adv. Mater. **7**, 551 (1995)
14. J. Kido, H. Shionoya and K. Nagai, Appl. Phys. Lett. **66** (16), 2281 (1995)

Poly(bithiazole)s: A New Class of Conjugated Polymers for Polymer-Based Light-Emitting Diodes

J. K. POLITIS*, J. NANOS*, YI HE**, J. KANICKI***, M.D. CURTIS*
*Department of Chemistry and the Macromolecular Science and Engineering Center, University of Michigan, Ann Arbor, Mi 48019-1055
**Department of Applied Physics, Unversity of Michigan, Ann Arbor, Mi 48109
***Department of Electrical Engineering and Computer Science, University of Michigan, Ann Arbor, Mi 48109-2108

ABSTRACT

A new class of conjugated polymers have been synthesized based on the bithiazole moiety. The photoluminescence (PL) of these polymers ranges throughout the visible, from blue-green to red. These materials also show high stability, both thermal and oxidative. Therefore, it appears that this class of polymers are good candidates for polymer-based light emitting diodes.

INTRODUCTION

Over the last decade, conducting polymers have come to the forefront of polymer science. This is due in large part to their potential applications in electroactive devices. Of these devices, the most widely studied has been polymer-based light-emitting diodes (LEDs).

There has been much progress since the first crude device, characterized by low efficiency, was reported by the Cambridge group.[1] The most significant improvement is the introduction of either an electron transport/hole blocking layer,[2-4] a hole transport/electron blocking layer[5,6] or both.[7] These multilayer devices have shown improved efficiencies up to 40 times better than the simple three layer device[3]. Also, the effects of the polymer-metal interface has been greatly reduced. In the simple three layer devices, changing from a high work function metal, aluminum, to a low work function metal, calcium, increases the luminescence efficiency by a factor of 50. In the more complex device, these changes are less dramatic, only a factor of 17.[6]

This improvement, however, focuses on the workings of the device itself. Because electroluminescence efficiency can only be 25% of the photoluminescence efficiency,[6] it is important to also improve the emitting material. Up to now, the two most studied polymers are poly p-phenylene vinylene and polythiophene, along with their derivatives. We have gained some understanding about quantum efficiency from studies on these materials. It is clear that by creating a more amorphous polymer, the electroluminescence efficiency increases. This is evidenced by the fact that as the length of the side chains on these polymers increases, so does the efficiency.[8,9] Also, in poly(3-alkyl thiophene), the degree of order may be by controlled by the polymer regioregularity, and it is observed that the intensity of the fluorescence decreases with increasing order[10]. This phenomenon is also observed in our first material, poly(5,5'-4,4'-dinonyl-2,2'-bithiazole) (PNBT)[11]. PNBT exhibits three distinct morphologies: each with differing degrees of crystallinity. Importantly, the fluorescence quenches as the more crystalline phase is obtained. This is presumably due to increased exciplex formation that favor nonradiative decay pathways in the more crystalline phase.

Our laboratory has focused its attention on creating a series of new polymers based on the bithiazole moiety. Three materials are currently being studied for their potential in LEDs. Using simply one base unit we have created a new class of fluorescent polymers with band gaps ranging from 2-3 eV.

EXPERIMENTAL

Synthesis of 4,4'-dinonyl-2,2'-bithiazole (NBT): The procedure described by Nanos was employed.[11] Yield = 6.17 g (72.9 %). ^{1}H-NMR (CDCl$_3$): d 0.88 (6 H, t, J = 6.5 Hz), 1.24-1.31 (24 H, m) 1.73 (4 H, m, J = 7.1 Hz), 2.81 (4 H, t, J = 7.7 Hz), 6.92 (2 H, s).

Synthesis of 5,5'-diiodo-4,4'-dinonyl-2,2'-bithiazole (I$_2$-NBT): A dry 250 mL Schlenk flask equipped with a stir bar and N$_2$ inlet/outlet was charged with dry THF (50 mL). The flask was cooled to -78 °C and BuLi (2.45 M in hexanes, 4.26 mL, 10.46 mmol) was added. NBT (2.00 g, 4.75 mmol) in THF (75 mL) was added dropwise via cannula. The dark solution was stirred at room temperature for 2.5 hrs and I$_2$ (2.77 g, 10.93 mmol) in THF (75 mL) was added dropwise over 20 min. The reaction was stirred for an additional 1 hr and washed with Na$_2$S$_2$O$_5$ (sat'd, 2 X 100 mL). The light brown solution was dried over MgSO$_4$ and decolorized with carbon. The solvent was removed under reduced pressure to give an oil which solidifies upon standing. The solid was recrystallized twice from CH$_3$CN/CHCl$_3$ (2:1) to give yellow crystals. Yield = 1.60 g (50.2 %). ^{1}H-NMR (CDCl$_3$): δ 0.88 (6 H, t, J = 6.4 Hz), 1.26-1.34 (24 H, m), 1.70 (4 H, m, J = 6.6 Hz), 2.77 (4 H, t, J = 7.6 Hz). λ_{max} (ex) = 244 nm, 362 nm. λ_{max} (em) = 444 nm). Anal.: Calcd. for C$_{24}$H$_{38}$I$_2$N$_2$S$_2$: C, 42.86; H, 5.71; N, 4.17. Found: C, 43.12; H, 5.51; 4.17.

Synthesis of 1,3-bis(2-(4-nonylthiazole)-5-[t]butylbenzene (BBNBT): A dry 100 mL Schlenk flask equipped with a stir bar and condenser was charged with the 1,3-bis(thioamide)-5-[t]butylbenzene (0.816 g, 3.23 mmol) and 1-bromo undecanone (1.65 g, 6.63 mmol) in dry DMF (30 mL). The yellow solution was refluxed for 23.5 hrs. The reaction mixture was allowed to cool and placed into the freezer. The tan crystals were collected and dried under vacuum. Yield = 1.505 g (84.3%). ^{1}H-NMR (CDCl$_3$): δ 0.88 (6 H, t, J = 6.7 Hz), 1.28-1.39 (24 H, m), 1.43 (10 H, s), 1.78 (4 H, m, J = 7.1 Hz), 2.85 (4 H, t, J = 7.7 Hz), 6.89 (2 H, s), 8.02 (2 H, d, J = 1.5 Hz), 8.24 (1 H, t, J = 1.5 Hz).

Synthesis of 1,3-bis(2-(5-bromo-4-nonylthiazole))-5-[t]butylbenzene (Br$_2$-BBNBT): A 250 mL 3-neck flask equipped with a condenser, addition funnel, stir bar, and N$_2$ inlet/outlet was charged with BBNBT (5.51 g, 0.01 mole) in CHCl$_3$ (50 mL). Bromine (3.42 g, 0.0214 moles) in CHCl$_3$ (50 mL) was added dropwise and the dark solution was refluxed for 22 hrs. The dark solution was cooled and washed with Na$_2$CO$_3$ (sat'd, 100 mL) and NaHSO$_3$ (sat'd, 100 mL). The solution was heated over decolorizing carbon and dried over MgSO$_4$. The solvent was removed in a rotary evaporator to yield a brown oil. The material was recrystallized twice from ethanol to give a white solid. Yield = 2.94 g (41.4 %). ^{1}H-NMR (CDCl$_3$): δ 0.88 (6 H, t, J = 5.9 Hz), 1.27-1.41 (24 H, m), 1.42 (10 H, s), 1.76 (4 H, m, J = 7.3 Hz), 2.79 (4 H, t, J = 7.6 Hz), 7.91 (2 H, d, J = 1.6 Hz), 8.07 (1 H, t, J = 1.5 Hz). λ_{max} (ex) = 307 nm, 232 nm. λ_{max} (em) = 373 nm. Anal.: Calcd. for C$_{34}$H$_{50}$Br$_2$N$_2$S$_2$: C, 57.45; H, 7.10; N, 3.94. Found: C, 57.49; H, 7.17; N, 3.74.

Synthesis of 4,4'-bis(p-dodecylphenyl)-2,2'-bithiazole (DPBT): A 100 mL Schlenk flask equipped with a stir bar, condenser, and N$_2$ inlet/outlet was charged with 1-(bromoacetyl)-4-dodecylbenzene (6.89 g, 18.9 mmol) and dithiooxamide (1.07 g, 8.9 mmol) in DMF (80 mL). The reaction was heated to reflux for 1 hr and at 60 °C for 5 hrs. The reaction was allowed to cool and placed into a freezer. The brown crystals were collected and recrystallized from CH$_3$CN/CHCl$_3$ (3:1) to give tan crystals. Yield = 4.48 g (76.6 %). ^{1}H-NMR (CDCl$_3$): δ 0.88 (6 H, t, J = 5.6 Hz), 1.26-1.32 (36 H, m), 1.65 (4

H, m, J = 7.4 Hz), 2.65 (4 H, t, J = 7.6 Hz), 7.27 (4 H, d, J = 6.9 Hz), 7.53 (2 H, s), 7.89 (4 H, d, J = 7.8 Hz).

Synthesis of 5,5'-dibromo-4,4'-bis(p-dodecylphenyl)-2,2'-bithiazole (Br₂-DPBT: The procedure for Br₂-BBNBT was followed with some modifications. The reaction was carried out at room temperature and the solid was recrystallized twice from $CH_3CN/CHCl_3$ (2:1). Yield = 3.39 g (79.2 %). ¹H-NMR ($CDCl_3$): δ 0.88 (6 H, t, 6.7 Hz), 1.26-1.32 (36 H, m), 1.66 (4 H, m, J = 7.1 Hz), 2.67 (4 H, t, J = 7.6 Hz), 7.29 (4 H, d, J = 8.2 Hz), 7.90 (4 H, d, J = 8.2 Hz). λ_{max} (ex) = 262 nm, 367 nm. λ_{max} (em) = 432 nm. Anal.: Calcd. for $C_{42}H_{58}Br_2N_2S_2$: C, 61.90; H, 7.19; N, 3.44. Found: C, 62.10; H, 7.21; N, 3.41.

Synthesis of poly(ethynylene NBT) (PENBT): A dry 100 mL Schlenk flask equipped with a stir bar and N_2 inlet/outlet was charged with Pd(dba)₂ (0.04 g, 0.07 mmol) and triphenyl phosphine (0.092 g, 0.35 mmol). Dry toluene (60 mL) was added and the system was freeze-pump thawed. The mixture was allowed to warm to room temperature and stirred until the yellow catalyst color was formed. The solution was transferred via cannula to a dry 100 mL Schlenk flask equipped with a stir bar and N_2 inlet/outlet containing I₂-NBT (1.5721 g, 2.34 mmol) and 1,2-bis(trimethyltin) acetylene (0.8220, 2.34 mmol) and a condenser was placed on the flask. The reaction was heated to reflux for 26 hrs during which the fluorescence changed from violet to yellow. The red mixture was precipitated into 5% HCl in MeOH (450 mL). The red solid was collected and reprecipitated from toluene into 5% HCl in MeOH (600 mL). The red powder was collected and dried under vacuum at 50 °C. Yield = 0.62 g (59.6 %). ¹H-NMR ($CDCl_3$): δ 0.80-0.90 (6 H, br), 1.20-1.48 (24 H, m) 1.75-1.86 (4 H,br), 2.88-2.95 (4 H, br). λ_{max} (ex) = 503 nm (film), 475 nm (soln). λ_{max} (em) = 600 nm (film), 532 nm (soln). Anal.: Calcd. for $C_{26}H_{38}N_2S_2$: C, 70.52; H, 8.87; N, 6.33. Found: C, 65.62; H, 8.01; N, 5.44.

Synthesis of poly(5,5'-(4,4'-bis(p-dodecylphenyl)-2,2'-bithiazole)) (PDPBT): A dry 100 mL Schlenk flask equipped with a stir bar and N_2 inlet/outlet was charged with 5,5'-dibromo-4,4'-bis(p-dodecylphenyl)-2,2'-bithiazole (1.597 g, 1.96 mmol), Ni(COD)₂ (0.72 g, 2.55 mmol), and 2,2'-bithiazole (0.41 g, 2.65 mmol) in dry toluene (20 mL). A condenser was placed on the flask and the reaction was heated to reflux overnight. The dark, viscous mixture was precipitated into 5% HCl in MeOH (400 mL). The light green solid was collected and redissovled in hot toluene (100 mL). The mixture was filtered through a 1/4" plug of Celite and the solvent reduced to 50 mL. The solution was precipitated into 5% HCl in MeOH (400 mL). The yellow solid was collected and stirred in warm 7% EDTA disodium salt in H_2O (300 mL). The solid was filtered and washed with warm H_2O. The polymer was redissolved in hot toluene (100 mL) and filtered through a 1/4" plug of Celite. The solvent was reduced to 50 mL and precipitated into MeOH (400 mL). The fluffy yellow solid was dried under vacuum at 50 °C. Yield = 0.806 g (62.8 %). ¹H-NMR ($CDCl_3$): δ 0.80-0.88 (6 H, br), 1.10-1.45 (36 H, m), 1.55-1.67 (4 H, br), 2.53-2.65 (4 H, br), 7.02-7.15 (4 H, br), 7.67-7.75 (4 H, br). λ_{max} (ex) = 420 nm (film), 417 nm (soln). λ_{max} (em) = 511 nm (film), 504 nm (soln). Anal.: Calcd. for $C_{42}H_{50}N_2S_2$: C, 77.00; H, 8.94; N, 4.28. Found: C, 77.04; H, 9.02; N, 4.19.

Synthesis of poly(1,3-(5-'butylphenylene)-2,2'-(5,5'-(4,4'-dinonyl) bithiazole)) (PBBNBT): A 100 mL Schlenk flask equipped with a stir bar and N_2 inlet/outlet was charged with Br₂-BBNBT (1.01 g, 1.42 mmol), Ni(COD)₂ (0.523 g, 1.85 mmol), and 2,2'-bypyridyl (0.300 g, 1.92 mmol) in dry toluene (14 mL). A condenser was placed on the flask and the reaction was heated to reflux. After 20 hrs, the dark solution was precipitated into 5% HCl in MeOH (400 mL). The light green solid was collected and redissolved in toluene (60 mL). The mixture was filtered through a 1/4" plug of Celite and precipitated into 5% HCl in MeOH (400 mL). The light yellow solid was collected and stirred in a warm solution of 7% EDTA disodium salt in H_2O (300 mL). The solid was collected and dissolved in $CHCl_3$ (75 mL) and the mixture filtered through a 1/4" plug of Celite. The solvent was reduced to 50 mL and the polymer was precipitated into MeOH (400 mL). The slightly stringy solid was collected and dried under vacuum to give a pale yellow material. Yield = 0.423 g (54.0 %). ^{1}H-NMR ($CDCl_3$): δ 0.82 (6 H, t, J = 5.8 Hz), 1.20-1.40 (24 H, m), 1.46 (10 H, s), 1.76-1.86 (4 H, m), 2.76 (4 H, t, J = 7.0 Hz), 8.05 (2 H, d, J = 1.5 Hz), 8.30 (1 H, t, 1.5 Hz). λ_{max} (ex) = 338 nm (film), 337 nm (soln). λ_{max} (em) = 452 nm (film), 457 nm (soln). Anal.: Calcd. for $C_{34}H_{50}N_2S_2$: C, 74.11; H, 9.17; N, 5.09. Found: C, 74.69; H, 9.30; N, 5.05.

RESULTS AND DISCUSSION

Poly(bithiazole)s, both homopolymers and copolymers (Figure 1), have proven to be relatively simple to synthesize by Pd or Ni mediated coupling reactions. The halogen functionalized thiazole rings are good precursors for many polymers. Polymers can be altered in a number of ways, as demonstrated in the three materials shown here. Changing the side chain on the bithiazole rings is a trivial task and therefore a series of homopolymers, like PDPBT, can be synthesized via the Ni(COD)$_2$ coupling of aryl bromides developed by Yamamoto[12]. Copolymers can be created by two different methods. The first is by copolymerization of two different monomers such as the Stille coupling employed in the PENBT synthesis. Another route is in the monomer synthesis. By incorporating the two separate monomers into a single molecule, a simple homocoupling can produce a copolymer as shown for PBBNBT. The chemistry behind the poly(bithiazole)s is straightforward and allows one to tailor the polymers to provide desirable characteristics, e.g. good luminescence efficiency, good film forming ability, etc.

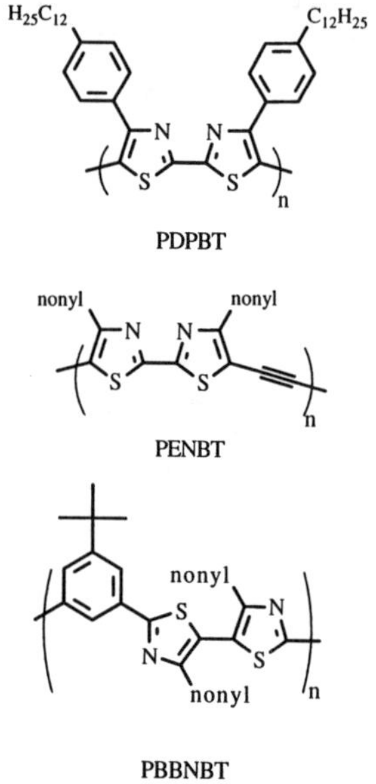

It is important to note that although true molecular weights are not known, with the exception of PENBT, microanalysis does give evidence that the materials are of substantial molecular weight. The end groups of these polymers should be a halogen, or in the case of PENBT potentially trimethyl tin. The effect of the end group on the elemental analysis is significant for low molecular weight materials, but as the molecular weight increases, the end groups become less significant. Analysis based on the repeat unit implies that the end groups are negligible and the materials are of substantial molecular weight. PENBT is currently under investigation for the low results. Because of the rigidity of the backbone, the polymer falls out of solution during the polymerization reaction. However, end group analysis should prove the material to oligomeric. Nonetheless, all the polymers form reasonable spin coated films.

The violet emission of the bithiazole fluorophore allows access to the full visible spectrum. Higher energy wavelengths can be obtained by trapping the fluorophore into a nonconjugated backbone, whereas the longer wavelengths, down to the red, can be obtained by controlling the effective conjugation lengths. These three polymers illustrate this characteristic (Figure 2). The PBBNBT emits a blue-green light while a low energy,

Figure 1. Poly(bithiazole)s

deep red color is emitted from PENBT. Further, from the absorbance spectra, optical band gaps between 2-3 eV are observed. The optical data are summarized in Table I.

Both PBBNBT and PDPBT give essentially identical spectra in the solid state and in solution, respectively. As noted earlier, high degree of order effectively quenches the fluorescence. Hence, completely amorphous polymers should be ideal for obtaining the highest efficiency possible. A red shift in the emission and absorption spectra, as in

Table I. Polymer Properties.

POLYMER	λ_{max} (ex)	λ_{max} (em)	band gap	TGA 5% wt.loss
PENBT	503 nm (film) 475 nm (soln)	600 nm (film) 532 nm (soln)	2.00 eV	310 °C
PDPBT	420 nm (film) 417 nm (soln)	511 nm (film) 504 nm (soln)	2.65 eV	542 °C
PBBNBT	338 nm (film) 337 nm (soln)	452 nm (film) 457 nm (soln)	2.91 eV	545 °C

PENBT, is often observed when going from solution to a film. This is due to a lower energy conformation in the solid state than in solution, either by extended conjugation or by interchain stacking. The optical data of PBBNBT and PDPBT suggests that both polymers have the same conformation in the films and in solution.

Further studies have shown poly(bithiazole)s to be quite robust. Both PBBNBT and PDPBT have thermal decomposition temperatures above 500 °C while PENBT is also high at 310 °C (Figure 3). The lower decomposition temperature of PENBT has not yet been explained, however, could by caused by a crosslinking reaction with the loss of a small molecule byproduct. The materials are also oxidative stable. Each polymer was studied closely over time by FT-IR. Subtraction of the spectra over a one month period exhibits no new peaks and no change in relative intensities of peaks.

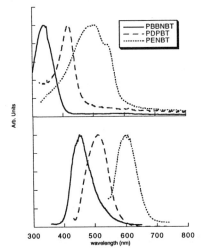

Figure 2. Top: UV-vis spectra of the polymer films. Bottom: Emission spectra of the polymer films.

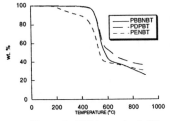

Figure 3. TGA results in N_2.

This information implies that all three polymers should be stable under the harsh conditions present in LEDs.

CONCLUSIONS

The results thus far are promising for incorporating poly(bithiazole)s into LEDs. We can obtain the full spectrum of colors simply by adjusting the conjugation length. Also, it appears that we can to some extent control the degree of order by altering the side chains or by copolymerization. Finally, these polymers show a high degree of thermal stability. As the conditions for LEDs are quite harsh, thermal and oxidative stability is vital in producing materials for LEDs. Therefore, poly(bithiazole)s are good candidates for multicolor polymer-based LEDs.

ACKNOWLEDGMENTS

The authors thank the donors of the Petroleum Research Fund, administered by the American Chemical Society, and the National Science Foundation (DMR-9510274) for support for this research.

REFERENCES

1. J. H. Burroughs, D. D. C. Bradley, A. R. Brown, R. N. Marks, R. H. Friend, P. L. Burn, A. B. Holmes, *Nature,* **347,** 539 (1990).

2. C. Adachi, T. Tsutsui, and S. Saito, *Appl. Phys. Lett.,* **56,** 799 (1990).

3. Q. Pei and Y. Yang, *Chem. Mater.,* **7,** 1568-1575 (1995).

4. M. Strukelj, F. Papadimitrakopoulos, T. M. Miller, and L. J. Rothberg, *Science,* **267,** 1969-1971 (1995).

5. P. E. Burrows and S. R. Forrest, *Appl. Phys. Lett.,* **64,** 2285-2287 (1994).

6. A. B. Holmes, D. D. C. Bradley, A. R. Brown, P. L. Burn, J. H. Burroughs, R. H. Friend, N. C. Greenham, R. W. Gymer, D. A. Halliday, R. W. Jackson, A. Kraft, J. H. F. Martens, K. Pichler, and I. D. W. Samuel, *Synthetic Metals,* **55-57,** 4031 (1993).

7. I. H. Campbell, M. D. Joswick, and I. D. Parker, *Appl. Phys. Lett.,* **67,** 3137-3173 (1995).

8. M. Uchida, Y. Ohmori, C. Morishima and K. Yoshino, *Synthetic Metals,* **55-57,** 4168 (1993).

9. S. Doi, M. Kuwabara, T. Noguchi, and T. Ohnishi, *Synthetic Metals,* **55-57,** 4174-4179 (1993).

10. B. Xu and S. Holdcroft, *SPIE,* **1910,** 65-68 (1993).

11. J. I. Nanos, J. W. Kampf, M. D. Curtis, L. Gonzalez, and D. C. Martin, *Chem. Mater.,* **1995,**7, 2232-2234.

12. T. Yamamoto, T. Maruyama, Z. Zhou, T. Ito, T. Fukuda, Y. Yoneda, F. Begum, T. S. Ikeda, H. Tahezoe, A. Fukuda, K. Kubota, *J. Am. Chem. Soc.,* **116,** 4832 (1994).

OPTICAL STUDIES OF ZnSe NANOCRYSTALS IN POTASSIUM BOROSILICATE GLASS MATRICES

CHRISTINE A. SMITH[1], SUBHASH H. RISBUD[1], HOWARD W. H. LEE[2], J. DIANE COOKE[2]

[1]Department of Chemical Engineering and Materials Science, University of California, Davis, CA 95616, csmith@ucdphy.ucdavis.edu

[2]Photonics Group, Lawrence Livermore National Laboratory, Livermore, CA 94550, hwhlee@llnl.gov

ABSTRACT

We have prepared ZnSe nanocrystals embedded in potassium borosilicate glass matrices in concentrations up to 14 wt%. The optical and electronic properties of these nanocrystals as a function of the fabrication process were investigated with High Resolution Transmission Electron Microscopy (HRTEM) and various optical techniques including absorption, photoluminescence, and excitation spectroscopy, and photoluminescence lifetime measurements. HRTEM confirms the presence of nanocrystals and indicates an average diameter size of 50 Å. The majority of the photoluminescence extends throughout the visible and is highly red-shifted from the absorption-edge, indicating emission from trap states such as Se vacancies. Contributions from the ZnSe nanocrystals as well as the borosilicate glass matrix are also observed. Exponential and bi-exponential photoluminescence decays are observed with decay times ranging from a few nanoseconds to a few hundred nanoseconds. The significance of these results to the optical and electronic properties of the ZnSe nanocrystals and their relevance as luminescent materials for light emitting applications will be discussed.

INTRODUCTION

The promising optical properties and potential applications of II-VI material systems such as ZnSe-based quantum well devices for light emitting applications have led us to explore the properties of ZnSe nanocrystals. Recent developments in ZnSe quantum wells have led to the successful fabrication of blue and green light emitting diodes (LED) and diode lasers [1]. A green (515 nm) diode laser based on a ZnSe quantum well system has recently been reported to achieve room temperature continuous wave (CW) operation for over 100 hours [2]. ZnSe, which has a direct band gap of approximately 2.65 eV at room temperature, has seen intense competition recently from III-V systems such as GaN. Though ZnSe is the more mature of the two technologies and was the first material system to demonstrate blue-green and blue semiconductor lasing operation, GaN has yielded the shortest wavelength laser output reported at 417 nm. The major shortcoming in the GaN system has been a limited lifetime and pulsed operation. However, it's wider bandgap (3.36 eV at room temperature) allows potentially shorter wavelength lasing operation, which is highly desirable for applications in optical data storage.

Our study of ZnSe quantum dots is motivated by the potential for shorter wavelength laser operation enabled by blue-shifted quantum confined energy levels, for longer operational lifetimes due to reduced defect concentrations, and by the inherent interest in quantum confined systems. The optical and electronic properties of these nanocrystals as a function of the fabrication process were investigated with High Resolution Transmission Electron Microscopy (HRTEM) and various optical techniques including absorption, photoluminescence and excitation spectroscopy, and photoluminescence lifetime measurements.

Mat. Res. Soc. Symp. Proc. Vol. 424 © 1997 Materials Research Society

EXPERIMENTAL

The ZnSe nanoparticles in borosilicate glass samples were prepared by first melting a base glass composition formulated specifically for compatibility with chalcogenide semiconductors. The base glass consisted of (in wt. % of each oxide): 56% silica, 24% potassium oxide, 9% barium oxide, 8% boron oxide, and 3% calcium oxide. Twenty-five to thirty gram batches of this oxide powder mixture were melted in alumina crucibles and refined to remove bubbles at 1400°C for several hours. Next, the materials were ground into a fine powder. ZnSe powder was then added to the base glass powder and the blended mixture was re-melted again at 1400°C for approximately one and a half hours before casting the melt into small slabs. The mold of the as-cast sample appeared reddish-orange in color. A portion of this as-cast sample was then melted and rapidly quenched. The quenching process involved squeezing of the melt between two metal plates. Examination of thin flakes of this quenched material revealed an apparent color change to a combination of orange-yellow. In all cases, excess ZnSe was added to compensate for the expected volatilization losses from the melting process.

Optical excitation for photoluminescence studies was provided by three sources: (1) the ~100 fs output from a self-modelocked Ti:sapphire laser was frequency-doubled with an appropriate KDP crystal to give excitation pulses varying from 355-400 nm and at a 82 MHz repetition rate, (2) the ~150 fs output from a Ti:sapphire regenerative amplifier was frequency-doubled to give 400 nm excitation pulses at a 1 kHz repetition rate, and (3) a picosecond nitrogen laser. Photoluminescence spectra were recorded with a 0.25 m monochromator and an intensified optical multichannel analyzer. The photoluminescence spectra are not corrected for the spectral response of the optical system. A photomultiplier tube (1.5 ns resolution) was used to monitor the photoluminescence decay. The photoluminescence decays were recorded at various emission wavelengths using appropriate bandpass filters. Absorption spectra were recorded with a commercial spectrometer. All experiments were performed at room temperature.

RESULTS

HRTEM performed on the samples revealed fringes consistent with the lattice sturcture of ZnSe and an average nanoparticle size of 50 Å in diameter [3]. Since the ZnSe exciton Bohr diameter is approximately 90 Å [4], quantum confinement of the particle excitations is likely. The absorption spectrum of the ZnSe quantum dots, shown in figure 1, supports this likelihood.

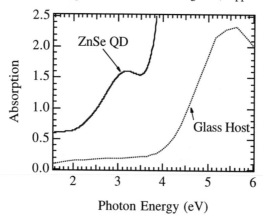

Figure 1. Absorption spectrum of ZnSe quantum dots and the borosilicate glass matrix.

The spectrum differs in shape from bulk ZnSe and shows an absorption peak near 3.1 eV, which is blue-shifted by ~ 0.45 eV from the bulk bandgap of 2.65 eV. This absorption peak likely originates from discrete quantum confined electron-hole pair states in the ZnSe nanocrystals and may correspond to the $1S_h$-$1S_e$ transition. The absorption peak is very broad (~ 1 eV) and suggests a wide distribution of particle sizes in the sample. Figure 1 also contrasts the absorption of the borosilicate host glass to the ZnSe quantum dots to understand their respective contributions. The onset of the glass absorption begins at a much higher energy than the ZnSe quantum dots. However, the glass has a small absorption tail that extends into the visible region and overlaps with the quantum dot absorption, an undesirable property for a host material.

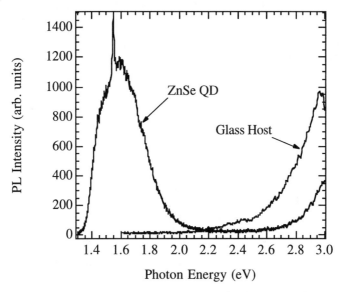

Figure 2. Photoluminescence spectrum of ZnSe quantum dots.

Figure 2 shows the photoluminescence spectrum of the ZnSe quantum dots excited at a wavelength of 400 nm (3.1 eV). Although a weak photoluminescence is observed near the absorption edge, most of the photoluminescence is highly red-shifted from the absorption edge. This suggests rapid trapping of the initial excitation and subsequent emission from low lying impurities or trap states. Evidence for the presence of traps is revealed in the resemblance of part of the red-shifted photoluminescence that we observed to photoluminescence from Se vacancies in ZnSe nanocrystals reported recently [5]. This would suggest that traps comprised of Se vacancies may be responsible for part of the photoluminescence in our samples. Consequently, the weak nature of absorption-edge luminescence may be attributed to rapid photoluminescence decay near the absorption-edge, which will be shown later to be the case, as well as rapid trapping of the responsible excitation. Note that the narrow peak at 800 nm (1.55 eV) is due to the second order of the laser excitation.

As the absorption spectrum in figure 1 shows, excitation at 400 nm (3.1 eV) may complicate the analysis by exciting the borosilicate glass to a small degree. In order to establish the relative contribution of the photoluminescence that is attributable only to the ZnSe quantum dots, a comparison was made between the two. Figure 2 illustrates that the photoluminescence from the borosilicate glass occurs mainly at higher energies than that from the ZnSe quantum dots. However, the glass host has a small photoluminescence tail extending to lower energies into the visible region and overlapping with that from the quantum dots.

The decay of the ZnSe quantum dot photoluminescence is shown in figure 3 and occurs in the nanosecond regime. Both exponential and bi-exponential decays are observed. Figure 3 also reveals that as the energy of the emission increases, the decay rate of the nanoparticles increases. absorption-edge recombination occurs primarily in tens of nanoseconds. Lower energy emission shows a long-lived component in the hundreds of nanoseconds regime. This is indicative of transport and trapping of the initial excitation, which is consistent with the red-shift observed in the photoluminescence spectra. Preliminary results also show that trapping of the initial excitation occurs in the sub-nanosecond regime. All of this is consistent with quantum confinement and trapping. In addition, the decay rate of the photoluminescence can be related to the overlap of the electron-hole wave functions via a simple model as $1/\tau \propto \omega^3 (\text{e-h overlap})^2$. This model would suggest that smaller particles have faster decay rates, a result consistent with our wavelength dependent lifetime data and quantum confinemnet of the excitations.

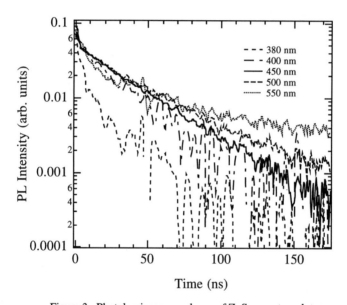

Figure 3. Photoluminescence decay of ZnSe quantum dots.

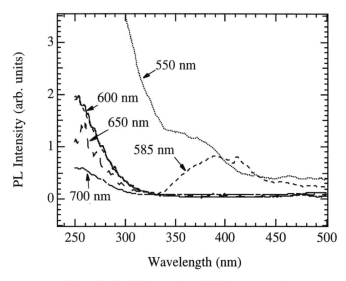

Figure 4. Excitation spectra of ZnSe quantum dots.

Finally, the excitation spectra of the ZnSe quantum dots were examined and is shown in figure 4. The excitation spectra reveal greater and more defined structure than the absorption spectrum. The excitation spectra from the 550 and 585 nm emission show contributions mainly from the ZnSe quantum dots. These spectra show a similar peak near 400 nm as seen in the absorption spectrum. However, the peaks in the excitation spectra are considerably narrower (~0.4 eV) than in the absorption spectrum (~1.0 eV). In addition, the peak in the 550 nm excitation spectrum occurs at higher energies than that observed in the 585 nm excitation spectrum. These observations are consistent with the quantum confinement model that smaller nanocrystals absorb and emit at shorter wavelengths. The excitation spectra taken at specific PL wavelengths monitor a smaller distribution of particle sizes that emit within that specific wavelength range, hence accounting for the narrower linewidths in the excitation spectra. Furthermore, the excitation spectra from the 550 nm emission monitors smaller particles than that from the 585 nm emission, thus accounting for the higher energy peak in the 550 nm excitation spectrum.

Another noteworthy point is that the wavelengths monitored in the excitation spectra originate not from bandedge states but from trap states. This would indicate that the energy levels of the trap(s) shift in a similar fashion as the quantum confined levels.

Examination of the spectra from 600-700 nm emission reveals contributions from both the ZnSe quantum dots and the borosilicate glass host. Overall, the excitation spectra indicate a strongly blue-shifted absorption edge due to quantum confinement of the excitations in the ZnSe nanocrystals. This observation is consistent with the absorption spectral data which show a similar blue shift.

CONCLUSIONS

The absorption and excitation spectra of the ZnSe nanocrystals reveal a peak near 3.1 eV and higher that is ~0.45 eV blue-shift from the bandedge of bulk ZnSe. The photoluminescence spectra show a weak emission near the absorption-edge as well as strong luminescence from trap states in the visible-infrared region. Preliminary evidence indicates that the trapping of the initial excitation occurs very rapidly in the sub-nanosecond regime. The absorption-edge emission occurs in the nanosecond regime while trap emission occurs in hundreds of nanoseconds. Examination of the photoluminescence decay rates reveals a faster rate of decay as the emission energy increases. Furthermore, the excitation and absorption spectra indicate that smaller particles absorb and emit at shorter wavelengths. All these observations confirm that quantum confinement is present in these ZnSe nanocrystals. In addition, the trap states appear to shift in energy in the same fashion as the quantum confined levels of the ZnSe nanocrystals. These results point to the dominant role traps play in these quantum dots. Possible applications for this luminescent material system include flat panel displays, LEDs, or lasers. However, better control over size distributions and impurity or trap states needs to occur before such applications can be realized.

ACKNOWLEDGMENTS

We gratefully acknowledge NSF support through the Division of Materials Research, Electronic Materials Program (Dr. LaVerne Hess, Program Director) under grant DMR 94-11179. Work at LLNL was performed under the auspices of the U.S. Department of Energy under contract number W-7405-ENG-48.

REFERENCES

1. M. Haase, J. Qiu, J. M. DePuydt, and H. Cheng, Appl. Phys. Lett., **59**, p. 1272(1991); H. Jeon, J. Ding, W. Patterson, A. Nurmikko, W. Xie, D. Grillo, M. Kobayashi, and R. L. Gunshor, Appl. Phys. Lett., 59, p. 3619(1991).
2. "Sony ZnSe Laser Breaks the 100 Hour Barrier," Compound Semiconductor 2, 7 (1996).
3. Valorie J. Leppert, private communication.
4. F. Henneberger, S. Schmitt-Rink, and E. O. Gobel, eds. "Optics of Semiconductor Nanostructures," (Akademie Verlag; Berlin; 1993).
5. G. Li and M. Nogami, J. Appl. Phys., **75**, p. 4276(1994).

AUTHOR INDEX

Abelson, J.R., 53
Ahn, Byung Tae, 225
Alessandri, A., 355
Anderson, W.A., 219
Ando, Masahiko, 341
Arai, Toshiaki, 37
Awadelkarim, Osama, 237, 249

Bae, Byung-Sung, 171
Bae, S.S., 273
Baik, Hong Koo, 375, 403
Baik, Young-Joon, 399
Bally, A.R., 471
Barthou, C., 471
Bawendi, M.G., 477
Beccherelli, R., 355
Benalloul, P., 471
Benoit, J., 471
Bhargava, R.N., 441
Bhat, Navakanta, 281, 287
Braunger, D., 453

Campoli, F., 355
Caruso, J., 433
Chakhovskoi, A.G., 415
Chamberlain, S.G., 25
Chen, Chun-ying, 43, 77
Chen, Yu, 103
Cheng, Huang-Chung, 177
Chi, Eung Joon, 375, 403
Cho, K.I., 243
Choi, Hong-Seok, 115, 119
Choi, Woong Sik, 85
Chung, In-Jae, 371
Cooke, J. Diane, 501
Couillard, J. Greg, 261
Curtis, M.D., 495

Dabbousi, B.O., 477
D'Alessandro, A., 355
Dankoski, P., 267
Deane, S.C., 91
Degertekin, F.L., 267
Dye, R.C., 459

Feenstra, K.F., 31
Fehlner, Francis P., 261
Felter, Thomas E., 387, 393, 415
Foglietti, V., 355
Fonash, Stephen J., 207, 237, 249
Fortunato, E., 59
Frank, C.W., 311

Galbato, A., 355
Givargizov, E.I., 393
Glesková, Helena, 71
Globus, T., 97

Goldburt, E.T., 441
Gudem, P.S., 25

Hack, M., 91, 213
Hampden-Smith, M.J., 433
Han, Min-Koo, 115, 119, 171, 183
Hascard, M.R., 371
Hatalis, Miltiadis K., 159, 165
Hautala, J.J., 9, 19
He, S.S., 109
He, Yi, 495
Hess, Dennis W., 165
Hiromasu, Yasunobu, 37
Holleman, J., 31
Holloway, Paul H., 425
Hong, Chan Hee, 85
Hovagimian, Howard, 237
Howell, Robert S., 165
Huang, Chun-Yao, 177
Huet, O., 127
Hunt, C.E., 415

Ihn, Tae-Hyung, 189, 195

Jacunski, M.D., 213
Jang, Keun-Ho, 115, 119
Jastrzebski, L., 201
Jensen, K.F., 477
Jeon, B.S., 421
Jeon, Yoo-Chan, 189
Jeong, E.J., 329
Jones, Sean, 425
Joo, Seung-Ki, 153, 189, 195
Ju, Byeong-Kwon, 371, 399
Jun, Jae-Hong, 115, 119, 171
Jung, D.K., 329
Jung, Jae-Hoon, 371, 399
Jung, M.Y., 273
Jung, Y.H., 273

Kaluri, Sita R., 165
Kang, S.W., 421
Kang, S-Y., 243
Kang, Young-Tae, 195
Kanicki, Jerzy, 43, 77, 347, 495
Kenyon, A.J., 483
Khuri-Yakub, B.T., 267
Kim, C.W., 329
Kim, Chang-Soo, 231
Kim, Iee Gon, 323, 335
Kim, J.H., 243
Kim, Jeong Hyun, 85
Kim, Jin-Won, 231, 255
Kim, Ki-Bum, 231, 255
Kim, S.Y., 329
Kim, Sang-Joo, 255
Kim, Seong-Jin, 399

SUBJECT INDEX